AN ATLAS OF LOCAL GROUP GALAXIES

ASTROPHYSICS AND SPACE SCIENCE LIBRARY

VOLUME 221

A C.I.P. Catalogue record for this book is available from the Library of Congress.

ISBN 1-4020-0673-X

Published by Kluwer Academic Publishers,
P.O. Box 17, 3300 AA Dordrecht, The Netherlands.

Sold and distributed in North, Central and South America
by Kluwer Academic Publishers,
101 Philip Drive, Norwell, MA 02061, U.S.A.

In all other countries, sold and distributed
by Kluwer Academic Publishers,
P.O. Box 322, 3300 AH Dordrecht, The Netherlands.

Printed on acid-free paper

Cover image supplied by Travis Rector, Knute Olsen and Chris Smith
of the National Optical Astronomical Observatories.

Printed in the Netherlands.

AN ATLAS OF LOCAL GROUP GALAXIES

by

PAUL W. HODGE

BROOKE P. SKELTON

and

JOY ASHIZAWA

University of Washington,
Seattle, U.S.A.

KLUWER ACADEMIC PUBLISHERS
DORDRECHT / BOSTON / LONDON

CONTENTS

DESCRIPTION OF THE ATLAS

INTRODUCTION

The Local Group of Galaxies is the small, loose cluster to which our Milky Way Galaxy belongs. It consists of approximately 36 known members, most of which are dwarf galaxies. This Atlas includes identification charts for 32 members for which identification atlases have not hitherto been published. There are numerous atlases of various sorts for the Milky Way Galaxy and there is a published atlas for M31 and its two close companions, M32 and NGC 205 (Hodge 1981). The Magellanic Clouds have also been covered by identification atlases (Hodge and Wright 1967 and 1977). This atlas, then, completes the set for the remaining members of the Local Group.

ORGANIZATION

The atlas is organized as follows: each Local Group member is presented by charts that include identifications of important known objects within them. For easy positional reference and for scale we have reproduced for each galaxy its image from the Digitized Sky Survey together with a grid of J2000 coordinates. Because of the large number of charts that are needed to cover all of M33, it is given first (Part I). The remaining galaxies are given in Part II in alphabetical order, starting with Andromeda I (abbreviated And I) and ending with the Wolf-Lundmark-Melotte galaxy (abbreviated WLM). Table 1 provides a complete listing of the charts. Each galaxy in Part II has a page that summarizes its basic data and gives some of its aliases, if any. A list of references follows. Except for M33, we have attempted to include all references to papers (up to early 1998) that deal with the galaxies of the Atlas.

As explained in the "Note added regarding publication" at the end of this text, a long delay has allowed us to add images of four objects that have been added to the Local Group list after the Atlas was originally submitted to be published. These galaxies are included at the end of the Atlas in Part III.

MEMBERSHIP

There is some uncertainty about Local Group membership for several galaxies. Either the needed facts are not yet available or certain criteria are not fulfilled. We have primarily followed van den Bergh's (1994) selection, based on the velocity arguments used earlier by Yahil, Tammann and Sandage (1977). More generous criteria have been adopted by others, most notably by Mateo (1998), who included several interesting galaxies that are not included here. Mateo's review is strongly recommended for its very thorough coverage, not only of possible members beyond those included here, but also for its extremely useful collection of uniform data for each galaxy; we have used many of his data for our summary pages.

Galaxies that Mateo describes but that are not included here are: DDO 210, GR8, IC 5152, Leo A, NGC 55, NGC 3109, UGC A86, and UKS 2323-326. Some of these have uncertain distances or unknown velocities and some we consider to be too distant to be included by us as members of the Local Group.

SOURCES

Our choices in selecting images to be used in this Atlas have been based on the appropriateness of the image for showing particular components of each galaxy, as well as on the ease with which the images could be made available to us and used. In the early years of this project, which dates back to the 1980's, we were generously helped by S. Tritton of the Royal Observatory, Edinburgh, who provided a splendid set of images taken by the U.K. Schmidt telescope. These have since been supplemented by images taken by us at various observatories and images obtained from others. The source in each case is listed in Table 1. All images are oriented with north at the top of the page and east to the left.

We note particular gratitude to W. Freedman, who granted permission to use her plates of M33 taken with the CFHT on Mauna Kea. However, we must warn users that these plates, used for Charts 3-19 and 21-32, were obtained with a Racine wedge in place, so that each bright star has a faint secondary image to its south-southeast.

ACKNOWLEDGMENTS

This Atlas was prepared with the help of many individuals and two funding sources. The National Science Foundation supported the preparation of the atlas of M33 (as well as the spin-off in the form of several scientific papers published elsewhere) under grant AST-9215821. The National Geographic Society helped financially with the preparation of the atlases of the other Local Group galaxies. The Royal Observatory, Edinburgh, generously supplied many photographs taken by the U.K. Schmidt telescope.

Many individuals also provided important materials, especially images, and other help. We are indebted to Taft Armandroff, Antonio Aparicio, William Blair, Ronald Canterna, Karen Fisher, Phillip Flower, Wendy Freedman, Jay Gallagher, S. M. Gordon, Rodrigo Ibata, Robert Kennicutt, Thomas Kinman, Robert Kirschner, M. G. Lee, Knox Long, Mario Mateo, James Nemec, Ata Sarajedini, Toby Smith, Sue Tritton, Alistair Walker, and Daniel Zucker. Many thanks to all of these and to the others who helped us along the way. We are especially grateful to our editors at Kluwer, most recently Dr. J. J. Blom, who have shown remarkable patience and have provided excellent help and advice.

Special Acknowledgment for Use of UK Schmidt images, the Digitized Sky Survey and SkyView
Fields with southern declinations are based on photographic data obtained using the UK Schmidt Telescope. The UK Schmidt Telescope was operated by the Royal Observatory Edinburgh, with funding from the UK Science and Engineering Research Council, until 1988 June, and thereafter by the Anglo-Australian Observatory. Original plate material is copyright the Royal Observatory Edinburgh and the Anglo-Australian Observatory. The plates were processed into the present compressed digital form with their permission. The Digitized Sky Survey was produced at the Space Telescope Science Institute under US Government grant NAG W-2166.
Fields with northern declinations are based on photographic data obtained using the Oschin Schmidt Telescope on Palomar Mountain. The Palomar Observatory Sky Survey was funded by the National Geographic Society. The Oschin Schmidt Telescope is operated by the California Institute of Technology and Palomar Observatory. The plates were processed into the present compressed digital format with their permission. The Digitized Sky Survey was produced at the Space Telescope Science Institute under U.S. Government grant NAG W-2166.

SkyView was developed and is maintained under NASA ADP Grant NAS5-32068 with P.I. Thomas A. McGlynn under the auspices of the High Energy Astrophysics Science Archive Research Center (HEASARC) at the GSFC Laboratory for High Energy Astrophysics.

REFERENCES

Hodge, P. W. 1981, Atlas of the Andromeda Galaxy (U. Washington Press, Seattle)
Hodge, P. W. and Wright, F. W. 1967, The Large Magellanic Cloud (Smithsonian Press, Washington)
Hodge, P. W. and Wright, F. W. 1977, The Small Magellanic Cloud (U. Washington Press, Seattle)
Mateo, M. L. 1998, "Dwarf Galaxies of the Local Group", ARAA, 36, 435
Van den Bergh, S. 1994, "The Outer Fringes of the Local Group", AJ, 107, 1320
Yahil, A., Tammann, G. and Sandage, A. R. 1977, "The Local Group: The Solar Motion Relative to its Centroid", ApJ, 217, 903

NOTE ADDED REGARDING PUBLICATION

Because of a series of computer failures, this Atlas has seen an unusual delay in publication. The Atlas was completed in July, 1998, and sent in paper form to the Publisher. When, several months later, we attempted to send the computer files for the charts, it was found that the computer used for the Atlas had failed and was irreparably damaged. We had backed up the files on another computer, however, so we turned to it. While we were reformatting the files from that computer preparatory to sending them to the Publisher, it crashed and the files were all lost. However, we still had versions of the charts on Bernoulli disks, made from the original computer when we were carrying out the identifications and marking the charts. However, by then we had no computers in the Department that could read Bernoulli disks, so they were sent out to be read. Unfortunately, it was found that the Bernoulli disks were damaged and could not be read. They were sent to the Publisher who was able to locate a company that offered to try to recapture the files on the damaged disks. Finally in December of 2000, the Publisher received the files from that company, which had been able to read all but one of the Bernoulli disks, the one exception being termed hopelessly damaged. By that time one of the authors had obtained her PhD and had left the University and another had completed her bachelors degree and had left both the University and astronomy. The third author therefore spent the next two months attempting to re-do the charts for the missing galaxies, eight in number, in spite of the fact that all of the original hardware, as well as all of the software, used for the original version of the charts, was no longer available.

This series of calamities delayed publication an unfortunate amount of time. However, during that time four galaxies were added to the list of Local Group members and we have taken the opportunity to include them in a special section at the back of the Atlas. The rest of the Atlas remains as it was originally submitted. Readers who wish to find more recent references to a galaxy are urged to consult one of the on-line sources of reference, such as the NASA Extragalactic Database.

LIST OF TABLES

The Charts

Table 1. The Charts

Chart	Galaxy	Objects Displayed	Telescope[a]	Filter[b]	Image Source[c]
1	M33	Coordinates	DSS	E	STScI
2	M33	Master chart, clusters	DSS	E	STScI
3	M33	Clusters and associations	CFHT	V	W. Freedman
4	M33	Clusters and associations	CFHT	V	W. Freedman
5	M33	Clusters and associations	CFHT	V	W. Freedman
6	M33	Clusters and associations	CFHT	V	W. Freedman
7	M33	Clusters and associations	CFHT	V	W. Freedman
8	M33	Clusters and associations	CFHT	V	W. Freedman
9	M33	Clusters and associations	CFHT	V	W. Freedman
10	M33	Clusters and associations	CFHT	V	W. Freedman
11	M33	Clusters and associations	CFHT	V	W. Freedman
12	M33	Clusters and associations	CFHT	V	W. Freedman
13	M33	Clusters and associations	CFHT	V	W. Freedman
14	M33	Clusters and associations	CFHT	V	W. Freedman
15	M33	Clusters and associations	CFHT	V	W. Freedman
16	M33	Clusters and associations	CFHT	V	W. Freedman
17	M33	Clusters and associations	CFHT	V	W. Freedman
18	M33	Clusters and associations	CFHT	V	W. Freedman
19	M33	Clusters and associations	CFHT	V	W. Freedman
20	M33	Master chart, variables	DSS	E	W. Freedman
21	M33	Variable stars	CFHT	B	W. Freedman
22	M33	Variable stars	CFHT	B	W. Freedman
23	M33	Variable stars	CFHT	B	W. Freedman
24	M33	Variable stars	CFHT	B	W. Freedman
25	M33	Variable stars	CFHT	B	W. Freedman
26	M33	Variable stars	CFHT	B	W. Freedman
27	M33	Variable stars	CFHT	B	W. Freedman
28	M33	Variable stars	CFHT	B	W. Freedman
29	M33	Variable stars	CFHT	B	W. Freedman
30	M33	Variable stars	CFHT	B	W. Freedman
31	M33	Variable stars	CFHT	B	W. Freedman
32	M33	Variable stars	CFHT	B	W. Freedman

33	M33	Master chart, emission regions	KPNO 4m	H	W. Freedman
34	M33	Emission regions	KPNO 4m	H	S. Gordon et al.
35	M33	Emission regions	KPNO 4m	H	S. Gordon et al.
36	M33	Emission regions	KPNO 4m	H	S. Gordon et al.
37	M33	Emission regions	KPNO 4m	H	S. Gordon et al.
38	M33	Emission regions	KPNO 4m	H	S. Gordon et al.
39	M33	Emission regions	KPNO 4m	H	S. Gordon et al.
40	M33	Emission regions	KPNO 4m	H	S. Gordon et al.
41	M33	Emission regions	KPNO 4m	H	S. Gordon et al.
42	M33	Emission regions	KPNO 4m	H	S. Gordon et al.
43	M33	Emission regions	KPNO 4m	H	S. Gordon et al.
44	M33	Emission regions	KPNO 4m	H	S. Gordon et al.
45	M33	Emission regions	KPNO 4m	H	S. Gordon et al.
46	M33	Emission regions	KPNO 4m	H	S. Gordon et al.
47	M33	Emission regions	KPNO 4m	H	S. Gordon et al.
48	M33	Emission regions	KPNO 4m	H	S. Gordon et al.
49	M33	Emission regions	KPNO 4m	H	S. Gordon et al.
50	M33	Emission regions	KPNO 4m	H	S. Gordon et al.
51	M33	Emission regions	KPNO 4m	H	S. Gordon et al.
52	M33	Emission regions	KPNO 4m	H	S. Gordon et al.
53	M33	Emission regions	KPNO 4m	H	S. Gordon et al.
54	M33	Emission regions	KPNO 4m	H	S. Gordon et al.
55	M33	Emission regions	KPNO 4m	H	S. Gordon et al.
56	M33	Emission regions	KPNO 4m	H	S. Gordon et al.
57	M33	Emission regions	KPNO 4m	H	S. Gordon et al.
58	M33	Emission regions	KPNO 4m	H	S. Gordon et al.
59	M33	HII region sub-components	KPNO 4m	H	T. Wyder
60	M33	HII region sub-components	KPNO 4m	H	T. Wyder
61	M33	HII region sub-components	KPNO 4m	H	T. Wyder
62	M33	HII region sub-components	KPNO 4m	H	T. Wyder
63	M33	HII region sub-components	KPNO 4m	H	T. Wyder
64	M33	HII region sub-components	KPNO 4m	H	T. Wyder
65	M33	HII region sub-components	KPNO 4m	H	T. Wyder
66	M33	HII region sub-components	KPNO 4m	H	T. Wyder
67	M33	HII region sub-components	KPNO 4m	H	T. Wyder

68	M33	HII region sub-components	KPNO 4m	H	T. Wyder
69	M33	HII region sub-components	KPNO 4m	H	T. Wyder
70	M33	HII region sub-components	KPNO 4m	H	T. Wyder
71	M33	HII region sub-components	KPNO 4m	H	T. Wyder
72	M33	HII region sub-components	KPNO 4m	H	T. Wyder
73	M33	HII region sub-components	KPNO 4m	H	T. Wyder
74	M33	HII region sub-components	KPNO 4m	H	T. Wyder
75	M33	HII region sub-components	KPNO 4m	H	T. Wyder
76	M33	HII region sub-components	KPNO 4m	H	T. Wyder
77	M33	HII region sub-components	KPNO 4m	H	T. Wyder
78	M33	HII region sub-components	KPNO 4m	H	T. Wyder
79	M33	HII region sub-components	KPNO 4m	H	T. Wyder
80	M33	HII region sub-components	KPNO 4m	H	T. Wyder
81	M33	HII region sub-components	KPNO 4m	H	T. Wyder
82	M33	HII region sub-components	KPNO 4m	H	T. Wyder
83	M33	HII region sub-components	KPNO 4m	H	T. Wyder
84	And I	Coordinates	DSS	E	STScI
85	And I	--	KPNO 4m	V	T. Kinman
86	And II	Coordinates	DSS	E	STScI
87	And II	--	KPNO 4m	V	T. Kinman
88	And III	Coordinates	DSS	E	STScI
89	And III	--	KPNO 4m	V	T. Kinman
90	Antlia	Coordinates	DSS	J	STScI
91	Antlia	--	CTIO 4m	V	A. Sarajedini
92	Carina	Coordinates	DSS	J	STScI
93	Carina	Carbon Stars	UKS	V	ROE
94	Draco	Coordinates	DSS	E	STScI
95	Draco	Variables, carbon stars	Pal. Schmidt	V	D. Malin
96	Fornax	Coordinates	DSS	J	STScI
97	Fornax	Globular clusters	UKS	V	ROE
98	IC 10	Coordinates	DSS	E	STScI
99	IC 10	Associations, dust clouds	APO 3.5m	B	D. Zucker
100	IC 10	Variables	APO 3.5m	I	D. Zucker
101	IC 10	Master chart, emission regions	APO 3.5m	H	D. Zucker
102	IC 10	Emission regions	APO 3.5m	H	D. Zucker

103	IC 10	Emission regions	APO 3.5m	H	D. Zucker
104	IC 10	Emission regions	APO 3.5m	H	D. Zucker
105	IC 10	Emission regions	APO 3.5m	H	D. Zucker
106	IC 10	Emission regions	APO 3.5m	H	D. Zucker
107	IC 10	Emission regions	APO 3.5m	H	D. Zucker
108	IC 1613	Coordinates	DSS	J	STScI
109	IC 1613	Master, stellar objects	Lick 3m	B	P. Hodge
110	IC 1613	Associations, variables, stars	Lick 3m	B	P. Hodge
111	IC 1613	Associations, variables, stars	Lick 3m	B	P. Hodge
112	IC 1613	Associations, variables, stars	Lick 3m	B	P. Hodge
113	IC 1613	Associations, variables, stars	Lick 3m	B	P. Hodge
114	IC 1613	Star clusters, dust clouds	Lick 3m	V	P. Hodge
115	IC 1613	Star clusters, dust clouds	Lick 3m	V	P. Hodge
116	IC 1613	Star clusters, dust clouds	Lick 3m	V	P. Hodge
117	IC 1613	Star clusters, dust clouds	Lick 3m	V	P. Hodge
118	IC 1613	Master, emission regions	Lick 3m	V	P. Hodge
119	IC 1613	Emission regions	KPNO 2.1m	H	P. Hodge
120	IC 1613	Emission regions	KPNO 2.1m	H	P. Hodge
121	IC 1613	Emission regions	KPNO 2.1m	H	P. Hodge
122	IC 1613	Emission regions	KPNO 2.1m	H	P. Hodge
123	IC 1613	Emission regions	KPNO 2.1m	H	P. Hodge
124	IC 1613	Emission regions	KPNO 2.1m	H	P. Hodge
125	IC 1613	Emission regions	KPNO 2.1m	H	P. Hodge
126	IC 1613	Emission regions	KPNO 2.1m	H	P. Hodge
127	IC 1613	Emission regions	KPNO 2.1m	H	P. Hodge
128	IC 1613	Emission regions	KPNO 2.1m	H	P. Hodge
129	IC 1613	Emission regions	KPNO 2.1m	H	P. Hodge
130	IC 1613	Emission regions	KPNO 2.1m	H	P. Hodge
131	IC 1613	Emission regions	KPNO 2.1m	H	P. Hodge
132	IC 1613	Emission regions	KPNO 2.1m	H	P. Hodge
133	IC 1613	Emission regions	KPNO 2.1m	H	P. Hodge
134	IC 1613	Emission regions	KPNO 2.1m	H	P. Hodge
135	Leo I	Coordinates	DSS	E	STScI
136	Leo I	Carbon stars	UKS	V	ROE
137	Leo I	Variable stars (outer)	UKS	V	ROE

138	Leo I	Variable stars (core)	UKS	V	ROE
139	Leo II	Coordinates	DSS	E	STScI
140	Leo II	Carbon stars	UKS	V	ROE
141	LGS 3	Coordinates	DSS	E	STScI
142	LGS 3	--	NOT	I	A. Aparicio
143	NGC 147	Coordinates	DSS	E	STScI
144	NGC 147	Globular clusters, planetaries	UKS	V	ROE
145	NGC 185	Coordinates	DSS	E	STScI
146	NGC 185	Globulars, PNe, dust clouds	UKS	V	ROE
147	NGC 6822	Coordinates	DSS	J	STScI
148	NGC 6822	Master, variables, stars, clusters	Lick 3m	B	P. Hodge
149	NGC 6822	Variables, stars	Lick 3m	B	P. Hodge
150	NGC 6822	Variables, stars	Lick 3m	B	P. Hodge
151	NGC 6822	Variables, stars	Lick 3m	B	P. Hodge
152	NGC 6822	Variables, stars	Lick 3m	B	P. Hodge
153	NGC 6822	Associations, clusters	Lick 3m	V	P. Hodge
154	NGC 6822	Associations, clusters	Lick 3m	V	P. Hodge
155	NGC 6822	Associations, clusters	Lick 3m	V	P. Hodge
156	NGC 6822	Associations, clusters	Lick 3m	V	P. Hodge
157	NGC 6822	Master, emission regions	Lick 3m	V	P. Hodge
158	NGC 6822	Emission regions	KPNO 2.1m	H	P. Hodge
159	NGC 6822	Emission regions	KPNO 2.1m	H	P. Hodge
160	NGC 6822	Emission regions	KPNO 2.1m	H	P. Hodge
161	NGC 6822	Emission regions	KPNO 2.1m	H	P. Hodge
162	NGC 6822	Emission regions	KPNO 2.1m	H	P. Hodge
163	NGC 6822	Emission regions	KPNO 2.1m	H	P. Hodge
164	NGC 6822	Emission regions	KPNO 2.1m	H	P. Hodge
165	NGC 6822	Emission regions	KPNO 2.1m	H	P. Hodge
166	NGC 6822	Emission regions	KPNO 2.1m	H	P. Hodge
167	NGC 6822	Emission regions	KPNO 2.1m	H	P. Hodge
168	NGC 6822	Emission regions	KPNO 2.1m	H	P. Hodge
169	NGC 6822	Emission regions	KPNO 2.1m	H	P. Hodge
170	NGC 6822	Emission regions	KPNO 2.1m	H	P. Hodge
171	NGC 6822	Dust clouds	Lick 3m	B	P. Hodge

172	Pegasus	Coordinates	DSS	E	STScI
173	Pegasus	Variables, associations, HII regions, PNe	UKS	V	ROE
174	Phoenix	Coordinates	DSS	J	STScI
175	Phoenix	Globulars, associations	CTIO 4m	B	R. Canterna
176	Sag DIG	Coordinates	DSS	J	STScI
177	Sag DIG	Stellar associations		V	I. Karachentsev
178	Sag DIG	Emission regions	KPNO 0.9m	H	P. Hodge
179	Sagittarius	Isopleths	--	--	R. Ibata
180	Sculptor	Coordinates	DSS	J	STScI
181	Sculptor	Carbon stars	UKS	V	ROE
182	Sculptor	Type II Cepheids	UKS	V	ROE
183	Sextans I	Coordinates	DSS	E	STScI
184	Sextans I	Variables - East half	--	--	M. Mateo
185	Sextans I	Variables – West half	--	--	M. Mateo
186	Sextans A	Coordinates	DSS	J	STScI
187	Sextans A	Variables, associations, PNe	UKS	V	ROE
188	Sextans A	Emission regions	KPNO 0.9m	H	P. Hodge
189	Sextans B	Coordinates	DSS	E	STScI
190	Sextans B	Variables, associations	KPNO 0.9m	R	P. Hodge
191	Sextans B	Emission regions	KPNO 0.9m	H	P. Hodge
192	Tucana	Coordinates	DSS	J	STScI
193	Tucana	Coordinates	CTIO 4m	I	A. Walker
194	UGC A92	Coordinates	DSS	E	STScI
195	UGC A92	Coordinates	KPNO 0.9m	H	B. Miller
196	UGC A92	Emission regions	KPNO 0.9m	H	B. Miller
197	Ursa Minor	Coordinates	DSS	E	STScI
198	Ursa Minor	Carbon stars	Lick 3m	V	P. Hodge
199	Ursa Minor	Variables	Palomar 1.5m	B	J. Nemec
200	Ursa Minor	Variables	Palomar 1.5m	B	J. Nemec
201	WLM	Coordinates	DSS	J	STScI
202	WLM	Variables, associations, globulars	UKS	V	ROE
203	WLM	Emission regions	KPNO 0.9m	H	B. Miller
204	And V		KPNO 4m	V	T. Armandroff
205	And VI		WIYN	Comp.	T. Armandroff

206	And VII		Keck II	V	E. Grebel
207	Cetus		INT	V	A. Whiting

Notes to table:

[a]APO	Apache Point Observatory
CFHT	Canada France Hawaii Telescope
CTIO	Cerro Tololo Interamerican Observatory
DSS	Digitized Sky Survey
KPNO	Kitt Peak National Observatory
NOT	Nordic Optical Telescope
UKS	United Kingdom Schmidt Telescope
[b]E	Palomar Observatory Sky Survey red (103a-E) image
J	SERC yellow (103a-J) image
[c]ROE	Royal Observatory Edinburgh
STScI	Space Telescope Science Institute

Part I

The Atlas of M33

Table 2a. M33 Stellar Associations[a]

Association	Chart
1	11
2	11, 12
3	12
4	11, 12
5	11, 12
6	12
7	12
8	12
9	12
10	12
11	12
12	12
13	12
14	12
15	12
16	12
17	12
18	12
19	12
20	12
21	12, 13
22	13
23	13
24	13
25	12, 13
26	13
27	13
28	9
29	9
30	9
31	9
32	9
33	9
34	9

35	9
36	8, 9
37	8
38	8
39	8
40	8
41	4
42	4
43	4
44	4
45	4
46	4
47	4, 8
48	12
49	12
50	12
51	12
52	12
53	12
54	12
55	12
56	12
57	12
58	12
59	12
60	12
61	9, 12
62	9
63	9
64	9, 12
65	9
66	9
67	9
68	8, 9
69	8, 9
70	8, 9
71	8

72	8
73	8
74	8
75	8
76	8
77	8
78	8
79	8
80	8
81	8
82	8
83	8
84	8
85	8
86	8
87	8
88	8
89	7
90	8, 11
91	8
92	8, 11
93	8, 11
94	8
95	11
96	11
97	11
98	11
99	11
100	11
101	11
102	11
103	11
104	11
105	11
106	11, 12
107	11,12,14,15
108	12, 15

109	14, 15
110	15
111	15
112	15
113	15
114	15
115	12, 15
116	15
117	15, 16
118	16
119	16
120	16
121	16
122	12,13,15,16
123	13, 16
124	13
125	13
126	13
127	12
128	12
129	12
130	13
131	13
132	13
133	13
134	10
135	9
136	9
137	9
138	5
139	5
140	5
141	5
142	12
143	12

[a]All identifications are from Humphreys and Sandage (1980).

Table 2b. M33 Star Clusters[a]

Cluster	Other Name	Chart
MO 1	--	13
2	--	16
3	--	16
4	--	18
5	--	18
6	--	15
7	--	15
8	--	12
9[b]	CS 37	15
10	--	12
11	CS 38	15
12	--	12
13	CS 36	15
14	--	12
15	--	12
16	--	12
17	--	9
18	--	5
19	--	12
20	CS 24	9
21	CS 1	5
22	CS 23	9
23	--	15
24	--	9
25	--	12
26	--	12
27	--	9
28	--	5
29	CS 35	11
30	--	0

31	CS 17	9
32	--	14
33	--	8
34	--	8
35	--	11
36	CS 33	11
37	--	8
38	--	11
39	--	8
40	--	8
41	--	11
42	--	11
43	--	14
44	--	8
45	--	8
46	--	11
47	--	11
48	--	11
49	CS 18	8
50	CS 16	8
51	CS 39	14
52	CS 13	8
54	CS 27	8
55	CS 20	8
56	--	8
58	--	3
CS 2	--	4
3	--	4
4	--	4
6	--	5
7	--	6
8	--	5
9	--	9
10	--	9
11	--	9
12	--	8

14	--	4
15	--	8
19	--	8
21	--	9
22	--	7
25	--	9
26	--	9
28	--	9
29	--	8
30	--	8
31	--	9
32	--	11
34	--	13

[a]MO = Melnick and D'Odorico (1978); CS = Christian and Schommer (1982).
[b]background galaxy

Table 3. M33 Variable Stars[a]

Variable	Other Name	Chart
KMW 377	--	23
402	--	21
486	--	21
502	--	23
546	--	21
594	--	21
641	--	21
1902	--	22
1264	--	22
1322	--	22
1439	--	24
1600	--	21
1737	--	21
2003	--	24
2022	--	21
2034	--	23
2085	--	21, 22
2101	--	22
2426	VHK 47	21
2481	--	21
2732	4439	24
2897	--	22
2929	--	23
3014	--	23
3017	--	24
3039	--	23
3090	--	24
3282	--	21
3482	VHK 50	21
3581	--	24
3618	--	23
3698	--	23

3894	--	21
3936	VHK 52	24
3955	--	24
4054	--	21
4072	--	23
4092	--	24
4189	--	21
4439	2732	24
4593	--	22
4698	--	24
4709	--	22
4799	--	24
5086	--	24
5205	--	24
5724	8663	24
5863	--	24
5866	8862	24
5900	--	24
6048	--	24
6071	--	24
6088	--	22
6231	--	24
6283	VHK 71	24
6365	--	22
6419	--	24
6436	VHK 75	24
6451	--	24
6508	VHK 81	24
6512	--	24
6628	--	24
6788	--	24
6834	--	24
6836	--	24
6884	--	24
6892	--	25

6924	--	24
6981	--	24
7008	--	24
7021	--	24
7022	--	22
7050	--	23
7065	--	24
7126	--	23
7341	--	24
7438	--	24
7460	--	24
7538	--	24
7548	--	22
7567	--	22
7585	--	24
7772	--	24
7908	--	24
8090	--	22
8323	--	24
8480	--	24
8521	--	24
8600	11739	24
8663	5724	24
8784	--	24
8811	VHK 65	24
8862	5866	24
8911	--	25
8979	--	24
9149	--	24
9169	--	27
9219	--	24
9310	--	24
9492	--	25
9495	--	24
9498	--	24

9686	--	24
9738	--	24
9780	--	24
9820	--	25
10025	--	24
10036	--	22
10077	--	24
10138	--	24
10169	VHK 26	24
10246	--	24
10256	H 28	24
10275	--	24
10395	--	24
10504	--	25
10801	--	27
10980	--	24
11172	--	25
11255	--	27
11290	--	24
11343	--	24
11444	--	24
11629	--	24
11721	H 36	24
11736	--	25
11739	8600	24
11853A	--	25
11853B	--	27
11988	--	24
11994	--	27
12104	--	25
12170	H 7	24
12191	--	25
12254	VHK 72	24
12300	--	27
12510	--	27

12788	16339	24
12865	--	25
12989	--	24
13013	--	24
13113	H 11	24
13148	--	24
13352	--	24
13363	--	25
13399	--	27
13587	--	25
13943	--	27
13953	--	27
14022	H 5	27
14098	--	27
14161	--	25
14261	--	27
14347	--	27
14364	--	25
14476	--	25
14523	--	25
14651	--	27
14783	--	27
14837	--	25
14890	H 3, VHK 3	25
14906	--	27
15043	H 34	25
15069	--	24
15151	--	25
15195	--	27
15383	--	27
15476	--	27
15687	--	25
15718	--	24
15772	--	27
15834	--	25

16046	--	25
16050	--	25
16071	--	25
16182	--	27
16280	--	25
16339	12788	24
16481	--	27
16606	--	25
16673	H 11	24
16692	--	27
16747	--	25
16876	--	25
16905	--	28
16942	--	25
17032	--	28
17068	--	27
17088	--	25
17163	--	27
17255	--	25
17502	--	25
17675	--	27
17869	--	27
18004	--	25
18095	21091	28
18180	--	27
18231	--	27
18443	--	28
18508	--	27
18561	--	28
18568	VHK 51	25
18569	--	27, 28
18675	H 16	28
18694	VHK 48	25
18714	--	27
18755	--	28

18797	H 17	28
18860	--	25
18901	--	25
18972	--	25
18892	--	28
19019	--	25
19247	22263	28
19237	--	28
19548	--	28
19721	--	28
19844	--	25
19860	--	27
20084	23190	28
20111	--	25
20311	--	27, 28
20385	--	25
20402	--	25
20415	--	25
20512	--	28
20522	--	28
20549	--	28
20597	--	25
20807	--	28
20822	--	28
20917	--	28
21091	18095	28
21158	--	28
21204	H 38	27
21253	--	26
21272	--	28
21308	--	30
21312	--	27
21314	--	27
21333	--	28
21516	--	26

21671	--	25
21736	--	28
21740	VHK 56	30
21765	--	28
21837	--	28
21979	--	28
22208	--	27
22263	19247	28
22289	--	28
22304	--	28
22312	H 39, VHK 79	28
22330	--	28
22359	H 29	28
22366	VHK 62	28
22526	--	28
22689	--	28
22861	--	28
22921	--	28
23020	H 31	28
23152	--	28
23162	H41	28
23173	HS C	28
23190	20084	28
23496	--	28
23551	--	28
23770	--	28
23835	H 44	28
23908	--	28
24113	--	28
24174	--	28
24180	--	30
24212	--	28
24289	--	28
24293	--	26
24355	--	28

24385	VHK 77?	--
24427	--	30
24472	28325	28
24501	--	28
24533	--	31
24599	--	28
24798	--	30
24838	SC G14	28
24858	H 33	28
24887	--	30
25017	--	28
25060	VHK 80	31
25061	--	31
25201	--	28
25255	--	28, 31
25615	--	26
25620	--	28
25933	--	31
25987	--	28
26066	--	30
26163	--	31
26166	H 35,VHK 35, SC E7	28
26175	--	28
26195	--	31
26212	--	28
26348	H 32	28
26414	--	31
26428	--	31
26477	--	30
26524	SC B1	26
26647	--	31
26729	--	28
26788	--	30
26850	--	31
26910	--	26

27065	--	31
27330	--	31
27544	--	28
27608	--	31
27639	--	30
27689	--	31
27697	--	28
27777	--	29
27830	--	28
27840	SC E34	26
28021	SC E17	26
28065	--	28
28168	VHK 58	28
28325	24472	28
28367	--	31
28500	--	26
28578	--	31
28632	--	31
28639	--	28
28664	--	28
28706	--	31
28768	--	28
28785	SC F12	29
28834	SC G14	28
28986	VHK 55	26
28988	--	31
29032	--	26
29104	--	31
29131	--	31
29211	--	31
29250	--	31
29280	--	29
29334	--	31
29397	--	28
29476	--	28

29618	--	28
29631	--	31
29799	--	28
29926	--	28
30001	--	31
30125	--	28
30133	--	28
30173	--	28
30352	--	31
30422	--	31
30459	--	31
30526	VHK 69	31
30718	--	31
30804	--	29
31100	--	31
31267	--	31
31396	--	28
31458	--	31
31668	VHK 57	31
31786	--	28
31805	--	28
31929	--	26
31953	SC G6	31
31986	--	28
32023	--	31
32117	--	28
32202	--	28
32473	SC F9	29
33070	SC F8	29
33350	--	26
34052	--	31
34497	--	31
35284	--	31
35369	--	31
36117	--	26

36568	--	31
36781	--	32
36806	--	32
36841	--	26
37040	--	26

[a]The variable stars are all based on the Kinman, Mould and Wood (1987) survey and are identified by the numbers in that catalog (designation: KMW), which also provided the alternate names given here (designation VHK: van den Bergh, Herbst, Kowal (1975); H: Hubble (1926); HS: Hubble and Sandage (1953); SC: Sandage and Carlson (1983)). We note that the fainter variables were often very difficult to identify with certainty and users should be aware of possible misidentifications.

Table 4a. M33 Emission Regions Identified by Boulesteix et al. (1974)

ID		Chart	RA (2000)		DEC (2000)	Flux[a]	Area[b]	Comments
BCLMP 1		50	1 34 1.8		30 35 49	160	157	
2		45	1 33 54.3		30 38 7	59	204	
3a		45,59	1 33 55.4		30 37 46	9.2	25	
3b		45,59	1 33 55.6		30 37 38	37	66	
4a		45,59	1 33 59.5		30 35 44	880	494	
4b		45,59	1 33 59.4		30 35 58	24	24	
4c		50,59	1 33 59.7		30 35 59	100	63	
4d		50,59	1 34 0.1		30 35 52	49	47	
4e		50,59	1 33 58.3		30 35 54	9.7	28	
5		50,51	1 33 58.7		30 35 27	440	315	
8Aa		50,59	1 33 58.7		30 33 56	43	101	
8Ab		50,59	1 33 59.3		30 33 53	74	84	Filament to SW is SNR G62[c]
8Ac		50,59	1 33 60.0		30 33 52	340	277	SNR S17 is located near 8A
8Ba		50,59	1 33 59.0		30 33 23	13	49	
8Bb		50,59	1 33 59.0		30 33 34	290	313	
8Bc		50,59	1 33 59.6		30 33 41	45	39	
8Bd		50,59	1 33 60.0		30 33 38	80	215	
9		45	1 33 51.8		30 38 48	100	242	
10a		50,51,59	1 33 53.7		30 35 42	27	95	
10b		50,51,59	1 33 54.0		30 35 29	200	304	
10c		50,51,59	1 33 55.1		30 35 22	61	121	
10d		50,51,59	1 33 53.4		30 35 18	120	226	
10e		50,51,59	1 33 53.8		30 35 11	21	105	
10f		50,51,59	1 33 54.6		30 35 6	27	105	
10g		50,51,59	1 33 53.4		30 35 7	7.2	37	
11a		50,51,59	1 33 55.5		30 34 23	120	257	
11b		50,51,59	1 33 55.9		30 34 18	160	245	
12		51	1 33 52.0		30 36 33	83	232	
13a		50,51,59	1 33 53.4		30 33 8	71	165	
13b		50,51,59	1 33 54.2		30 32 57	350	128	
13c		50,51,59	1 33 54.0		30 33 6	420	145	
13d		50,51,59	1 33 54.8		30 33 9	180	107	Also identified as SNR G55

13e		50,51,59	1 33 54.8		30 32 59	100	95	
14a		45,59	1 33 50.8		30 37 11	7.6	15	
14b		45,59	1 33 50.6		30 37 5	34	51	
15a		51,59	1 33 50.3		30 33 50	280	200	
15b		51,59	1 33 50.2		30 33 41	130	135	
15c		51,59	1 33 49.2		30 33 37	7.4	17	
15d		51,59	1 33 49.8		30 33 33	23	49	
15e		51,59	1 33 50.8		30 33 42	2.6	9	
16		45	1 33 49.8		30 37 30	300	246	
17a		51,60	1 33 47.1		30 33 1	110	153	
17b		51,60	1 33 47.4		30 32 48	160	195	
17c		51,60	1 33 47.9		30 32 51	47	63	Also identified as SNR G47
17d		51,60	1 33 47.7		30 33 3	590	277	
17e		51,60	1 33 48.2		30 33 13	11	19	
18		51	1 33 47.5		30 33 35	200	255	
19		51	1 33 46.6		30 33 33	280	278	
20a		45,60	1 33 49.2		30 38 0	37	74	
20b		45,60	1 33 50.1		30 37 56	120	323	
21a		51,60	1 33 45.2		30 32 36	92	234	
21b		51,60	1 33 45.4		30 32 20	180	33	
21c		51,60	1 33 45.4		30 32 27	130	420	
22		51	1 33 44.9		30 33 9	230	196	
23		51	1 33 42.7		30 33 28	110	147	
25		46,51	1 33 44.7		30 36 22	--	--	Contains SNR G46
26		46,51	1 33 44.6		30 35 57	160	164	
27a		46,51,60	1 33 45.4		30 36 45	140	64	
27b		46,51,60	1 33 45.7		30 36 50	130	123	
27c		46,51,60	1 33 45.8		30 37 0	26	70	
28		46,51	1 33 43.6		30 36 44	330	252	
29		45,46	1 33 47.6		30 38 37	940	657	
30a		45,46,60	1 33 40.3		30 37 22	120	137	
30b		45,46,60	1 33 41.0		30 37 13	69	120	
31		46,51	1 33 36.9		30 36 35	190	310	
32a		46,51,60	1 33 36.0		30 36 25	440	308	Also identified as SNR G35
32b		46,51,60	1 33 34.8		30 36 30	140	232	

33a		46,60	1 33 34.7		30 36 59	7.2	12	
33b		46,60	1 33 34.9		30 37 4	550	356	
34		46	1 33 32.9		30 36 52	30	51	
35a		46,60	1 33 38.5		30 38 4	180	280	
35b		46,60	1 33 39.2		30 38 7	260	199	
36		46	1 33 39.0		30 37 57	400	350	Area E of 36 is probably SNR S10
37		45,46	1 33 48.0		30 39 15	220	167	
38a		45,46,60	1 33 43.2		30 38 44	160	229	
38b		45,46,60	1 33 43.9		30 38 48	61	40	
38c		45,46,60	1 33 44.2		30 38 46	150	231	
38d		45,46,60	1 33 45.5		30 38 39	46	114	
39a		45,46,61	1 33 41.0		30 39 9	14	31	
39b		45,46,61	1 33 41.8		30 38 55	120	267	
39c		45,46,61	1 33 42.2		30 39 18	71	245	
39d		45,46,61	1 33 42.7		30 38 58	280	294	
40		45,46	1 33 43.6		30 39 5	820	567	
41		46	1 33 34.6		30 39 18	310	383	
42a		46,61	1 33 35.1		30 39 24	310	249	
42b		46,61	1 33 35.5		30 39 33	25	99	
43a		45,46,61	1 33 48.6		30 39 38	20	29	
43b		45,46,61	1 33 48.2		30 39 33	53	49	Also identified as SNR G48
43c		45,46,61	1 33 48.2		30 39 28	29	49	
44		45,46	1 33 42.0		30 39 37	58	131	
45		46,47	1 33 28.6		30 40 21	1400	877	
46		46,47	1 33 29.5		30 40 43	340	281	
47		45,46	1 33 49.0		30 39 50	430	233	
48		45,46	1 33 41.2		30 40 41	140	380	
49		40,41,46	1 33 33.6		30 41 32	--	--	NGC 595, contains SNR G32
50		46	1 33 39.7		30 41 2	71	142	
51		45,46	1 33 43.9		30 40 43	260	445	
52		45,46	1 33 43.3		30 40 55	120	105	
53		41,45,46	1 33 40.7		30 41 30	420	696	
54		40,41,46	1 33 35.6		30 42 27	350	656	
55a		40,41,46,61	1 33 36.4		30 42 55	160	399	

55b		40,41,46,61	1 33 35.7		30 42 43	63	229	
56a		45,46,61	1 33 40.7		30 41 30	380	589	
56b		45,46,61	1 33 39.1		30 41 34	36	94	
57		40,41	1 33 36.4		30 43 18	89	159	
58		40,41	1 33 39.2		30 42 58	75	230	
59		40,41,45,46	1 33 40.4		30 42 49	240	349	
60		40,45,46	1 33 45.8		30 42 43	300	566	
61		40	1 33 42.5		30 44 36	530	329	
62a		40,61	1 33 44.6		30 44 47	620	557	
62b		40,61	1 33 43.8		30 44 35	190	142	
62c		40,61	1 33 44.2		30 44 36	820	266	
62d		40,61	1 33 43.5		30 44 22	30	79	
62e		40,61	1 33 43.9		30 44 13	22	80	
62f		40,61	1 33 45.2		30 44 27	32	50	
62g		40,61	1 33 44.9		30 44 14	9.8	48	
63		40,45	1 33 47.6		30 43 47	680	1074	
64a		45,61	1 33 50.1		30 41 34	61	121	
64b		45,61	1 33 49.9		30 41 19	120	203	
64c		45,61	1 33 50.3		30 41 21	65	156	
64d		45,61	1 33 50.1		30 41 15	84	73	
65a		40,45,61	1 33 51.7		30 43 50	51	107	
65b		40,45,61	1 33 52.5		30 43 46	380	360	
66a		45,61	1 33 51.3		30 40 50	450	434	
66b		45,61	1 33 50.5		30 40 40	43	111	
66c		45,61	1 33 51.0		30 40 36	26	86	
69		45	1 33 50.9		30 39 46	210	158	
70		45	1 33 51.1		30 40 17	62	96	
71a		45,62	1 33 52.5		30 40 24	15	22	
71b		45,62	1 33 52.0		30 40 19	56	103	
72		39,40,45	1 34 1.0		30 43 51	310	176	
73		40,45	1 33 57.4		30 42 11	610	309	
74a		40,45,62	1 33 57.9		30 42 27	150	186	
74b		40,45,62	1 33 57.8		30 42 18	120	106	
75		45	1 33 56.0		30 41 12	79	112	
76a		45,62	1 33 56.4		30 40 58	110	94	

76b		45,62	1 33 56.0		30 40 56	9.2	26	
77a		44,45,62	1 34 7.7		30 41 41	95	184	
77b		44,45,62	1 34 6.5		30 41 24	17	66	
77c		44,45,62	1 34 5.9		30 41 42	1300	950	
78		44,45	1 34 10.7		30 42 32	42	157	
79a		45,62	1 33 59.7		30 40 47	200	166	
79b		45,62	1 34 0.1		30 40 43	180	98	
79c		45,62	1 34 0.4		30 40 50	77	97	
80		45	1 34 0.8		30 40 22	71	207	
81		45	1 34 1.3		30 39 59	390	505	
83		44,45	1 34 9.7		30 39 7	410	442	
84		45	1 33 56.8		30 39 17	26	72	
85a		44,45,62	1 34 6.8		30 39 22	12	16	
85b		44,45,62	1 34 7.4		30 39 24	250	346	
86		45	1 33 55.8		30 39 0	58	159	
87a		45,62	1 34 1.8		30 38 13	75	99	
87b		45,62	1 34 1.7		30 38 20	150	96	
87c		45,62	1 34 2.3		30 38 19	110	104	
87d		45,62	1 34 2.0		30 38 26	110	71	
87e		45,62	1 34 0.9		30 38 45	46	256	
87f		45,62	1 34 1.8		30 38 38	160	73	
87g		45,62	1 34 2.5		30 38 36	1200	457	
87h		45,62	1 34 3.3		30 38 42	130	131	
87i		45,62	1 34 3.3		30 39 0	110	110	
87j		45,62	1 34 2.6		30 38 58	36	40	
87k		45,62	1 34 2.2		30 39 6	50	121	
87l		45,62	1 34 2.6		30 39 9	21	85	
88a		44,50,62	1 34 17.2		30 37 13	140	295	
88b		44,50,62	1 34 16.4		30 37 12	700	274	
88c		44,50,62	1 34 15.1		30 37 22	100	93	
88d		44,50,62	1 34 16.0		30 37 6	330	184	
88e		44,50,62	1 34 15.5		30 37 9	1300	624	
88f		44,50,62	1 34 13.7		30 37 8	25	111	
89Aa		45,62	1 34 1.2		30 38 1	47	115	
89Ab		45,62	1 34 1.2		30 38 7	29	64	

89Ac		45,62	1 34 0.8		30 38 18	29	58	
89Ad		45,62	1 34 0.4		30 38 21	66	148	
89Ba		45,63	1 34 1.8		30 37 29	21	40	
89Bb		45,63	1 34 2.2		30 37 36	46	119	
89Bc		45,63	1 34 2.5		30 37 40	79	203	
89Bd		45,63	1 34 2.8		30 37 39	12	12	
90		44,45	1 34 4.1		30 38 7	94	103	
91		44,45	1 34 6.5		30 37 47	120	214	
92		45	1 33 55.2		30 38 56	--	--	
93a		45,63	1 33 52.5		30 39 7	420	478	
93b		45,63	1 33 52.7		30 39 14	260	71	
93c		45,63	1 33 52.3		30 39 20	540	116	
93d		45,63	1 33 52.2		30 39 28	160	151	
93e		45,63	1 33 52.8		30 39 23	110	76	
94		45	1 34 0.2		30 38 10	--	--	
95a		50,63	1 34 10.9		30 36 13	420	334	
95b		50,63	1 34 10.6		30 36 25	52	87	
95c		50,63	1 34 11.3		30 36 25	36	69	
95d		50,63	1 34 12.1		30 36 12	12	40	
95e		50,63	1 34 10.2		30 36 23	1.4	5	
95Aa		50,63	1 34 10.7		30 35 53	16	32	
95Ab		50,63	1 34 11.2		30 35 46	120	103	
96		45	1 33 56.6		30 38 19	130	399	
97a		45,63	1 34 1.7		30 37 18	84	108	
97b		45,63	1 34 1.6		30 37 22	75	147	
99a		45,63	1 33 53.9		30 38 37	18	51	
99b		45,63	1 33 53.2		30 38 49	1000	585	
100a		50,63	1 34 2.6		30 36 32	47	91	
100b		50,63	1 34 3.2		30 36 33	62	124	
100c		50,63	1 34 1.7		30 36 30	650	456	
100d		50,63	1 33 59.4		30 36 19	140	444	
100e		50,63	1 34 1.9		30 36 8	800	893	
200		51	1 33 39.4		30 32 36	560	439	
201		51	1 33 40.9		30 32 6	220	469	
202		51	1 33 39.2		30 32 6	130	134	

203		51	1 33 37.6		30 31 58	56	93	
204		51	1 33 39.1		30 31 32	--	--	
206		51	1 33 37.7		30 30 7	100	205	Also identified as SNR S9
207a		51,56,63	1 33 35.1		30 29 53	21	57	
207b		51,56,63	1 33 35.1		30 29 50	37	79	
208a		51,63	1 33 33.1		30 32 17	5.5	12	
208b		51,63	1 33 32.5		30 32 12	9.7	28	
208c		51,63	1 33 32.7		30 32 6	23	49	
208d		51,63	1 33 33.2		30 32 11	150	83	
208e		51,63	1 33 33.2		30 32 1	290	255	
208f		51,63	1 33 33.8		30 32 13	11	19	
208g		51,63	1 33 33.8		30 32 8	52	26	
208h		51,63	1 33 34.2		30 32 8	390	131	
208i		51,63	1 33 34.9		30 32 10	100	91	
208j		51,63	1 33 34.5		30 31 58	59	92	
209a		51,64	1 33 31.9		30 33 52	19	78	
209b		51,64	1 33 33.3		30 33 43	81	263	
209c		51,64	1 33 33.3		30 33 36	130	229	
209d		51,64	1 33 33.2		30 33 30	19	44	
209e		51,64	1 33 33.6		30 33 23	21	56	
210		51,52	1 33 31.2		30 33 31	190	124	Also identified as SNR G31
211a		51,52,64	1 33 30.6		30 32 1	130	119	
211b		51,52,64	1 33 30.1		30 32 4	44	61	
211c		51,52,64	1 33 29.9		30 31 58	32	44	
212a		51,52,64	1 33 29.1		30 31 51	30	35	
212b		51,52,64	1 33 29.1		30 31 54	57	52	
212c		51,52,64	1 33 29.3		30 31 54	27	27	
213a		51,52,64	1 33 27.5		30 31 59	14	35	
213b		51,52,64	1 33 27.7		30 31 54	280	259	
213c		51,52,64	1 33 28.1		30 31 48	72	59	
213d		51,52,64	1 33 28.1		30 31 40	56	89	
214a		51,52,64	1 33 29.6		30 31 50	47	20	
214b		51,52,64	1 33 30.1		30 31 46	890	278	
215		52	1 33 14.6		30 32 26	280	287	
216		52	1 33 11.0		30 29 57	--	--	

217a	52,64	1 33 9.3	30 29 49	710	583	
217b	52,64	1 33 7.9	30 29 43	410	549	
218	52,53	1 33 0.5	30 30 43	2400	2648	Contains SNRs G10, G12, G13
220	52	1 33 11.0	30 27 42	550	481	
221	52	1 33 9.9	30 27 21	600	700	
222	57	1 33 24.5	30 25 33	--	--	IC 135
223	52	1 33 24.5	30 30 53	--	--	
224a	52,64	1 33 20.5	30 32 0	82	190	
224b	52,64	1 33 21.1	30 32 4	49	148	
225	52	1 33 21.1	30 31 25	60	166	Also identified as SNR G23
226	52	1 33 12.0	30 30 26	170	315	
227	52	1 33 11.5	30 30 16	--	--	
229	47	1 33 1.8	30 34 48	--	--	
230a	47,48,64	1 33 0.8	30 34 18	170	252	
230b	47,48,64	1 33 1.1	30 34 8	120	187	
231	47,48	1 32 59.6	30 34 43	170	246	
232a	48,64	1 32 57.8	30 34 40	79	118	
232b	48,64	1 32 58.0	30 34 49	27	53	
233a	47,48,65	1 32 57.3	30 35 43	16	42	
233b	47,48,65	1 32 56.5	30 35 46	76	209	
233c	47,48,65	1 32 57.2	30 35 55	370	771	
234a	48,65	1 32 55.7	30 35 34	67	167	
234b	48,65	1 32 56.6	30 35 36	58	108	
234c	48,65	1 32 56.8	30 35 29	22	72	
235	48	1 32 55.1	30 35 46	100	259	
236a	48,65	1 32 51.7	30 35 55	71	213	
236b	48,65	1 32 52.4	30 35 47	94	283	
237	48	1 32 52.5	30 34 58	370	426	
238	48	1 32 44.7	30 34 58	270	317	
239	53	1 32 57.0	30 32 51	100	115	
240a	53,65	1 32 56.1	30 32 22	130	166	
240b	53,65	1 32 56.1	30 32 28	140	144	
241	53	1 32 57.5	30 31 59	190	295	
243	53	1 32 56.5	30 27 17	220	495	
245a	52,57,65	1 33 18.5	30 26 45	96	282	

245b		52,57,65	1 33 20.9		30 27 8	4.0	16	
255a		57,65	1 33 10.0		30 23 5	590	652	
255b		57,65	1 33 11.0		30 23 3	150	202	
256a		57,65	1 33 11.8		30 23 18	800	751	
256b		57,65	1 33 11.7		30 23 33	370	443	
256c		57,65	1 33 13.3		30 23 29	30	103	
257		57	1 33 9.6		30 23 24	190	194	
258a		57,65	1 33 7.1		30 23 31	58	132	
258b		57,65	1 33 6.0		30 23 26	230	340	
259a		57,65	1 33 7.3		30 23 14	60	83	
259b		57,65	1 33 7.3		30 22 57	220	333	
259c		57,65	1 33 8.1		30 23 4	40	124	
261a		58,66	1 32 53.6		30 23 21	68	156	
261b		58,66	1 32 54.6		30 23 19	280	389	
262a		58,66	1 32 45.7		30 20 16	92	207	
262b		58,66	1 32 46.2		30 20 28	95	290	
263a		58,66	1 32 43.6		30 21 7	20	92	
263b		58,66	1 32 43.2		30 21 0	56	219	
263c		58, 66	1 32 42.4		30 20 54	58	208	
263d		58,66	1 32 42.1		30 20 37	2.7	12	Ring surrounding 263d is SNR G3
263e		58,66	1 32 40.9		30 20 43	8.1	35	
264		58	1 32 41.0		30 22 26	210	385	
265a		58,66	1 32 41.7		30 24 5	87	135	
265b		58,66	1 32 41.9		30 24 13	67	152	
265c		58,66	1 32 43.6		30 24 6	180	427	
265d		58,66	1 32 44.5		30 23 42	4.7	19	
266		58	1 32 40.7		30 24 23	270	497	
268		58	1 32 43.4		30 24 36	390	1249	
269		58	1 32 41.9		30 25 12	470	935	
270		58	1 32 46.5		30 25 3	160	301	
272		53	1 32 34.7		30 27 45	520	606	
273a		53,66	1 32 35.3		30 30 27	280	362	
273b		53,66	1 32 34.1		30 30 43	77	207	
274a		48,66	1 32 31.6		30 35 24	290	338	North part of 274 is SNR G2

274b		48,66	1 32 31.8		30 35 16	340	254	
274c		48,66	1 32 32.5		30 35 18	260	194	
274d		48,66	1 32 30.6		30 35 10	590	1668	
275a		48,66	1 32 29.7		30 36 7	80	187	
275b		48,66	1 32 29.3		30 36 18	24	64	
275c		48,66	1 32 29.9		30 36 25	4.9	17	
277		47	1 33 11.4		30 38 50	1100	760	NGC 592
278a		46,47,66	1 33 18.7		30 39 34	27	103	
278b		46,47,66	1 33 18.7		30 39 25	31	95	
278c		46,47,66	1 33 19.3		30 39 24	20	76	
279a		48,66	1 32 55.1		30 39 28	23	69	
279b		48,66	1 32 55.3		30 39 26	21	58	
279c		48,66	1 32 56.1		30 39 27	21	111	
280		48	1 32 45.5		30 39 4	--	--	NGC 588
283		48	1 32 56.7		30 40 38	120	556	Also identified as SNR G8
284		47	1 33 3.5		30 39 49	96	144	Also identified as SNR G15
285		42	1 33 2.7		30 41 4	92	67	
286		41,42	1 33 14.4		30 41 28	140	212	Also identified as SNR S2
287a		41,46,67	1 33 17.8		30 41 17	17	15	
287b		41,46,67	1 33 18.1		30 41 17	20	26	
287c		41,46,67	1 33 17.9		30 41 10	72	119	
288a		42,67	1 33 6.9		30 42 29	470	824	North part is SNR G16
288b		42,67	1 33 9.0		30 42 53	74	346	
289		42	1 32 58.3		30 44 25	72	172	
301a		39,40,67	1 33 55.0		30 45 38	140	230	IC 142
301b		39,40,67	1 33 54.9		30 45 33	64	93	IC 142
301c		39,40,67	1 33 55.6		30 45 17	710	471	IC 142
301d		39,40,67	1 33 55.1		30 45 17	940	241	IC 142
301e		39,40,67	1 33 54.1		30 45 15	140	227	IC 142, also identified as SNR G54
301f		39,40,67	1 33 55.6		30 45 8	43	38	IC 142
301g		39,40,67	1 33 54.8		30 45 4	49	129	IC 142
301h		39,40,67	1 33 53.3		30 45 16	120	455	IC 142
302		39	1 34 6.8		30 47 33	--	--	
601		41,42	1 33 14.1		30 45 53	180	374	
602a		41,42,67	1 33 16.8		30 45 52	38	57	

602b		41,42,67	1 33 16.5		30 45 58	120	499	
603a		41,42,67	1 33 16.4		30 46 46	67	94	
603b		41,42,67	1 33 15.9		30 46 44	58	110	
603c		41,42,67	1 33 16.5		30 46 36	53	132	
603d		41,42,67	1 33 15.0		30 46 48	3.2	18	
604a		41,42,67	1 33 16.1		30 47 5	74	111	
604b		41,42,67	1 33 16.4		30 47 8	21	103	
606		37,41	1 33 19.7		30 47 40	200	731	
607a		41,67	1 33 21.1		30 47 34	33	146	
607b		41,67	1 33 21.5		30 47 39	86	278	
607c		41,67	1 33 22.5		30 47 50	8.4	40	
608a		41,67	1 33 25.2		30 47 24	95	402	
608b		41,67	1 33 26.2		30 47 32	33	143	Area just NE is SNR G26
608c		41,67	1 33 25.8		30 48 1	4.2	22	
609a		41,67	1 33 28.2		30 47 39	100	273	
609b		41,67	1 33 27.6		30 47 33	30	80	
609c		41,67	1 33 27.2		30 47 38	15	41	
609d		41,67	1 33 27.3		30 47 42	9.1	24	
609e		41,67	1 33 27.1		30 47 49	2.2	14	
609f		41,67	1 33 27.5		30 47 41	14	30	
609g		41,67	1 33 29.9		30 47 41	120	578	Also identified as SNR G30
610		40,41	1 33 33.4		30 47 2	180	387	
611a		40,41,68	1 33 33.4		30 47 38	59	268	
611b		40,41,68	1 33 34.6		30 47 35	5.1	36	
612a		40,41,68	1 33 40.4		30 45 38	110	225	
612b		40,41,68	1 33 40.3		30 45 51	150	134	
612c		40,41,68	1 33 40.2		30 45 56	26	43	
612d		40,41,68	1 33 39.8		30 45 53	39	71	
612e		40,41,68	1 33 39.4		30 45 57	31	95	
613		40	1 33 41.6		30 47 31	64	77	
614		36,40	1 33 43.9		30 48 29	5.7	24	
615a		36,40,68	1 33 36.1		30 48 24	28	107	
615b		36,40,68	1 33 36.2		30 48 36	7.2	29	
615c		36,40,68	1 33 36.1		30 48 31	6.7	29	
615d		36,40,68	1 33 36.7		30 48 28	15	56	

615e		36,40,68	1 33 36.7	30 48 21	5.0	25	
615f		36,40,68	1 33 36.9	30 48 29	4.7	27	
615g		36,40,68	1 33 37.5	30 48 21	10	68	
615h		36,40,68	1 33 34.5	30 48 42	42	259	
616		37	1 32 54.3	30 50 25	84	224	
618a		37,68	1 33 12.9	30 50 27	130	101	
618b		37,68	1 33 13.6	30 50 28	15	35	
618c		37,68	1 33 13.6	30 50 23	6.6	39	
619		37	1 33 14.2	30 51 4	13	23	
620		37	1 33 14.4	30 51 32	110	66	
621		37	1 33 13.3	30 52 5	47	207	
622		37	1 33 15.2	30 52 7	22	38	
623a		37,68	1 33 15.1	30 52 37	83	185	IC 133
623b		37,68	1 33 16.1	30 52 42	900	436	IC 133
623c		37,68	1 33 15.8	30 52 52	200	332	IC 133
623d		37,68	1 33 16.6	30 52 54	110	334	IC 133
624		37	1 33 14.5	30 53 15	470	586	IC 133
625		37	1 33 14.5	30 53 34	300	690	IC 133
626		37	1 33 16.2	30 53 59	470	617	IC 133
627		36	1 33 24.0	30 50 22	32	138	
628		36	1 33 34.3	30 50 3	86	140	
629		36	1 33 36.1	30 50 31	360	238	
630		36	1 33 36.4	30 50 52	690	2040	
631a		36,68	1 33 47.1	30 50 53	44	87	
631b		36,68	1 33 47.8	30 50 49	300	435	
631c		36,68	1 33 47.6	30 50 38	9.9	8	
632a		36,68	1 33 49.5	30 51 36	350	827	
632b		36,68	1 33 48.7	30 51 49	100	355	
633a		36,68	1 33 46.0	30 52 9	43	19	
633b		36,68	1 33 46.0	30 52 7	15	13	
634		36	1 33 41.6	30 52 37	--	--	Contains SNR G40
635a		36,68	1 33 42.9	30 53 9	43	225	
635b		36,68	1 33 42.2	30 53 6	20	111	
636		36	1 33 47.3	30 54 2	--	--	
651a		34,69	1 34 29.0	30 57 0	85	370	

651b		34,69	1 34 29.1		30 57 22	96	152	
651c		34,69	1 34 29.5		30 57 9	460	315	
651d		34,69	1 34 30.3		30 57 13	200	349	
652a		34,69	1 34 33.0		30 56 38	12	61	
652b		34,69	1 34 32.5		30 56 33	25	142	
652c		34,69	1 34 31.3		30 56 48	13	68	
652d		34,69	1 34 31.1		30 56 44	8.9	50	
652e		34,69	1 34 30.5		30 56 35	4.2	24	
653a		34,69	1 34 38.6		30 56 12	94	241	
653b		34,69	1 34 38.2		30 56 10	34	83	
654		34	1 34 40.2		30 56 3	26	132	
655a		34,69	1 34 42.6		30 55 46	4.5	10	
655b		34,69	1 34 42.5		30 55 42	21	28	
655c		34,69	1 34 41.9		30 55 34	7.0	45	
655d		34,69	1 34 42.7		30 55 31	4.7	22	
656		34	1 34 46.1		30 55 17	49	214	
657		34	1 34 49.5		30 54 55	--	--	
659		34,35	1 34 23.8		30 57 1	150	224	
660a		35,83	1 34 15.6		30 54 55	58	343	See note d
660b		35,83	1 34 15.3		30 55 7	2.4	14	See note d
660c		35,83	1 34 16.1		30 56 1	4.8	21	See note d
661		35	1 34 21.4		30 54 20	59	66	
662		34,35	1 34 24.4		30 54 48	170	323	
663a		34,69	1 34 29.7		30 53 8	51	240	
663b		34,69	1 34 27.8		30 53 44	24	168	
663c		34,69	1 34 26.8		30 53 56	13	97	
663d		34,69	1 34 25.5		30 53 32	19	133	
664a		34,69	1 34 28.1		30 52 43	180	879	
664b		34,69	1 34 27.0		30 52 52	85	476	
664c		34,69	1 34 25.9		30 53 5	21	103	
664d		34,69	1 34 25.4		30 52 44	37	248	
664e		34,69	1 34 26.3		30 52 38	36	192	
665a		35,69	1 34 14.7		30 53 12	93	543	
665b		35,69	1 34 15.5		30 53 4	76	275	

666a	35,69	1 34 16.6	30 52 23	23	94	BCLMP 666 contains SNR G84
666b	35,69	1 34 17.0	30 52 34	42	172	
666c	35,69	1 34 16.3	30 52 41	92	375	
666d	35,69	1 34 14.8	30 52 36	71	289	
667a	35,69	1 34 13.7	30 52 30	86	184	BCLMP 667 contains SNR DDB12
667b	35,69	1 34 13.9	30 52 19	75	117	
667c	35,69	1 34 12.8	30 52 8	94	282	
667d	35,69	1 34 13.3	30 52 16	78	178	
667e	35,69	1 34 12.0	30 52 19	160	451	
668a	35,70	1 34 3.9	30 54 56	280	331	
668b	35,70	1 34 3.3	30 54 43	24	78	
669	35	1 34 58.8	30 355 28	39	61	
670	35	1 34 3.4	30 53 5	10	32	
671a	35,70	1 34 7.0	30 51 41	55	115	
671b	35,70	1 34 6.3	30 51 38	290	77	
672a	35,70	1 34 4.2	30 51 25	27	115	
672b	35,70	1 34 2.6	30 51 21	460	659	
673	35	1 34 1.0	30 51 8	61	153	
674	35	1 33 58.2	30 51 6	45	220	
675	39	1 34 2.2	30 50 24	55	80	
676	38	1 34 32.0	30 49 0	130	344	
677	38	1 34 40.4	30 48 13	130	367	
678	38	1 34 41.2	30 47 3	70	157	
679a	38,70	1 34 37.2	30 47 12	57	38	
679b	38,70	1 34 36.9	30 47 10	30	25	
680	38	1 34 32.8	30 47 6	--	--	NGC 604, contains SNR G94
681	38	1 34 36.5	30 46 39	--	--	
682a	39,70	1 34 23.3	30 46 52	34	108	
682b	39,70	1 34 23.3	30 46 46	27	90	
682c	39,70	1 34 22.9	30 46 52	79	265	
683a	39,70	1 34 23.3	30 48 48	49	111	
683b	39,70	1 34 23.3	30 48 39	36	175	
684a	39,70	1 34 22.1	30 48 41	22	24	

684b	39,70	1 34 21.7	30 48 38	250	353	
684c	39,70	1 34 19.9	30 48 23	100	244	
685a	39,70	1 34 20.2	30 49 3	180	219	
685b	39,70	1 34 19.8	30 48 49	240	436	
685c	39,70	1 34 19.8	30 48 56	18	15	
686	39	1 34 17.8	30 48 37	160	242	
687a	39,70	1 34 14.1	30 48 14	75	278	Also identified as SNR G77
687b	39,70	1 34 14.0	30 48 22	39	103	
687c	39,70	1 34 13.8	30 48 29	130	183	
687d	39,70	1 34 13.0	30 48 33	34	110	Also identified as SNR G75
687e	39,70	1 34 12.4	30 48 38	15	101	
688	39	1 34 11.1	30 46 39	--	--	
689	39	1 34 8.3	30 46 46	--	--	
690	39	1 34 6.9	30 47 1	35	135	
691a	35,71	1 34 17.6	30 52 16	130	480	
691b	35,71	1 34 17.2	30 51 53	240	219	
691c	35,71	1 34 16.6	30 52 2	42	51	
691d	35,71	1 34 16.4	30 51 49	1200	438	
691e	35,71	1 34 16.9	30 51 47	270	469	Also identified as SNR G83
692a	39,40,71	1 34 3.5	30 48 7	18	45	
692b	39,40,71	1 34 3.6	30 47 53	92	237	
693	39,40	1 34 4.1	30 46 53	240	341	
694	40	1 33 51.8	30 47 12	60	80	
695	39,40	1 33 58.8	30 49 7	90	112	
696	39,40	1 33 57.8	30 48 37	38	182	
697	40	1 33 52.2	30 46 26	64	86	
698	39,40	1 34 1.6	30 44 55	95	82	
699	39,40	1 34 0.4	30 44 34	24	68	
702a	50,71	1 34 10.1	30 31 55	83	108	
702b	50,71	1 34 10.1	30 32 3	7.9	18	
702c	50,71	1 34 9.8	30 32 1	6.5	12	
702d	50,71	1 34 9.6	30 31 55	69	274	
702e	50,71	1 34 10.7	30 32 1	3.3	18	
703	50	1 33 59.8	30 32 41	160	84	
704a	50,51,71	1 33 55.3	30 32 3	54	244	

704b	50,51,71	1 33 57.9	30 32 8	81	182	Also identified as SNR G59
704c	50,51,71	1 33 57.8	30 32 16	18	39	
704d	50,51,71	1 33 57.5	30 32 21	11	31	
704e	50,51,71	1 33 57.8	30 32 23	34	157	
704f	50,51,71	1 33 58.3	30 32 21	25	44	
704g	50,51,71	1 33 58.8	30 32 22	42	170	
705	49	1 34 40.3	30 31 18	110	95	
706	49	1 34 42.2	30 31 40	210	237	
707a	50,51,71	1 33 55.5	30 33 38	81	267	
707b	50,51,71	1 33 55.6	30 33 47	110	192	
707c	50,51,71	1 33 55.7	30 33 56	220	359	
708	50	1 34 8.2	30 34 2	160	195	
709a	50,71	1 34 8.7	30 34 26	7.4	38	
709b	50,71	1 34 9.3	30 34 25	45	115	
709c	50,71	1 34 9.8	30 34 38	63	254	
709d	50,71	1 34 10.4	30 34 51	24	78	
709e	50,71	1 34 11.5	30 34 50	19	78	
709f	50,71	1 34 10.6	30 34 43	12	32	
709g	50,71	1 34 10.9	30 34 47	14	30	
709h	50,71	1 34 11.4	30 34 42	34	82	
709i	50,71	1 34 11.0	30 34 35	55	109	
709j	50,71	1 34 12.5	30 34 43	24	108	
709k	50,71	1 34 11.2	30 34 28	30	52	
709l	50,71	1 34 11.2	30 34 20	61	89	
709m	50,71	1 34 11.7	30 34 20	24	76	
709n	50,71	1 34 10.8	30 34 24	77	252	
709o	50,71	1 34 10.3	30 34 9	12	53	
710	50	1 34 13.8	30 33 42	710	373	
711a	50,71	1 34 17.3	30 33 41	490	350	
711b	50,71	1 34 17.3	30 33 56	110	250	
711c	50,71	1 34 16.2	30 34 6	67	248	
711d	50,71	1 34 16.4	30 33 50	48	65	
711e	50,71	1 34 16.4	30 33 38	58	104	
711f	50,71	1 34 15.7	30 33 56	51	64	
711g	50,71	1 34 14.6	30 33 44	210	254	

711h	50,71	1 34 14.7	30 33 56	87	218	
712a	50,71	1 34 19.5	30 33 43	270	379	Also identified as SNR G86
712b	50,71	1 34 18.6	30 33 55	34	113	
712c	50,71	1 34 20.1	30 33 35	14	66	
713	50	1 34 18.7	30 34 28	260	410	
714a	50,71	1 34 16.0	30 35 5	39	122	
714b	50,71	1 34 16.4	30 34 56	18	53	
714c	50,71	1 34 16.7	30 34 49	31	105	
714d	50,71	1 34 15.5	30 35 17	14	78	
714e	50,71	1 34 14.8	30 35 9	12	28	
714f	50,71	1 34 15.1	30 35 2	41	94	
714g	50,71	1 34 13.7	30 34 47	180	130	
714h	50,71	1 34 13.8	30 34 38	36	87	
714i	50,71	1 34 14.3	30 34 28	180	430	Also identified as SNR G80
714j	50,71	1 34 14.4	30 34 38	92	116	
714k	50,71	1 34 15.5	30 34 19	170	447	
714l	50,71	1 34 14.3	30 34 51	230	429	
714m	50,71	1 34 15.2	30 34 45	77	160	
715a	50,72	1 34 22.4	30 33 17	290	719	
715b	50,72	1 34 23.2	30 33 3	180	493	
716	50	1 34 22.2	30 32 35	59	154	
717a	49,72	1 34 33.9	30 35 24	25	55	
717b	49,72	1 34 34.1	30 35 13	110	147	
718	49	1 34 35.0	30 35 49	59	222	
719	49	1 34 32.5	30 36 5	75	183	
720	50	1 34 16.3	30 36 42	240	359	
721	44	1 34 21.2	30 37 56	98	283	
722	44	1 34 18.5	30 38 39	--	--	
723	44	1 34 16.9	30 39 17	23	57	
724a	44,72	1 34 19.7	30 39 23	21	82	
724b	44,72	1 34 19.5	30 39 17	29	112	
724c	44,72	1 34 18.5	30 39 7	7.2	44	
725	43	1 34 46.5	30 38 13	62	129	
726	43	1 34 47.2	30 37 52	9.4	29	
729	43	1 34 38.5	30 37 49	63	341	Also identified as SNR G96

730	43	1 34 37.1	30 38 14	--	--	
731	44	1 34 25.0	30 39 48	37	84	
732	44,45	1 34 13.1	30 40 2	150	391	
733	44	1 34 13.1	30 40 14	69	270	
734	44	1 34 30.0	30 40 33	190	284	
735	44	1 34 30.5	30 41 20	70	163	
736	44	1 34 33.7	30 41 3	31	36	
737	43	1 34 38.0	30 40 53	74	151	
738a	43,72	1 34 39.0	30 41 4	93	113	
738b	43,72	1 34 39.5	30 41 3	41	83	
739	43	1 34 41.9	30 41 8	50	193	
740a	43,72	1 34 39.6	30 41 58	300	385	
740b	43,72	1 34 39.5	30 41 46	660	454	
740c	43,72	1 34 39.1	30 41 34	220	300	
740d	43,72	1 34 40.1	30 41 18	36	138	
741	43	1 34 36.8	30 41 39	19	95	
742	43	1 34 38.9	30 41 18	120	204	
743a	44,72	1 34 14.3	30 42 18	42	146	
743b	44,72	1 34 14.5	30 42 26	35	130	
743c	44,72	1 34 15.2	30 42 26	31	124	
743d	44,72	1 34 12.0	30 42 40	30	177	
743e	44,72	1 34 11.7	30 42 55	8.6	43	
744	44	1 34 13.2	30 41 56	63	93	
745	49	1 34 37.3	30 35 0	--	--	
746a	39,72	1 34 19.3	30 44 18	29	181	
746b	39,72	1 34 20.6	30 44 34	21	138	
747	39	1 34 12.7	30 45 5	83	135	
748	38,43	1 34 36.5	30 44 11	--	--	
749	38,43	1 34 38.9	30 44 6	--	--	
750a	43,72	1 34 40.5	30 43 36	77	156	
750b	43,72	1 34 40.9	30 43 32	140	187	See note e, also identified as SNR G97
750c	43,72	1 34 40.0	30 43 27	67	197	
750d	43,72	1 34 41.1	30 43 9	5.4	29	
750e	43,72	1 34 39.1	30 43 16	8.6	44	

751	50	1 34 6.9	30 36 30	120	462	Contains SNR G70
752	43	1 34 55.3	30 41 18	140	543	
753	43	1 34 59.2	30 41 35	--	--	
756	43	1 35 7.1	30 41 45	--	--	
758	55,56	1 33 56.5	30 27 16	250	849	
1001	46,51	1 33 45.9	30 35 58	240	269	
1002	51	1 33 43.6	30 35 32	--	--	
1003	51	1 33 38.8	30 35 4	150	274	
1004	51	1 33 36.6	30 35 41	100	244	
1005	46	1 33 33.4	30 37 21	430	1021	
1006a	46,72	1 33 31.5	30 38 1	63	107	
1006b	46,72	1 33 31.4	30 37 55	69	130	
1006c	46,72	1 33 30.9	30 38 2	150	307	
1006d	46,72	1 33 30.3	30 38 8	20	70	
1006e	46,72	1 33 29.6	30 37 44	30	159	
1007a	46,47,73	1 33 25.8	30 38 58	120	94	
1007b	46,47,73	1 33 26.3	30 38 53	170	191	
1007c	46,47,73	1 33 26.9	30 39 2	210	256	
1008a	46,73	1 33 35.9	30 38 43	53	34	
1008b	46,73	1 33 35.2	30 38 43	150	535	
1501a	51,73	1 33 50.3	30 32 30	71	174	
1501b	51,73	1 33 51.5	30 32 32	40	109	
1502a	51,73	1 33 43.4	30 32 12	76	213	
1502b	51,73	1 33 44.4	30 31 45	99	354	

BCLMP emission regions are labeled on the M33 charts with their number and no prefix. Separate sub-components of complex regions are given lowercase Arabic letters and are identified on Charts 59-83.

Notes to table:

[a]Flux is in units of 10^{-15} erg cm^{-2} s^{-1} (from Wyder et al., 1997).

[b]Area is in units of arcsec^2 (from Wyder et al., 1997).

[c]SNR candidates are from Sabbadin (1979, identified with prefix S), D'Odorico, Dopita, and Benvenuti (1980, prefix DDB), and Gordon et al. (1998, prefix G). Gordon et al. (1998) cross-reference various identifications of the SNRs.

[d]BCLMP 660 appears as a complete ring in Figure 7b of Courtès et al. (1987). We detect only a partial ring in our images.

[e]SNR DDB18 is contained within BCLMP 750. Gordon et al. (1998) do not identify it with G97, but we believe they are the same object.

Table 4b. M33 Emission Regions Identified by Courtès et al. (1987)

ID		Chart	RA(2000)		DEC(2000)	Flux[a]	Area[b]	Comments
CPSDP1		58	1 32 42.4		30 20 22	24	78	
1Aa		50,51,73	1 33 58.2		30 34 15	190	472	
1Ab		50,51,73	1 34 0.5		30 34 17	660	666	Contains SNR G65[c]
1Ac		50,51,73	1 34 0.5		30 34 36	37	95	
1Ad		50,51,73	1 33 59.9		30 34 27	240	128	
1Ae		50,51,73	1 33 59.7		30 34 34	480	394	
1Af		50,51,73	1 34 0.6		30 34 45	44	150	
1Ag		50,51,73	1 33 59.3		30 34 50	150	311	
2		58	1 32 43.0		30 21 58	30	46	
2Aa		51,73	1 33 51.7		30 33 34	69	130	
2Ab		51,73	1 33 51.3		30 33 21	13	40	
2Ac		51,73	1 33 51.8		30 33 23	36	65	
2Ad		51,73	1 33 52.0		30 33 29	42	84	
2Ae		51,73	1 33 52.6		30 33 30	28	75	
3a		58,73	1 32 38.9		30 25 4	11	20	
3b		58,73	1 32 38.1		30 24 49	140	246	
3c		58,73	1 32 37.7		30 24 39	62	158	
3A		58	1 32 39.3		30 24 51	130	265	
3B		58	1 32 40.2		30 24 43	62	132	
4		58	1 32 35.5		30 25 31	3.9	16	
4Aa		50,51,73	1 33 55.9		30 35 12	5.9	31	
4Ab		50,51,73	1 33 57.3		30 35 10	120	237	
5a		58,73	1 32 34.3		30 26 24	28	76	
5b		58,73	1 32 34.9		30 26 27	19	72	
6		53	1 32 34.3		30 27 5	280	316	
7		53	1 32 35.1		30 27 13	170	206	
8		58	1 32 44.7		30 22 14	7.6	33	Also identified as SNR G4
9		58	1 32 45.2		30 23 21	--	--	
10		51	1 33 42.3		30 33 11	250	192	
10A		51	1 33 42.1		30 32 56	480	340	
11		58	1 33 42.1		30 32 59	--	--	
13a		58,74	1 32 54.1		30 25 27	22	66	Also identified as SNR G7

13b		58,74	1 32 54.5		30 25 31	4.6	19	
14		50,51	1 33 54.6		30 36 46	34	100	
15		58	1 32 55.1		30 23 43	18	38	
16		58	1 32 57.5		30 23 43	8.1	37	
17		58	1 32 57.4		30 23 16	5.7	25	
18		58	1 32 59.2		30 24 16	38	87	
19a		58,74	1 32 59.8		30 23 42	18	70	
19b		58,74	1 33 0.0		30 23 16	3.3	12	
20a		58,74	1 32 56.5		30 25 56	12	28	
20b		58,74	1 32 56.1		30 25 52	36	123	
21		58	1 33 0.3		30 24 32	34	55	
22		57,58	1 33 1.4		30 24 31	130	101	
23		52	1 33 14.0		30 29 41	90	92	
23Aa		57,74	1 33 4.5		30 23 26	4.9	16	
23Ab		57,74	1 33 4.0		30 23 22	19	57	
23Ac		57,74	1 33 3.7		30 23 11	25	88	
23Ad		57,74	1 33 3.6		30 23 25	64	264	
24		57	1 33 3.4		30 22 1	1.4	5	
25		52	1 33 2.8		30 29 31	39	68	
26		52	1 33 4.0		30 29 46	4.0	14	
26Aa		52,74	1 33 6.8		30 30 11	79	206	
26Ab		52,74	1 33 6.2		30 30 6	51	179	
26Ac		52,74	1 33 5.6		30 30 2	9.4	40	
26Ad		52,74	1 33 8.4		30 30 22	34	113	
26Ae		52,74	1 33 7.8		30 30 21	69	272	
27		52	1 33 4.2		30 28 54	29	91	
27A		52	1 33 4.8		30 29 29	26	73	
28		52	1 33 8.7		30 27 53	100	362	
30		57	1 33 12.3		30 23 1	33	59	
30A		57	1 33 15.8		30 23 47	79	287	
31		52	1 33 13.0		30 28 9	45	176	
32		57	1 33 13.5		30 22 35	7.4	12	
33a		51,74	1 33 36.8		30 33 46	21	72	
33b		51,74	1 33 37.8		30 33 51	12	48	
33Aa		51,74	1 33 37.1		30 32 51	20	71	Also identified as SNR G36

33Ab		51,74	1 33 37.1		30 33 0	6.1	21	
33Ac		51,74	1 33 37.1		30 33 9	56	248	
33Ad		51,74	1 33 37.3		30 33 27	12	48	
34		52	1 33 14.1		30 27 10	130	117	
36		57	1 33 20.4		30 26 34	11	39	
37		57	1 33 21.8		30 25 59	19	76	
38		57	1 33 21.9		30 25 40	14	29	
40		52,57	1 33 22.5		30 27 3	9.1	25	Also identified as SNR G24
41		57	1 33 23.4		30 26 32	22	62	
42		57	1 33 23.9		30 26 13	33	53	Also identified as SNR G25
43		57	1 33 26.2		30 24 6	11	40	
43Aa		57,74	1 33 21.2		30 26 26	3.6	14	CPSDP 43A is possibly SNR DDB5
43Ab		57,74	1 33 20.0		30 26 20	8.1	35	
43Ac		57,74	1 33 20.0		30 26 7	6.7	29	
43Ad		57,74	1 33 20.8		30 25 55	1.4	5	
47		57	1 33 32.2		30 20 13	34	47	
48		57	1 33 33.0		30 20 27	56	85	
50		56	1 33 35.5		30 25 17	--	--	
51		56	1 33 35.4		30 25 54	21	36	
53a		51,74	1 33 46.7		30 32 42	96	147	
53b		51,74	1 33 45.6		30 32 44	28	74	
54		53	1 32 30.4		30 32 37	--	--	
58a		42,75	1 32 45.7		30 44 42	19	37	
58b		42,75	1 32 45.9		30 44 38	18	42	
60		48	1 32 46.0		30 40 45	7.7	43	
61		48	1 32 42.5		30 35 52	25	70	
62		48	1 32 44.3		30 36 1	81	125	
63		48	1 32 46.8		30 34 11	63	157	
64		48	1 32 51.8		30 38 7	45	219	Also identified as SNR G5
65		48	1 32 56.7		30 40 11	30	122	
65Aa		40,45,75	1 33 50.8		30 43 59	56	101	
65Ab		40,45,75	1 33 50.4		30 43 46	95	236	Also identified as SNR G50
65Ac		40,45,75	1 33 51.2		30 44 2	1.2	6	
66a		42,75	1 32 58.5		30 41 26	24	62	

66b		42,75	1 32 59.3		30 41 20	19	84	
67		42	1 32 58.8		30 41 5	24	131	
68		42	1 33 1.2		30 41 0	12	54	
69a		48,75	1 32 53.2		30 37 46	79	182	Ring is SNR G6
69b		48,75	1 32 53.9		30 37 52	26	91	
69c		48,75	1 32 53.0		30 37 40	19	63	
69d		48,75	1 32 53.6		30 37 46	40	120	
69Aa		48,75	1 32 52.6		30 37 4	31	115	
69Ab		48,75	1 32 52.8		30 36 49	9.0	32	
69Ac		48,75	1 32 53.6		30 36 42	13	56	
70		48	1 32 56.7		30 39 22	56	89	Also identified as SNR G9
71		48	1 32 57.0		30 39 7	20	105	
72		53	1 32 51.8		30 33 3	45	58	
73a		53,75	1 32 52.4		30 31 38	13	53	
73b		53,75	1 32 53.5		30 31 37	2.2	9	
73Aa		53,75	1 32 53.9		30 29 46	77	254	
73Ab		53,75	1 32 50.0		30 29 33	11	48	
74		53	1 32 56.3		30 31 48	270	668	
75		47	1 33 1.0		30 35 15	--	--	
75A		47	1 33 1.2		30 34 50	160	444	
76		53	1 32 56.9		30 32 13	62	168	
77a		47,52,53,75	1 33 0.5		30 33 47	37	112	
77b		47,52,53,75	1 33 1.1		30 33 29	170	309	
77c		47,52,53,75	1 33 0.4		30 33 32	60	204	
78		45	1 34 3.4		30 37 54	93	320	
78Aa		47,75	1 33 0.6		30 33 54	12	21	
78Ab		47,75	1 33 0.9		30 33 58	14	196	
79		47	1 33 6.3		30 38 39	200	1114	
80		46	1 33 35.5		30 40 51	64	167	
81		45,46	1 33 40.3		30 38 20	450	1512	
82		47	1 33 10.9		30 39 38	180	700	Also identified as SNR G20
83		47	1 33 11.0		30 39 10	200	560	
84		47	1 33 13.3		30 39 24	82	134	
85		47	1 33 12.9		30 39 15	180	303	
85A		47	1 33 26.6		30 38 35	130	265	

86		47	1 33 12.5		30 38 36	990	925	Contains SNR G21
86Aa		47,76	1 33 27.6		30 39 6	160	699	
86Ab		47,76	1 33 27.3		30 39 40	32	128	
87a		45,76	1 33 57.3		30 41 56	56	183	
87b		45,76	1 33 56.4		30 41 42	21	82	
88		52	1 33 4.9		30 31 54	91	103	
90		42	1 32 49.6		30 45 50	9.9	14	
91		37	1 32 52.2		30 50 13	20	16	
92		42	1 32 55.6		30 43 48	30	78	
93		42	1 33 6.3		30 45 53	52	176	
95		37	1 33 9.6		30 49 49	29	183	
96		37	1 33 9.5		30 49 13	70	386	
97		37	1 33 10.4		30 48 17	21	90	
98		44	1 34 14.1		30 37 32	220	443	
100a		41,42,76	1 33 13.5		30 45 23	100	234	
100b		41,42,76	1 33 12.3		30 45 38	190	577	
101a		37,76	1 33 11.7		30 48 58	320	878	
101b		37,76	1 33 12.4		30 48 33	500	1572	
102		37	1 33 12.5		30 49 12	200	255	
103		37	1 33 15.1		30 51 53	22	40	
104		37	1 33 12.9		30 53 35	60	289	
104A		37	1 33 16.5		30 53 43	87	404	
105a		37,76	1 33 17.2		30 54 26	8.5	44	
105b		37,76	1 33 17.4		30 54 19	14	71	
108		41	1 33 23.0		30 47 31	30	138	
109		36	1 33 29.1		30 50 6	11	37	
110		36	1 33 37.7		30 51 16	--	--	
111		36	1 33 29.7		30 50 17	13	64	
112		36	1 33 35.6		30 50 49	190	118	
113		47	1 33 7.3		30 35 12	27	45	
114a		47,76	1 33 9.3		30 34 53	95	210	
114b		47,76	1 33 10.0		30 34 41	15	56	
114c		47,76	1 33 8.9		30 34 42	12	49	
115a		47,76	1 33 10.5		30 34 22	19	62	
115b		47,76	1 33 11.8		30 34 11	38	95	Also identified as SNR G19

115c		47,76	1 33 12.1		30 34 17	65	133	Also identified as SNR G19
115d		47,76	1 33 11.5		30 34 36	31	114	
115e		47,76	1 33 11.0		30 34 20	64	151	
116		47	1 33 12.5		30 34 10	240	382	
117		52	1 33 11.9		30 31 38	43	145	
121		52	1 33 17.0		30 31 8	72	61	
122		52	1 33 18.9		30 33 14	80	181	
123a		52,76	1 33 20.6		30 32 52	16	79	
123b		52,76	1 33 20.6		30 32 46	74	144	
123A		52	1 33 16.3		30 32 29	37	155	
124		52	1 33 22.1		30 32 16	38	79	
125		52	1 33 20.9		30 30 31	15	56	
128		46	1 33 31.3		30 38 34	52	111	See note d
129a		52,76	1 33 23.4		30 31 33	39	71	
129b		52,76	1 33 23.1		30 31 29	4.1	10	
129Aa		51,52,77	1 33 28.0		30 31 26	64	169	Contains SNR G27
129Ab		51,52,77	1 33 28.8		30 31 29	31	95	
130a		46,77	1 33 31.0		30 37 15	5.2	8	
130b		46,77	1 33 31.1		30 37 21	23	64	
131		51,52	1 33 31.2		30 33 51	28	69	
131A		46	1 33 30.3		30 36 56	--	--	
132		51	1 33 34.6		30 33 14	88	300	
132A		51	1 33 35.1		30 32 53	37	90	
133a		51,77	1 33 36.9		30 34 31	60	229	
133b		51,77	1 33 35.7		30 34 24	18	69	
133A		--	1 33 37.5		30 35 10	44	104	See note e
134		46,51	1 33 37.4		30 36 11	41	116	
135		46,51	1 33 39.0		30 36 28	85	234	See note f
136		46	1 33 40.1		30 36 48	59	163	See note f
137		51	1 33 38.3		30 30 14	41	90	
138		46	1 33 37.5		30 39 30	48	40	
139		56	1 33 42.3		30 25 47	30	74	
140a		45,77	1 33 49.9		30 40 52	180	473	
140b		45,77	1 33 50.5		30 41 8	72	204	
141		51	1 33 48.6		30 33 55	87	207	

141Aa		50,51,77	1 33 52.4		30 33 58	1.7	5	
141Ab		50,51,77	1 33 52.8		30 33 57	15	55	
141B		50,51	1 33 54.1		30 33 51	68	195	Also identified as SNR G53
143		51	1 33 49.8		30 35 30	34	137	Also identified as SNR G49
143A		50,51	1 33 54.4		30 34 46	--	--	
144		51	1 33 49.3		30 30 50	30	180	
145a		56,77	1 33 49.5		30 29 31	66	316	
145b		56,77	1 33 51.0		30 29 36	9.8	59	
145c		56,77	1 33 50.9		30 29 22	28	164	
145d		56,77	1 33 50.4		30 29 6	3.3	21	
146		51	1 33 51.4		30 31 4	12	85	Also identified as SNR G52
147		51	1 33 51.5		30 30 39	17	115	Also identified as SNR G51
148		56	1 33 52.1		30 28 30	--	--	
148Aa		55,56,77	1 33 56.8		30 27 11	21	102	
148Ab		55,56,77	1 33 58.9		30 27 30	7.4	31	
148Ac		55,56,57	1 33 57.7		30 27 54	16	94	
148Ad		55,56,77	1 33 54.5		30 27 52	21	137	
149		50,51	1 33 54.8		30 32 44	88	224	
150		51	1 33 51.3		30 35 49	23	21	
160a		45,77	1 33 52.3		30 37 26	19	119	
160b		45,77	1 33 52.7		30 37 37	6.2	9	
160c		45,77	1 33 52.9		30 37 44	70	254	
160A		45	1 33 52.8		30 37 14	--	--	
171		45	1 33 59.2		30 40 56	130	258	
171Aa		45,77	1 34 0.6		30 40 35	110	138	
171Ab		45,77	1 34 0.5		30 40 41	19	24	
172		35	1 34 12.8		30 52 8	79	251	
173		35	1 34 8.8		30 51 44	32	130	
174		50	1 34 1.4		30 30 25	8.0	31	
175a		50,78	1 34 1.9		30 30 45	95	201	
175b		50,78	1 34 2.7		30 30 40	12	67	
176a		50,78	1 34 3.2		30 31 25	48	153	
176b		50,78	1 34 2.3		30 31 20	90	153	
176c		50,78	1 34 2.7		30 31 15	44	174	Area just south is SNR G67
180		50	1 34 2.7		30 36 49	95	280	

181a		50,78	1 34 0.7		30 32 1	25	106	
181b		50,78	1 34 1.7		30 32 0	2.5	12	
181c		50,78	1 34 1.2		30 31 58	11	49	
181d		50,78	1 34 1.1		30 31 54	9.5	39	
181e		50,78	1 34 0.9		30 31 51	9.4	47	
182a		50,78	1 34 7.8		30 32 46	4.3	8	
182b		50,78	1 34 8.1		30 32 46	12	24	
183		50	1 34 8.7		30 30 54	14	46	
184a		50,78	1 34 12.4		30 35 7	44	171	Also identified as SNR G74
184b		50,78	1 34 13.1		30 35 12	32	109	
184A		50	1 34 9.3		30 34 13	78	175	
185a		50,78	1 34 13.4		30 35 24	2.8	8	
185b		50,78	1 34 13.7		30 35 18	75	105	
187		50	1 34 15.4		30 30 59	17	105	
188a		50,78	1 34 18.3		30 31 40	12	60	
188b		50,78	1 34 19.0		30 31 31	33	144	
188c		50,78	1 34 19.5		30 31 17	2.1	14	
189		50	1 34 19.8		30 31 1	4.0	21	
190a		49,50,78	1 34 24.0		30 31 25	30	129	
190b		49,50,78	1 34 23.4		30 31 15	23	98	
190c		49,50,78	1 34 24.6		30 31 11	31	182	
192		50	1 34 19.6		30 33 54	73	119	Also identified as SNR G87
193		41,42	1 33 14.4		30 45 1	330	272	
194		41,42	1 33 13.5		30 45 23	500	1466	
196a		41,42,78	1 33 14.2		30 45 13	580	622	IC 131
196b		41,42,78	1 33 14.4		30 45 1	440	358	IC 131
197		41	1 33 16.9		30 45 4	120	175	
198		41	1 33 15.0		30 44 29	170	1791	
199		36	1 33 24.4		30 49 59	18	122	
200		41	1 33 25.3		30 43 28	9.4	54	
201		41	1 33 25.8		30 45 56	8.8	63	
202		41	1 33 26.9		30 44 59	13	37	
203a		36,79	1 33 29.3		30 49 3	5.4	17	Also identified as SNR G29
203b		36,79	1 33 29.4		30 49 5	2.0	7	
204a		44,45,79	1 34 8.3		30 39 17	130	318	

204b		44,45,79	1 34 7.8		30 39 12	36	19	
204c		44,45,79	1 34 7.6		30 39 4	37	135	
205		41,46	1 33 25.2		30 41 34	18	98	
206		41,46	1 33 28.7		30 42 11	83	51	Also identified as SNR G28
207		41	1 33 31.2		30 47 22	8.8	33	
208		44,45	1 34 10.4		30 39 13	120	89	
209a		40,41,79	1 33 33.1		30 43 14	68	208	
209b		40,41,79	1 33 33.8		30 43 7	12	58	
210		40,41	1 33 33.8		30 44 39	9.8	47	
211a		36,79	1 33 35.3		30 49 19	8.6	28	CPSDP 211 also identified as SNR G34
211b		36,79	1 33 35.5		30 49 16	2.6	9	
211c		36,79	1 33 35.3		30 49 16	4.0	15	
212		40,41	1 33 57.8		30 49 2	100	80	
213		40	1 33 39.1		30 44 7	25	92	
214		36,40	1 33 37.1		30 48 43	23	157	
215		40	1 33 40.9		30 47 48	30	107	
216		40	1 33 42.5		30 46 28	--	--	
217		40	1 33 43.6		30 47 17	--	--	
218		40	1 33 43.3		30 45 39	15	55	
219		40	1 33 44.8		30 47 20	23	110	
220		36	1 33 46.8		30 51 22	--	--	
221		36	1 33 48.4		30 52 4	9.9	58	
222a		36,79	1 33 49.6		30 52 24	21	86	
222b		36,79	1 33 49.0		30 52 18	3.8	18	
223		39,40	1 33 58.5		30 47 25	9.9	27	See note g
223C		45,46	1 33 43.5		30 41 24	99	284	See note g
224		39,40	1 33 57.8		30 46 36	35	94	See note g
224A		39,40	1 33 57.0		30 46 5	14	56	
224C		45,46	1 33 42.6		30 41 14	70	178	See note g
226		40	1 33 52.2		30 44 57	210	585	
226Aa		40,79	1 33 47.3		30 44 39	45	88	
226Ab		40,79	1 33 47.5		30 44 43	53	137	
227		40	1 33 52.9		30 45 47	240	431	
228		40	1 33 52.7		30 46 13	46	122	

229		39,40	1 33 55.9		30 47 56	9.9	58	
230		39,40	1 33 57.7		30 48 57	91	167	
231a		35,79	1 33 58.4		30 51 57	5.0	25	
231b		35,79	1 33 59.0		30 51 49	5.1	23	Also identified as SNR G61
232		39,40	1 33 55.5		30 44 56	180	525	
233		39,40	1 33 58.5		30 48 29	48	78	
233A		39,40	1 33 58.6		30 48 40	120	123	
234a		39,79	1 33 59.7		30 49 21	5.7	14	
234b		39,79	1 33 59.8		30 49 19	12	25	
235a		40,45,79	1 34 0.9		30 43 41	32	85	
235b		40,45,79	1 34 1.5		30 43 37	2.9	13	
236a		40,45,80	1 33 54.1		30 43 3	29	124	
236b		40,45,80	1 33 54.8		30 43 1	28	115	
236c		40,45,80	1 33 55.6		30 43 6	20	79	
236d		40,45,80	1 33 55.6		30 43 12	71	184	
236e		40,45,80	1 33 55.0		30 43 21	31	103	
236f		40,45,80	1 33 51.6		30 43 21	150	507	
236g		40,45,80	1 33 51.9		30 42 58	59	225	
236h		40,45,80	1 33 51.5		30 43 5	50	250	
236i		40,45,80	1 33 51.7		30 43 10	23	61	
237		40,45	1 33 55.7		30 42 48	8.0	15	
238		40,45	1 33 59.4		30 43 2	17	91	
239		39,40	1 34 1.3		30 44 30	23	96	
240a		40,45,80	1 34 0.3		30 42 19	11	28	CPSDP 240 also identified as SNR G64
240b		40,45,80	1 33 59.9		30 42 15	11	32	
241		39,40,45	1 34 1.5		30 44 0	--	--	
242		39,40	1 34 3.4		30 44 6	30	175	
243		39,40	1 34 3.4		30 44 42	12	75	Also identified as SNR G68
244		39,40	1 34 2.2		30 45 52	28	121	
245		39,40	1 34 1.3		30 46 38	--	--	
246a		39,80	1 34 5.8		30 48 3	17	40	
246b		39,80	1 34 5.5		30 47 48	40	145	
246A		39	1 34 7.0		30 47 53	79	211	
246B		39	1 34 8.3		30 47 46	--	--	
247		39	1 34 6.8		30 48 53	280	234	

248		35	1 34 3.7		30 50 47	10	71	
249		39	1 34 7.0		30 49 37	25	73	
250		39	1 34 10.2		30 49 5	18	85	
251		39	1 34 10.5		30 49 17	33	122	
252		39	1 34 11.2		30 49 21	17	93	
253		44,45	1 34 11.6		30 43 17	18	25	
254		44,45	1 34 11.6		30 43 28	13	59	
255		44,45	1 34 10.6		30 42 4	12	65	
256a		39,80	1 34 14.8		30 44 9	1.8	5	
256b		39,80	1 34 14.4		30 44 22	35	184	
256c		39,80	1 34 13.6		30 44 38	17	42	
256d		39,80	1 34 12.9		30 44 27	22	154	
256e		39,80	1 34 11.9		30 44 21	3.1	14	
257		39	1 34 12.7		30 46 4	31	93	
258a		39,80	1 34 13.6		30 47 23	13	24	
258b		39,80	1 34 13.8		30 47 19	25	53	
259		39	1 34 14.4		30 49 37	6.4	34	
260		39	1 34 14.2		30 49 30	17	101	
261a		39,80	1 34 19.2		30 48 33	72	176	
261b		39,80	1 34 19.8		30 48 37	62	141	
262		39	1 34 24.9		30 48 58	--	--	
263a		58,80	1 32 45.4		30 25 28	110	448	
263b		58,80	1 32 45.9		30 24 45	27	103	
263c		58,80	1 32 44.9		30 24 25	68	221	
263d		58,80	1 32 43.2		30 24 23	100	301	
265		51	1 33 36.5		30 31 7	66	164	
266a		51,80	1 33 35.6		30 32 22	24	69	
266b		51,80	1 33 36.1		30 32 17	16	43	
267a		51,80	1 33 32.2		30 31 44	40	112	
267b		51,80	1 33 33.0		30 31 39	71	198	
270		39	1 34 16.2		30 48 34	--	--	
271a		35,81	1 34 19.2		30 51 43	26	176	
271b		35,81	1 34 20.2		30 51 23	11	83	
272		44	1 34 14.5		30 38 53	--	--	
273		50	1 34 13.6		30 36 45	43	96	

273A		50	1 34 16.1		30 36 29	63	146	
274		44	1 34 17.3		30 41 16	30	144	Also identified as SNR G85
274A		44	1 34 21.5		30 40 33	7.7	26	
275		44	1 34 16.4		30 43 18	--	--	
276		39	1 34 19.2		30 44 4	68	232	
277a		39,81	1 34 19.0		30 45 1	20	83	
277b		39,81	1 34 19.1		30 44 55	7.5	32	
277c		39,81	1 34 18.4		30 44 50	2.2	16	
279		39	1 34 22.0		30 47 9	42	96	
280a		39,81	1 34 22.3		30 45 20	45	286	
280b		39,81	1 34 20.3		30 45 32	53	304	
280c		39,81	1 34 22.7		30 46 5	72	500	
281a		38,81	1 34 31.7		30 48 19	5.9	16	
281b		38,81	1 34 31.8		30 48 14	41	141	
282		38	1 34 32.9		30 49 22	6.4	36	
283		38	1 34 34.9		30 48 26	--	--	
284		38	1 34 37.2		30 46 50	59	235	
284A		38	1 34 37.8		30 46 28	--	--	
292		38	1 34 35.7		30 46 4	--	--	
295		35	1 34 3.7		30 54 23	19	98	
302		34,35	1 34 24.7		30 56 20	35	157	
303		34	1 34 26.0		30 56 25	7.0	44	
304a		43,81	1 34 40.9		30 43 50	9.7	49	
304b		43,81	1 34 39.9		30 43 46	25	116	
305a		35,81	1 34 15.8		30 53 41	68	279	
305b		35,81	1 34 16.2		30 53 26	68	316	
305A		35	1 34 14.2		30 53 46	50	197	Also identified as SNR G78
307		56	1 33 46.6		30 27 21	210	204	
311		55,56	1 33 56.8		30 28 45	38	95	
314		55,56	1 33 57.1		30 25 8	40	149	
320		55	1 34 5.3		30 23 45	--	--	
321		55	1 34 8.2		30 23 23	21	28	
322		55	1 34 10.4		30 25 56	48	151	
322A		55	1 34 14.2		30 25 47	45	112	
324		55	1 34 11.4		30 27 14	12	52	

325		55	1 34 12.9		30 27 29	1.4	7	
326		55	1 34 14.1		30 23 14	--	--	
327		55	1 34 15.3		30 27 48	4.6	17	
328		55	1 34 15.8		30 28 9	17	99	
329		55	1 34 19.0		30 23 28	64	110	
330		55	1 34 22.8		30 25 28	9.6	61	Also identified as SNR G88
331		54	1 34 28.5		30 24 53	1.8	13	
332		54	1 34 29.0		30 23 46	16	42	
334		54	1 34 42.2		30 23 58	20	14	
338a		49,81	1 34 29.9		30 35 16	2.3	9	
338b		49,81	1 34 29.4		30 35 15	7.2	47	
339a		49,81	1 34 29.7		30 35 41	11	37	Also identified as SNR G92
339b		49,81	1 34 30.1		30 35 37	3.2	15	
339c		49,81	1 34 30.6		30 35 42	3.3	20	
340		44	1 34 32.0		30 37 26	--	--	
341		49	1 34 32.5		30 35 22	22	93	Also identified as SNR G93
343		44	1 34 24.3		30 39 3	8.5	31	
344		44	1 34 26.7		30 39 41	6.4	36	
345		44	1 34 33.2		30 40 6	--	--	
347		43	1 34 39.5		30 39 8	--	--	Also identified as SNR DDB17
348a		43,81	1 34 44.1		30 42 40	20	65	
348b		43,81	1 34 44.7		30 42 32	17	53	Also identified as SNR G98
349		43	1 34 44.2		30 47 20	--	--	
350		43	1 34 45.4		30 42 22	48	62	
352		43	1 34 44.7		30 38 44	15	81	
353		49	1 34 38.0		30 32 28	12	36	
354		49	1 34 40.3		30 33 53	12	54	
356a		49,82	1 34 42.0		30 32 23	4.9	10	
356b		49,82	1 34 42.0		30 32 30	23	73	
357		49	1 34 40.6		30 30 52	42	73	
358		49	1 34 43.3		30 31 59	37	149	
358A		49	1 34 45.4		30 32 37	--	--	
359		49	1 34 50.8		30 36 8	--	--	
360a		43,82	1 34 42.8		30 37 19	52	354	

360b		43,82	1 34 42.7	30 37 43	6.5	47	
360c		43,82	1 34 44.8	30 38 1	17	90	
360d		43,82	1 34 44.0	30 37 54	9.2	63	
360e		43,82	1 34 45.3	30 37 47	15	105	
361		49	1 34 52.5	30 35 25	--	--	
366		43	1 35 5.4	30 41 25	--	--	
366A		43	1 35 5.9	30 41 5	--	--	
368		35	1 35 15.2	30 54 37	--	--	
370a		34,82	1 34 24.8	30 52 39	3.7	13	
370b		34,82	1 34 24.4	30 52 36	38	260	
371		43	1 35 0.0	30 41 48	9.2	36	
372		34,35	1 34 25.4	30 55 2	37	169	Also identified as SNR G89
373		34,35	1 34 25.5	30 54 33	21	126	
374		34	1 34 28.0	30 54 21	21	118	
377		34	1 34 31.7	30 57 22	130	164	
378		34	1 34 32.0	30 57 33	110	168	
379		34	1 34 35.0	30 56 48	--	--	
380		34	1 34 35.3	30 56 42	37	242	
381		34	1 34 35.0	30 52 1	20	44	
383		34	1 34 39.5	30 57 16	--	--	
384a		34,82	1 34 41.5	30 51 6	20	81	
384b		34,82	1 34 40.3	30 51 17	22	173	
385		35	1 34 42.9	30 49 23	--	--	
386		34	1 34 50.6	30 55 28	--	--	
387		34	1 34 50.9	30 54 20	--	--	
388		34	1 34 52.0	30 54 19	26	118	
389		34	1 34 52.6	30 53 52	51	44	
390		34	1 34 53.3	30 54 44	40	137	
394		53	1 32 33.8	30 27 28	51	83	
395		58	1 32 39.5	30 22 26	79	215	
396a		58,82	1 32 41.8	30 21 13	7.1	24	
396b		58,82	1 32 42.0	30 21 10	15	51	
398		48	1 32 43.6	30 35 15	23	18	
399		58	1 32 44.1	30 22 3	11	11	
400a		48,82	1 32 44.6	30 35 16	40	59	

400b	48,82	1 32 44.8	30 35 16	3.5	7	
401	42	1 32 45.4	30 41 29	10	26	
402	58	1 32 48.7	30 25 51	13	13	
403	52	1 33 7.4	30 30 57	42	75	
407	44,45	1 34 10.9	30 42 18	79	43	
409a	58,82	1 32 42.2	30 22 23	20	37	
409b	58,82	1 32 42.1	30 22 28	2.9	7	
410	58	1 32 38.5	30 25 12	9.2	41	

CPSDP emission regions are labeled on the M33 charts with the prefix Z. Separate sub-components of complex regions are given lowercase Arabic letters and are identified on Charts 59-83.

Notes to table:
[a]Flux is in units of 10^{-15} erg cm^{-2} s^{-1} (from Wyder et al., 1997).
[b]Area is in units of $arcsec^2$ (from Wyder et al., 1997).
[c]SNR candidates are from Sabbadin (1979, identified with prefix S), D'Odorico, Dopita, and Benvenuti (1980, prefix DDB), and Gordon et al. (1998, prefix G). Gordon et al. (1998) cross-reference various identifications of the SNRs.
[d]CPSDP 128 (Z128) has the same position as CPSDP 130A (Z130A) in Table 2 of Courtès et al. (1987). We have identified this HII region as Z128 because it is labeled as Z128 in their Figure 6b.
[e]The HII region we identify as CPSDP 133A (Z133A) is labeled in figure 5b of Courtès et al. (1987) but is absent from their Table 2.
[f]The regions CPSDP 135 (Z135) and CPSDP 136 (Z136) which are listed in Table 2 of Courtès et al. (1987) are labeled correctly in their Figure 5b. They are mislabeled as Z134A and Z134B in their Figure 6b.
[g]The labels CPSDP 223 (Z223) and CPSDP 224 (Z224) are each assigned to two different objects in Figure 6b of Courtès et al. (1987) but are only listed once in their Table 2. The positions in Table 2 match the positions of the northern-most Z223 and Z224 in their Figure 6b. We have labeled the southern Z223 and Z224 as CPSDP 223C (Z223C) and CPSDP 224C (Z224C), respectively.

Table 4c. M33 Emission Regions Identified by Hodge et al. (1998)

ID		Chart	RA(2000)	DEC(2000)	Flux[a]	Area[b]	Comments[c]
HBW1		48	1 32 28.7	30 36 43	2.5	9	
2		53	1 32 29.5	30 30 51	--	--	half off frame
3		53	1 32 30.1	30 32 34	33	195	
4		53	1 32 30.2	30 26 58	7.5	46	
5		58	1 32 30.5	30 24 55	--	--	
6		53	1 32 31.0	30 29 32	0.25	1	u
7		48	1 32 31.0	30 35 54	--	--	
8		53	1 32 31.2	30 27 17	--	--	linear
9		48	1 32 31.5	30 37 16	--	--	
10		53	1 32 31.6	30 32 32	2.4	14	
11		48	1 32 31.7	30 36 17	--	--	d
12		53,58	1 32 31.8	30 26 50	--	--	d
13		53	1 32 31.9	30 28 6	--	--	d, linear
14		58	1 32 32.1	30 23 19	--	--	u
15		53	1 32 32.1	30 28 1	--	--	d, linear
16		58	1 32 32.2	30 23 33	1.3	6	u
17		53	1 32 32.2	30 28 57	--	--	d
18		53	1 32 32.3	30 27 6	--	--	d, linear
19		53	1 32 32.6	30 32 37	--	--	
20		53	1 32 32.7	30 30 21	--	--	d
21		53	1 32 33.0	30 27 7	--	--	d, linear
22		53	1 32 33.2	30 31 59	20	19	u
23		58	1 32 33.3	30 24 42	--	--	u
24		48	1 32 33.6	30 36 44	1.8	11	
25		53	1 32 33.7	30 28 34	--	--	d, linear
26		58	1 32 34.2	30 22 39	1.3	6	u
27		58	1 32 34.2	30 22 42	4.9	12	u
28		53	1 32 34.6	30 30 26	370	716	
29		48	1 32 35.2	30 35 20	--	--	
30		53	1 32 36.0	30 28 44	--	--	u
31		53	1 32 36.2	30 27 11	1.6	9	
32		48	1 32 36.3	30 36 48	14	34	

33		53	1 32 36.6	30 28 55	1.1	6	u
34		53	1 32 36.9	30 30 44	1.1	6	u
35		53	1 32 36.9	30 31 1	14	83	
36		53	1 32 37.0	30 27 39	1.5	10	
37		53	1 32 37.3	30 28 50	--	--	u
38		53	1 32 37.5	30 28 57	--	--	d
39		58	1 32 37.6	30 23 20	--	--	
40		58	1 32 37.9	30 23 58	4.4	9	
41		58	1 32 38.1	30 22 43	--	--	
42		58	1 32 38.2	30 21 1	--	--	d
43		53	1 32 38.2	30 29 20	3.5	23	
44		58	1 32 38.8	30 25 46	0.28	1	
45		53	1 32 38.8	30 29 48	6.4	30	u
46		48	1 32 38.9	30 34 10	--	--	
47		48	1 32 39.0	30 37 28	--	--	
48		53	1 32 39.3	30 27 35	1.8	11	
49		58	1 32 39.7	30 23 41	--	--	
50		53	1 32 40.1	30 27 23	--	--	u
51		53	1 32 40.3	30 27 22	--	--	
52		58	1 32 40.6	30 23 55	--	--	d
53		58	1 32 40.6	30 37 28	0.64	3	
54		58	1 32 40.7	30 25 25	4.6	31	
55		53	1 32 41.0	30 32 56	--	--	
56		48	1 32 41.1	30 35 52	0.04	--	
57		53	1 32 41.4	30 27 15	0.54	4	
58		48	1 32 41.4	30 36 55	--	--	
59		48	1 32 41.6	30 36 0	2.4	15	
60		58	1 32 42.0	30 22 58	--	--	
61		48	1 32 42.2	30 36 34	--	--	u
62		48	1 32 42.2	30 37 26	1.7	9	
63		48	1 32 42.3	30 36 44	10	56	
64		48	1 32 42.5	30 36 22	--	--	d, arc or bubble fragment
65		48	1 32 42.5	30 36 34	26	37	
66		48	1 32 42.6	30 34 22	--	--	d, arc
67		48	1 32 43.1	30 37 8	--	--	d, arc

68		48	1 32 43.7	30 36 15	1.6	10	
69		58	1 32 43.9	30 22 45	--	--	
70		48	1 32 44.1	30 36 32	2.5	15	
71		58	1 32 45.0	30 21 45	1.2	6	u
72		48	1 32 45.2	30 35 13	3.6	15	
73		53	1 32 45.3	30 31 54	4.7	23	u
75		48	1 32 45.6	30 35 6	40	114	
76		48	1 32 45.7	30 35 18	6.0	23	
77		48	1 32 45.7	30 36 26	--	--	u
78		58	1 32 45.8	30 22 12	--	--	
79		53	1 32 45.8	30 28 53	18	59	
80		58	1 32 46.0	30 21 4	--	--	d
81		42	1 32 46.4	30 45 21	7.1	30	
82		48	1 32 46.5	30 39 30	20	207	
83		48	1 32 46.6	30 39 40	3.0	14	
84		58	1 32 46.7	30 24 22	--	--	d
85		48	1 32 46.8	30 34 35	--	--	bubble or arc
86		48	1 32 46.9	30 36 23	--	--	
87		58	1 32 47.1	30 22 31	--	--	u
88		48	1 32 47.8	30 39 44	--	--	
89		53	1 32 48.0	30 31 26	--	--	
90		48	1 32 48.4	30 34 19	--	--	d, filaments
91		48	1 32 48.4	30 35 0	0.23	1	
92		37	1 32 48.5	30 53 15	1.4	14	
93		48	1 32 49.0	30 36 46	--	--	d
94		58	1 32 49.3	30 22 11	4.7	22	
95		42	1 32 49.4	30 45 58	--	--	u
96		48	1 32 49.8	30 38 37	--	--	u
97		48	1 32 50.2	30 37 30	9.2	81	
98		48	1 32 50.3	30 35 51	--	--	d, circular arc
99		37	1 32 50.3	30 54 0	1.7	15	
100		48	1 32 50.4	30 34 32	1.6	10	
101		37	1 32 50.4	30 50 42	5.9	38	
102		48	1 32 50.5	30 37 48	1.6	14	
103		48	1 32 50.5	30 40 32	--	--	d

104		48	1 32 50.8	30 37 35	1.6	15	
105		37	1 32 50.8	30 51 27	2.3	10	u
106		48	1 32 50.9	30 35 31	--	--	d, arcs
107		48	1 32 51.0	30 37 30	--	--	u
108		48	1 32 51.1	30 35 10	--	--	u
109		48	1 32 51.1	30 37 40	1.3	10	
110		48	1 32 51.2	30 37 43	0.38	3	
111		37	1 32 51.3	30 51 2	1.1	13	d, isolated
112		48	1 32 51.4	30 39 47	--	--	
113		42	1 32 51.5	30 45 26	1.4	18	
114		48	1 32 51.8	30 37 8	17	54	
115		37	1 32 52.1	30 53 26	4.4	25	
116		37	1 32 52.5	30 49 34	1.2	12	
117		53	1 32 52.6	30 28 49	--	--	d
118		53	1 32 52.6	30 32 47	0.11	--	
119		48	1 32 52.8	30 40 12	--	--	d
120		48	1 32 53.0	30 35 31	12	33	
121		53	1 32 53.1	30 33 41	--	--	
122		37	1 32 53.2	30 48 18	--	--	
123		58	1 32 53.4	30 25 46	--	--	d
124		48	1 32 53.4	30 37 23	--	--	u
125		48	1 32 53.5	30 39 18	7.8	65	
126		48	1 32 53.8	30 37 23	3.6	8	
127		42	1 32 53.8	30 42 37	3.2	39	
128		58	1 32 54.0	30 26 14	--	--	
129		53	1 32 54.2	30 31 41	--	--	d, linear
130		53	1 32 54.2	30 33 52	--	--	d, linear
131		48	1 32 54.3	30 34 59	4.1	19	
132		58	1 32 54.5	30 25 45	2.4	13	
133		42	1 32 54.5	30 41 41	0.18	2	
134		48	1 32 54.6	30 38 40	3.0	25	
135		53	1 32 54.7	30 27 51	--	--	d
136		48	1 32 54.9	30 37 35	--	--	u
137		42	1 32 55.0	30 44 1	--	--	linear
138		53	1 32 55.3	30 28 50	--	--	

139		37	1 32 55.3	30 49 8	6.3	61	
140		53	1 32 55.4	30 30 4	--	--	
141		48	1 32 55.4	30 37 48	6.6	67	d
142		58	1 32 55.5	30 26 52	--	--	d
143		58	1 32 55.6	30 22 50	--	--	d
144		53	1 32 56.1	30 33 27	--	--	bubble
145		48	1 32 56.2	30 34 16	--	--	d, filaments
146		48	1 32 56.4	30 34 48	6.9	32	
147		42	1 32 56.4	30 42 18	--	--	
148		48	1 32 56.5	30 40 20	2.0	12	
149		48	1 32 56.7	30 35 10	1.6	9	
150		53	1 32 56.8	30 27 46	4.6	22	u
151		58	1 32 56.9	30 26 45	--	--	
152		53	1 32 57.1	30 31 29	--	--	
153		48	1 32 57.1	30 36 10	--	--	d, filaments
154		48	1 32 57.2	30 38 26	7.8	82	
155		42	1 32 57.3	30 41 48	1.6	17	
156		42	1 32 57.5	30 44 6	--	--	Includes BCLMP 289, CPSDP 92
157		48	1 32 57.7	30 35 26	--	--	u
158		48	1 32 57.9	30 39 21	--	--	
159		53	1 32 58.1	30 31 53	3.1	13	
160		53	1 32 58.2	30 31 43	--	--	d, linear
161		53	1 32 58.4	30 31 20	--	--	d
162		37	1 32 58.4	30 47 50	0.18	2	
163		58	1 32 58.7	30 22 51	--	--	
164		53	1 32 58.7	30 33 27	--	--	d, 3 parallel lines
165		47,48	1 32 58.7	30 36 36	--	--	
166		42	1 32 58.7	30 44 59	0.55	5	
167		53	1 32 58.8	30 32 38	1.1	6	
168		58	1 32 59.0	30 23 13	--	--	d, arc
169		58	1 32 59.0	30 25 45	--	--	u
170		58	1 32 59.1	30 21 47	--	--	
171		37	1 32 59.1	30 54 28	--	--	
172		58	1 32 59.3	30 23 2	3.3	20	

173		42	1 32 59.5	30 43 36	--	--	
174		53	1 32 59.6	30 31 40	44	227	
175		53	1 32 59.6	30 32 43	--	--	d
176		47,48	1 32 59.7	30 39 57	12	141	
177		47,48	1 32 59.9	30 38 56	16	76	
178		42	1 33 0.5	30 43 59	--	--	
179		53	1 33 0.1	30 30 11	--	--	bubble
180		53	1 33 0.4	30 32 1	9.3	38	
181		53	1 33 0.5	30 33 5	--	--	d
182		42	1 33 0.7	30 41 31	2.2	24	
183		37	1 33 0.7	30 54 32	1.1	12	
184		47	1 33 0.9	30 40 37	--	--	d
185		42	1 33 0.9	30 43 53	2.2	20	
186		53	1 33 1.0	30 28 51	--	--	d
187		47	1 33 1.0	30 39 22	4.4	29	
188		52	1 33 1.1	30 32 41	--	--	u
189		47	1 33 1.3	30 39 47	2.0	16	
190		57	1 33 1.4	30 23 36	2.0	10	
191		57,58	1 33 1.5	30 24 9	--	--	d
192		52	1 33 1.6	30 31 48	--	--	
193		42	1 33 1.8	30 46 50	0.02	2	
194		47	1 33 1.9	30 39 44	3.1	13	
195		47	1 33 2.0	30 39 48	6.0	17	
196		52	1 33 2.6	30 33 48	4.2	17	u
197		52	1 33 2.7	30 32 55	22	76	
198		42	1 33 2.7	30 45 5	4.3	48	
199		52	1 33 3.4	30 33 20	23	73	
200		57	1 33 3.5	30 23 55	--	--	u
201		57	1 33 3.5	30 25 31	1.2	7	
202		52	1 33 3.5	30 30 35	--	--	d, arc
203		52	1 33 3.5	30 32 20	--	--	d
204		52	1 33 3.6	30 28 39	--	--	d
205		52	1 33 3.6	30 30 48	--	--	d, arc
206		52	1 33 4.0	30 30 56	--	--	u
207		42	1 33 4.0	30 41 42	0.66	6	

208		57	1 33 4.1	30 25 46	--	--	d
209		52	1 33 4.3	30 31 42	16	69	
210		52	1 33 4.4	30 27 16	--	--	d
211		52	1 33 4.4	30 31 26	--	--	d, filaments
212		42	1 33 4.4	30 43 49	--	--	d
213		52	1 33 4.6	30 32 34	11	61	
214		52	1 33 4.7	30 30 57	8.5	22	u
215		42	1 33 5.0	30 45 26	--	--	
216		57	1 33 5.1	30 23 38	1.9	11	
217		52	1 33 5.1	30 33 20	2.1	10	
218		52	1 33 5.2	30 32 58	36	161	
219		52	1 33 5.8	30 33 22	--	--	
220		47	1 33 5.9	30 39 10	--	--	d, arc
221		52	1 33 6.0	30 31 1	--	--	u
222		52	1 33 6.2	30 31 30	15	54	
223		42	1 33 6.4	30 46 37	--	--	
224		42	1 33 6.5	30 45 25	--	--	u?
225		52	1 33 6.6	30 31 25	20	27	
226		52	1 33 6.7	30 27 50	--	--	d, arcs
227		52	1 33 6.7	30 31 13	--	--	u
228		52	1 33 6.8	30 28 23	--	--	d
229		52	1 33 6.9	30 31 54	--	--	d, filaments
230		47	1 33 7.0	30 35 22	--	--	
231		42	1 33 7.1	30 46 15	--	--	
232		37	1 33 7.2	30 54 17	--	--	u
233		57	1 33 7.4	30 24 9	--	--	d, linear arc
234		57	1 33 7.4	30 24 49	--	--	
235		42	1 33 7.4	30 42 18	--	--	u
236		52	1 33 7.6	30 31 34	--	--	d
237		42	1 33 7.6	30 46 34	--	--	
238		47	1 33 7.7	30 37 21	--	--	
239		52	1 33 7.8	30 31 15	--	--	d
240		47	1 33 7.8	30 35 16	5.8	37	
241		52	1 33 7.9	30 32 55	--	--	d
242		42	1 33 7.9	30 42 18	--	--	u

243		37	1 33 7.9	30 47 54	0.64	6	
244		52	1 33 8.0	30 33 10	--	--	d
245		42	1 33 8.1	30 45 3	--	--	u
246		37	1 33 8.1	30 48 59	3.1	27	
247		57	1 33 8.5	30 22 16	0.24	12	
248		42	1 33 8.5	30 47 39	0.95	11	
249		52	1 33 8.9	30 29 12	--	--	u
250		41,42	1 33 8.9	30 46 52	4.0	11	
251		57	1 33 9.5	30 26 26	--	--	d, arc
252		52	1 33 9.5	30 29 11	--	--	u
253		47	1 33 9.8	30 37 18	--	--	
255		42	1 33 9.9	30 46 46	--	--	part of giant arc
256		52	1 33 10.1	30 33 10	--	--	d
257		52	1 33 10.2	30 32 6	0.24	5	
258		57	1 33 10.3	30 22 22	--	--	d, arc
260		52	1 33 10.7	30 32 13	0.95	5	
261		47	1 33 10.8	30 38 19	42	125	
262		52	1 33 10.9	30 33 1	--	--	d
263		52	1 33 11.0	30 28 2	--	--	d, arc
264		57	1 33 11.2	30 24 0	--	--	d
265		47	1 33 11.2	30 40 13	--	--	d
266		52,57	1 33 11.5	30 27 2	--	--	arcs
267		47	1 33 11.5	30 39 16	12	35	
268		52	1 33 11.7	30 28 11	--	--	d, arc
269		47	1 33 11.8	30 39 20	12	36	
270		52	1 33 12.1	30 29 12	--	--	d, arcs
271		37	1 33 12.1	30 49 33	1.6	12	
272		41,42	1 33 12.2	30 43 36	--	--	u
273		37	1 33 12.2	30 50 34	5.1	30	dbl
274		47	1 33 12.3	30 40 0	--	--	
275		41,42	1 33 12.4	30 44 6	3.0	24	within large complex
276		41,42	1 33 12.7	30 42 14	--	--	d
277		57	1 33 12.8	30 22 24	2.6	12	
278		52	1 33 12.8	30 32 17	--	--	d, filaments
279		47	1 33 12.8	30 34 22	--	--	d

280		37	1 33 12.8	30 51 10	7.2	53	
281		37	1 33 12.9	30 49 43	2.8	16	
282		37	1 33 12.9	30 50 41	3.6	28	
283		57	1 33 13.0	30 22 55	--	--	d
284		57	1 33 13.1	30 21 16	1.1	6	
285		57	1 33 13.1	30 22 24	6.0	16	
286		41,42	1 33 13.1	30 43 9	3.3	21	
287		52	1 33 13.3	30 30 14	--	--	d, bright arcs
288		47	1 33 13.3	30 39 40	--	--	bubble (or horseshoe)
289		57	1 33 13.4	30 22 40	16	64	
290		41,42	1 33 13.5	30 43 9	--	--	u
291		47	1 33 13.6	30 34 38	0.79	5	
292		41,42	1 33 13.6	30 44 21	21	49	within large complex
293		41,42	1 33 13.7	30 43 15	--	--	u
294		37	1 33 13.7	30 47 50	--	--	d
295		52	1 33 13.8	30 28 4	--	--	u
296		47	1 33 13.8	30 35 55	0.57	4	
297		47	1 33 13.8	30 40 36	--	--	u
298		37	1 33 14.0	30 50 44	0.61	7	
299		37	1 33 14.0	30 51 18	--	--	u
300		37	1 33 14.0	30 52 44	--	--	u
301		52	1 33 14.1	30 33 37	--	--	d
302		47	1 33 14.1	30 34 23	--	--	u
303		41,42	1 33 14.1	30 45 41	4.9	39	
304		52	1 33 14.3	30 32 48	4.4	28	
305		57	1 33 14.4	30 22 32	2.2	12	u
306		37	1 33 14.4	30 51 44	1.3	8	
307		47	1 33 14.5	30 38 54	9.8	33	
308		57	1 33 14.6	30 23 24	--	--	d
309		52,57	1 33 14.6	30 26 59	--	--	d, arc
310		57	1 33 14.8	30 22 22	0.43	3	
311		52	1 33 14.8	30 29 24	--	--	d, arcs
312		37	1 33 14.8	30 54 7	4.8	36	
313		57	1 33 14.9	30 22 29	--	--	
314		52	1 33 14.9	30 27 3	--	--	u

315		47	1 33 14.9	30 35 47	--	--	
316		41,42	1 33 14.9	30 45 59	5.4	37	
317		41,42	1 33 14.9	30 46 47	--	--	u
318		47	1 33 15.0	30 38 45	--	--	d, arc
319		37	1 33 15.0	30 54 11	7.2	56	
320		37	1 33 15.1	30 52 15	--	--	d, linear
321		47	1 33 15.2	30 37 24	2.9	13	
322		47	1 33 15.2	30 39 2	6.2	31	
323		47	1 33 15.2	30 39 25	14	52	2 components
324		41,42	1 33 15.2	30 47 1	7.5	51	
325		52	1 33 15.3	30 31 26	5.8	90	
326		52	1 33 15.3	30 31 59	31	167	
327		47	1 33 15.3	30 36 36	--	--	u
328		47	1 33 15.3	30 36 40	--	--	u
329		47	1 33 15.3	30 37 20	1.8	10	
330		47	1 33 15.4	30 35 31	--	--	d, arc
331		52	1 33 15.5	30 28 26	--	--	d, arc
332		47	1 33 15.5	30 39 47	--	--	u
333		57	1 33 15.6	30 25 49	--	--	
334		52	1 33 15.7	30 31 45	1.5	10	
335		37	1 33 15.7	30 53 23	31	96	
336		37	1 33 15.7	30 54 12	2.2	20	
337		52	1 33 15.8	30 29 52	--	--	
338		47	1 33 15.8	30 35 0	--	--	u
339		47	1 33 15.9	30 36 13	--	--	
340		46,47	1 33 15.9	30 36 57	--	--	d, arc
341		47	1 33 15.9	30 39 12	--	--	
342		52	1 33 16.1	30 32 42	--	--	d
343		47	1 33 16.1	30 38 56	14	61	
344		57	1 33 16.4	30 25 11	--	--	
345		57	1 33 16.5	30 21 38	--	--	d
346		52	1 33 16.5	30 33 17	3.1	12	
347		41,42	1 33 16.5	30 47 20	13	66	
348		52	1 33 16.6	30 30 8	--	--	d, arcs
349		52	1 33 16.6	30 31 28	5.0	24	

350		46,47	1 33 16.6	30 36 15	1.8	11	
351		37	1 33 16.7	30 47 47	25	185	
352		52	1 33 16.8	30 32 5	1.6	10	
353		52	1 33 16.9	30 28 6	--	--	d, plus faint u
354		37	1 33 16.9	30 53 19	6.8	29	
355		57	1 33 17.0	30 22 54	--	--	d
356		52	1 33 17.0	30 32 41	7.0	31	
357		47	1 33 17.0	30 35 17	2.3	16	
358		41,46	1 33 17.0	30 41 28	14	67	
359		57	1 33 17.1	30 23 2	--	--	u
360		47	1 33 17.1	30 39 49	--	--	
361		37	1 33 17.1	30 54 25	8.1	50	
362		52	1 33 17.3	30 33 0	--	--	d
363		41	1 33 17.4	30 47 23	7.6	63	arc
365		46,47	1 33 17.5	30 39 17	--	--	
366		41	1 33 17.6	30 47 0	18	135	
367		37	1 33 17.6	30 52 41	4.2	27	
368		46,47	1 33 17.7	30 40 47	5.0	37	
369		52	1 33 17.8	30 29 2	--	--	d, arc
370		52	1 33 17.9	30 32 10	1.4	9	
371		47	1 33 17.9	30 35 37	6.8	43	
372		52	1 33 18.0	30 33 2	10	38	
373		52	1 33 18.0	30 33 35	--	--	
374		52	1 33 18.2	30 33 25	12	61	
375		57	1 33 18.4	30 26 3	--	--	
376		46,47	1 33 18.4	30 36 24	0.24	1	
377		47	1 33 18.5	30 35 31	--	--	
378		41	1 33 18.6	30 46 46	--	--	bubble
379		37	1 33 18.7	30 53 28	11	61	
380		57	1 33 18.8	30 25 38	--	--	u
381		46,47	1 33 18.9	30 36 37	1.2	7	
382		52	1 33 19.2	30 27 48	--	--	d
383		52	1 33 19.2	30 29 37	--	--	u
384		41	1 33 19.2	30 47 6	0.29	3	
385		46,47	1 33 19.3	30 40 40	--	--	d

386		46,47	1 33 19.4	30 39 24	17	63	
387		37	1 33 19.4	30 50 35	2.0	18	
388		46,47	1 33 19.5	30 39 37	5.1	27	
389		46,47	1 33 19.6	30 36 18	--	--	
390		37	1 33 19.6	30 53 8	--	--	d, linear
391		57	1 33 19.7	30 24 27	--	--	
392		41,46	1 33 19.7	30 42 14	--	--	d
393		52	1 33 19.9	30 27 38	--	--	u
394		46,47	1 33 19.9	30 36 31	0.42	2	
395		46,47	1 33 20.0	30 39 19	4.0	28	
396		47	1 33 20.3	30 34 46	0.48	3	
397		57	1 33 20.4	30 21 24	--	--	u
398		41	1 33 20.4	30 46 35	--	--	u
399		41	1 33 20.5	30 42 27	7.3	66	
400		57	1 33 21.3	30 23 58	3.0	12	
401		46,47	1 33 21.3	30 38 53	2.3	17	
402		36	1 33 21.8	30 50 9	--	--	
403		57	1 33 22.0	30 23 12	--	--	d
404		47	1 33 22.2	30 35 7	--	--	
405		47	1 33 22.4	30 35 42	--	--	d
406		57	1 33 22.7	30 25 10	--	--	
407		52	1 33 22.7	30 30 14	--	--	d
408		47	1 33 22.7	30 35 24	--	--	
409		36	1 33 22.7	30 49 53	1.2	13	
410		57	1 33 22.8	30 23 40	--	--	u
411		41,46	1 33 22.8	30 42 31	--	--	
412		36	1 33 22.9	30 50 5	2.2	20	
413		57	1 33 23.0	30 20 19	--	--	
414		52	1 33 23.0	30 32 58	--	--	
415		41	1 33 23.1	30 42 11	--	--	d
416		57	1 33 23.2	30 23 46	--	--	d
417		36	1 33 23.2	30 49 36	--	--	u
418		36	1 33 23.3	30 50 40	15	133	
419		36	1 33 23.3	30 54 23	--	--	d
420		46,47	1 33 23.4	30 38 38	4.6	24	

421		41	1 33 23.4	30 42 49	--	--	
422		41,46	1 33 23.7	30 44 4	1.6	15	
423		36	1 33 23.7	30 51 7	5.3	48	
424		41	1 33 23.8	30 43 6	--	--	
425		52	1 33 23.9	30 31 51	--	--	on bad columns
426		46,47	1 33 23.9	30 38 39	3.3	23	
427		36	1 33 24.0	30 49 4	3.9	35	
428		36	1 33 24.0	30 53 23	--	--	d
429		46,47	1 33 24.3	30 39 6	12	67	
430		36	1 33 24.3	30 49 44	--	--	d, linear
431		36	1 33 24.4	30 50 42	2.6	28	
432		57	1 33 24.5	30 23 26	--	--	d
433		36	1 33 24.6	30 49 9	1.2	11	
434		41	1 33 24.7	30 44 58	--	--	u
435		46,47	1 33 24.8	30 37 50	4.2	39	
436		36	1 33 24.8	30 50 28	6.8	46	
437		36	1 33 25.0	30 49 42	--	--	d
438		46,47	1 33 25.4	30 38 53	3.3	18	
439		46,47	1 33 25.4	30 39 41	7.5	49	
440		57	1 33 25.5	30 24 59	--	--	
441		36	1 33 25.5	30 48 14	--	--	d
442		46,47	1 33 25.9	30 40 39	--	--	u
443		36	1 33 25.9	30 53 27	--	--	d
444		46,47	1 33 26.0	30 36 56	12	66	
445		46,47	1 33 26.3	30 37 54	4.0	29	
446		41	1 33 26.3	30 42 21	--	--	d
447		46,47	1 33 26.5	30 37 35	8.9	41	
448		46,47	1 33 26.5	30 37 47	14	72	
449		41	1 33 26.9	30 43 41	7.7	55	
450		57	1 33 27.2	30 23 55	3.5	24	u, cluster
451		57	1 33 27.2	30 24 19	--	--	d
452		52	1 33 27.3	30 30 26	--	--	on defect
453		51,52	1 33 27.4	30 33 40	1.7	9	
454		47	1 33 27.5	30 34 19	0.93	5	
455		47	1 33 27.7	30 34 25	--	--	

456		52	1 33 27.8	30 29 31	0.85	5	
457		46,47	1 33 28.0	30 36 48	1.1	7	
458		46,47	1 33 28.0	30 37 5	3.5	21	
459		41	1 33 28.0	30 43 22	29	207	
460		47	1 33 28.1	30 34 43	1.2	5	
461		51,52	1 33 28.5	30 33 11	4.4	20	
462		46,47	1 33 28.5	30 40 41	22	42	
463		46	1 33 28.5	30 41 4	11	51	
464		46,47	1 33 28.6	30 39 59	14	45	
465		41,46	1 33 28.7	30 41 57	7.1	44	
466		41	1 33 28.8	30 43 50	4.6	44	
467		41	1 33 28.8	30 46 11	--	--	d
468		36	1 33 28.8	30 52 54	--	--	d
469		41	1 33 28.9	30 43 29	--	--	d (structured)
470		41	1 33 29.1	30 46 31	0.51	6	
471		57	1 33 29.2	30 23 56	--	--	
472		41	1 33 29.4	30 43 19	--	--	
473		51	1 33 29.8	30 30 30	--	--	
475		51,52	1 33 29.9	30 33 9	--	--	d
476		51	1 33 29.9	30 34 28	--	--	d
477		51,52	1 33 30.0	30 32 34	--	--	u
478		46	1 33 30.2	30 41 9	10	32	
479		41	1 33 30.2	30 44 16	--	--	
480		57	1 33 30.5	30 21 1	--	--	
481		51,52	1 33 30.5	30 33 39	2.3	12	
482		57	1 33 30.6	30 23 56	--	--	
483		51,52	1 33 30.8	30 31 29	--	--	d
484		57	1 33 30.9	30 20 25	3.2	18	
485		41	1 33 31.0	30 45 1	0.46	6	
486		52	1 33 31.1	30 27 32	--	--	
487		51	1 33 31.2	30 30 1	--	--	u
488		51,52	1 33 31.3	30 31 21	--	--	d
489		56,57	1 33 31.5	30 26 16	1.8	9	u
490		51	1 33 31.7	30 30 24	1.9	11	
491		51	1 33 31.7	30 30 57	--	--	

492		41	1 33 31.7	30 45 29	1.5	15	
493		51	1 33 31.8	30 30 29	1.6	5	
494		36	1 33 31.9	30 50 41	--	--	d
495		57	1 33 32.1	30 21 56	2.2	12	
496		56	1 33 32.1	30 28 19	2.2	6	
497		51	1 33 32.1	30 30 15	0.81	5	
498		57	1 33 32.2	30 22 3	0.29	1	
499		46	1 33 32.3	30 37 49	6.5	35	
500		56	1 33 32.5	30 24 42	--	--	
501		56,57	1 33 32.5	30 26 39	--	--	u
502		56	1 33 32.6	30 24 32	--	--	u
503		36	1 33 32.6	30 49 35	1.0	8	
504		36	1 33 32.8	30 50 29	--	--	d
505		36	1 33 32.8	30 50 43	--	--	d
506		56	1 33 33.0	30 24 30	0.33	1	
507		36	1 33 33.0	30 49 6	0.25	3	
508		36	1 33 33.0	30 49 44	4.4	18	
509		51	1 33 33.1	30 30 37	--	--	d
510		46	1 33 33.1	30 38 14	--	--	d
511		36	1 33 33.1	30 49 22	5.2	44	
512		51	1 33 33.3	30 30 7	4.1	15	
513		46	1 33 33.3	30 37 53	6.7	24	
514		46	1 33 33.6	30 39 27	15	53	
515		51	1 33 33.7	30 30 27	--	--	
516		51	1 33 33.8	30 30 17	--	--	
517		51	1 33 33.8	30 34 29	0.95	5	
518		40,41	1 33 33.9	30 47 20	22	77	
519		36	1 33 34.1	30 49 49	--	--	d, linear
520		36	1 33 34.2	30 50 27	20	104	
521		56	1 33 34.3	30 28 25	1.8	5	
522		51	1 33 34.3	30 34 24	2.6	13	
523		46	1 33 34.3	30 37 49	7.2	43	
524		40,41	1 33 34.3	30 44 25	1.6	14	
525		51	1 33 34.4	30 31 23	1.6	11	
526		46	1 33 34.4	30 40 51	70	155	

527		56	1 33 34.5	30 26 20	1.6	9	
528		40,41	1 33 34.7	30 44 39	--	--	u
529		51	1 33 34.8	30 30 14	--	--	d
530		56	1 33 34.9	30 25 37	--	--	
531		51	1 33 34.9	30 35 5	--	--	d
532		40,41	1 33 34.9	30 47 10	7.4	57	
533		51	1 33 35.1	30 31 17	7.5	39	
534		51	1 33 35.4	30 31 35	--	--	d, arc
535		46	1 33 35.4	30 39 33	13	43	
536		51	1 33 35.6	30 30 51	--	--	
537		40,41,46	1 33 35.9	30 42 44	16	62	
538		51	1 33 36.1	30 31 47	7.3	27	
539		46	1 33 36.1	30 39 22	28	87	
540		36	1 33 36.1	30 50 10	--	--	d, linear
541		51	1 33 36.3	30 30 42	--	--	defect?
542		51	1 33 36.4	30 30 57	5.5	24	
543		56	1 33 36.5	30 28 59	--	--	
544		51	1 33 36.5	30 31 54	--	--	d, arc
545		56	1 33 36.6	30 28 49	--	--	d, double arc
546		56	1 33 36.6	30 29 11	--	--	
547		51	1 33 36.6	30 29 57	1.0	5	
548		56	1 33 36.8	30 29 40	--	--	d
549		51	1 33 36.8	30 30 4	13	61	
550		56	1 33 36.9	30 29 26	3.9	14	
551		46	1 33 37.0	30 39 22	--	--	u
552		51	1 33 37.2	30 31 8	--	--	d
553		51	1 33 37.2	30 32 6	2.9	13	
554		51	1 33 37.4	30 30 44	11	54	
555		40,41,46	1 33 37.4	30 42 38	25	120	
556		56	1 33 37.5	30 28 3	1.8	11	
557		51	1 33 38.0	30 31 31	--	--	d
558		51	1 33 38.0	30 31 46	--	--	d, linear (parallel lines)
559		40,41	1 33 38.0	30 45 45	1.7	16	
560		51	1 33 38.2	30 30 52	1.9	10	
561		51	1 33 38.2	30 31 16	4.9	17	

562		56	1 33 38.3	30 28 6	5.3	20	
563		51	1 33 38.7	30 31 24	--	--	d
564		56	1 33 38.9	30 27 28	1.6	12	
565		56	1 33 38.9	30 28 21	--	--	d
566		51	1 33 38.9	30 34 38	--	--	d
567		40,41	1 33 38.9	30 45 34	34	50	
568		56	1 33 39.1	30 28 17	8.4	14	
570		51	1 33 39.2	30 31 39	--	--	d
571		40,41	1 33 39.2	30 45 36	36	84	
572		46	1 33 39.3	30 40 4	17	35	
573		40,41,46	1 33 39.3	30 42 11	--	--	u, two?
574		56	1 33 39.4	30 28 5	1.7	9	
575		46	1 33 39.4	30 38 16	21	43	
576		36	1 33 39.4	30 50 9	1.2	10	
577		46	1 33 39.5	30 37 23	9.8	45	
578		46	1 33 39.6	30 38 19	14	41	
579		56	1 33 39.8	30 28 43	--	--	
580		56	1 33 39.9	30 27 22	--	--	
582		40,41	1 33 40.0	30 44 14	--	--	u
583		51	1 33 40.1	30 34 15	--	--	
584		40,41	1 33 40.1	30 44 16	--	--	u
585		51	1 33 40.5	30 36 1	--	--	d
586		45,46	1 33 40.7	30 39 4	12	27	
587		45,46	1 33 40.8	30 39 10	6.0	24	
588		45,46	1 33 40.8	30 39 12	3.3	18	
590		51	1 33 41.2	30 34 15	16	42	
591		45,46	1 33 41.3	30 40 12	--	--	
592		45,46	1 33 41.5	30 37 20	--	--	
593		51	1 33 41.6	30 35 22	--	--	d
594		45,46	1 33 41.7	30 39 13	--	--	
595		36	1 33 41.7	30 50 23	1.2	11	
596		56	1 33 41.8	30 26 47	--	--	
597		36	1 33 41.8	30 49 24	2.9	26	
598		40	1 33 41.9	30 45 6	16	81	
599		40	1 33 41.9	30 45 15	--	--	linear, u-shaped

600		40	1 33 41.9	30 46 30	--	--	u
601		56	1 33 42.1	30 27 18	5.8	65	
602		45,46	1 33 42.1	30 39 5	--	--	
603		51	1 33 42.3	30 34 18	8.8	48	
604		40	1 33 42.3	30 45 5	12	49	
605		40	1 33 42.4	30 44 55	--	--	linear arc
606		45,46	1 33 42.5	30 40 5	--	--	u
607		44,45	1 34 10.3	30 39 29	--	--	
608		51	1 33 42.6	30 30 9	3.7	34	
609		45,46	1 33 42.6	30 40 10	5.8	38	
610		45,46	1 33 42.8	30 42 8	--	--	u
611		56	1 33 43.1	30 23 14	--	--	
612		51	1 33 43.2	30 31 45	13	22	
613		51	1 33 43.2	30 33 14	16	38	
614		45,46	1 33 43.7	30 38 22	28	70	
615		45,46	1 33 43.7	30 39 28	--	--	d, arc
616		51	1 33 43.8	30 33 14	20	61	
617		56	1 33 43.9	30 26 49	0.78	9	
618		40	1 33 44.0	30 44 12	19	66	
619		45,46	1 33 44.2	30 37 4	--	--	
620		45,46	1 33 44.4	30 41 36	--	--	u
621		45,46	1 33 44.5	30 42 4	--	--	u
622		51	1 33 44.6	30 34 27	2.8	22	
623		40	1 33 44.7	30 44 14	9.4	59	
624		56	1 33 44.8	30 23 41	--	--	
625		45,46	1 33 45.0	30 40 5	8.1	32	
626		45,46	1 33 45.0	30 40 10	23	65	
627		56	1 33 45.1	30 27 37	--	--	u
628		56	1 33 45.4	30 27 41	1.6	13	
629		45,46	1 33 45.5	30 39 42	54	220	
630		45,46	1 33 46.0	30 37 51	13	46	
631		56	1 33 46.2	30 26 53	--	--	u
632		45,46	1 33 46.2	30 41 6	22	86	
633		51	1 33 46.9	30 35 19	--	--	d
634		36	1 33 47.0	30 51 29	4.8	44	

635		45,46	1 33 47.1	30 39 40	36	124	
636		40	1 33 47.2	30 45 19	6.7	44	
637		36	1 33 47.2	30 52 1	3.4	32	
638		51	1 33 47.3	30 34 55	--	--	d
639		40	1 33 47.3	30 45 29	2.6	17	
640		56	1 33 47.4	30 29 40	--	--	d
641		40	1 33 47.4	30 44 44	42	98	
642		40	1 33 47.4	30 44 50	13	44	
643		40, 45,46	1 33 47.5	30 42 54	11	58	
644		40	1 33 47.5	30 44 37	51	120	
645		40	1 33 47.5	30 45 3	52	137	
646		51	1 33 47.6	30 33 54	12	42	
647		51	1 33 47.7	30 30 26	--	--	d
648		40	1 33 47.7	30 44 54	6.3	25	
649		56	1 33 47.8	30 29 12	--	--	
650		51	1 33 48.0	30 35 54	4.2	26	
651		40	1 33 48.0	30 44 59	21	127	
652		40	1 33 48.0	30 45 15	16	87	
653		40,45	1 33 48.1	30 43 24	--	--	linear arc
654		40	1 33 48.3	30 44 34	5.8	26	
655		45	1 33 48.4	30 39 16	2.7	13	
656		56	1 33 48.5	30 24 14	--	--	
657		51	1 33 48.5	30 35 45	--	--	u
658		40	1 33 48.6	30 44 37	--	--	u
659		56	1 33 48.7	30 23 50	--	--	
660		45	1 33 48.9	30 38 4	98	52	
661		56	1 33 49.0	30 23 24	--	--	
662		51	1 33 49.0	30 30 24	--	--	u
663		56	1 33 49.1	30 28 5	--	--	d
664		45	1 33 49.1	30 39 53	--	--	u
665		40	1 33 49.2	30 44 34	9.3	39	
666		36	1 33 49.2	30 52 24	39	254	
667		56	1 33 49.3	30 27 38	--	--	
668		51	1 33 49.3	30 32 4	--	--	u
669		51	1 33 49.3	30 32 26	--	--	u, 2 jets

670		45	1 33 49.3	30 39 50	--	--	u
671		51	1 33 49.4	30 33 25	8.5	33	
672		40	1 33 49.4	30 44 42	--	--	d
673		51	1 33 49.5	30 33 14	--	--	
674		40	1 33 49.5	30 45 42	6.2	40	
675		51	1 33 49.7	30 30 13	--	--	bubble
676		40	1 33 49.9	30 44 8	--	--	u
677		45	1 33 50.0	30 39 50	35	47	
678		51	1 33 50.1	30 33 17	34	65	
679		45	1 33 50.2	30 39 10	17	32	
680		40	1 33 50.2	30 45 24	16	87	
681		40	1 33 50.2	30 45 55	6.7	35	
682		45	1 33 50.4	30 40 39	61	65	
683		56	1 33 50.5	30 28 31	--	--	d
684		45	1 33 50.5	30 38 57	21	71	
685		45	1 33 50.6	30 39 31	--	--	defect on one frame?
686		45	1 33 50.7	30 40 0	47	52	
687		40	1 33 50.7	30 44 49	--	--	linear arc
688		40	1 33 50.7	30 45 28	2.9	15	
689		40	1 33 50.7	30 45 34	2.6	8	
690		51	1 33 50.9	30 30 6	0.46	5	
691		45	1 33 51.0	30 40 36	6.9	25	
692		51	1 33 51.1	30 34 53	--	--	d
693		40	1 33 51.1	30 46 19	0.83	6	
694		56	1 33 51.2	30 27 24	--	--	
695		51	1 33 51.2	30 33 14	7.4	18	
696		40	1 33 51.2	30 45 18	7.7	25	
697		36	1 33 51.2	30 52 46	--	--	d
698		51	1 33 51.3	30 31 23	--	--	d
699		45	1 33 51.3	30 40 33	5.5	21	
700		45	1 33 51.4	30 39 55	32	58	
701		56	1 33 51.5	30 23 38	--	--	
702		56	1 33 51.5	30 28 4	12	116	
703		45	1 33 51.5	30 40 6	71	88	
704		51	1 33 51.7	30 32 59	--	--	u, 2 jets

705		56	1 33 51.8	30 27 36	1.3	13	
706		56	1 33 52.0	30 29 2	1.8	14	
707		51	1 33 52.0	30 34 48	11	38	
708		45	1 33 52.1	30 41 18	8.5	29	
709		45	1 33 52.2	30 40 19	13	25	
711		45	1 33 52.4	30 39 33	--	--	linear (thick)
712		56	1 33 52.6	30 23 39	--	--	
713		45	1 33 52.6	30 38 10	7.6	41	
714		51	1 33 52.8	30 33 12	2.1	12	
715		45	1 33 53.0	30 40 9	--	--	near defect
716		51	1 33 53.3	30 33 21	12	27	
717		51	1 33 53.4	30 33 27	6.9	30	
718		45	1 33 53.6	30 39 9	--	--	linear arc
719		45	1 33 53.6	30 39 24	--	--	linear arc
720		51	1 33 53.7	30 33 22	5.6	15	
721		45	1 33 53.7	30 40 35	36	122	
722		45	1 33 53.9	30 38 33	9.0	35	
723		45	1 33 54.0	30 39 16	110	49	
724		50,51	1 33 54.3	30 32 30	--	--	linear, part of bubble
725		51	1 33 54.3	30 33 32	1.3	6	
726		35	1 33 54.4	30 54 47	3.6	27	
727		50,51	1 33 54.6	30 36 4	--	--	u
728		45	1 33 54.6	30 37 57	--	--	
729		45	1 33 54.7	30 37 27	--	--	defect?
730		45	1 33 54.8	30 37 40	--	--	
731		45	1 33 54.8	30 40 9	--	--	
732		45	1 33 54.8	30 41 33	1.4	8	
733		39,40	1 33 54.9	30 45 4	21	62	
734		35	1 33 54.9	30 53 35	--	--	
735		45	1 33 55.0	30 40 44	13	92	
736		45	1 33 55.1	30 41 26	2.7	15	
737		39,40	1 33 55.1	30 45 44	96	147	
738		45	1 33 55.3	30 37 40	--	--	
739		45	1 33 55.4	30 40 8	11	62	
740		45	1 33 55.4	30 40 21	--	--	

741		35	1 33 55.4	30 54 13	11	87	
742		55,56	1 33 55.6	30 26 9	--	--	
743		45	1 33 55.6	30 41 27	1.4	10	
744		35	1 33 55.6	30 54 25	0.82	8	
745		50,51	1 33 55.8	30 36 42	--	--	d, linear
746		39,40	1 33 55.8	30 45 10	35	32	
747		55,56	1 33 55.9	30 25 38	1.9	41	
748		56	1 33 55.9	30 28 39	--	--	
749		45	1 33 56.3	30 40 6	4.2	26	
750		39	1 33 56.3	30 49 26	0.87	6	
751		39,40	1 33 56.4	30 45 47	2.8	12	
752		35	1 33 56.4	30 55 32	0.61	7	
753		45	1 33 56.5	30 40 18	7.2	66	
754		39,40	1 33 56.6	30 45 40	5.8	31	
755		45	1 33 56.8	30 38 51	--	--	
756		50,51	1 33 56.9	30 31 34	0.44	3	
757		35	1 33 56.9	30 54 10	2.6	9	
758		50,51	1 33 57.0	30 31 20	4.4	23	
759		45	1 33 57.4	30 41 55	6.0	54	
760		39	1 33 57.4	30 49 33	1.4	14	
761		45	1 33 57.5	30 38 57	4.5	31	
762		55,56	1 33 57.6	30 26 13	--	--	
763		55,56	1 33 57.6	30 29 48	1.2	13	
764		45	1 33 57.6	30 41 5	7.2	35	
765		55,56	1 33 57.7	30 29 37	0.73	5	
766		50,51	1 33 57.7	30 35 38	--	--	d, complex
767		39,40	1 33 57.9	30 47 47	12	52	
768		50	1 33 58.0	30 33 45	--	-	d, knotty
769		55,56	1 33 58.1	30 28 44	--	-	u
770		45	1 33 58.2	30 38 56	12	67	
771		45	1 33 58.3	30 41 17	--	--	dbl
772		39	1 33 58.4	30 49 35	0.75	8	
773		35	1 33 58.4	30 55 2	6.8	40	
774		35	1 33 58.5	30 52 11	5.1	51	
775		55,56	1 33 58.6	30 23 9	--	--	u

776		50	1 33 58.6	30 34 56	36	96	
777		45	1 33 58.6	30 41 10	58	98	
778		39,40	1 33 58.7	30 46 20	2.0	11	
779		50	1 33 58.8	30 30 17	--	--	double bubble
780		45	1 33 58.8	30 39 52	12	31	
781		50	1 33 58.9	30 33 21	15	49	
782		45	1 33 58.9	30 41 36	30	43	
783		50	1 33 59.0	30 31 42	--	--	d, linear
784		50	1 33 59.2	30 36 13	11	37	
785		45	1 33 59.2	30 41 13	2.9	12	
786		39	1 33 59.3	30 49 10	1.3	8	
787		50	1 33 59.4	30 32 5	25	69	
788		50	1 33 59.4	30 36 19	12	48	
789		45	1 33 59.4	30 40 30	22	118	
790		45	1 33 59.4	30 41 19	1.8	9	
791		39	1 33 59.4	30 49 32	5.7	38	
792		55,56	1 33 59.5	30 27 25	0.38	4	
793		35	1 33 59.6	30 53 20	--	--	u
794		45	1 33 59.7	30 39 23	10	42	
795		35	1 33 59.7	30 52 11	7.6	79	
796		45	1 33 59.8	30 40 25	1.4	7	
797		45	1 33 59.9	30 40 18	2.6	29	dbl
798		39	1 33 59.9	30 49 14	0.77	5	
799		45	1 34 0.0	30 39 37	11	61	
800		50	1 34 0.1	30 33 2	--	--	d, linear
801		55,56	1 34 0.2	30 27 25	--	--	
802		45	1 34 0.2	30 37 14	--	--	d, filaments
803		56	1 34 0.3	30 23 19	0.80	16	
804		45	1 34 0.3	30 39 23	8.1	39	
805		39,40	1 34 0.3	30 46 5	0.78	5	
806		55,56	1 34 0.4	30 28 39	0.28	2	
807		39,40	1 34 0.5	30 49 1	0.85	6	
808		39	1 34 0.5	30 50 39	1.5	16	
809		35	1 34 0.5	30 53 12	8.3	92	
810		55,56	1 34 0.6	30 24 51	--	--	

811		50	1 34 0.6	30 35 3	12	51	
812		39	1 34 0.7	30 47 25	2.3	9	
813		39,40	1 34 0.8	30 46 12	6.5	40	
814		35	1 34 0.8	30 52 40	0.93	8	
815		39,40	1 34 0.9	30 48 45	2.5	20	
816		39	1 34 0.9	30 49 14	0.74	6	
817		35	1 34 0.9	30 51 57	8.1	91	
818		35	1 34 1.0	30 53 47	5.3	55	
819		50	1 34 1.1	30 32 43	14	77	
820		39	1 34 1.1	30 50 24	3.6	9	
821		39	1 34 1.2	30 49 27	1.1	10	
822		35	1 34 1.3	30 52 28	--	--	u
823		40,45	1 34 1.4	30 43 37	2.8	14	
824		35	1 34 1.6	30 52 32	--	--	u
825		50	1 34 1.8	30 33 39	--	--	d
826		39	1 34 1.8	30 49 18	0.19	2	
827		35	1 34 1.8	30 52 42	17	94	
828		55,56	1 34 2.0	30 28 32	2.7	28	
829		35	1 34 2.0	30 52 34	--	--	u
830		39	1 34 2.1	30 50 40	0.64	4	
831		55,56	1 34 2.4	30 28 17	0.57	5	
832		39,40	1 34 2.5	30 48 36	2.0	8	
833		50	1 34 2.9	30 32 51	0.63	7	
834		50	1 34 2.9	30 33 57	11	82	
835		55	1 34 3.0	30 28 48	2.9	22	
836		45	1 34 3.0	30 39 20	--	--	u
837		55	1 34 3.1	30 27 55	--	--	
838		35	1 34 3.1	30 52 31	--	--	u
839		39,40	1 34 3.2	30 46 27	4.3	23	
840		50	1 34 3.4	30 35 45	4.8	30	
841		45	1 34 3.4	30 39 13	--	--	u
842		40,45	1 34 3.4	30 43 30	2.5	22	
843		39,40,45	1 34 3.4	30 43 48	3.2	25	
844		39,40	1 34 3.5	30 46 39	15	83	
845		45	1 34 3.6	30 38 14	--	--	d

846		39,40	1 34 3.6	30 45 56	--	--	u
847		35	1 34 3.6	30 52 37	--	--	u
848		50	1 34 3.7	30 33 28	16	124	
849		40,45	1 34 3.7	30 42 45	5.5	50	
850		44,45	1 34 3.9	30 39 28	--	--	d
851		55	1 34 4.0	30 24 44	--	--	d
852		55	1 34 4.1	30 27 22	2.8	24	u
853		50	1 34 4.3	30 30 46	--	--	d
854		45	1 34 4.3	30 37 45	--	--	d, filaments
855		44,45	1 34 4.3	30 42 0	--	--	half-ring
856		39	1 34 4.3	30 45 12	0.88	8	
857		44,45	1 34 4.4	30 39 28	--	--	d
858		39	1 34 4.5	30 49 21	1.3	15	
859		44,45	1 34 4.6	30 41 11	11	35	
860		35	1 34 4.7	30 53 25	1.3	9	
861		44,45	1 34 4.8	30 39 20	--	--	d
862		35	1 34 4.9	30 51 54	7.4	72	
863		35	1 34 4.9	30 54 2	0.56	6	
864		50	1 34 5.2	30 35 22	1.6	15	
865		39	1 34 5.2	30 46 14	--	--	u
866		39	1 34 5.3	30 47 36	28	108	
867		55	1 34 5.5	30 30 6	--	--	
868		35	1 34 5.6	30 52 1	--	--	u
869		55	1 34 6.0	30 24 52	--	--	u
870		39	1 34 6.1	30 48 24	0.71	3	
871		35	1 34 6.1	30 52 25	38	399	
872		44,45	1 34 6.2	30 39 15	--	--	u
873		35	1 34 6.3	30 52 9	--	--	u
874		55	1 34 6.4	30 28 45	--	--	
875		44,45	1 34 6.4	30 37 24	--	--	d
876		39	1 34 6.4	30 47 53	8.8	24	
877		39	1 34 6.4	30 50 40	0.94	7	
878		55	1 34 6.5	30 23 44	--	--	d, linear
879		55	1 34 6.5	30 28 1	--	--	
880		50	1 34 6.5	30 35 54	--	--	d, arc

881		55	1 34 6.6	30 24 53	--	--	
882		55	1 34 6.6	30 26 33	9.3	73	
883		44,45	1 34 6.6	30 42 5	8.3	41	
884		39	1 34 6.6	30 48 20	1.2	5	
885		55	1 34 7.0	30 25 14	--	--	
886		50	1 34 7.0	30 34 9	--	--	d, linear
887		35	1 34 7.0	30 54 30	6.5	38	
888		50	1 34 7.1	30 33 55	--	--	d
889		50	1 34 7.1	30 35 20	1.9	14	
890		44,45	1 34 7.2	30 42 14	--	--	
891		44,45,50	1 34 7.3	30 37 3	--	--	
892		50	1 34 7.7	30 34 44	--	--	d, arc
893		55	1 34 7.9	30 29 22	0.33	3	
894		55	1 34 8.0	30 29 4	--	--	
895		44,45	1 34 8.0	30 38 47	--	--	d, filament
896		44,45	1 34 8.0	30 42 49	8.9	73	
897		55	1 34 8.1	30 25 42	0.82	10	
898		44,45	1 34 8.1	30 41 56	9.1	53	
899		44,45	1 34 8.3	30 38 14	--	--	
900		55	1 34 8.4	30 25 35	--	--	u
901		44,45	1 34 8.4	30 43 1	6.2	33	
902		35	1 34 8.4	30 53 25	--	--	
903		35	1 34 8.5	30 52 38	1.9	17	
904		44,45	1 34 8.6	30 41 47	7.9	33	
905		44,45	1 34 8.6	30 42 7	--	--	bubble
906		35	1 34 8.7	30 53 11	5.6	45	
907		39	1 34 8.8	30 49 30	--	--	
908		35	1 34 8.8	30 52 57	1.7	14	
909		35	1 34 8.8	30 53 51	--	--	d
910		44,45	1 34 9.0	30 42 53	4.1	30	
911		44,45	1 34 9.1	30 41 47	2.9	19	
912		44,45	1 34 9.1	30 42 45	6.2	33	
913		39	1 34 9.1	30 49 4	0.81	4	
914		55	1 34 9.2	30 23 33	--	--	d, linear
915		55	1 34 9.3	30 26 14	--	--	u

916		39	1 34 9.3	30 49 49	0.73	6	
917		39	1 34 9.5	30 47 34	0.94	7	
918		44,45	1 34 9.6	30 39 53	13	58	
919		55	1 34 9.9	30 23 23	--	--	u
920		55	1 34 9.9	30 25 33	--	--	double nucleus
921		35	1 34 9.9	30 52 53	1.3	13	
922		39	1 34 10.0	30 47 28	1.4	10	
923		55	1 34 10.1	30 23 14	2.3	27	
924		44,45,50	1 34 10.1	30 37 5	25	69	
925		35	1 34 10.1	30 52 44	0.60	6	
926		45	1 34 10.3	30 37 33	--	--	u
927		39	1 34 10.3	30 44 50	--	--	u
928		35	1 34 10.4	30 52 52	0.63	7	
929		55	1 34 10.5	30 25 30	0.60	4	
930		55	1 34 10.6	30 25 21	--	--	u
931		50	1 34 10.6	30 32 50	--	--	d, circular (bubble?)
932		55	1 34 10.7	30 25 35	3.4	13	u
933		55	1 34 10.7	30 25 46	9.0	37	u
934		50	1 34 10.7	30 31 14	3.5	19	
935		39	1 34 10.7	30 44 50	1.8	10	
936		55	1 34 10.8	30 25 38	4.3	20	u
937		50	1 34 10.9	30 34 3	8.0	42	
938		50	1 34 11.0	30 32 59	--	--	d
939		44,45	1 34 11.0	30 38 29	--	--	d
940		44,45	1 34 11.1	30 37 18	--	--	d
941		39	1 34 11.1	30 47 32	2.2	16	
942		39	1 34 11.1	30 49 47	0.99	8	
943		39	1 34 11.1	30 50 35	0.97	9	
944		55	1 34 11.2	30 25 35	--	--	superimposed star?
945		55	1 34 11.2	30 25 55	10	69	
946		55	1 34 11.3	30 24 14	--	--	
947		44,45	1 34 11.4	30 39 23	--	--	d
948		44,45,50	1 34 11.6	30 37 3	--	--	d
949		55	1 34 11.7	30 25 50	1.6	13	
950		55	1 34 11.7	30 26 26	0.33	3	

951		39	1 34 11.7	30 46 59	--	--	d
952		50	1 34 11.8	30 33 59	25	125	
953		44,45,50	1 34 11.8	30 37 5	4.3	14	
954		44,45	1 34 11.8	30 37 23	7.0	43	
955		44,45	1 34 12.0	30 38 22	4.3	30	
956		44,45	1 34 12.0	30 39 8	16	37	
957		39	1 34 12.0	30 44 58	--	--	u
958		55	1 34 12.1	30 26 13	0.35	2	
959		55	1 34 12.1	30 27 16	--	--	
960		55	1 34 12.3	30 26 16	1.5	9	
961		55	1 34 12.3	30 29 35	--	--	u
962		50	1 34 12.3	30 36 1	--	--	d, linear
963		44,45	1 34 12.3	30 37 26	--	--	d
964		35	1 34 12.3	30 51 45	190	1537	
965		50	1 34 12.6	30 34 4	--	--	d, linear
966		39	1 34 12.7	30 47 3	8.0	19	
967		55	1 34 12.8	30 23 21	5.1	57	
968		55	1 34 12.8	30 26 7	--	--	
969		39	1 34 12.8	30 47 15	1.9	10	
970		35	1 34 12.8	30 53 5	0.84	8	
971		55	1 34 13.0	30 29 55	--	--	u
972		50	1 34 13.0	30 31 39	--	--	d
973		50	1 34 13.0	30 32 39	--	--	d
974		39	1 34 13.0	30 47 17	1.7	9	
975		55	1 34 13.2	30 26 55	0.72	7	
976		55	1 34 13.3	30 27 7	--	--	d, linear
977		39	1 34 13.3	30 44 58	--	--	u
978		55	1 34 13.4	30 26 1	--	--	
979		55	1 34 13.6	30 25 38	1.7	15	
980		50	1 34 13.7	30 30 36	2.5	14	
981		39	1 34 13.7	30 45 53	--	--	d
982		55	1 34 13.9	30 29 10	0.94	10	
983		55	1 34 14.1	30 24 51	--	--	d, linear
984		50	1 34 14.1	30 30 13	0.59	6	
985		39	1 34 14.1	30 46 14	--	--	d

986		55	1 34 14.2	30 30 6	--	--	u
987		55	1 34 14.3	30 27 39	--	--	
988		44	1 34 14.3	30 38 18	--	--	
989		39	1 34 14.3	30 50 2	2.8	30	
990		55	1 34 14.5	30 29 52	--	--	u
991		55	1 34 14.6	30 29 9	--	--	
992		50	1 34 14.6	30 31 47	--	--	bubble with point source
993		55	1 34 14.8	30 28 48	0.83	9	
994		44	1 34 14.8	30 40 4	--	--	
995		50	1 34 14.9	30 35 49	2.7	17	u
996		50	1 34 14.9	30 36 19	--	--	d, linear
997		55	1 34 15.1	30 26 47	0.38	4	
998		55	1 34 15.1	30 28 43	--	--	u
999		39	1 34 15.1	30 44 55	--	--	u
1000		44	1 34 15.3	30 38 43	--	--	
1001		44	1 34 15.3	30 41 7	19	154	
1002		50	1 34 15.4	30 32 48	--	--	d, linear
1003		44	1 34 15.5	30 37 35	--	--	
1004		55	1 34 15.6	30 24 51	--	--	u
1005		55	1 34 15.8	30 29 42	--	--	
1006		44	1 34 15.8	30 40 51	0.34	3	
1007		44	1 34 16.0	30 39 3	--	--	u
1008		44	1 34 16.0	30 40 47	0.44	4	
1009		50	1 34 16.1	30 30 15	1.5	11	
1010		44	1 34 16.1	30 38 43	--	--	u
1011		44	1 34 16.1	30 39 45	--	--	
1012		44	1 34 16.2	30 38 26	--	--	
1013		44	1 34 16.4	30 41 7	1.5	8	
1014		50	1 34 17.3	30 34 58	--	--	d, linear, kinked
1015		55	1 34 17.6	30 25 20	--	--	u
1016		44,50	1 34 17.6	30 37 4	51	89	
1017		35	1 34 17.6	30 56 21	3.7	46	
1018		50	1 34 17.9	30 35 31	--	--	defect?
1019		50	1 34 17.9	30 35 46	2.0	11	

1020		44	1 34 17.9	30 38 0	--	--	d, large complex of filaments
1021		35	1 34 17.9	30 52 45	13	30	
1022		55	1 34 18.0	30 26 57	0.64	5	
1023		55	1 34 18.0	30 29 32	--	--	d
1024		44	1 34 18.1	30 41 36	--	--	
1025		55	1 34 18.2	30 28 3	--	--	d
1026		55	1 34 18.5	30 23 11	--	--	u
1027		39	1 34 18.5	30 47 53	23	213	
1028		39	1 34 18.5	30 48 18	6.9	18	
1029		55	1 34 18.6	30 28 44	0.77	8	
1030		55	1 34 18.7	30 26 51	--	--	u
1031		35	1 34 18.8	30 53 42	--	--	d
1032		35	1 34 18.8	30 55 12	0.72	4	
1033		55	1 34 18.9	30 25 13	--	--	u
1034		35	1 34 19.0	30 55 44	--	--	d
1035		55	1 34 19.1	30 25 21	--	--	u
1036		44	1 34 19.1	30 40 1	--	--	u
1037		35	1 34 19.4	30 52 31	--	--	d
1038		55	1 34 19.5	30 29 1	--	--	d
1039		50	1 34 19.5	30 32 59	9.1	55	
1040		35	1 34 19.5	30 56 44	--	--	d, linear
1041		39	1 34 19.6	30 46 25	2.8	21	
1042		55	1 34 19.8	30 25 14	--	--	u
1043		55	1 34 19.9	30 25 3	--	--	u
1044		50	1 34 20.0	30 34 39	4.4	18	
1045		44	1 34 20.0	30 40 8	--	--	
1046		55	1 34 20.2	30 26 37	--	--	u
1047		50	1 34 20.2	30 35 45	--	--	d, linear
1048		55	1 34 20.6	30 29 28	--	--	d
1049		50	1 34 20.6	30 33 3	3.2	20	
1050		50	1 34 20.7	30 30 37	--	--	u, double
1051		50	1 34 20.8	30 32 28	2.3	22	
1052		35	1 34 21.0	30 56 11	1.6	13	
1053		50	1 34 21.2	30 32 40	--	--	u, multiple
1054		44	1 34 21.2	30 38 17	--	--	d, arc

1055		44	1 34 21.2	30 43 34	26	59	
1056		55	1 34 21.4	30 23 24	--	--	
1057		55	1 34 21.4	30 29 25	--	--	
1058		44	1 34 21.4	30 37 39	--	--	
1059		39	1 34 21.4	30 46 46	4.7	40	
1060		39	1 34 21.5	30 44 59	--	--	d
1061		35	1 34 21.6	30 56 59	--	--	
1062		50	1 34 21.8	30 31 53	3.9	38	
1063		39	1 34 21.8	30 44 43	4.4	41	
1064		39	1 34 22.0	30 48 13	1.4	7	
1065		35	1 34 22.3	30 56 28	3.1	34	
1066		55	1 34 22.4	30 29 31	0.20	2	
1067		44	1 34 22.4	30 39 59	1.7	8	
1068		39	1 34 22.4	30 49 52	8.6	74	
1069		50	1 34 22.7	30 34 8	21	69	
1070		50	1 34 22.7	30 35 13	15	45	
1071		50	1 34 22.9	30 31 48	1.3	11	
1072		39	1 34 22.9	30 46 29	12	79	
1073		39	1 34 23.2	30 46 23	2.5	12	
1074		34,35	1 34 23.2	30 56 50	5.6	57	
1075		34,35	1 34 23.4	30 54 20	0.96	7	
1076		55	1 34 23.5	30 29 21	--	--	
1077		55	1 34 23.6	30 28 18	0.35	3	
1078		44	1 34 23.6	30 39 13	--	--	
1079		44	1 34 23.7	30 39 3	--	--	u
1080		55	1 34 24.1	30 27 52	--	--	u
1081		50	1 34 24.1	30 34 30	1.2	9	
1082		34,35	1 34 24.2	30 56 50	2.8	14	
1083		38,39	1 34 24.3	30 48 43	1.1	8	
1084		38,39	1 34 24.6	30 48 45	1.6	12	
1085		34,35	1 34 24.8	30 56 52	2.5	20	
1086		55	1 34 24.9	30 28 40	--	--	u
1087		44	1 34 25.2	30 36 53	--	--	
1088		44	1 34 25.2	30 39 37	--	--	
1089		44	1 34 25.3	30 40 7	2.5	9	

1090		44	1 34 25.5	30 38 26	--	--	
1091		34,35	1 34 25.6	30 56 57	--	--	
1092		49	1 34 25.7	30 33 12	0.96	11	u
1093		54	1 34 26.0	30 28 57	0.91	9	
1094		49	1 34 26.0	30 34 21	0.44	6	
1095		44	1 34 26.0	30 37 19	--	--	
1096		38,39	1 34 26.1	30 48 58	14	36	
1097		54	1 34 26.5	30 24 14	--	--	
1098		38,39	1 34 26.9	30 44 44	--	--	
1099		34	1 34 27.0	30 55 53	--	--	d, s-shaped
1100		49	1 34 27.1	30 36 41	2.6	28	
1101		49	1 34 27.2	30 34 13	--	--	
1102		38	1 34 27.7	30 49 12	--	--	u
1103		34	1 34 27.8	30 57 6	3.4	19	
1104		49	1 34 28.4	30 34 50	0.68	7	
1105		38	1 34 28.4	30 49 29	1.4	16	
1106		44	1 34 28.8	30 38 1	--	--	
1107		34	1 34 28.9	30 56 59	64	367	
1108		44	1 34 29.1	30 38 22	--	--	
1109		44	1 34 29.4	30 40 55	400	2395	incorporates BCLMP 734 and 735
1110		38	1 34 29.4	30 48 50	--	--	bubble
1111		54	1 34 29.6	30 29 37	--	--	u
1112		49	1 34 29.7	30 34 9	1.4	14	
1113		49	1 34 30.0	30 33 44	0.48	3	
1114		44	1 34 30.0	30 37 41	--	--	
1115		49	1 34 30.4	30 35 29	--	--	u
1116		49	1 34 30.5	30 36 33	--	--	u
1117		44	1 34 30.5	30 38 24	--	--	u
1118		44	1 34 30.6	30 39 15	--	--	
1119		44	1 34 30.8	30 37 42	--	--	
1120		38	1 34 30.8	30 49 40	--	--	u
1121		54	1 34 31.2	30 24 6	--	--	
1122		38	1 34 31.2	30 48 43	--	--	d
1123		49	1 34 31.3	30 34 13	2.2	21	

1124		49	1 34 31.3	30 36 2	--	--	d, arc
1125		49	1 34 31.9	30 34 12	--	--	
1126		38	1 34 31.9	30 45 2	12	100	
1127		38	1 34 31.9	30 49 36	--	--	d, linear
1128		49	1 34 32.1	30 35 1	--	--	bubble
1129		54	1 34 32.2	30 28 49	--	--	
1130		34	1 34 32.3	30 52 56	--	--	d
1131		44	1 34 32.4	30 38 0	--	--	d
1132		38	1 34 32.7	30 44 42	7.3	50	
1133		44	1 34 32.8	30 40 53	2.8	27	
1134		44	1 34 32.9	30 38 39	--	--	d
1135		38	1 34 33.1	30 44 42	5.6	38	
1136		49	1 34 33.3	30 34 55	6.6	49	
1137		49	1 34 33.3	30 35 15	--	--	u
1138		49	1 34 33.3	30 35 51	7.2	38	
1139		34	1 34 33.3	30 57 14	--	--	d, structured
1140		44	1 34 33.4	30 39 21	--	--	u
1141		38	1 34 34.4	30 45 21	7.0	54	
1142		44	1 34 34.6	30 42 8	--	--	bubble
1143		38	1 34 34.7	30 45 48	12	68	
1144		44	1 34 34.9	30 38 39	--	--	
1145		38	1 34 34.9	30 45 44	7.4	62	
1146		34	1 34 34.9	30 52 9	0.49	5	
1147		38	1 34 35.1	30 45 16	16	104	
1148		38	1 34 35.1	30 45 32	14	115	
1149		38	1 34 35.4	30 44 59	9.1	56	
1150		38	1 34 35.4	30 45 7	22	127	
1151		38	1 34 35.5	30 46 2	--	--	
1152		34	1 34 35.6	30 52 5	2.1	17	
1153		34	1 34 35.6	30 52 11	3.1	19	
1154		44	1 34 35.7	30 40 22	3.8	23	
1155		44	1 34 35.9	30 40 2	4.1	16	
1156		38	1 34 35.9	30 44 52	19	87	
1157		38	1 34 35.9	30 45 56	--	--	image defect?
1158		44	1 34 36.0	30 37 20	--	--	

1159		44	1 34 36.0	30 38 56	--	--	
1160		38	1 34 36.5	30 45 14	6.7	44	
1161		38	1 34 36.6	30 45 43	4.6	34	
1162		43,44	1 34 36.7	30 42 51	--	--	
1163		38	1 34 36.8	30 45 52	--	--	image defect?
1164		43	1 34 37.0	30 42 15	--	--	bubble
1165		38	1 34 37.0	30 44 51	2.7	19	
1166		34	1 34 37.0	30 55 59	--	--	d, arc
1167		38	1 34 37.1	30 45 26	2.2	25	
1168		38	1 34 37.3	30 48 0	0.44	3	u
1169		38	1 34 37.4	30 44 50	2.5	19	
1170		38	1 34 37.4	30 48 10	--	--	u
1171		34	1 34 37.4	30 54 35	--	--	
1172		38	1 34 37.5	30 48 13	1.1	6	
1173		34	1 34 37.6	30 56 28	--	--	d, linear
1174		38	1 34 37.7	30 44 57	--	--	d
1175		49	1 34 37.8	30 35 16	--	--	bad column
1176		34	1 34 37.8	30 53 4	2.0	24	
1177		49	1 34 38.0	30 31 19	--	--	next to bad column
1178		49	1 34 38.0	30 32 59	--	--	next to bad column
1179		49	1 34 38.1	30 31 24	--	--	next to bad column
1180		49	1 34 38.3	30 32 50	14	36	u
1181		38	1 34 38.4	30 46 58	3.9	13	
1182		38	1 34 38.6	30 50 4	--	--	
1183		38	1 34 38.8	30 46 20	13	71	
1184		34	1 34 38.8	30 55 45	5.1	48	
1185		49	1 34 39.0	30 35 52	0.93	10	
1186		49	1 34 39.1	30 33 43	--	--	
1187		49	1 34 39.2	30 31 12	--	--	u
1188		38	1 34 39.4	30 44 46	6.0	40	
1189		54	1 34 39.5	30 26 48	--	--	d
1190		38	1 34 39.6	30 47 49	110	873	
1191		49	1 34 39.7	30 31 13	--	--	u
1192		38	1 34 39.8	30 44 46	10	56	
1193		34	1 34 39.8	30 52 53	3.5	30	

1194		38	1 34 39.9	30 44 40	--	--	d
1195		38	1 34 40.0	30 44 49	25	128	
1196		49	1 34 40.1	30 35 49	0.83	8	
1197		34	1 34 40.1	30 55 46	--	--	d
1198		34	1 34 40.4	30 52 50	--	--	
1199		43	1 34 40.5	30 39 1	--	--	
1200		54	1 34 40.6	30 26 53	2.5	24	
1201		49	1 34 40.6	30 36 13	0.11	1	
1202		34	1 34 40.6	30 52 6	2.1	7	
1203		54	1 34 40.8	30 27 10	--	--	d
1204		38	1 34 40.8	30 46 10	0.58	4	
1205		49	1 34 41.0	30 30 20	--	--	u
1206		49	1 34 41.1	30 31 1	--	--	u
1207		54	1 34 41.2	30 28 9	--	--	
1208		54	1 34 41.6	30 25 58	--	--	
1209		38	1 34 41.6	30 47 44	6.6	58	
1210		49	1 34 41.7	30 31 59	--	--	d
1211		49	1 34 41.7	30 32 13	4.6	26	
1212		34	1 34 41.8	30 56 3	24	46	
1213		49	1 34 41.9	30 32 23	--	--	u
1214		49	1 34 42.0	30 32 57	--	--	u
1215		38	1 34 42.0	30 47 6	3.5	17	
1216		43	1 34 42.5	30 40 47	3.7	30	
1217		43	1 34 42.7	30 40 38	15	34	
1218		38	1 34 42.7	30 46 54	--	--	u
1219		54	1 34 42.8	30 26 35	--	--	d
1220		43	1 34 42.9	30 40 30	0.96	8	
1221		54	1 34 43.1	30 26 25	5.8	22	
1222		54	1 34 43.4	30 26 31	1.2	13	
1223		34	1 34 43.5	30 51 37	8.6	32	
1224		38	1 34 44.2	30 46 14	--	--	
1225		38	1 34 44.5	30 49 36	3.7	36	
1226		43	1 34 44.8	30 43 22	4.2	15	
1227		49	1 34 44.9	30 33 31	--	--	
1228		49	1 34 45.3	30 35 38	--	--	

1229		43	1 34 45.4	30 38 32	--	--	d, arcs
1230		34	1 34 46.0	30 54 47	2.8	21	
1231		38	1 34 46.1	30 44 39	--	--	u
1232		34	1 34 46.3	30 54 50	3.8	14	u
1233		49	1 34 46.6	30 35 44	1.0	12	
1234		49	1 34 46.9	30 36 9	0.93	8	
1235		43	1 34 47.0	30 37 42	--	--	
1236		49	1 34 47.1	30 34 23	5.0	52	
1237		43	1 34 47.1	30 37 30	--	--	
1238		34	1 34 47.2	30 54 54	2.2	24	
1239		38	1 34 48.1	30 50 15	0.69	8	
1240		54	1 34 48.6	30 28 26	51	369	
1241		43	1 34 49.0	30 37 25	--	--	
1242		49	1 34 50.5	30 33 4	--	--	bad column
1243		49	1 34 51.2	30 32 46	--	--	bad column
1244		49	1 34 51.2	30 32 52	--	--	bad column
1245		49	1 34 51.3	30 33 7	--	--	bad column
1246		34	1 34 51.9	30 54 32	1.3	11	
1247		38,43	1 34 53.5	30 43 56	1.4	9	
1248		34	1 34 53.9	30 54 5	0.19	2	
1249		43	1 34 54.2	30 40 18	1.8	16	
1250		38	1 34 54.3	30 50 14	--	--	
1251		49	1 34 54.9	30 34 48	16	29	u
1252		49	1 34 55.1	30 31 39	--	--	
1253		43	1 34 58.4	30 43 2	--	--	d
1254		43	1 34 59.0	30 42 7	52	481	
1255		43	1 34 59.1	30 42 24	--	--	d
1256		43	1 34 59.8	30 40 3	--	--	d
1257		43	1 35 0.1	30 42 27	0.65	3	
1258		43	1 35 0.3	30 41 55	3.3	28	
1259		43	1 35 0.7	30 40 38	--	--	d
1260		43	1 35 0.9	30 40 9	--	--	d
1261		43	1 35 1.1	30 41 15	2.6	9	
1262		43	1 35 1.4	30 42 20	0.34	3	
1263		43	1 35 2.9	30 43 54	3.1	14	

1264		43	1 35 4.3	30 40 13	--	--	d
1265		43	1 35 4.3	30 41 29	3.3	19	
1266		43	1 35 4.5	30 41 7	4.0	30	
1267		43	1 35 4.7	30 41 26	4.8	22	
1268		58	1 32 39.5	30 21 58	--	--	
1269		58	1 32 40.2	30 22 3	--	--	
1270		57	1 33 10.5	30 22 42	--	--	
1273		48	1 32 55.3	30 38 59	--	--	
1274		48	1 32 55.7	30 39 2	--	--	
1275		48	1 32 56.1	30 39 5	--	--	
1276		44,45	1 34 6.4	30 36 23	--	--	
1277		43	1 34 37.5	30 40 10	--	--	
1278		43	1 34 39.0	30 39 54	--	--	
1279		43	1 35 2.5	30 41 5	--	--	Partial border of large diffuse object
1280		40	1 33 42.3	30 47 47	--	--	
1281		39,40	1 33 58.5	30 44 42	--	--	
1282		38	1 34 26.6	30 48 51	--	--	
1283		35	1 33 56.5	30 54 7	--	--	

HBW emission regions are labeled on the M33 charts with the prefix H.

Notes to table:

[a]Flux is in units of 10^{-15} erg cm^{-2} s^{-1} (from Wyder et al., 1998). Low surface brightness objects were not measured.

[b]Area is in units of $arcsec^2$ (from Wyder et al., 1998).

[c]Regions of diffuse emission are noted with a "d" in the comments; unresolved emission regions are noted with a "u".

Table 4d. M33 Supernova Remnants

ID	Chart	RA (2000)	DEC (2000)
G1	53	1 32 30.8	30 27 39
G11	42	1 33 0.0	30 44 5
G14	52	1 33 3.5	30 31 16
G17	47	1 33 10.0	30 39 10
G18	42	1 33 9.9	30 42 18
G22	52	1 33 17.3	30 31 25
G37	46	1 33 37.4	30 40 4
G38	40,41,46	1 33 37.5	30 42 13
G39	46	1 33 40.8	30 39 40
G41	40,41,46	1 33 40.3	30 42 28
G42	51	1 33 41.2	30 32 28
G44	45,46	1 33 42.5	30 41 48
G45	45,46	1 33 42.9	30 41 1
G56	45	1 33 57.0	30 40 44
G57	50,51	1 33 57.0	30 34 59
G58	45	1 33 57.9	30 37 52
G60	50,51	1 33 58.3	30 36 22
G63	39,40	1 34 0.2	30 47 19
G66	50	1 34 1.3	30 35 21
G69	50	1 34 4.2	30 32 54
G71	39	1 34 10.1	30 47 11
G72	44,45	1 34 11.0	30 39 1
G73	44,45	1 34 10.6	30 42 21
G76	44	1 34 13.6	30 43 25
G79	44	1 34 14.2	30 41 52
G81	44	1 34 14.3	30 39 39
G82	50	1 34 15.5	30 33 0
G90	38,39	1 34 25.3	30 48 27
G91	44	1 34 29.4	30 41 32
G95	49	1 34 36.0	30 36 20
S14	45,46	1 33 46.9	30 37 18

SNRs with prefix G from Gordon et al. (1998) and prefix S from Sabbadin (1979). Gordon et al. (1998) cross-reference their identifications with previously catalogued supernova remnants.

M33

Chart 1

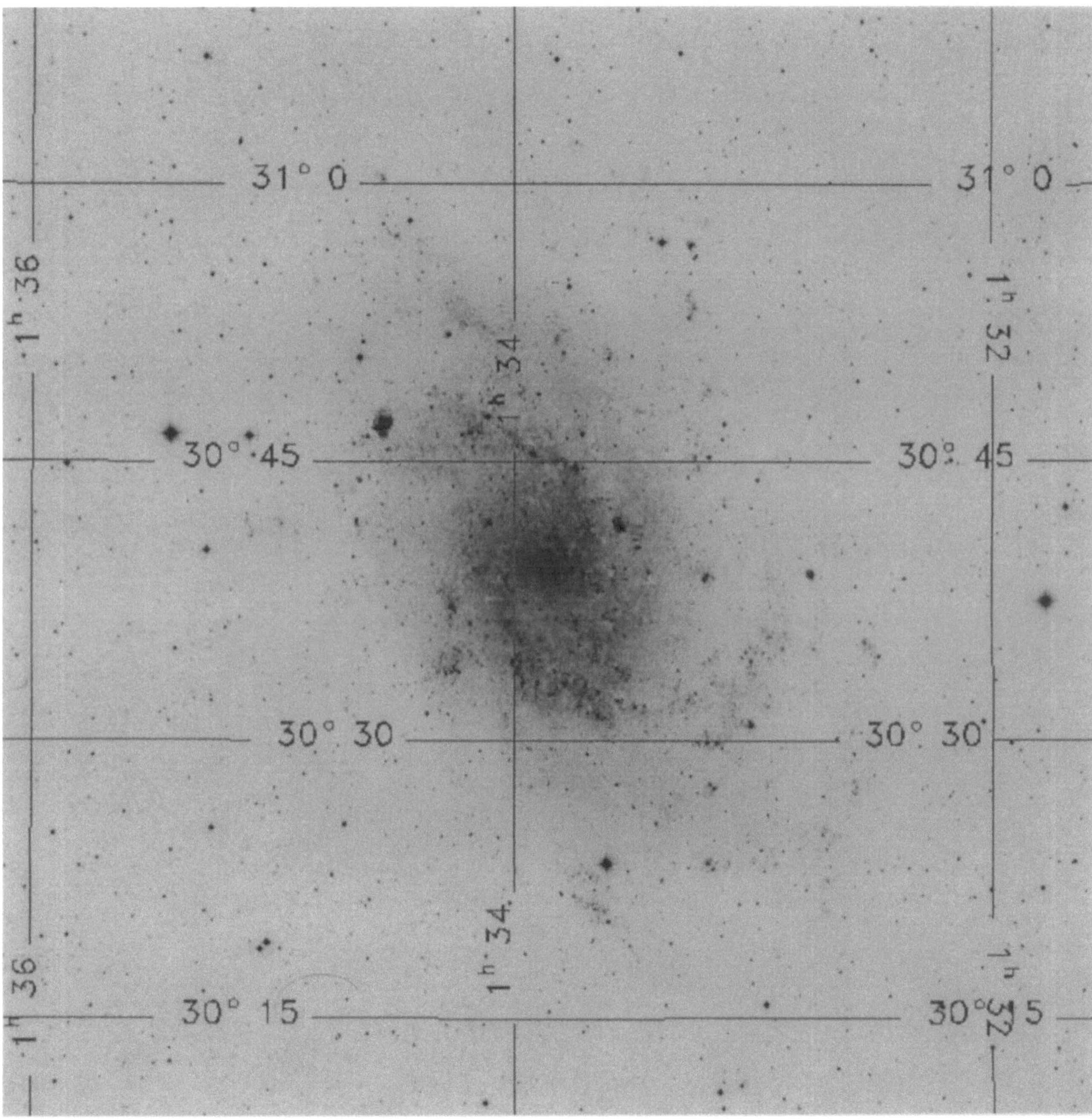

M33

Chart 2

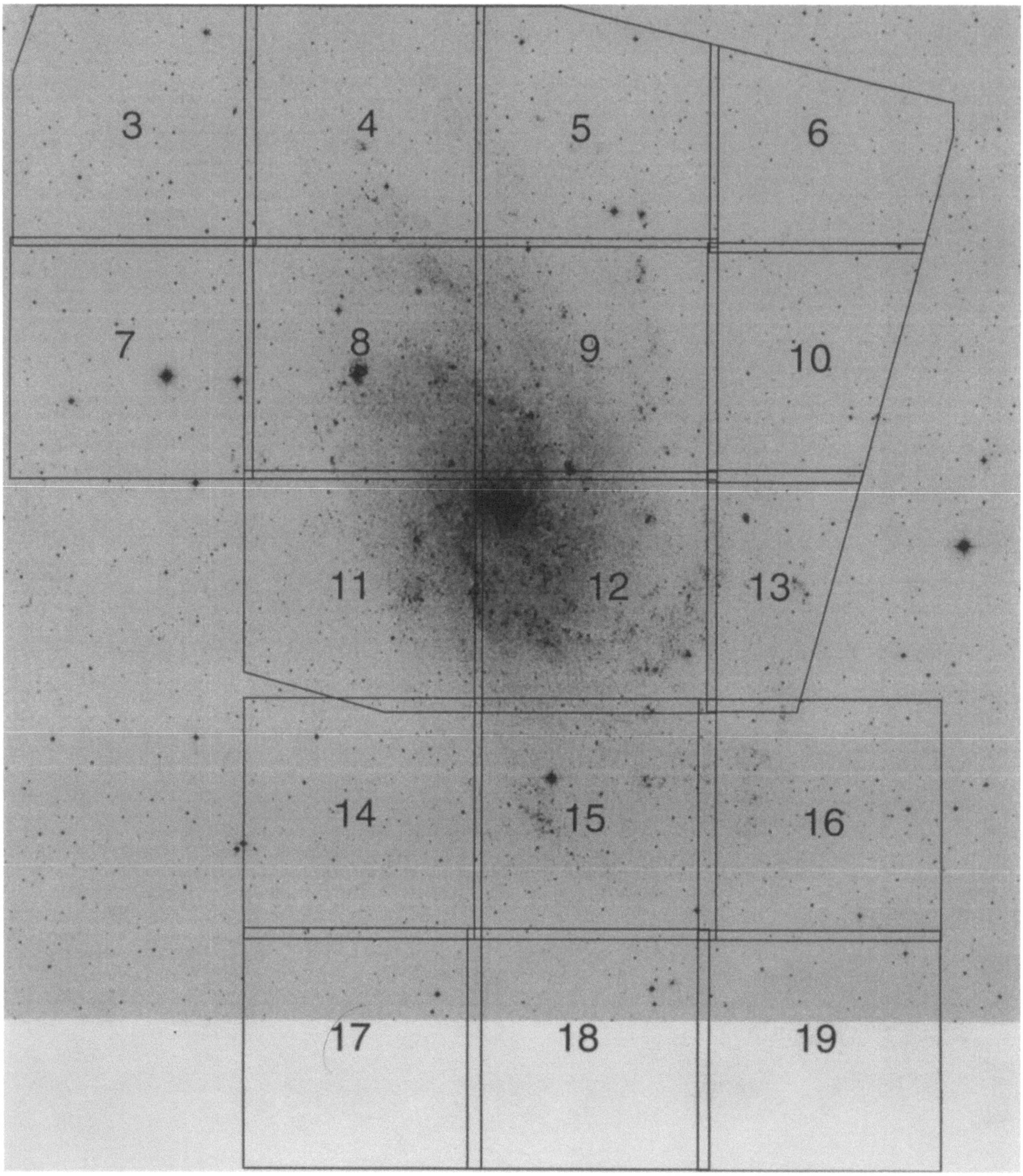

M33

Chart 3

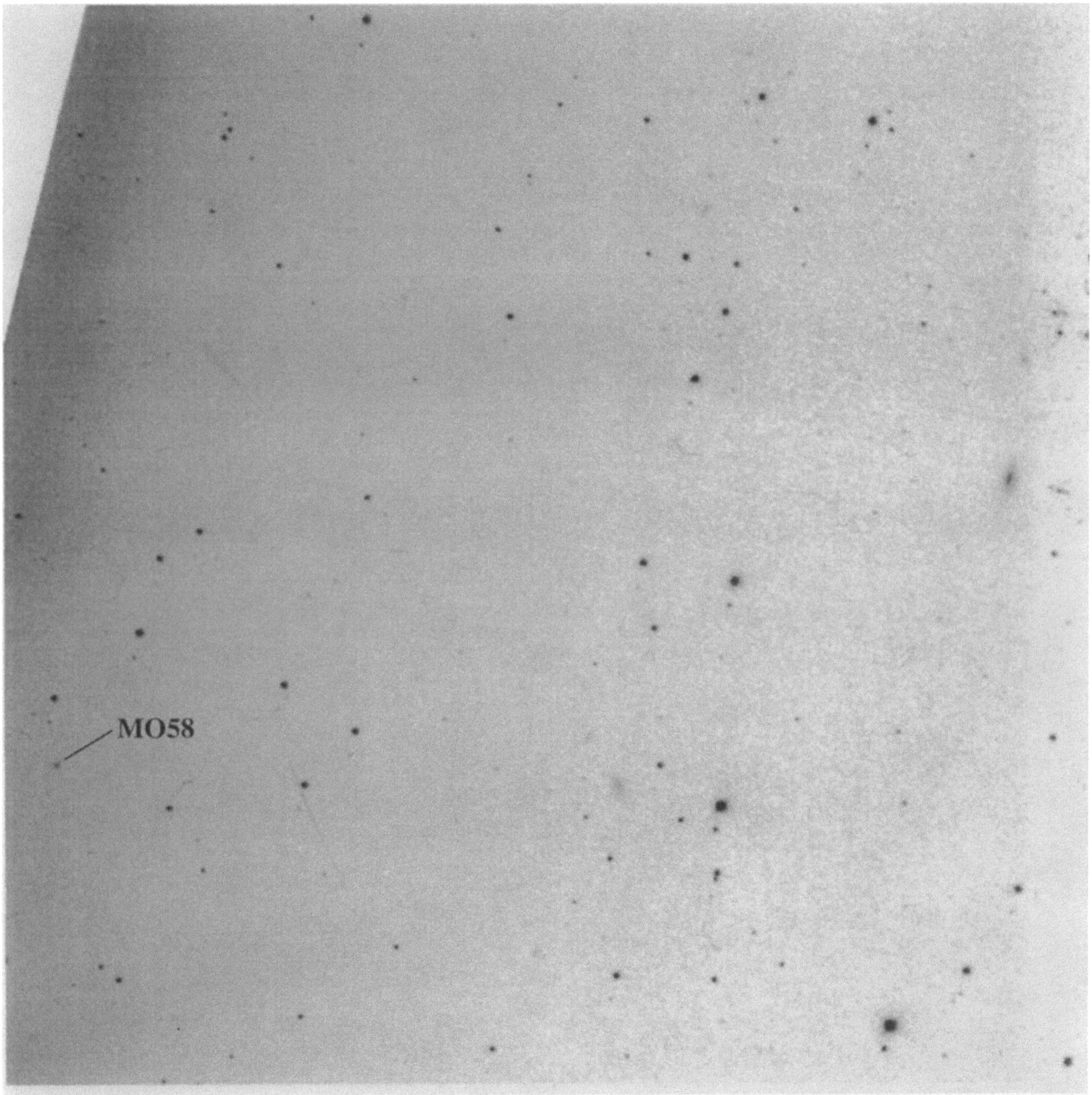

M33

Chart 4

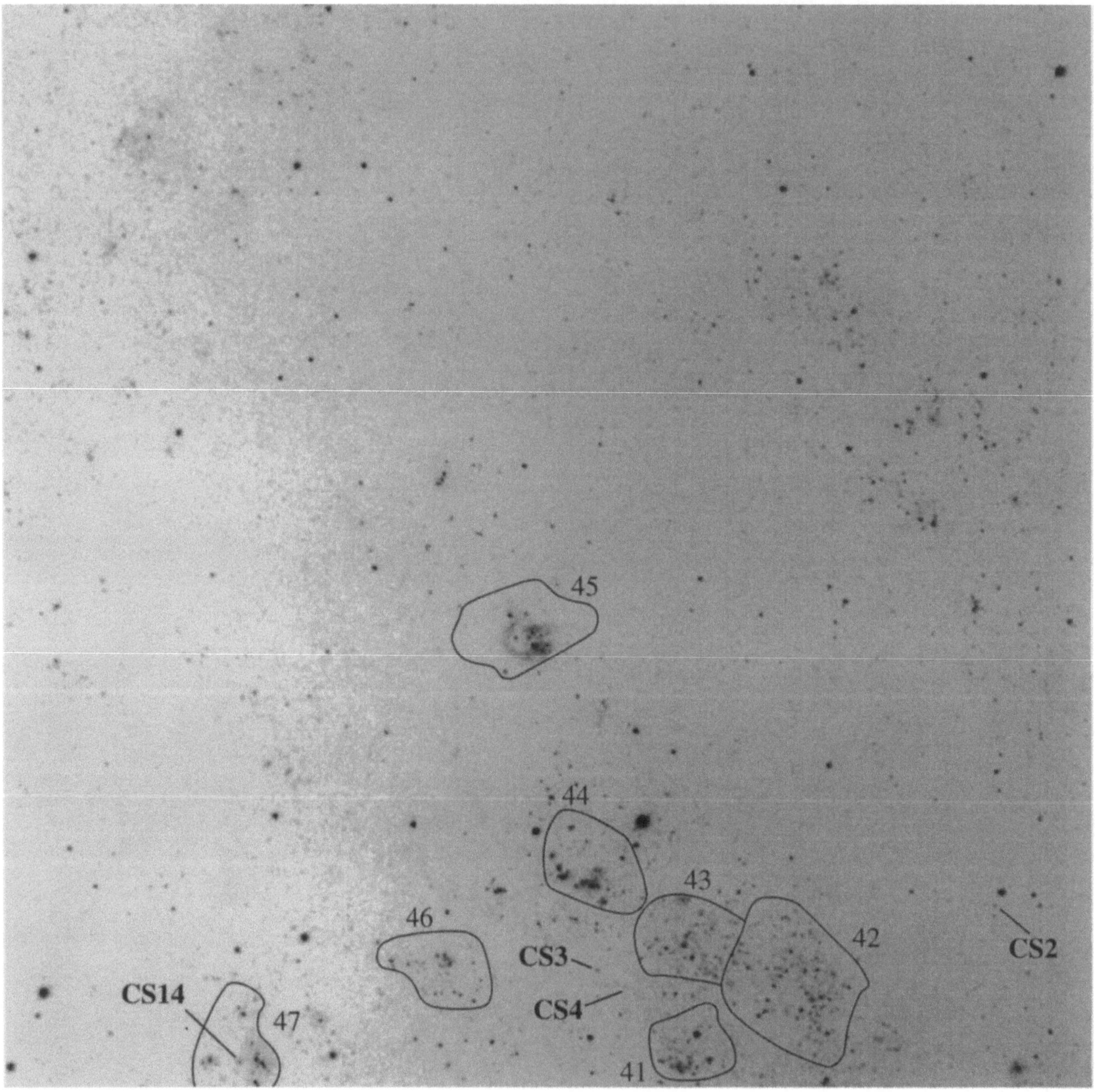

M33

Chart 5

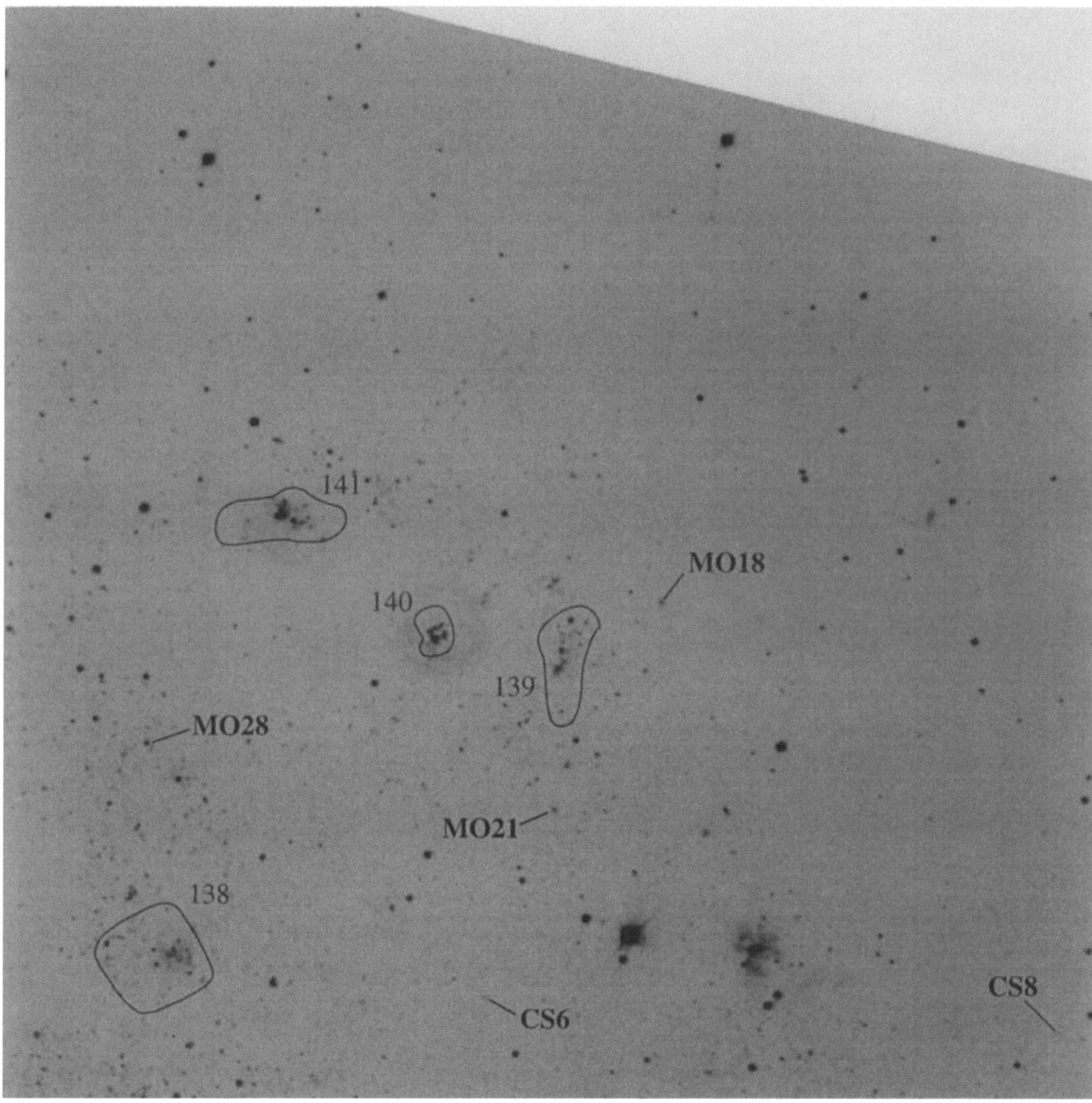

M33

Chart 6

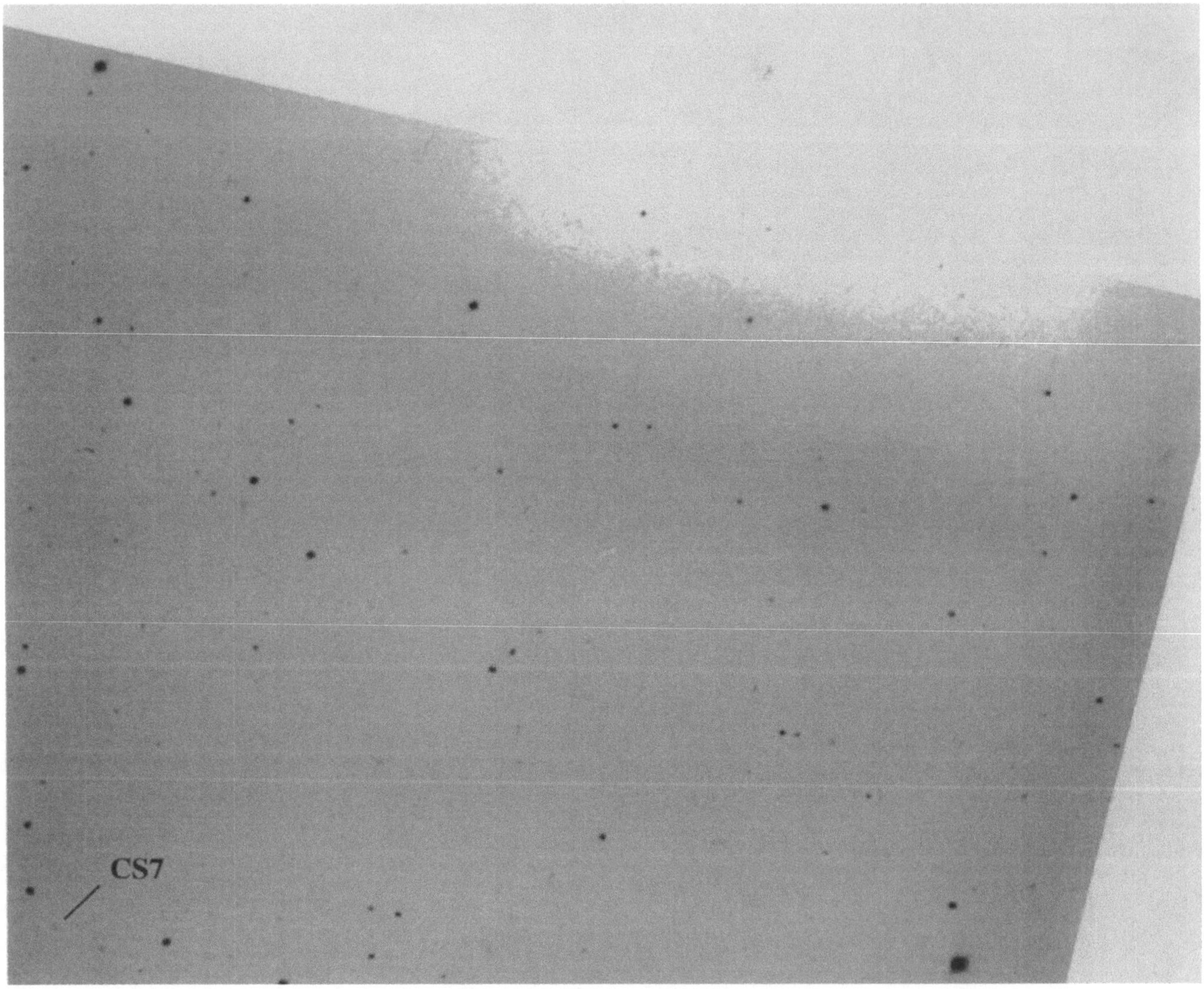

M33

Chart 7

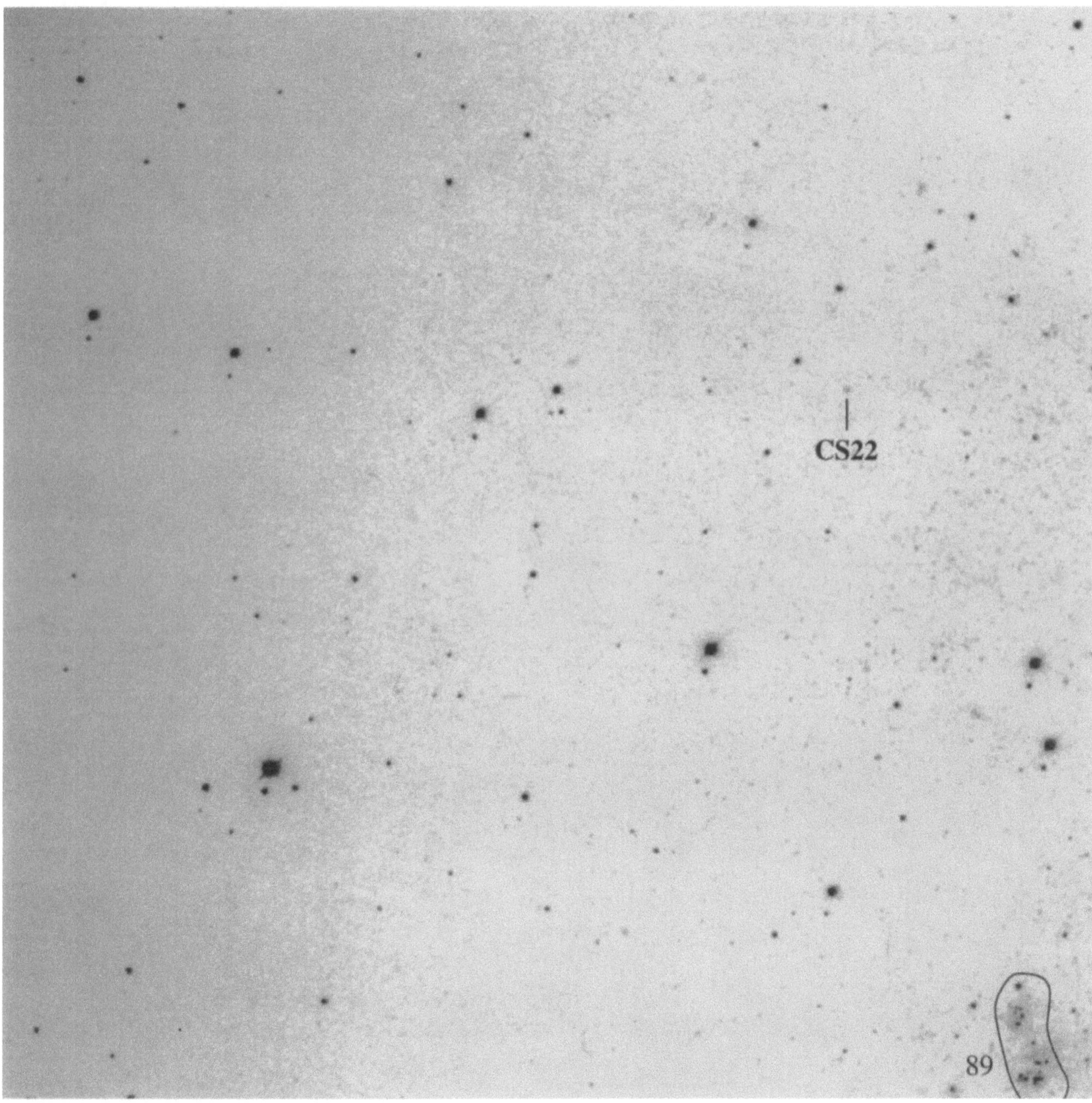

M33

Chart 8

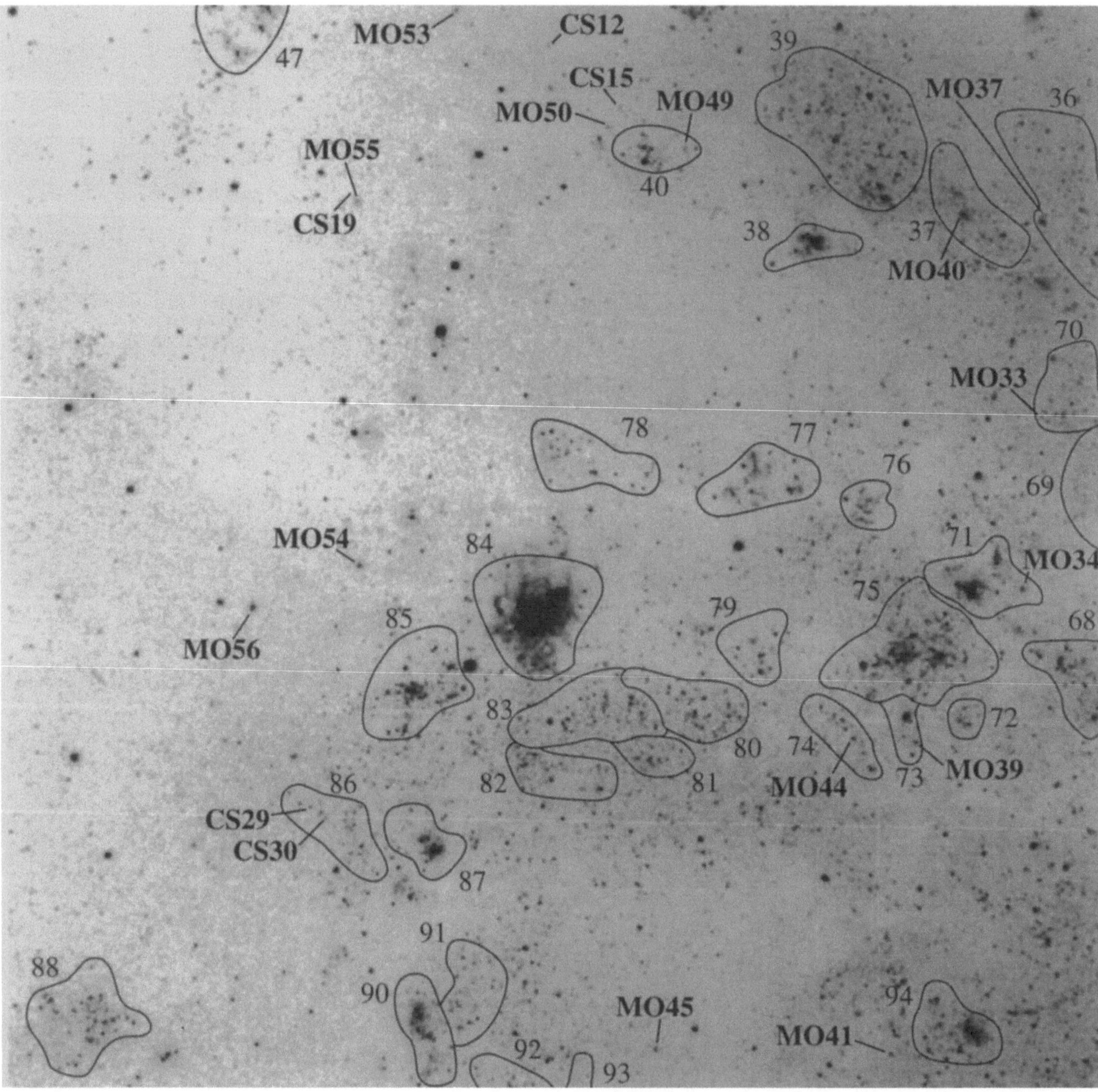

47
MO53
CS12
39
CS15
MO49
MO37
36
MO50
MO55
40
CS19
38
37
MO40
70
MO33
78
77
76
69
MO54
84
71
MO34
75
79
85
MO56
68
83
72
80
74
81
MO44
73
MO39
82
86
CS29
CS30
87
91
88
90
MO45
94
MO41
92
93

M33

Chart 9

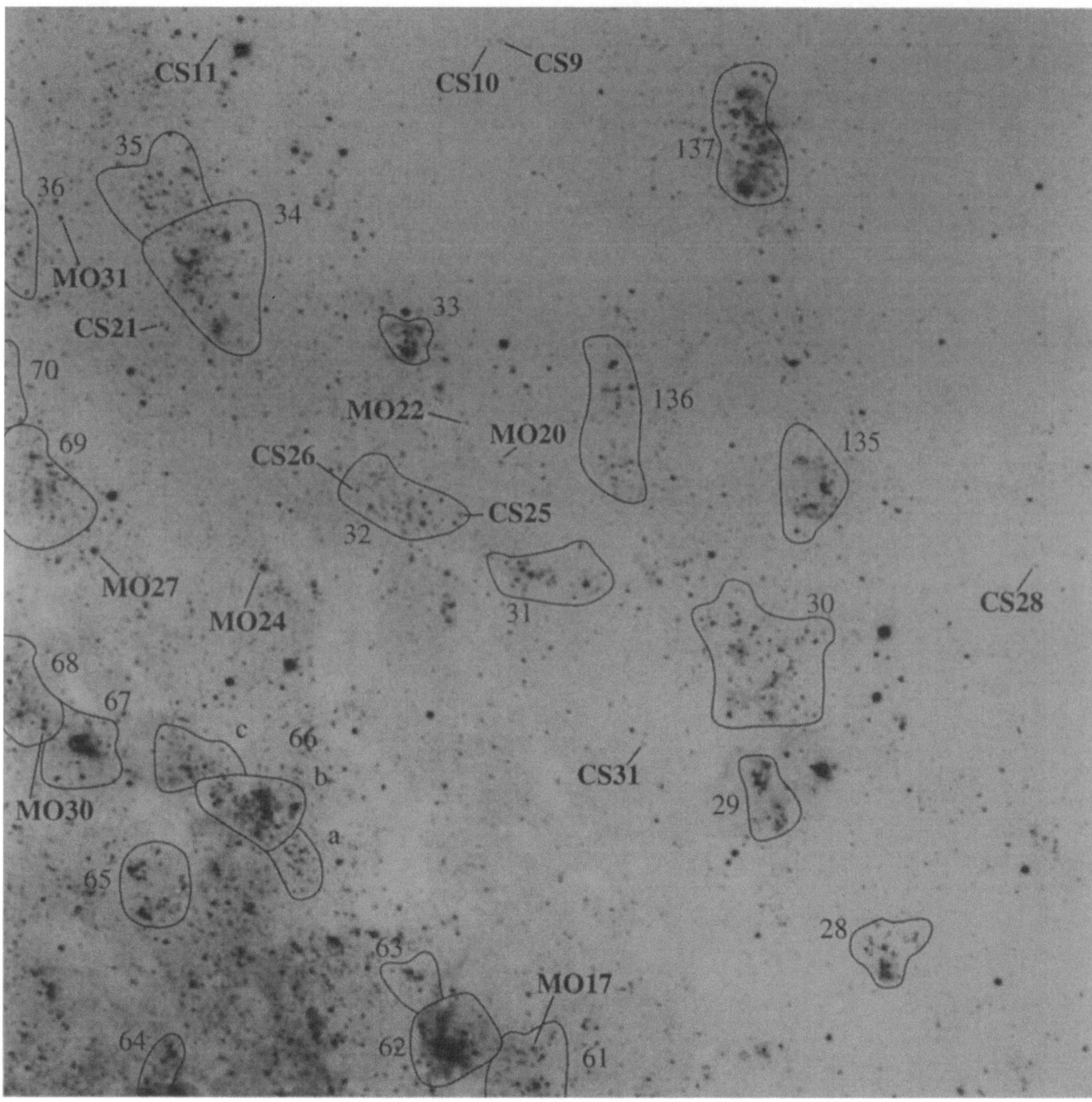
CS11
CS10
CS9
137
35
36
34
MO31
CS21
33
70
136
MO22
MO20
135
69
CS26
CS25
32
31
MO27
MO24
30
CS28
68
67
c
66
b
CS31
MO30
29
a
65
28
63
MO17
64
62
61

M33

Chart 10

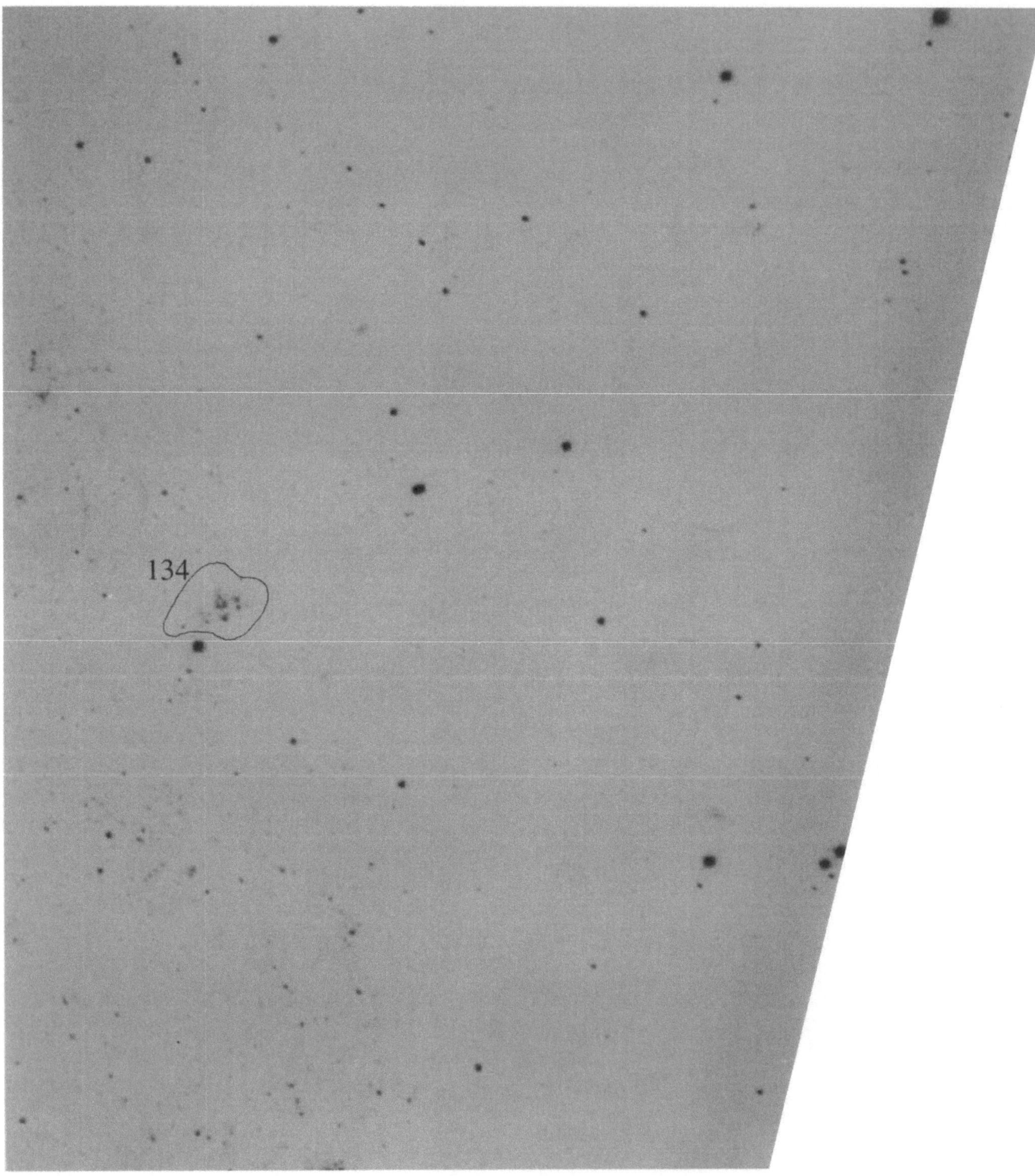

M33

Chart 11

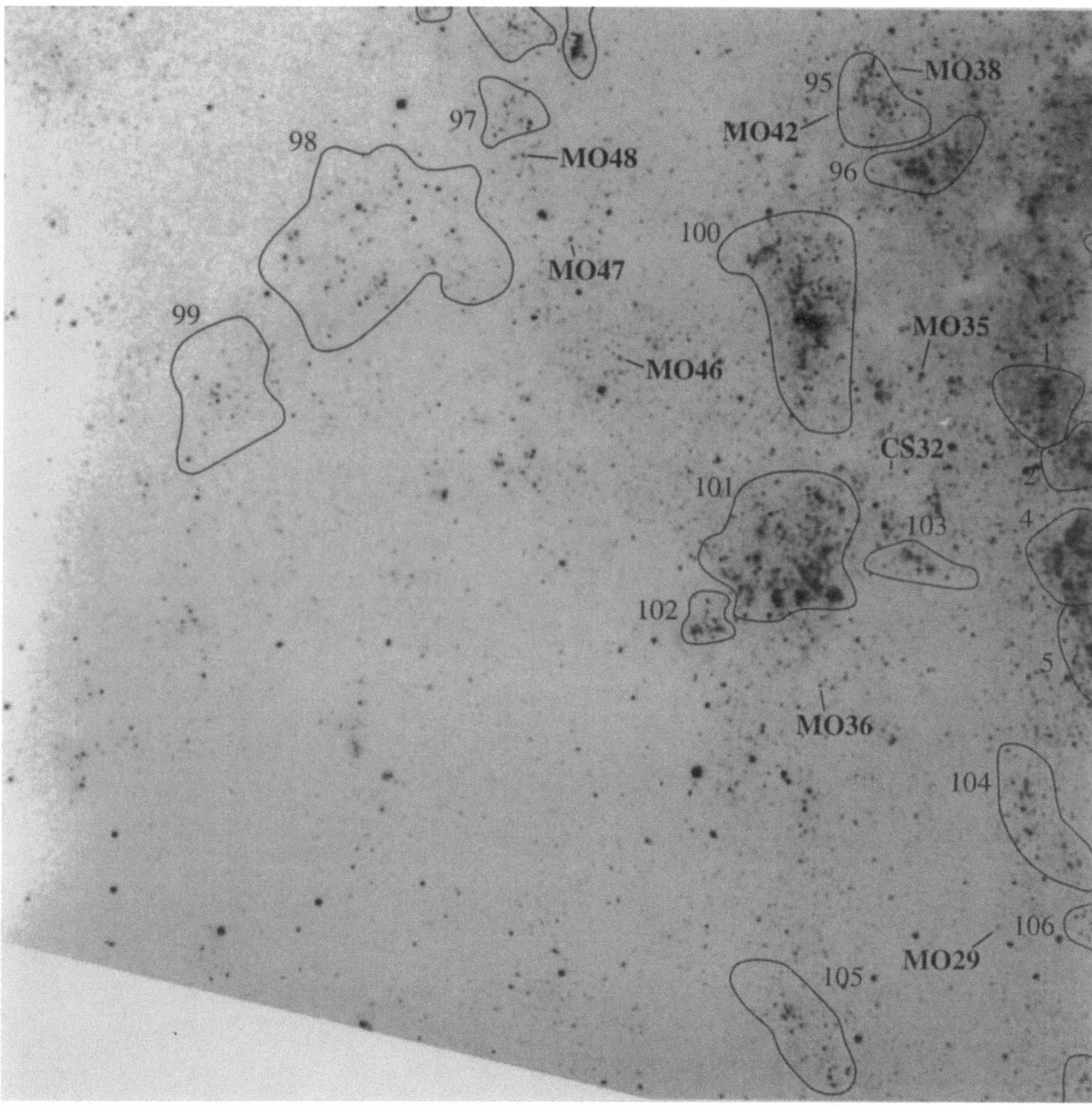
95
MO38
97
MO42
98
MO48
96
100
MO47
99
MO35
MO46
1
CS32
2
101
103
4
102
5
MO36
104
106
MO29
105

M33

Chart 12

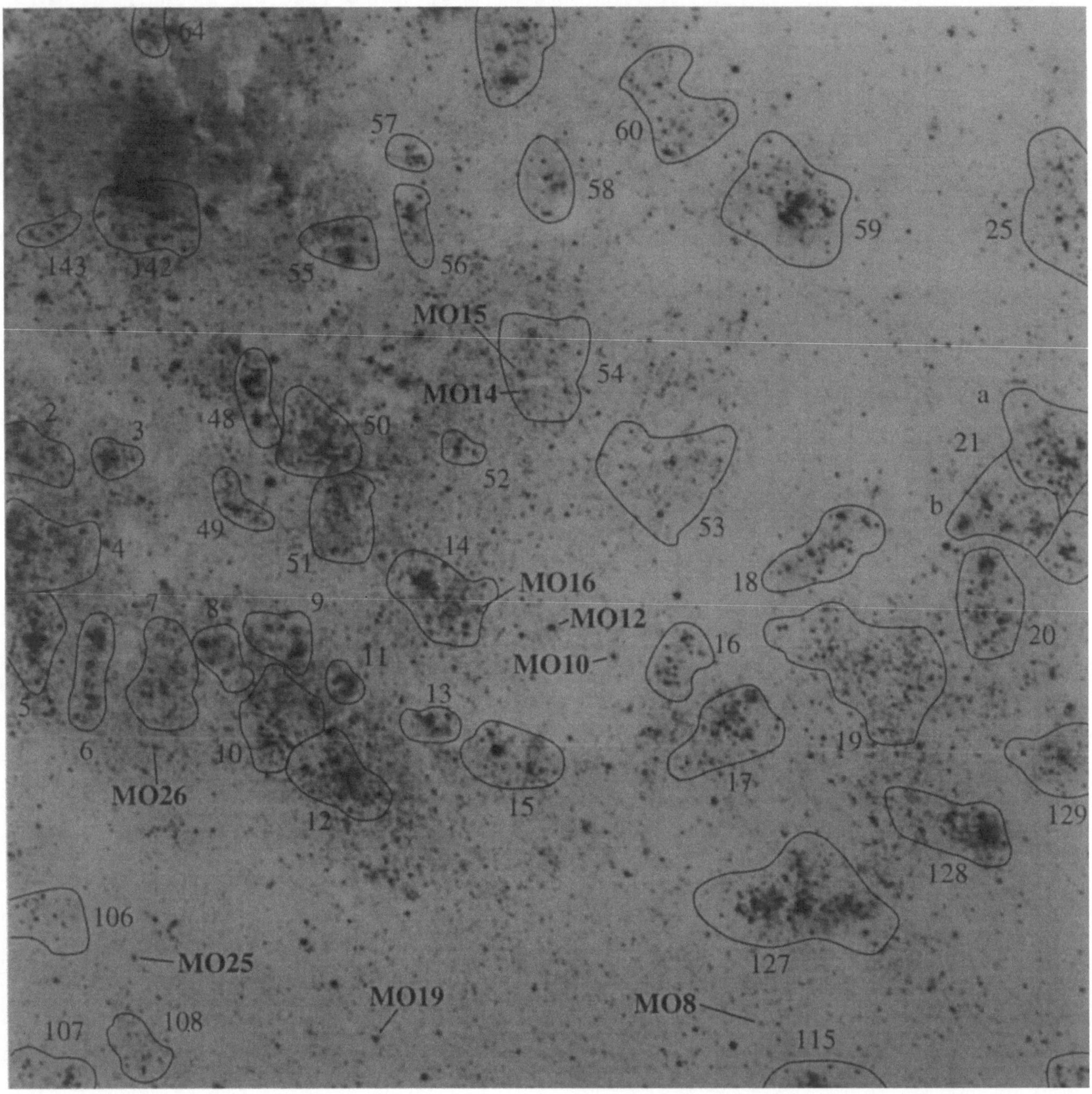
64
57
60
58
59
25
143
142
55
56
MO15
54
MO14
2
3
48
50
a
21
52
53
b
49
4
51
14
MO16
18
MO12
7
8
9
11
MO10
16
20
5
13
19
6
10
17
MO26
12
15
129
128
106
127
MO25
MO19
MO8
107
108
115

M33

Chart 13

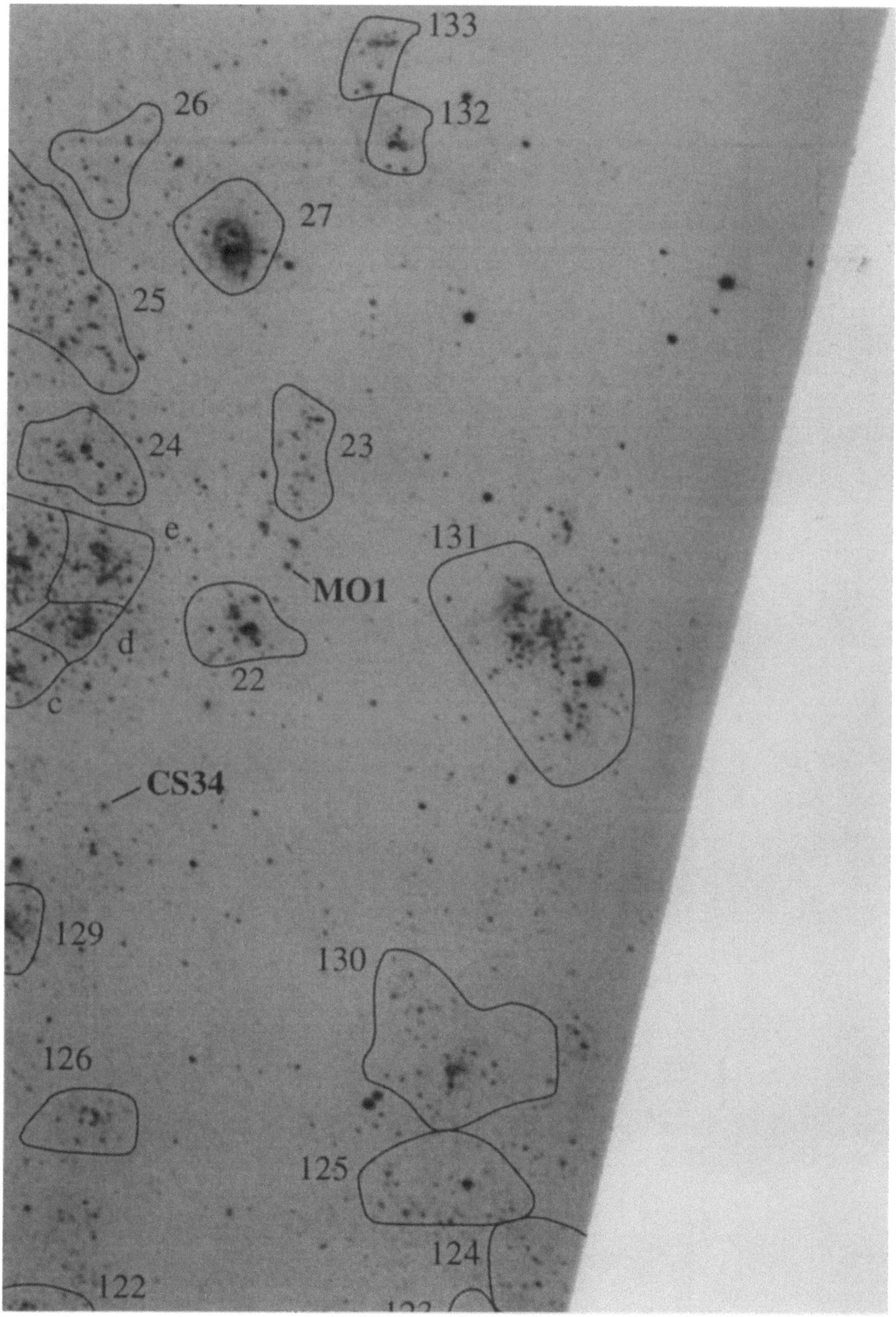

M33

Chart 14

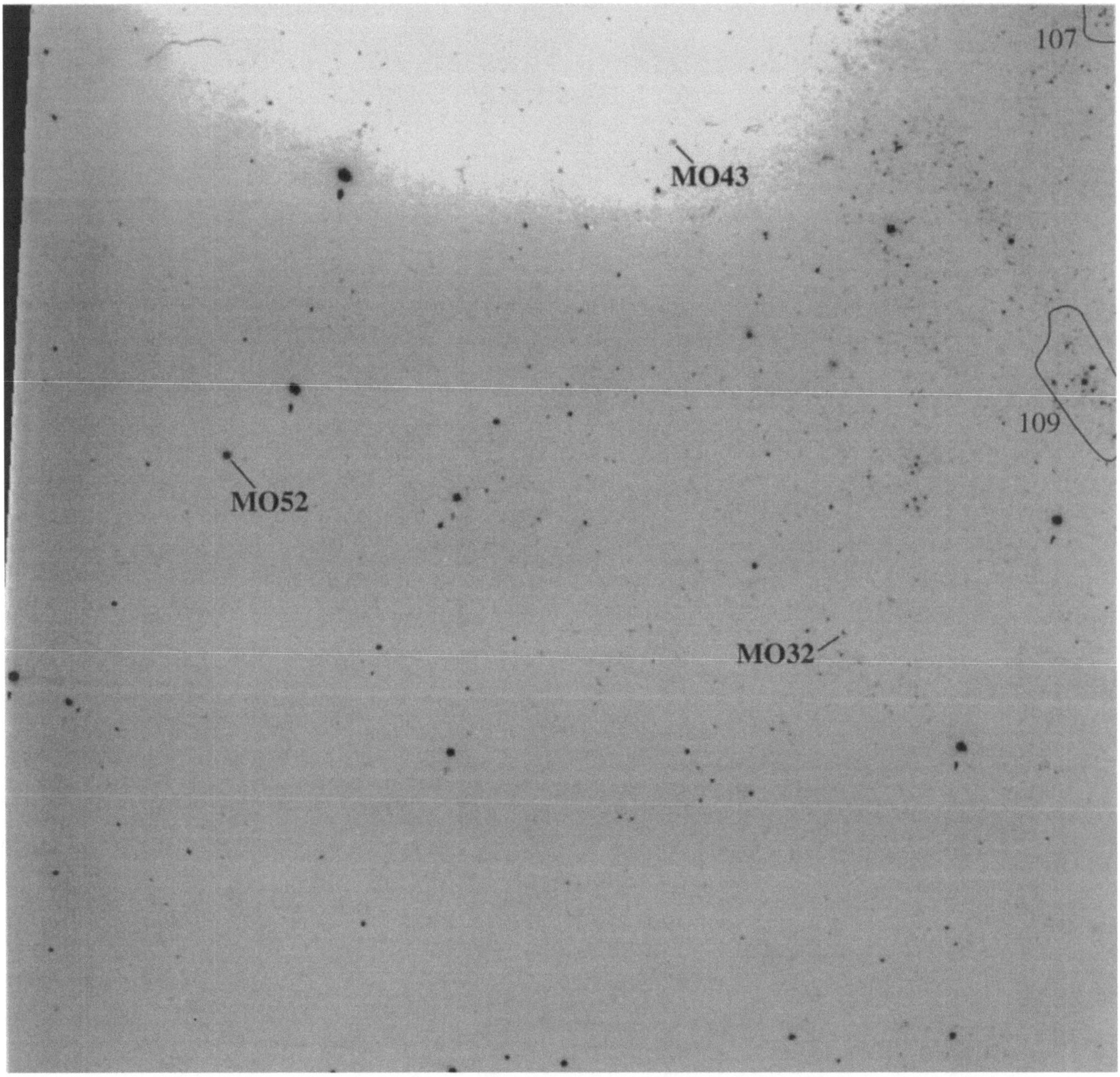

M33

Chart 15

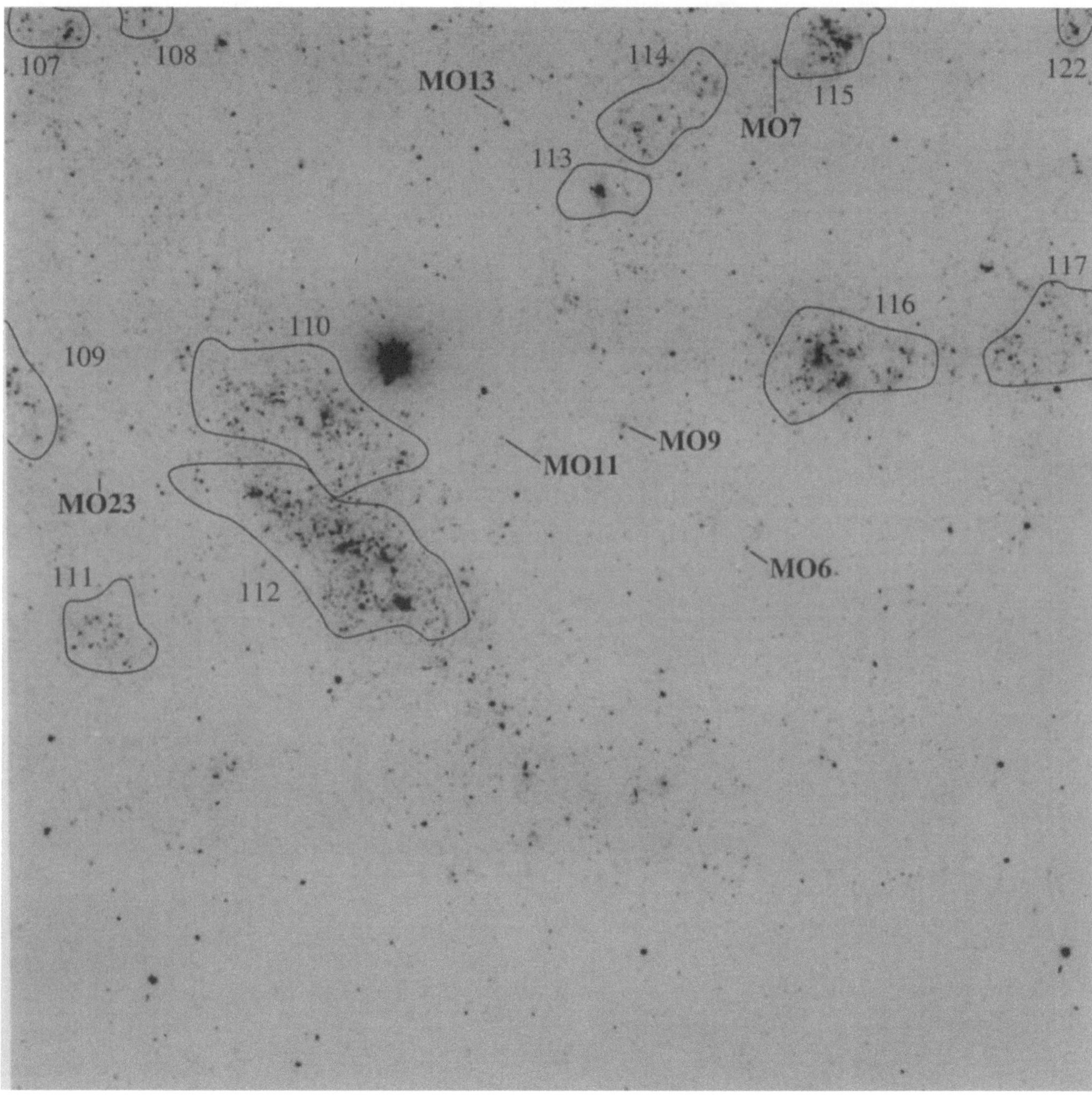
107
108
MO13
114
115
MO7
122
113
117
116
110
109
MO9
MO11
MO23
MO6
111
112

M33

Chart 16

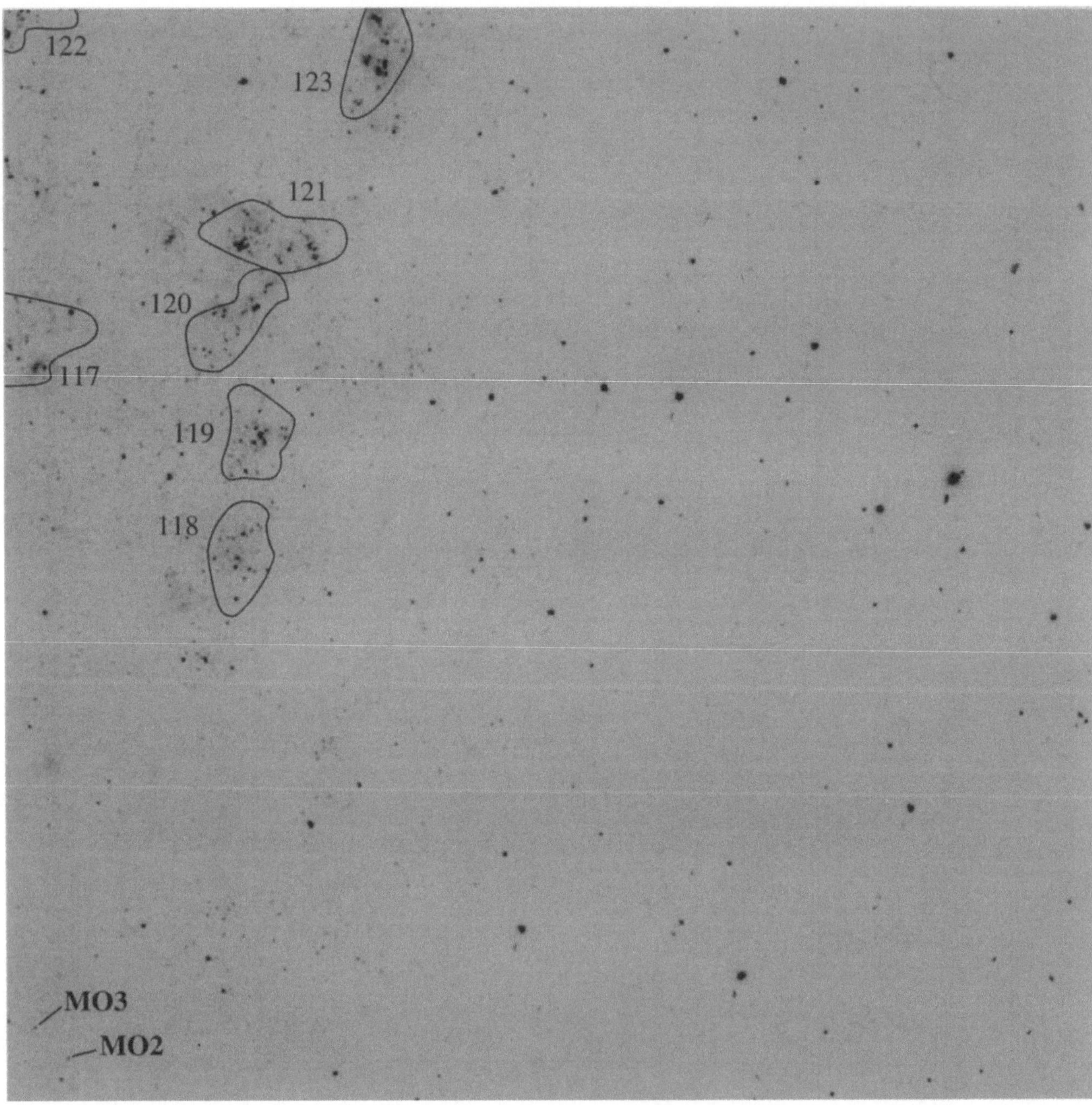

M33

Chart 17

M33

Chart 18

M33

Chart 19

M33

Chart 20

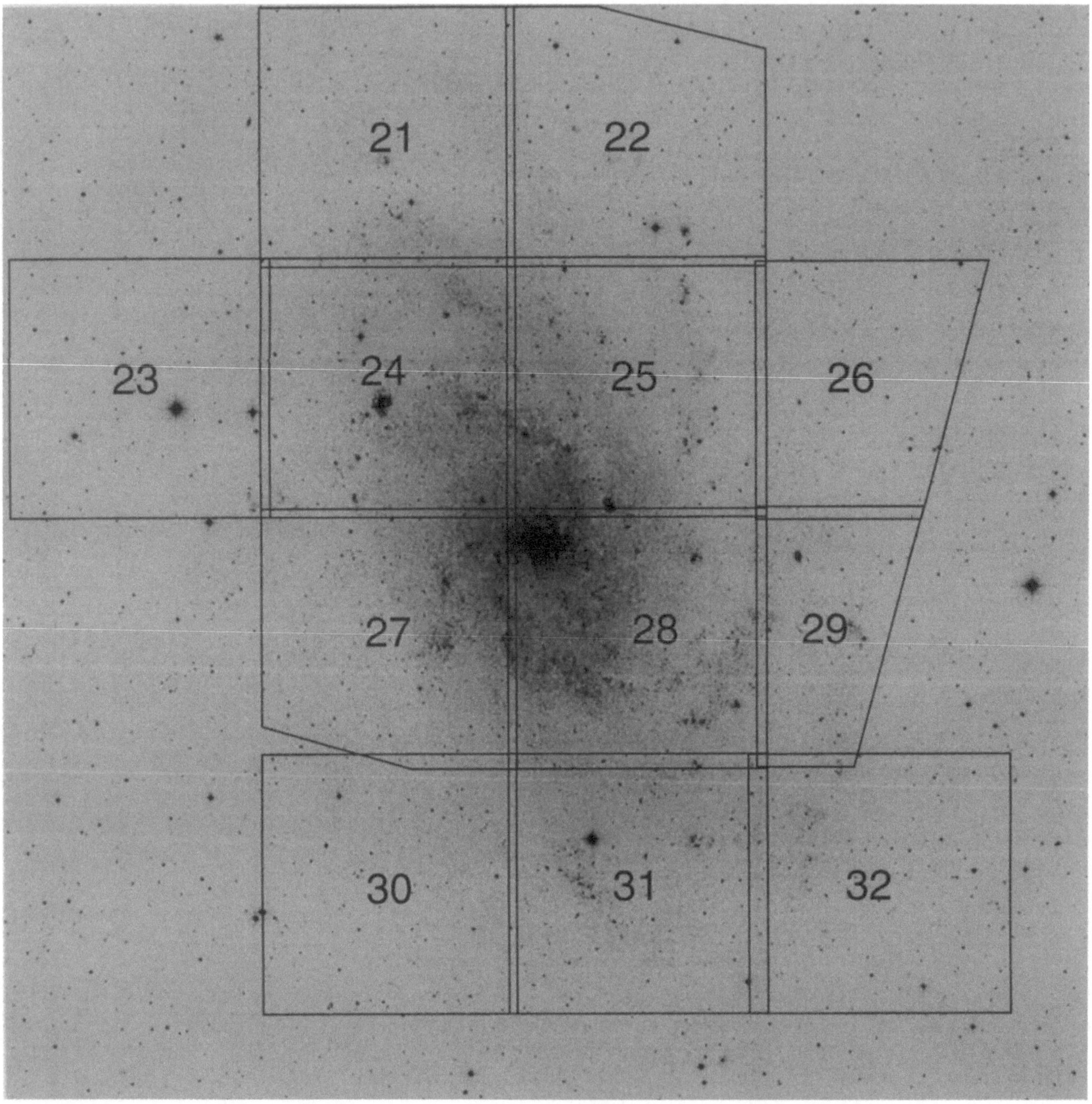

M33

Chart 21

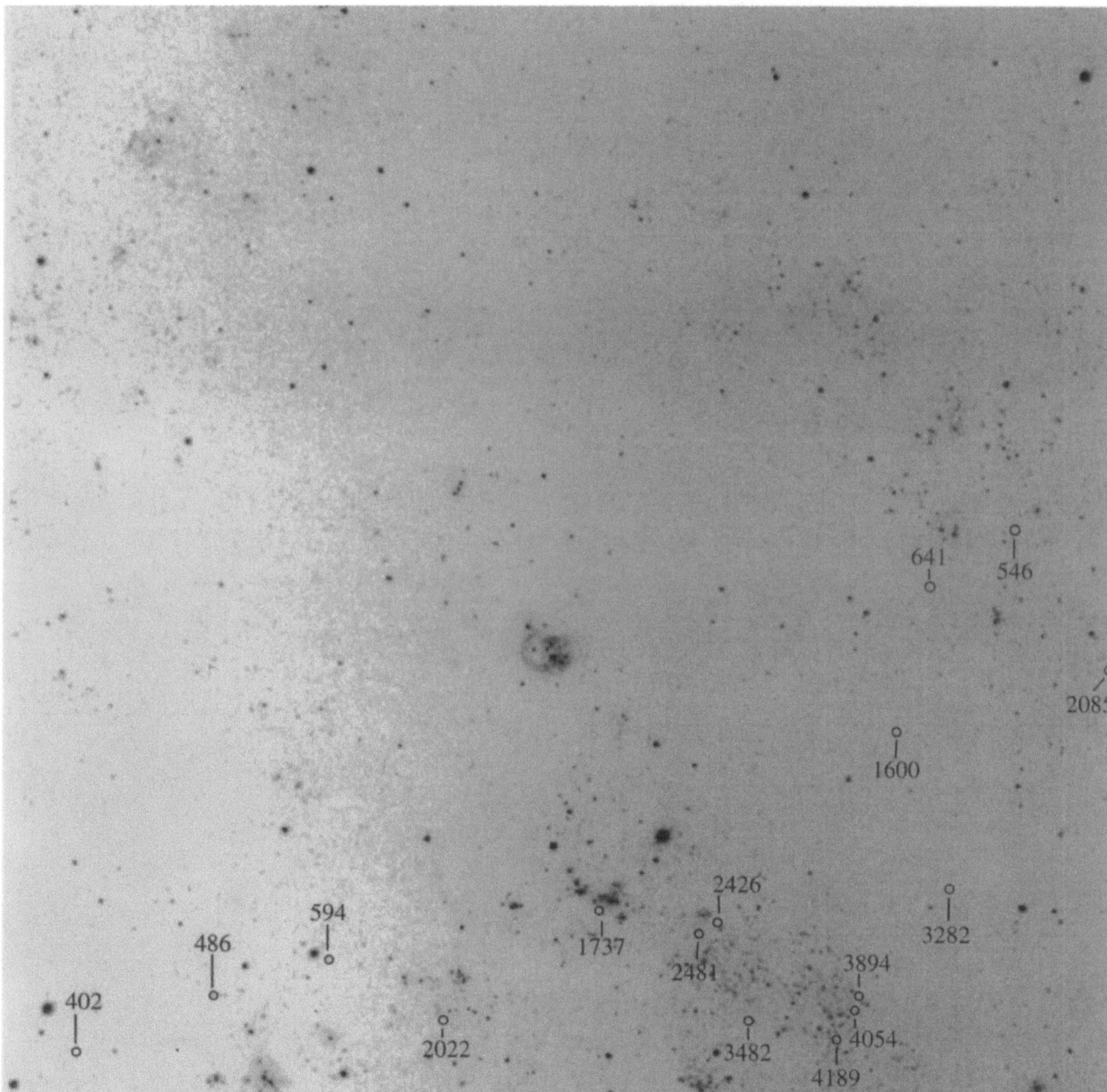

641
546
2085
1600
2426
594
486
1737
2481
3282
3894
402
4054
2022
3482
4189

M33

Chart 22

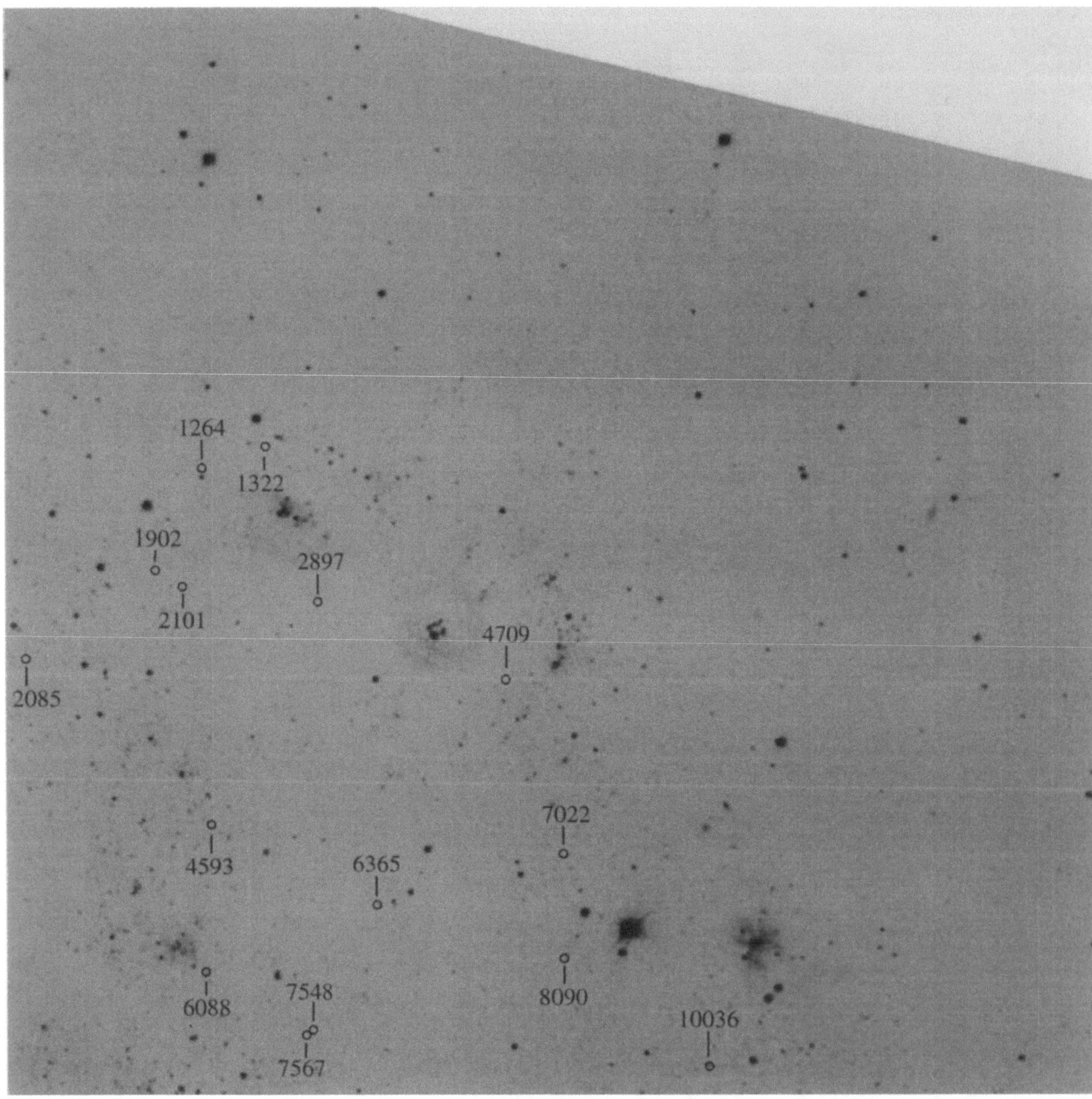
1264
1322
1902
2897
2101
4709
2085
7022
4593
6365
8090
6088
7548
10036
7567

M33

Chart 23

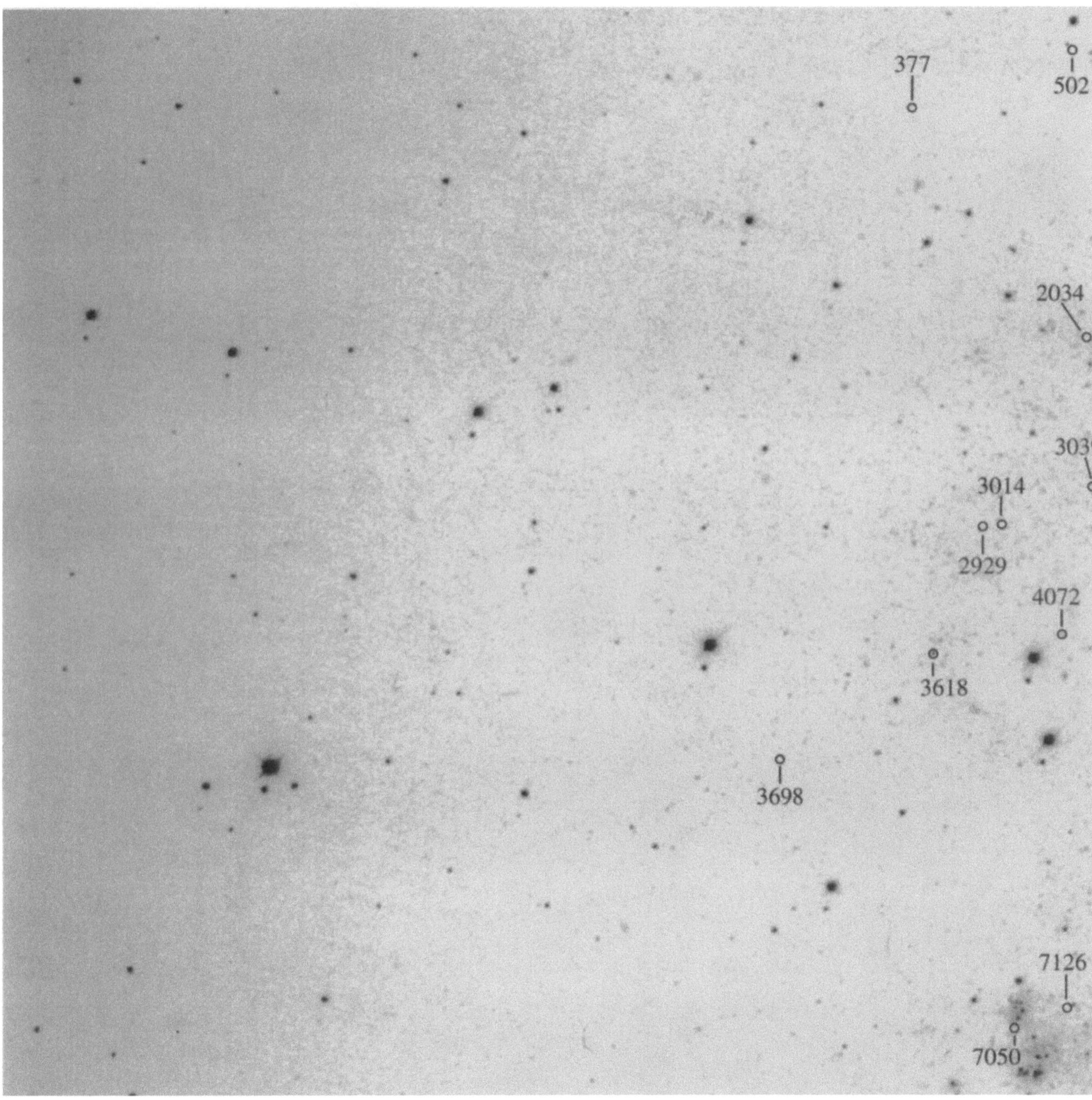

M33

Chart 24

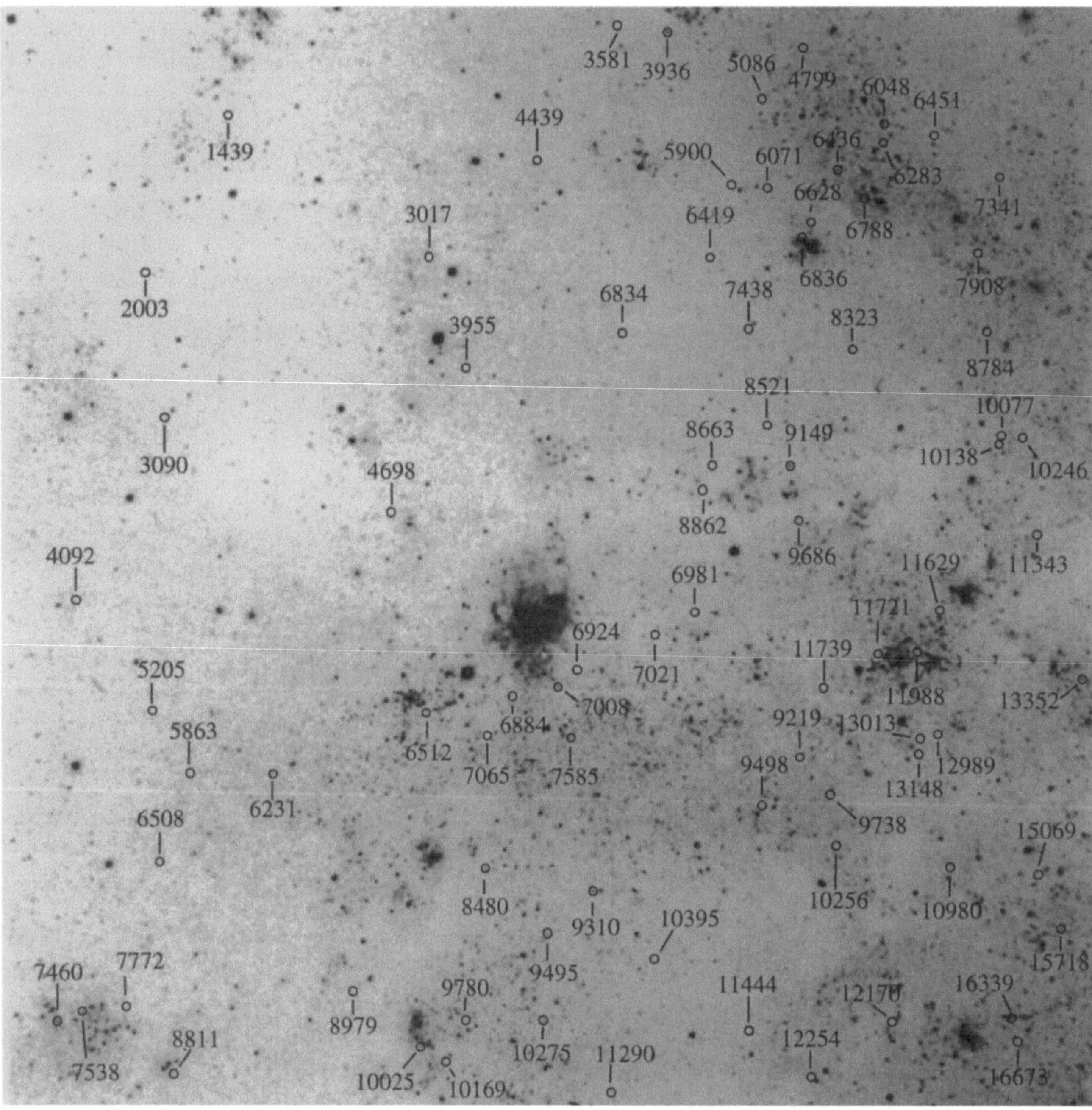
3581
3936
5086
4799
6048
6451
1439
4439
5900
6071
6436
6283
6628
7341
3017
6419
6788
6836
7908
2003
6834
7438
8323
3955
8784
8521
10077
3090
8663
9149
10138
10246
4698
8862
9686
4092
6981
11629
11343
11721
6924
7021
11739
5205
11988
13352
7008
9219
13013
6884
5863
6512
12989
7065
7585
9498
13148
6231
6508
9738
15069
10256
10980
8480
9310
10395
9495
15718
7460
7772
9780
11444
12170
16339
8979
12254
8811
10275
11290
7538
10025
10169
16673

M33

Chart 25

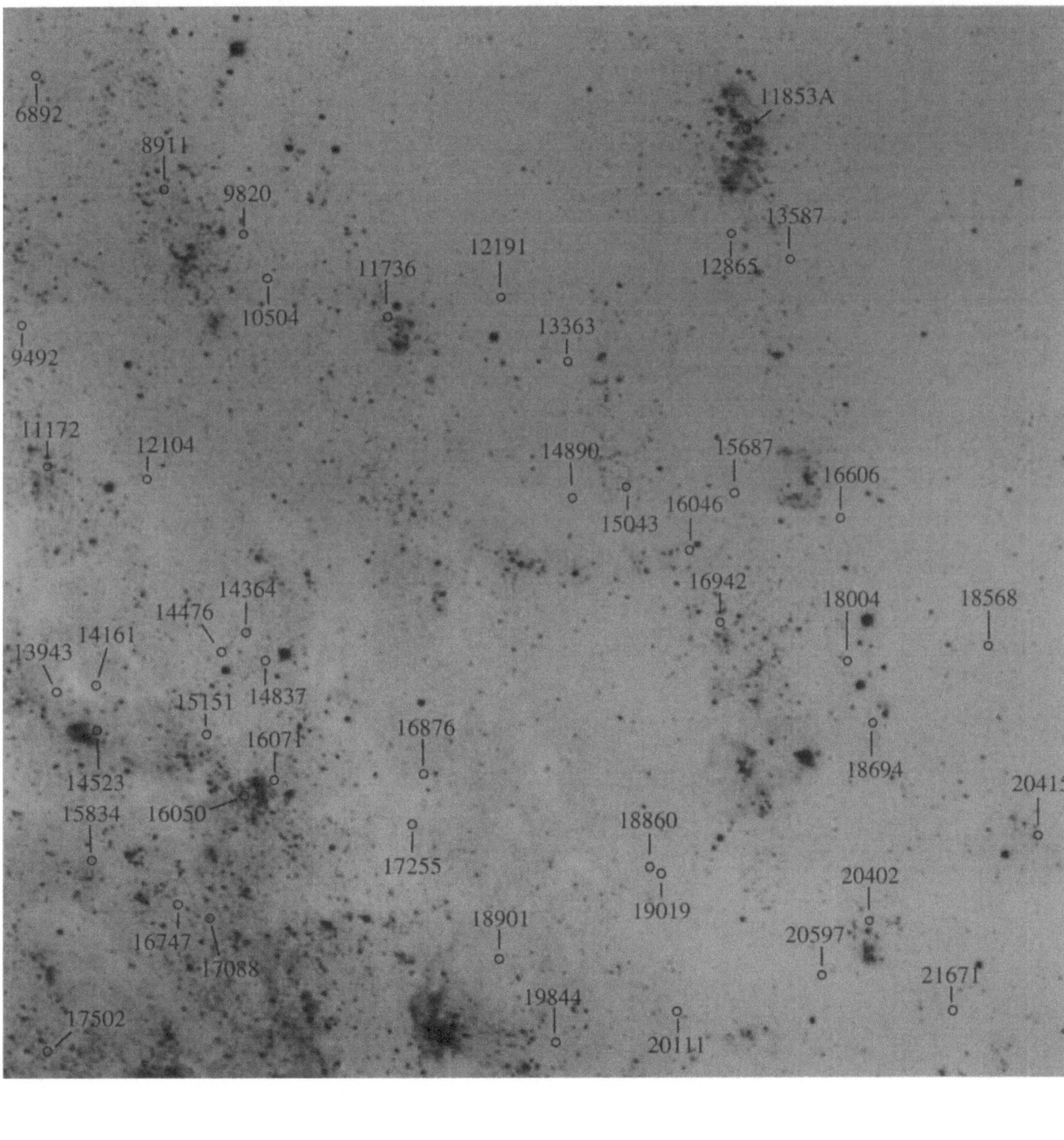
6892
8911
9820
11853A
13587
12865
12191
11736
10504
9492
13363
11172
12104
14890
15687
16606
15043
16046
16942
14364
14476
14161
13943
18004
18568
15151
14837
16876
16071
14523
18694
20415
15834
16050
18860
17255
19019
20402
16747
17088
18901
20597
21671
19844
17502
20111

M33

Chart 26

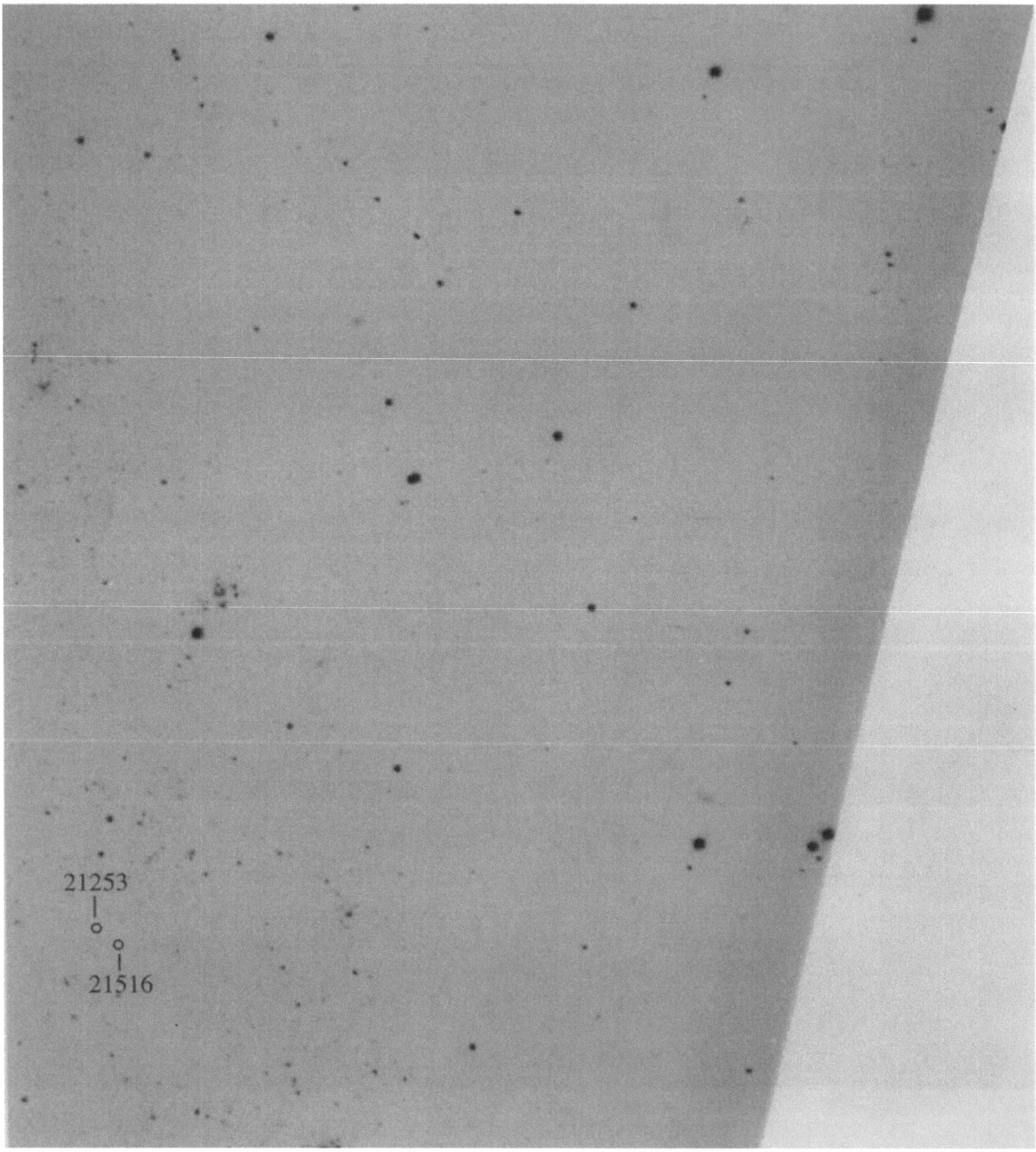

M33

Chart 27

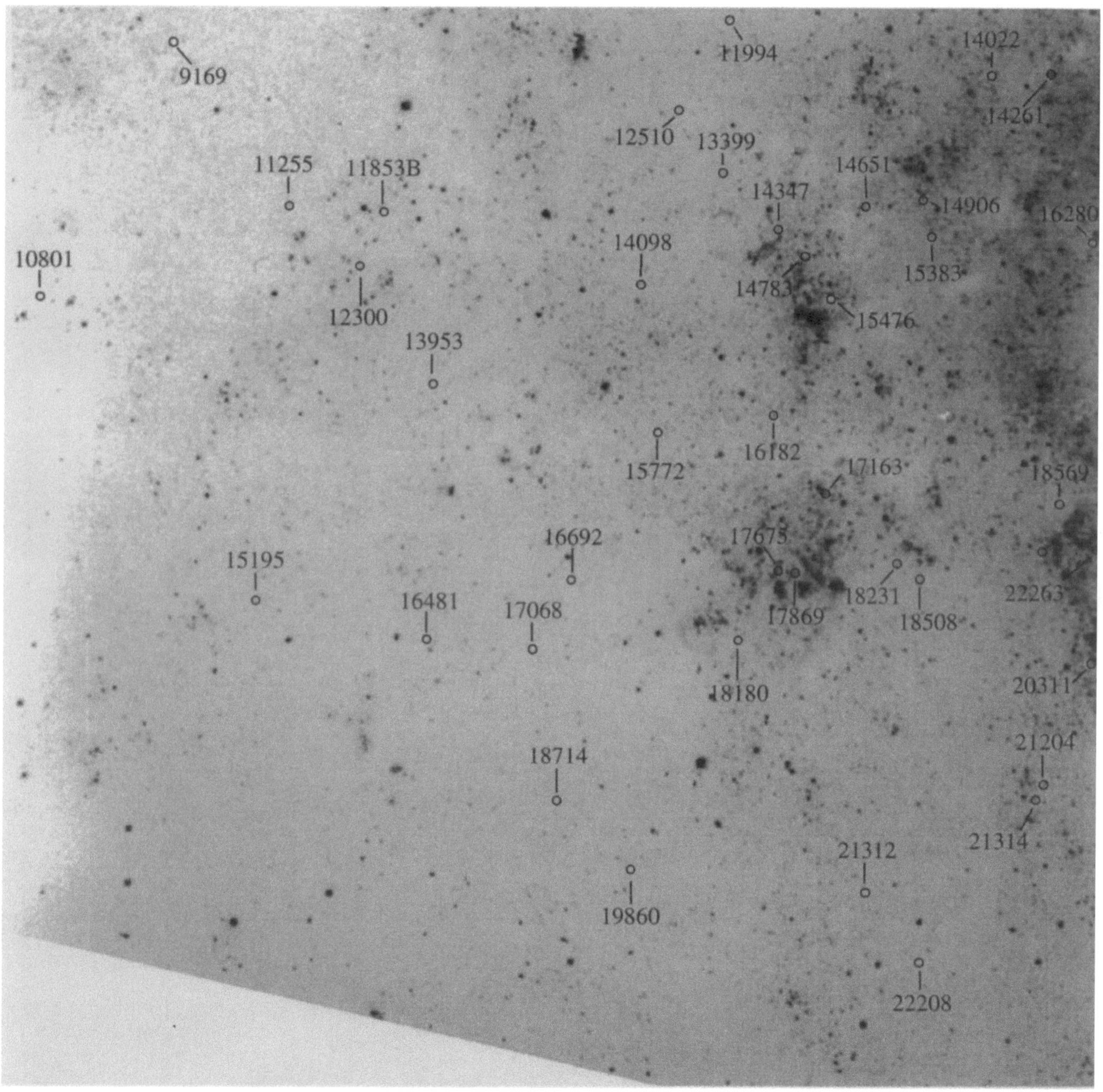
9169
11994
14022
14261
12510
13399
11255
11853B
14651
14347
14906
16280
10801
14098
15383
14783
12300
15476
13953
16182
15772
17163
18569
16692
17675
15195
18231
22263
17869
16481
17068
18508
18180
20311
21204
18714
21314
21312
19860
22208

M33

Chart 28

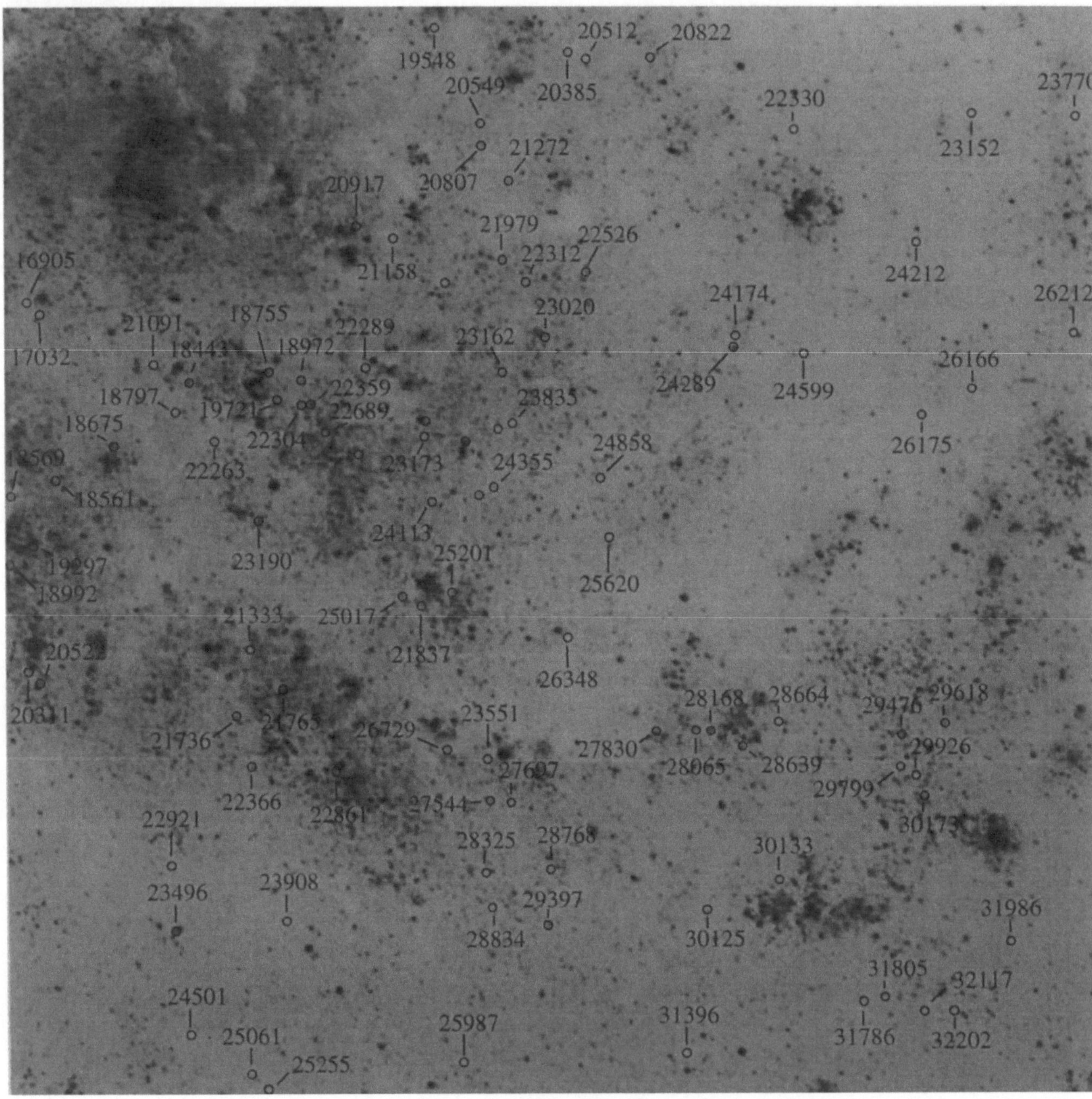
19548
20512
20822
20549
20385
22330
23770
23152
21272
20807
20917
21979
22526
22312
21158
16905
24212
24174
26212
23020
17032
21091
18755
18443
18972
22289
23162
26166
24289
24599
18797
19721
22359
22689
23835
18675
22304
24858
26175
18569
22263
23173
24355
18561
23190
24113
25201
19297
18992
25620
21333
25017
21837
20522
26348
20311
28168
28664
29470
29618
21736
21765
26729
23551
27830
28065
28639
29926
27607
29799
22366
27544
22921
22861
30173
28325
28768
30133
23496
23908
29397
31986
28834
30125
31805
32117
24501
31396
25987
31786
32202
25061
25255

M33

Chart 29

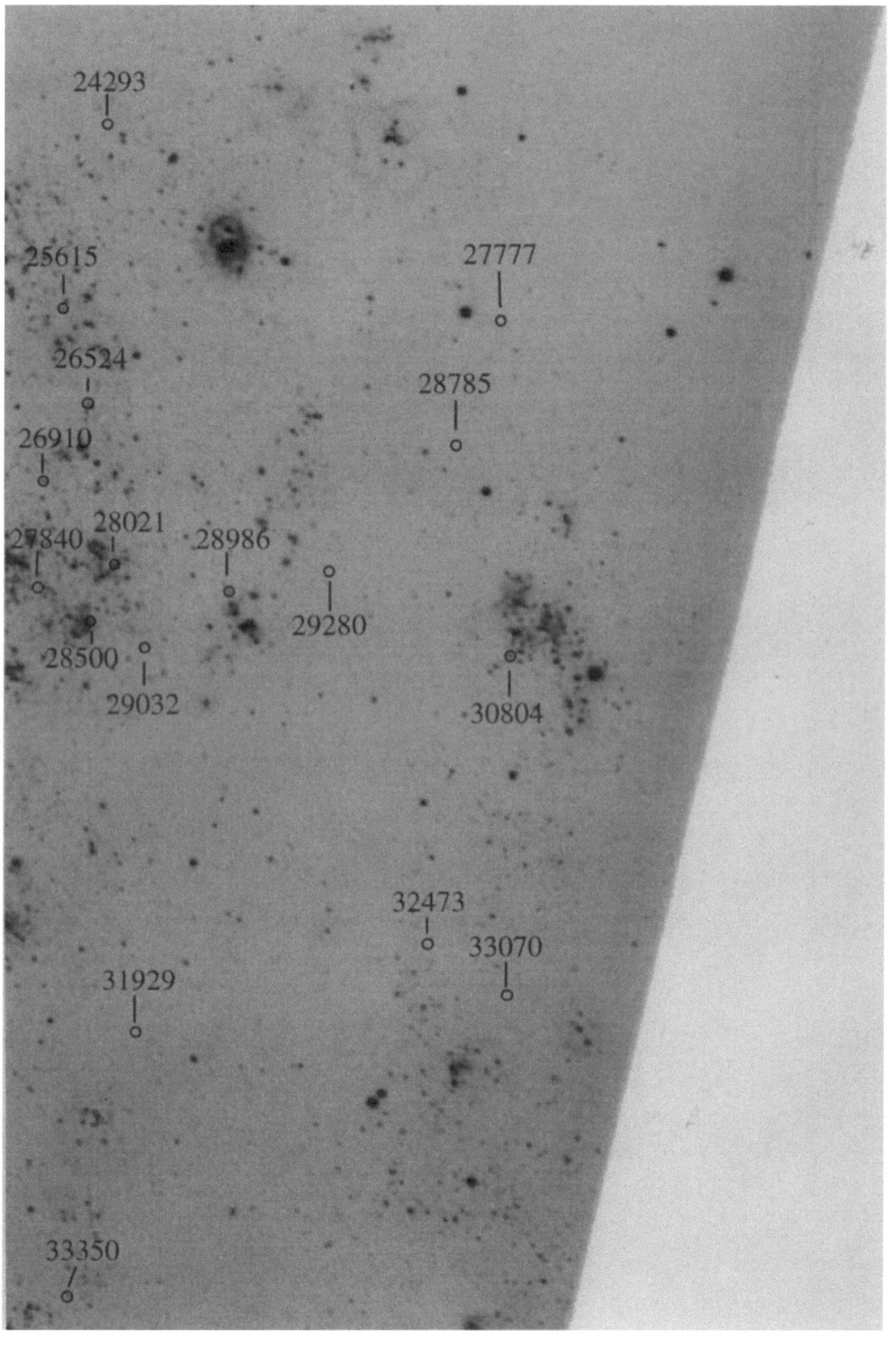
24293
25615
27777
26524
28785
26910
27840
28021
28986
29280
28500
29032
30804
32473
33070
31929
33350

M33

Chart 30

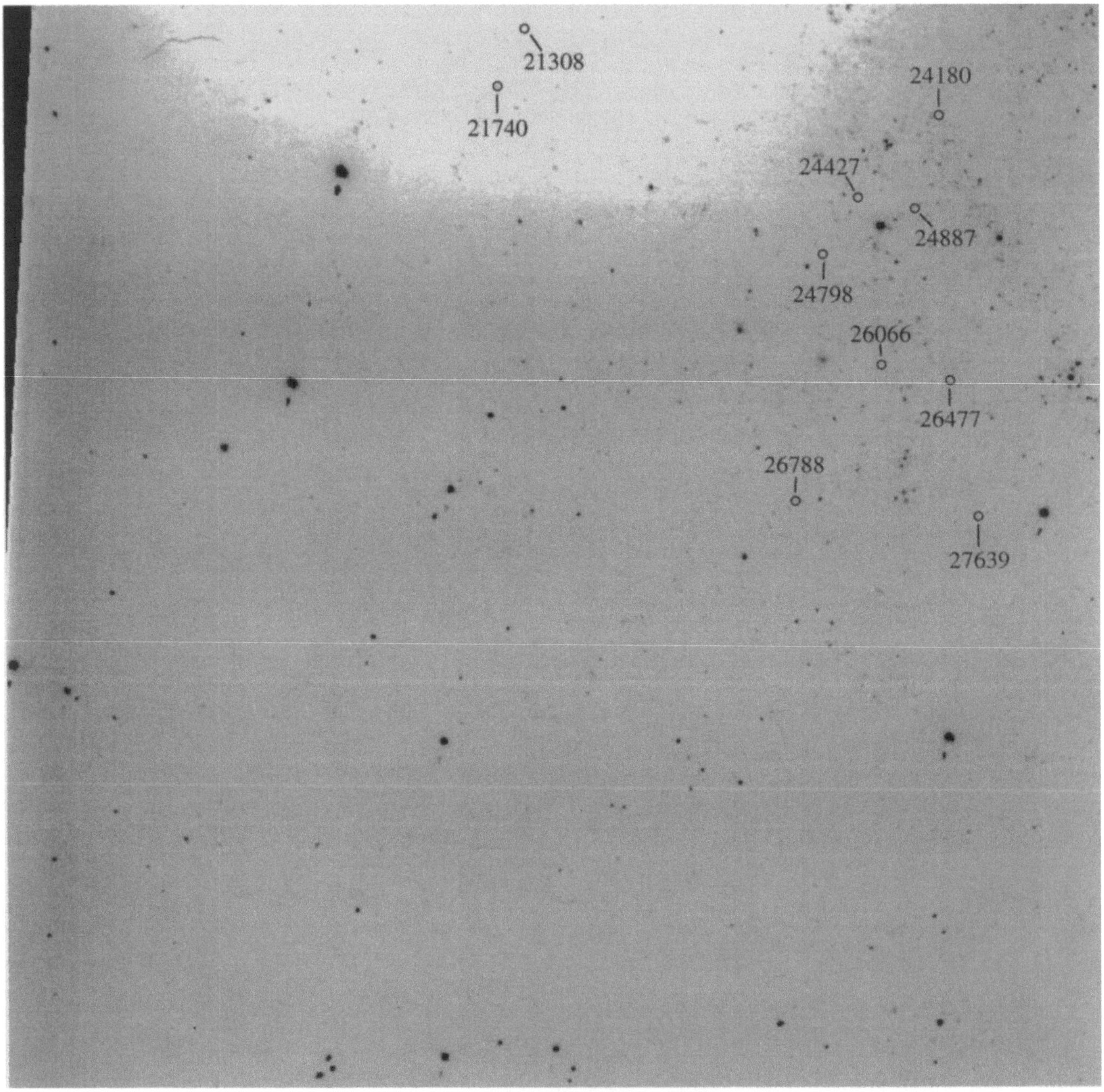
21308
21740
24180
24427
24887
24798
26066
26477
26788
27639

M33

Chart 31

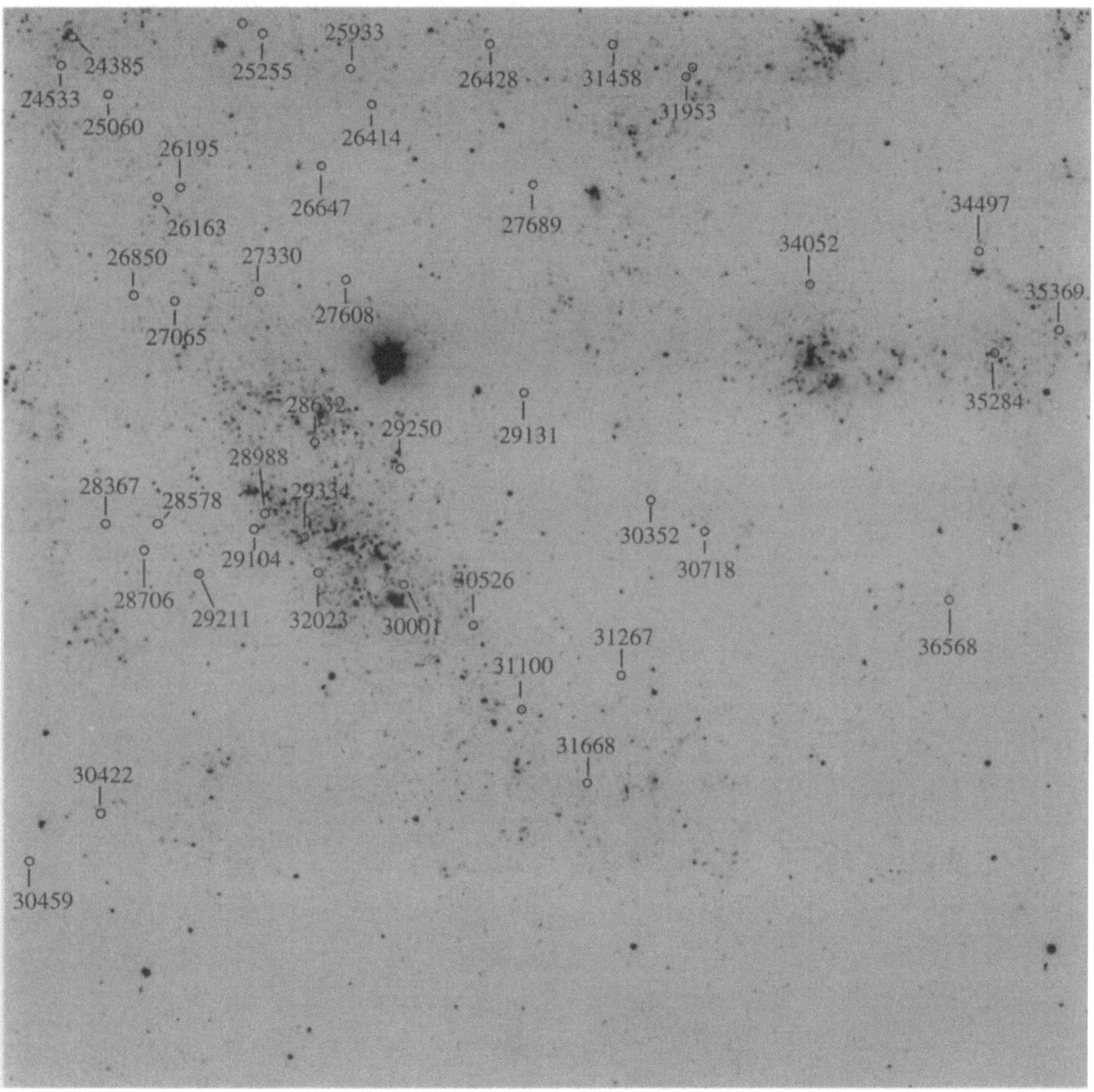

24385
24533
25255
25933
26428
31458
31953
25060
26414
26195
26647
27689
34497
26163
34052
26850
27330
35369
27608
27065
35284
28632
29250
29131
28988
28367
28578
29334
30352
30718
29104
30526
28706
29211
32023
30001
31267
36568
31100
31668
30422
30459

M33

Chart 32

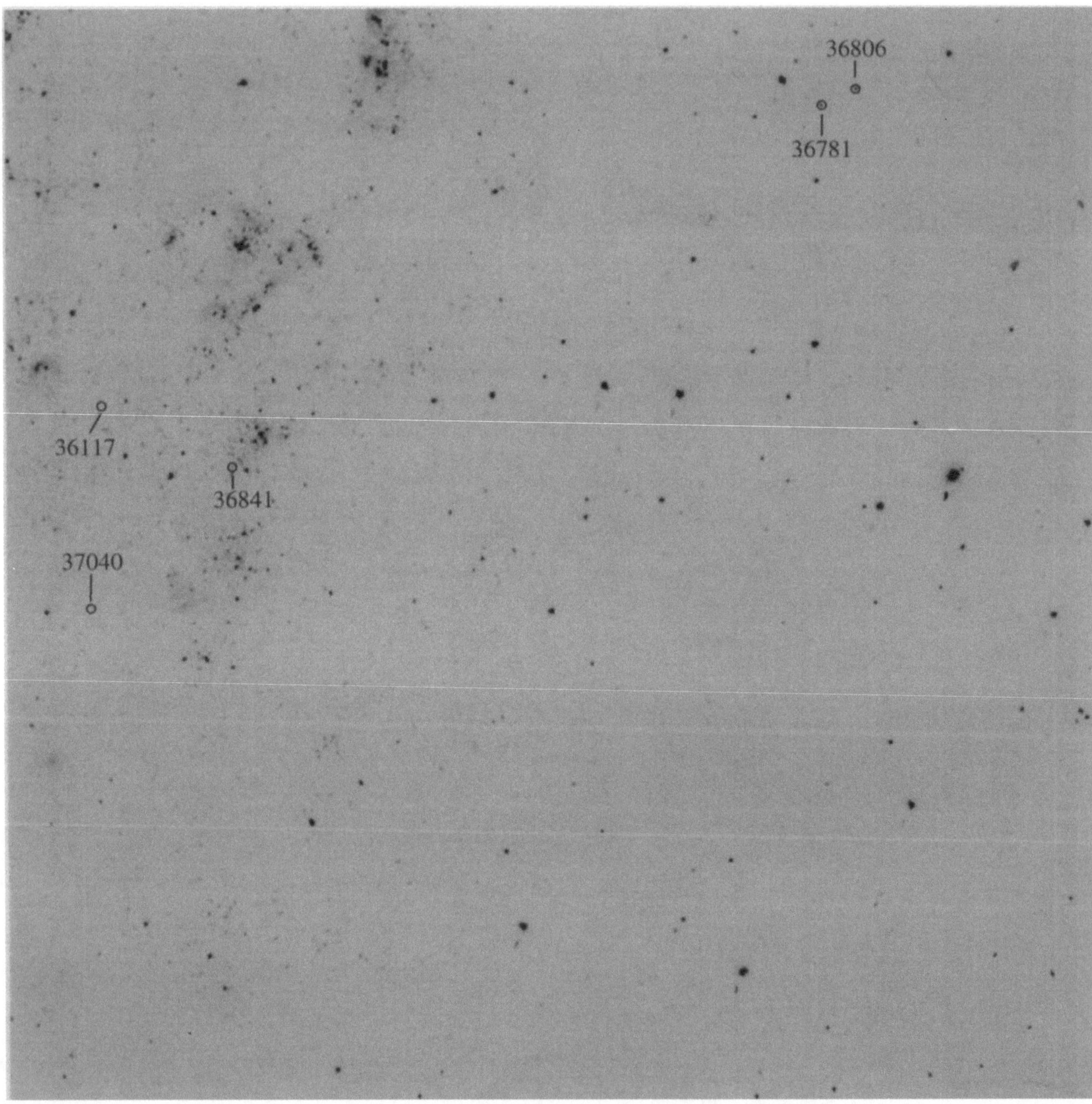

M33

Chart 33

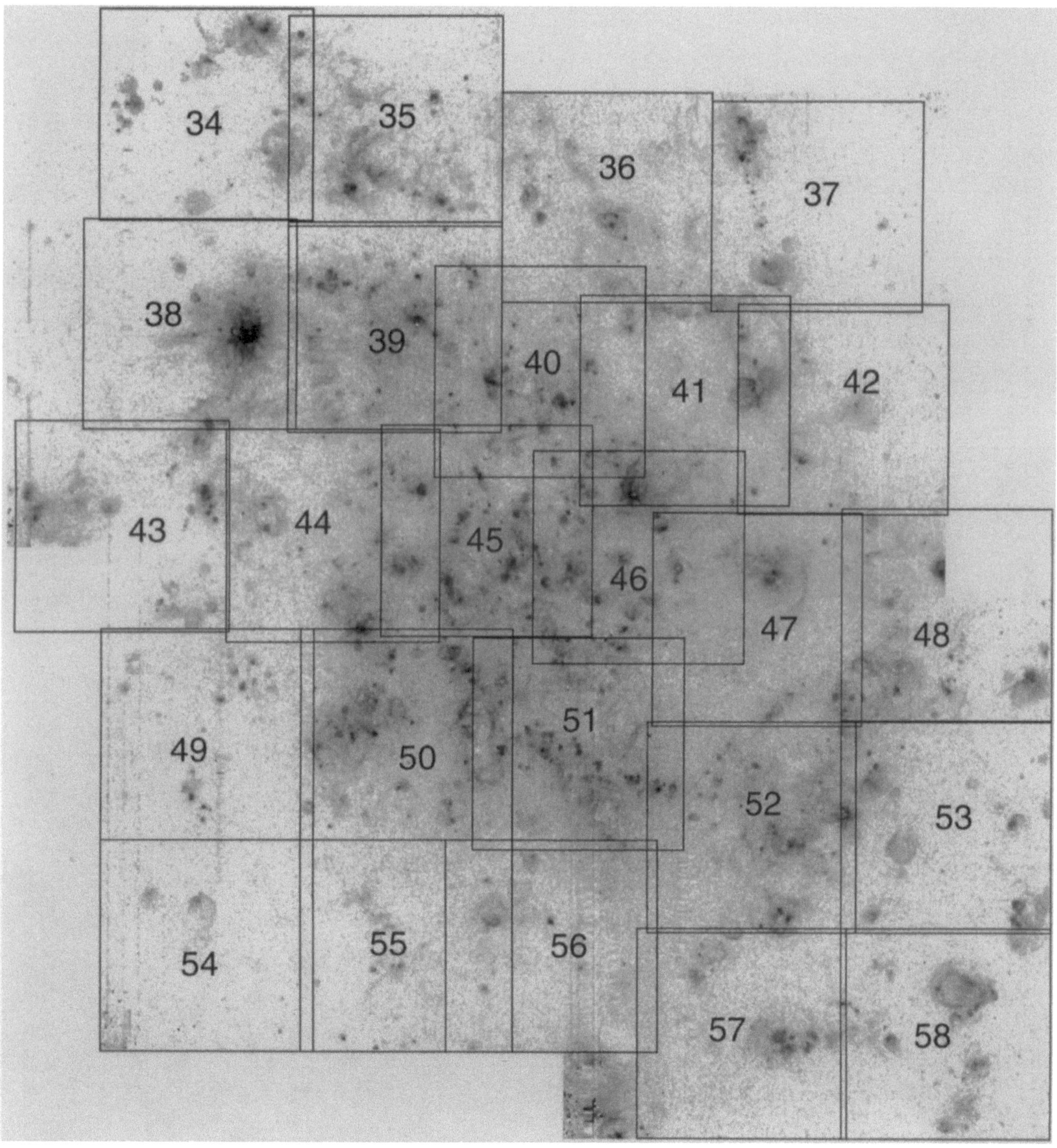
34
35
36
37
38
39
40
41
42
43
44
45
46
47
48
49
50
51
52
53
54
55
56
57
58

M33

Chart 34

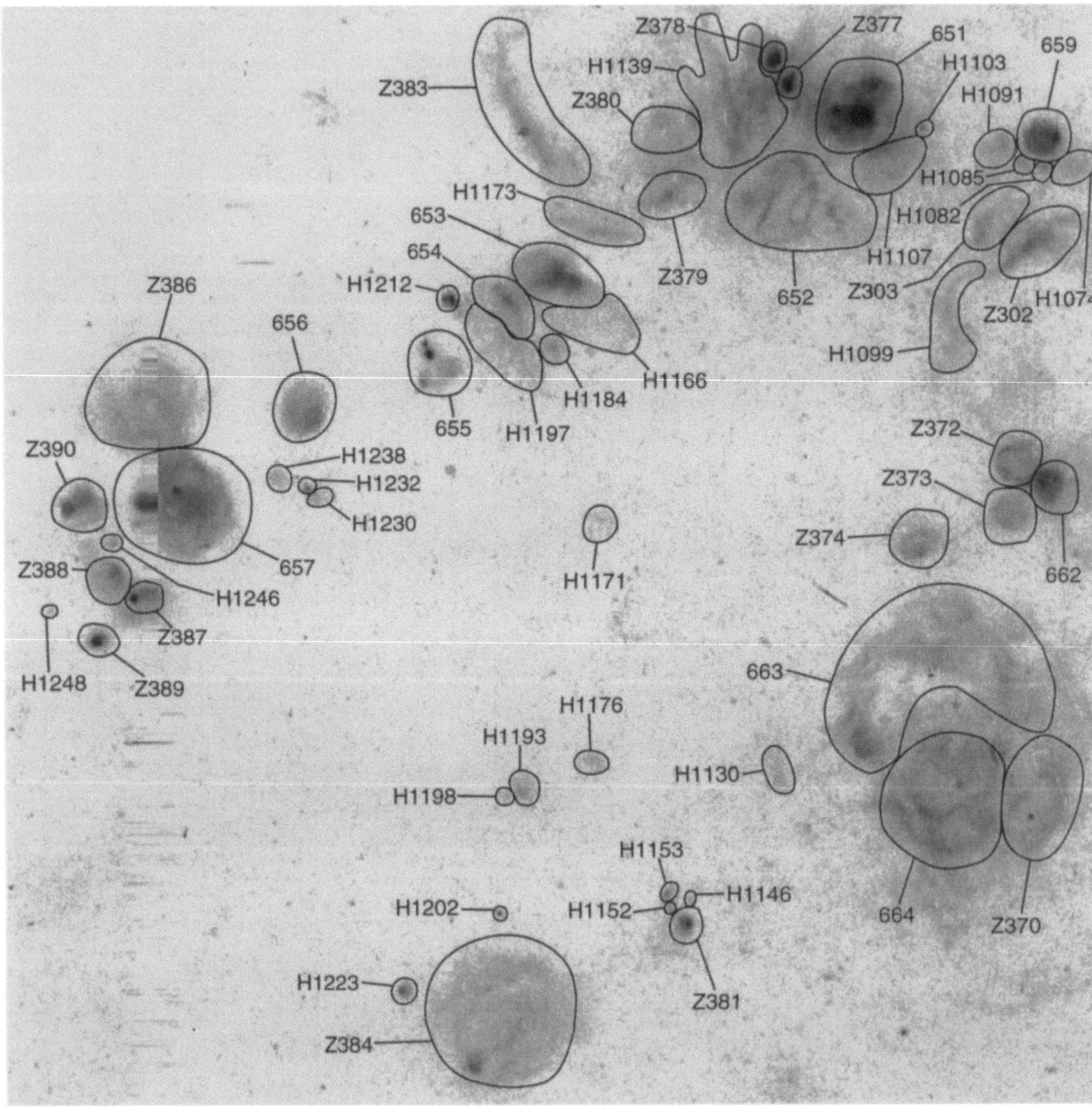
Z378
Z377
651
659
H1139
H1103
Z383
Z380
H1091
H1085
H1173
H1082
653
Z379
H1107
654
652
Z303
Z386
H1212
H1074
Z302
656
H1099
H1166
H1184
655
H1197
Z372
Z390
H1238
H1232
Z373
H1230
Z374
657
H1171
662
Z388
H1246
Z387
H1248
Z389
663
H1176
H1193
H1130
H1198
H1153
H1146
H1202
H1152
664
Z370
H1223
Z381
Z384

M33

Chart 35

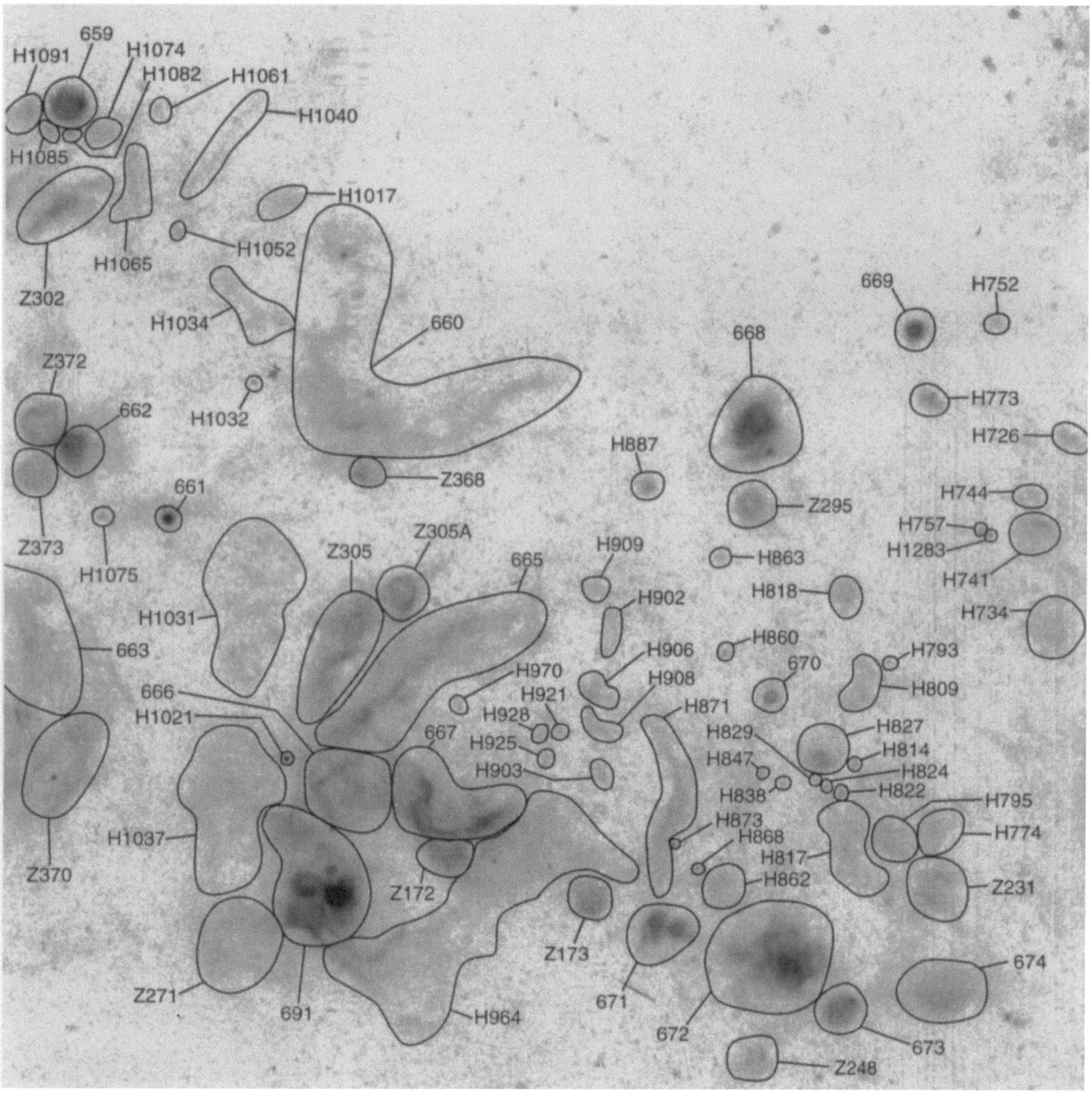
659
H1091
H1074
H1082
H1061
H1040
H1085
H1017
H1052
H1065
Z302
H1034
660
669
H752
668
Z372
662
H1032
H773
H726
H887
Z368
661
Z295
H744
H757
H1283
Z373
H1075
Z305
Z305A
665
H909
H863
H741
H818
H902
H1031
H734
663
H860
H906
H793
670
H970
H908
H809
666
H921
H871
H1021
H928
H827
H829
667
H925
H814
H847
H824
H903
H838
H822
H795
H873
H868
H774
H1037
H817
H862
Z370
Z172
Z231
Z173
674
Z271
671
691
H964
672
673
Z248

M33

Chart 36

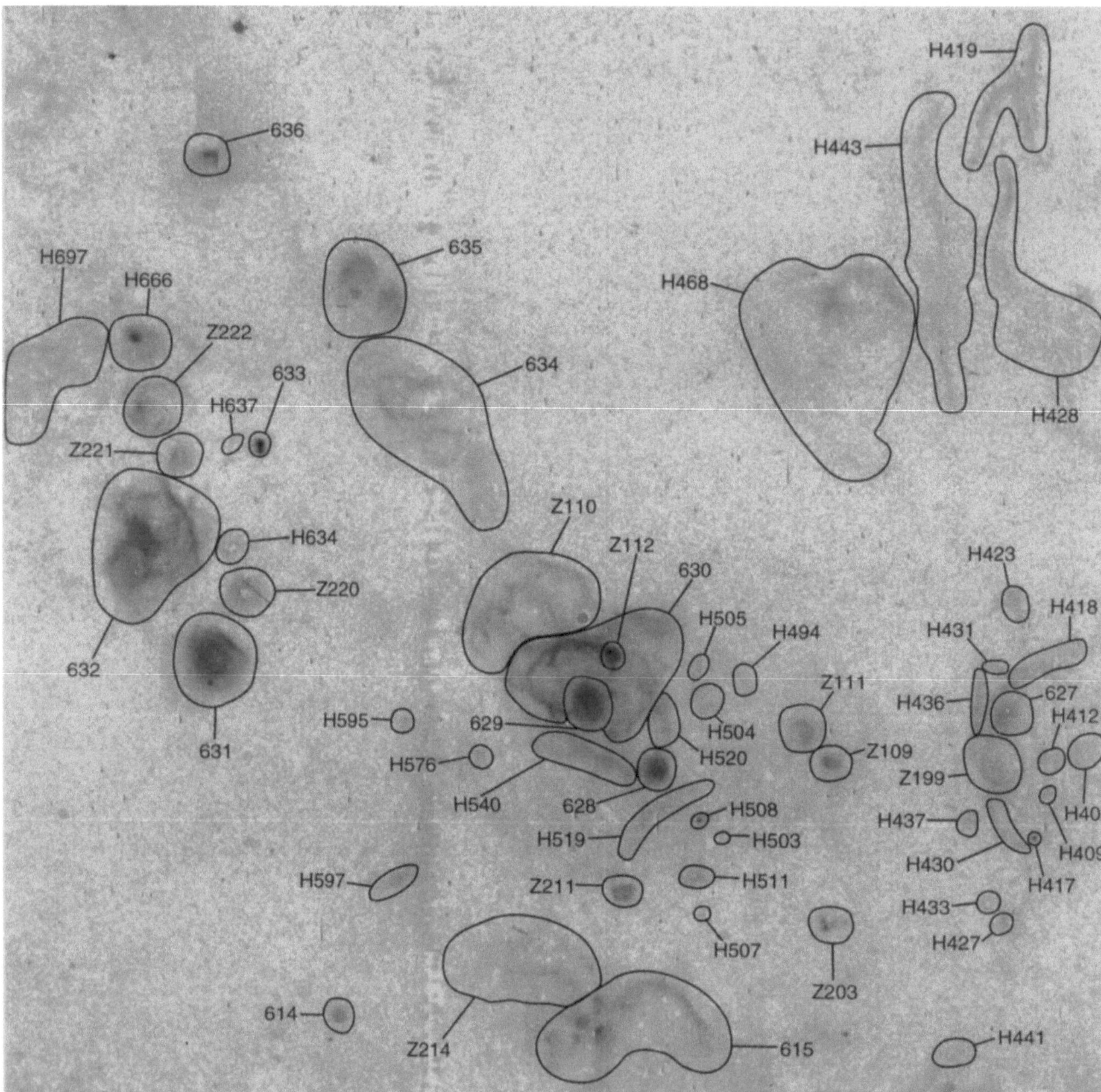
H419
H443
636
635
H697
H666
H468
Z222
634
633
H637
H428
Z221
Z110
H634
Z112
H423
630
Z220
H505
H494
H418
H431
632
Z111
627
H436
H595
629
H504
H412
H520
Z109
H576
Z199
H402
H540
628
H508
H437
H519
H503
H409
H430
H417
H597
Z211
H511
H433
H427
H507
Z203
614
Z214
615
H441
631

M33

Chart 37

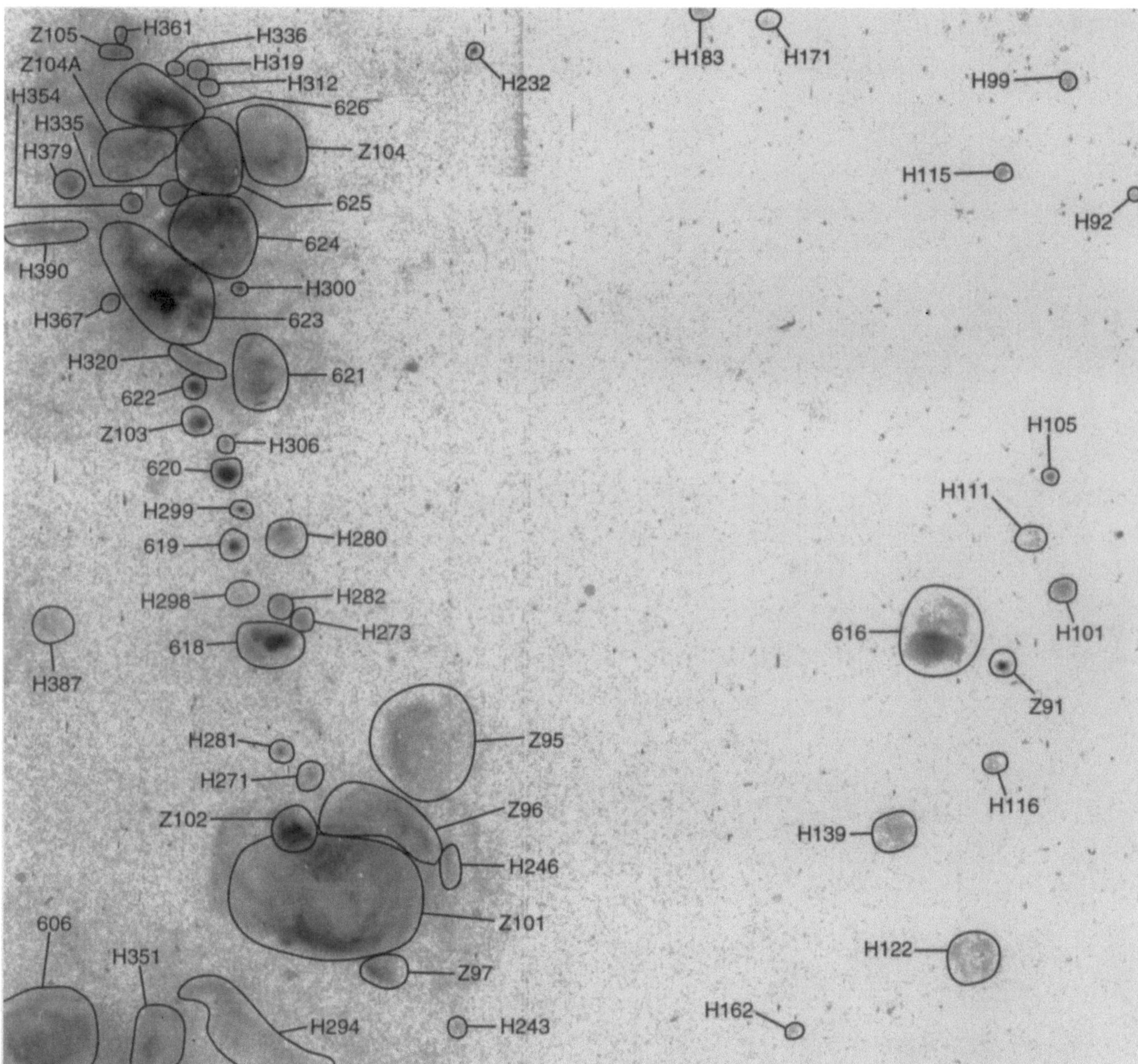
Z105
H361
H336
Z104A
H319
H312
H354
626
H335
H379
Z104
625
624
H390
H300
H367
623
H320
621
622
Z103
H306
620
H299
619
H280
H298
H282
618
H273
H387
H281
Z95
H271
Z102
Z96
H246
Z101
606
H351
Z97
H294
H243
H232
H183
H171
H99
H115
H92
H105
H111
H101
616
Z91
H116
H139
H122
H162

M33

Chart 38

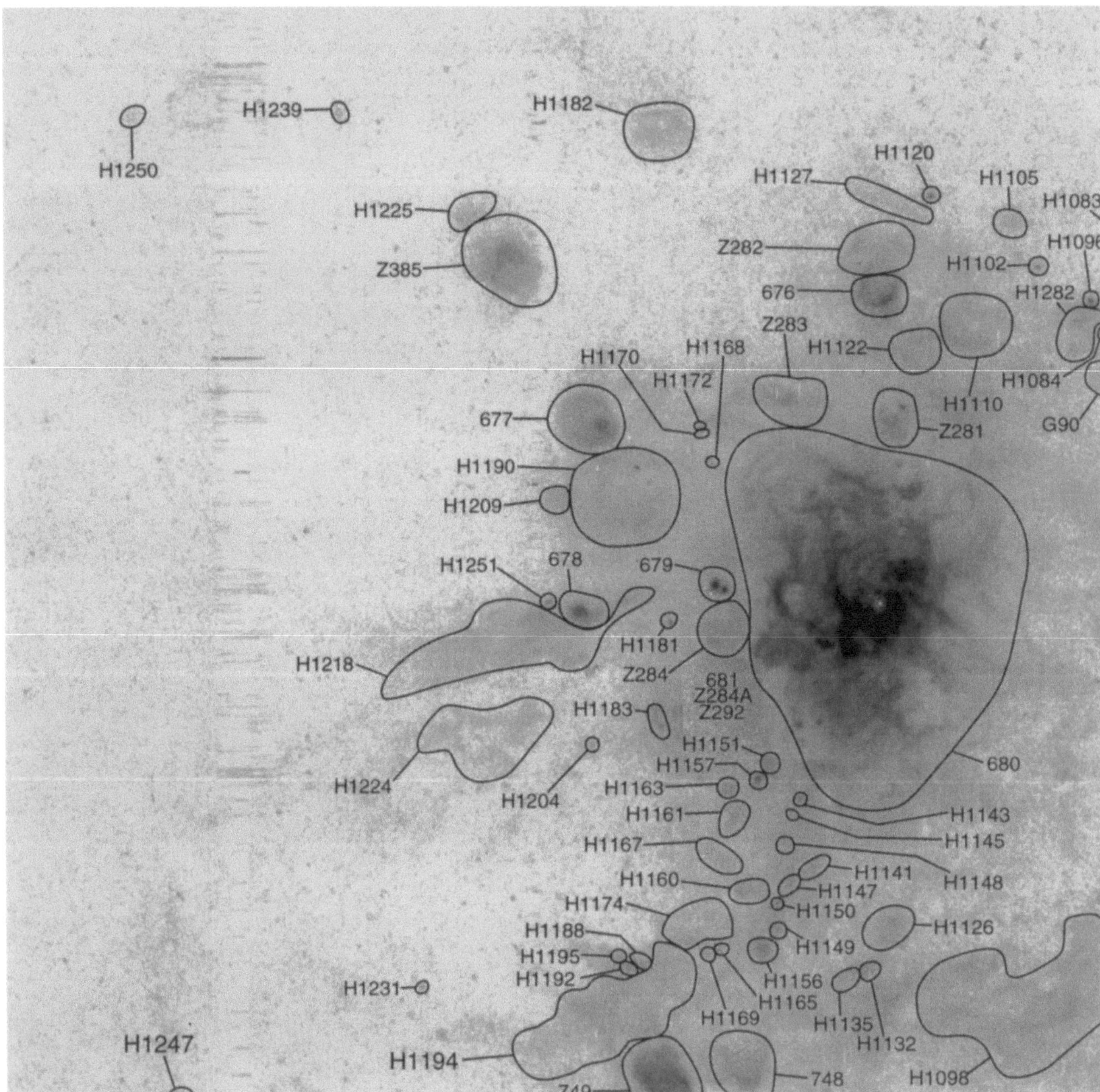
H1250
H1239
H1182
H1120
H1127
H1105
H1083
H1225
Z282
H1096
Z385
H1102
676
H1282
Z283
H1168
H1122
H1170
H1172
H1084
H1110
677
Z281
G90
H1190
H1209
H1251
678
679
H1181
H1218
Z284
681
Z284A
Z292
H1183
H1151
H1157
680
H1224
H1204
H1163
H1143
H1161
H1145
H1167
H1141
H1148
H1160
H1147
H1174
H1150
H1126
H1188
H1149
H1195
H1156
H1231
H1192
H1165
H1169
H1135
H1132
H1247
H1194
748
H1098
749

M33

Chart 39

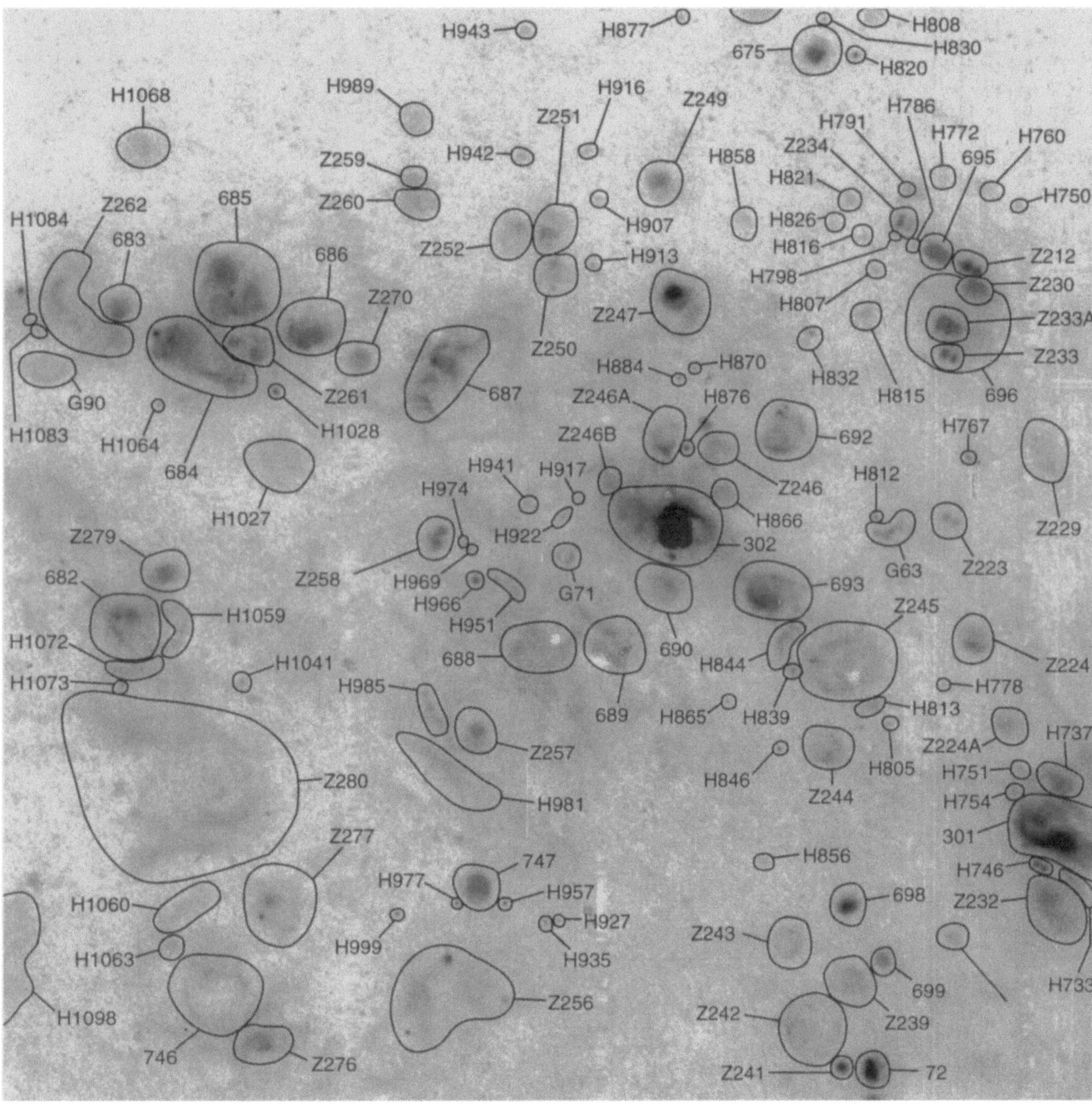
H943
H877
H808
H830
675
H820
H1068
H989
H916
Z249
H786
H791
H772
H760
Z251
H942
Z234
H858
695
Z259
H821
H750
Z262
685
Z260
H907
H826
H1084
683
Z252
H816
686
H913
H798
Z212
Z270
Z230
Z247
H807
Z233A
Z250
H884
H870
Z233
G90
Z261
687
H832
H815
696
Z246A
H876
H1028
H1083
H1064
Z246B
692
H767
684
H941
H917
H812
H1027
H974
Z246
H866
H922
Z229
Z279
302
Z258
H969
G63
Z223
682
H966
G71
693
H1059
H951
Z245
H1072
688
690
H844
Z224
H1041
H1073
H985
689
H865
H839
H778
H813
Z257
H805
Z224A
H737
H846
Z244
H751
Z280
H981
H754
301
Z277
H856
H746
747
H977
H957
698
Z232
H1060
H927
H999
Z243
H1063
H935
699
H733
Z256
Z239
H1098
Z242
746
Z276
Z241
72

M33

Chart 40

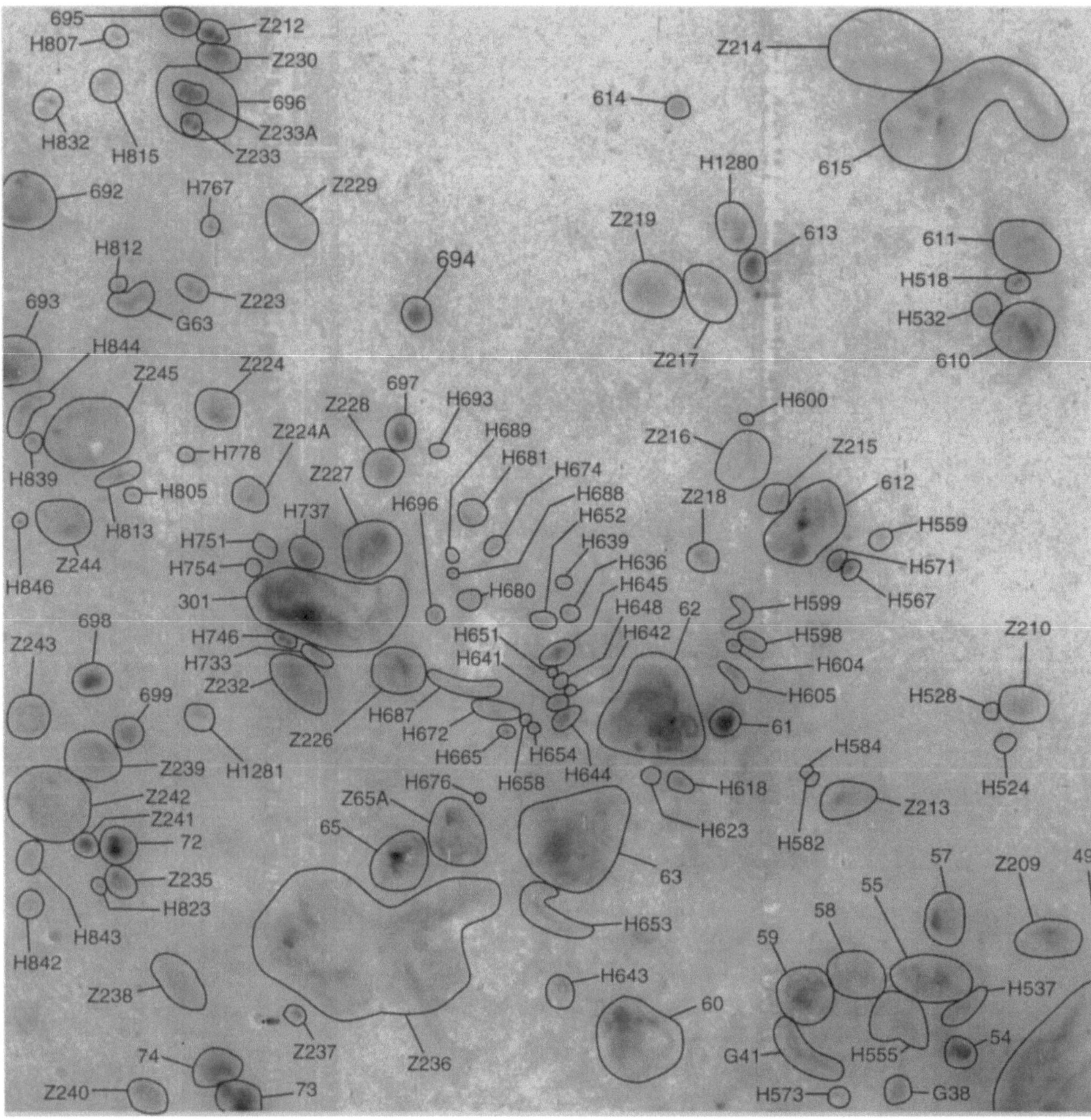

695
Z212
H807
Z230
Z214
614
696
Z233A
H832
H815
Z233
H1280
615
692
H767
Z229
Z219
613
611
H812
694
H518
693
Z223
G63
H532
H844
Z217
610
Z224
Z245
697
H600
Z228
H693
Z224A
H689
Z216
Z215
H778
H839
H681
H674
Z227
H805
H688
612
Z218
H813
H737
H696
H652
H559
H751
Z244
H639
H636
H571
H754
H846
H680
H645
H567
301
H599
H648
62
698
Z210
H746
H651
H642
H598
Z243
H733
H641
H604
Z232
699
H605
H528
H687
H672
61
Z226
H665
H654
H584
Z239
H1281
H658
H644
H618
H524
H676
Z242
Z65A
Z213
Z241
H623
72
65
H582
63
57
Z209
49
Z235
55
H823
58
H653
H843
59
H842
H643
H537
Z238
60
54
Z237
74
G41
H555
Z236
73
Z240
H573
G38

M33

Chart 41

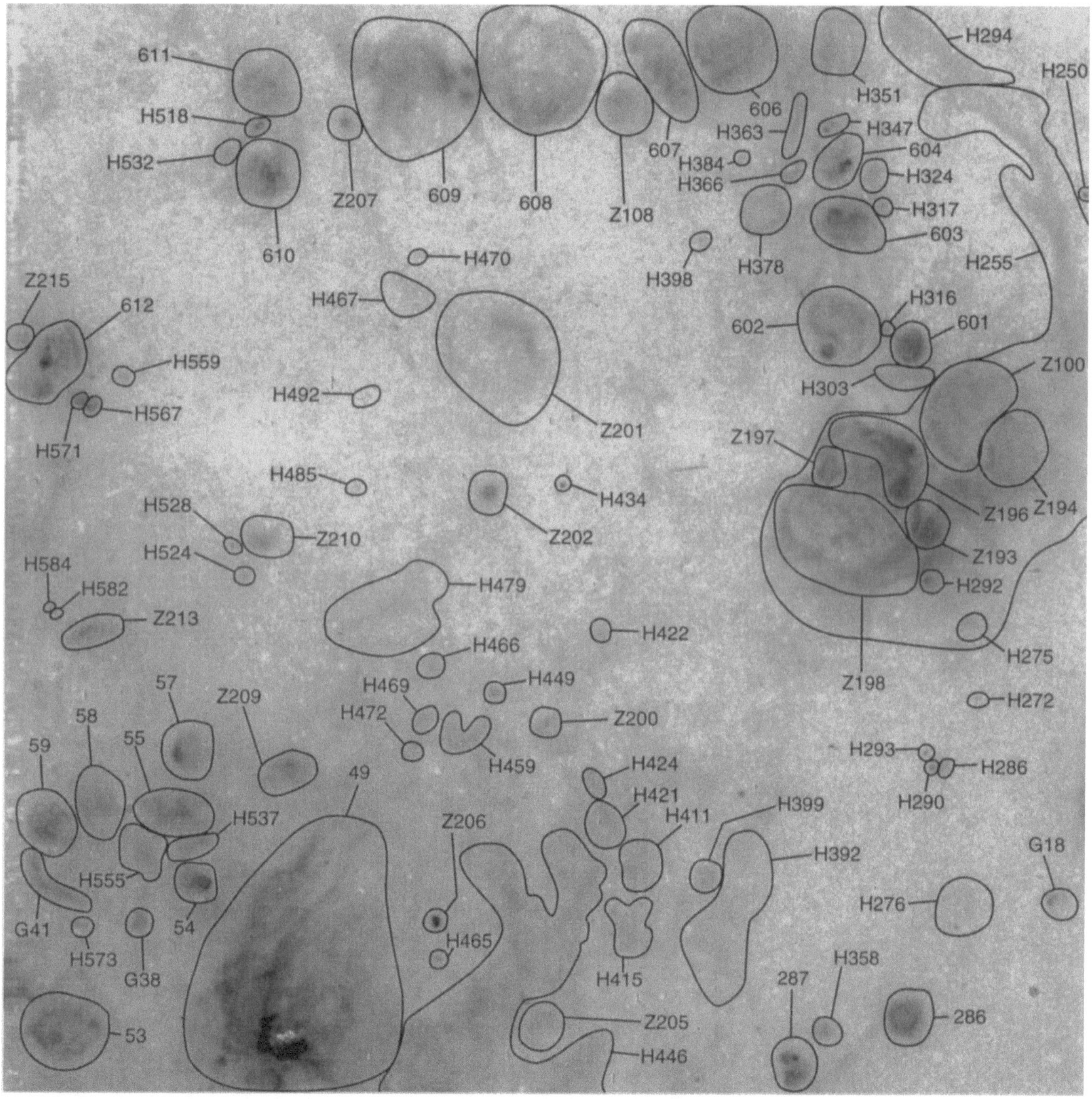
611
H518
H532
610
Z207
609
608
Z108
607
606
H363
H384
H366
H351
H347
604
H324
H317
603
H378
H398
H294
H250
H255
H470
Z215
612
H467
H559
H567
H571
H492
Z201
602
H316
601
H303
Z100
Z197
Z196
Z194
Z193
H292
H275
Z198
H272
H485
H434
Z202
H528
Z210
H524
H584
H582
Z213
H479
H466
H422
H469
H449
H472
Z200
H459
57
Z209
58
55
59
H537
H555
G41
H573
G38
54
49
53
Z206
H465
H424
H421
H411
H415
Z205
H446
H399
H392
H293
H286
H290
G18
H276
H358
287
286

M33

Chart 42

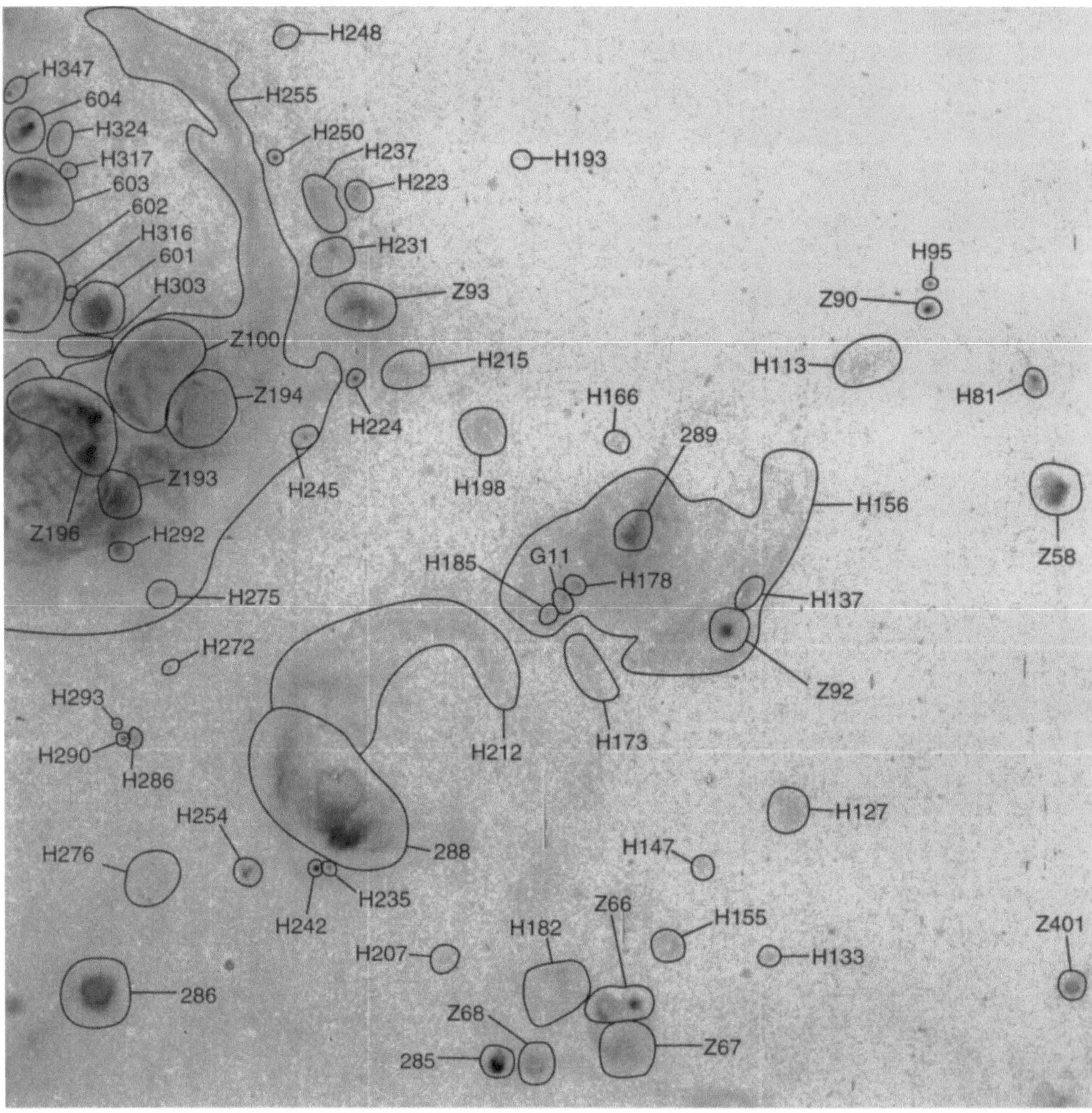
H248
H347
604
H255
H324
H250
H317
H237
H193
603
H223
602
H316
H231
601
H95
H303
Z93
Z90
Z100
H215
H113
Z194
H224
H166
H81
289
Z193
H245
H198
H156
Z196
H292
Z58
H185
G11
H178
H275
H137
H272
Z92
H293
H212
H173
H290
H286
H127
H254
H276
288
H147
H235
Z66
H242
H155
Z401
H182
H207
H133
286
Z68
Z67
285

M33

Chart 43

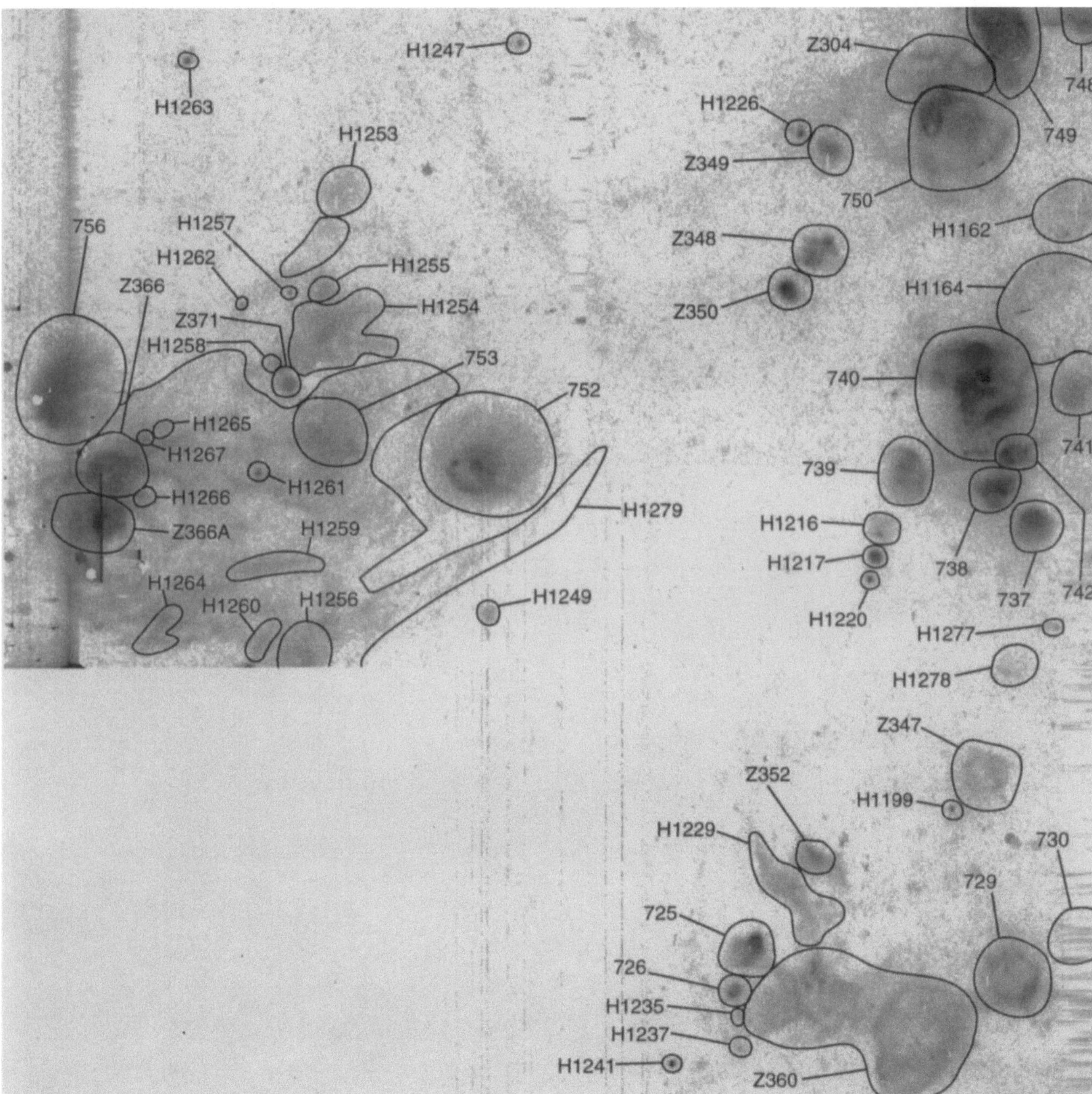
H1247
Z304
748
H1263
H1226
749
H1253
Z349
750
756
H1257
Z348
H1162
H1262
H1255
Z366
H1254
H1164
Z371
Z350
H1258
753
740
752
H1265
741
H1267
739
H1261
H1266
H1279
H1216
Z366A
H1259
H1217
738
H1264
742
H1260
H1256
H1249
737
H1220
H1277
H1278
Z347
Z352
H1199
H1229
730
729
725
726
H1235
H1237
H1241
Z360

M33

Chart 44

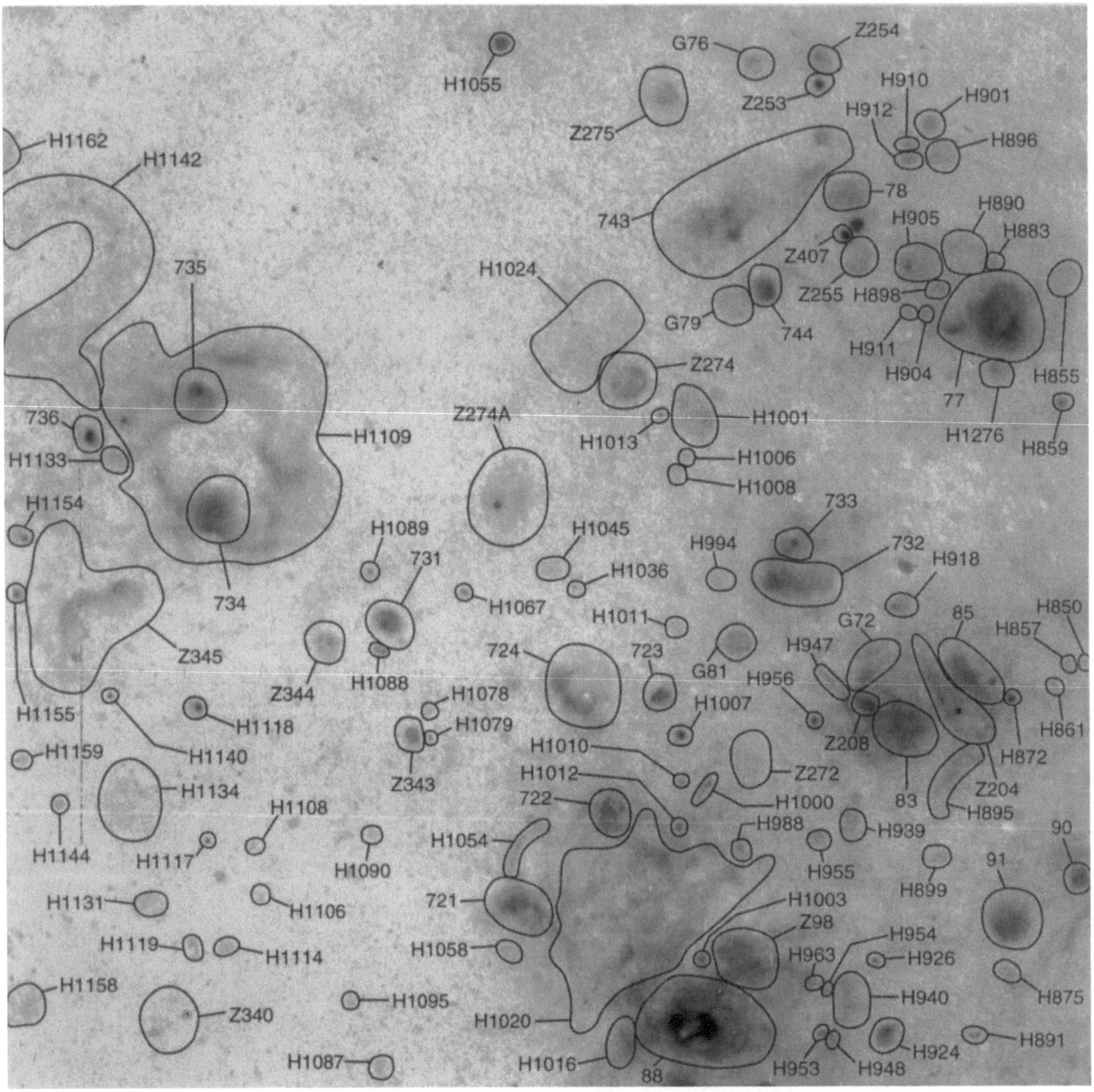
H1055
G76
Z254
Z253
H910
H901
H912
H896
Z275
H1162
H1142
78
743
H890
H905
H883
Z407
735
H1024
Z255
H898
G79
744
H911
H904
Z274
H855
77
H1276
736
Z274A
H1001
H1013
H1109
H1133
H1006
H859
H1008
733
H1154
H1089
H1045
731
H994
732
H918
H1036
734
H1067
H1011
G72
85
H850
H857
Z345
724
723
H947
G81
H956
Z344
H1088
H1078
H1155
H1118
H1007
H861
H1079
Z208
H872
H1159
H1140
H1010
Z272
H1134
Z343
H1012
83
Z204
H1000
H1108
722
H895
H988
H939
90
H1144
H1117
H1054
H1090
H955
91
H899
H1003
H1131
721
H1106
Z98
H954
H963
H926
H1119
H1114
H1058
H1158
H940
H875
Z340
H1095
H1020
H924
H891
H1087
H1016
88
H953
H948

M33

Chart 45

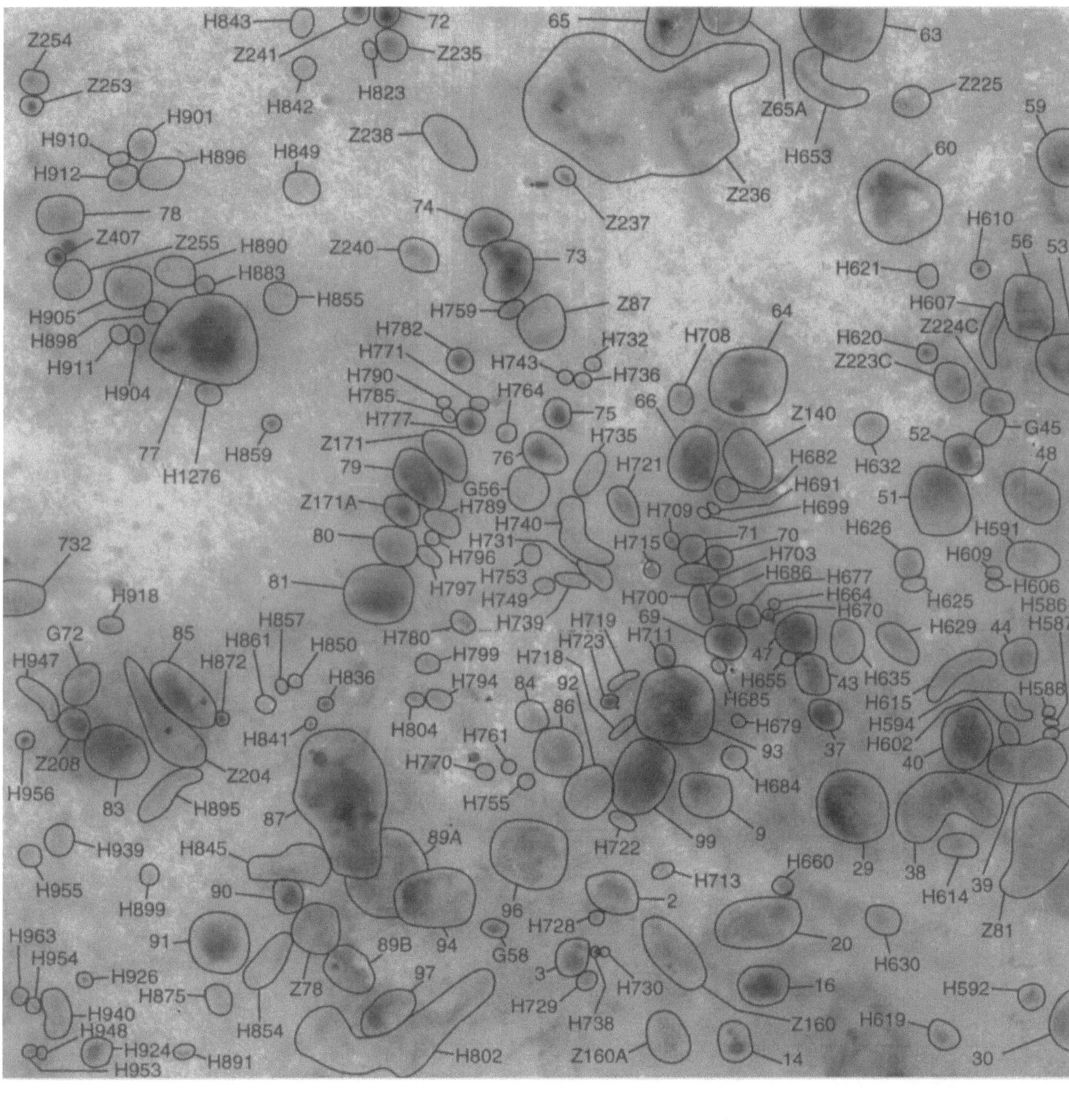
H843
72
65
63
Z254
Z241
Z235
Z253
H842
H823
Z225
H901
Z65A
59
H910
H896
H849
Z238
H653
60
H912
Z236
78
74
Z237
H610
Z407
Z255
H890
Z240
73
56
53
H883
H621
H905
H855
H759
Z87
64
H607
H898
H782
H732
H708
H620
Z224C
H911
H771
H743
H736
Z223C
H904
H790
H785
H764
H777
75
66
Z140
G45
77
H859
Z171
H735
52
H632
48
H1276
79
76
H721
H682
G56
H691
51
Z171A
H789
H709
H699
H740
71
70
H626
H591
732
80
H731
H715
H703
H609
H796
H753
H686
H677
H606
81
H797
H749
H700
H664
H625
H918
H857
H739
H719
69
H670
H586
G72
85
H861
H850
H780
H723
H711
H629
44
H587
H947
H872
H799
H718
47
H635
H836
H794
84
92
H655
43
H615
H588
H804
86
H685
H594
H841
H761
H679
37
H602
Z208
93
37
40
H956
Z204
H770
H684
H895
H755
87
89A
H722
99
9
29
38
H614
39
H939
H845
H660
H955
H899
90
96
H713
2
H728
20
Z81
H963
91
89B
94
G58
H630
H954
H926
97
3
H730
16
H875
H940
Z78
H729
H592
H948
H854
H738
Z160
H619
H924
H891
H802
Z160A
14
30
H953

M33

Chart 46

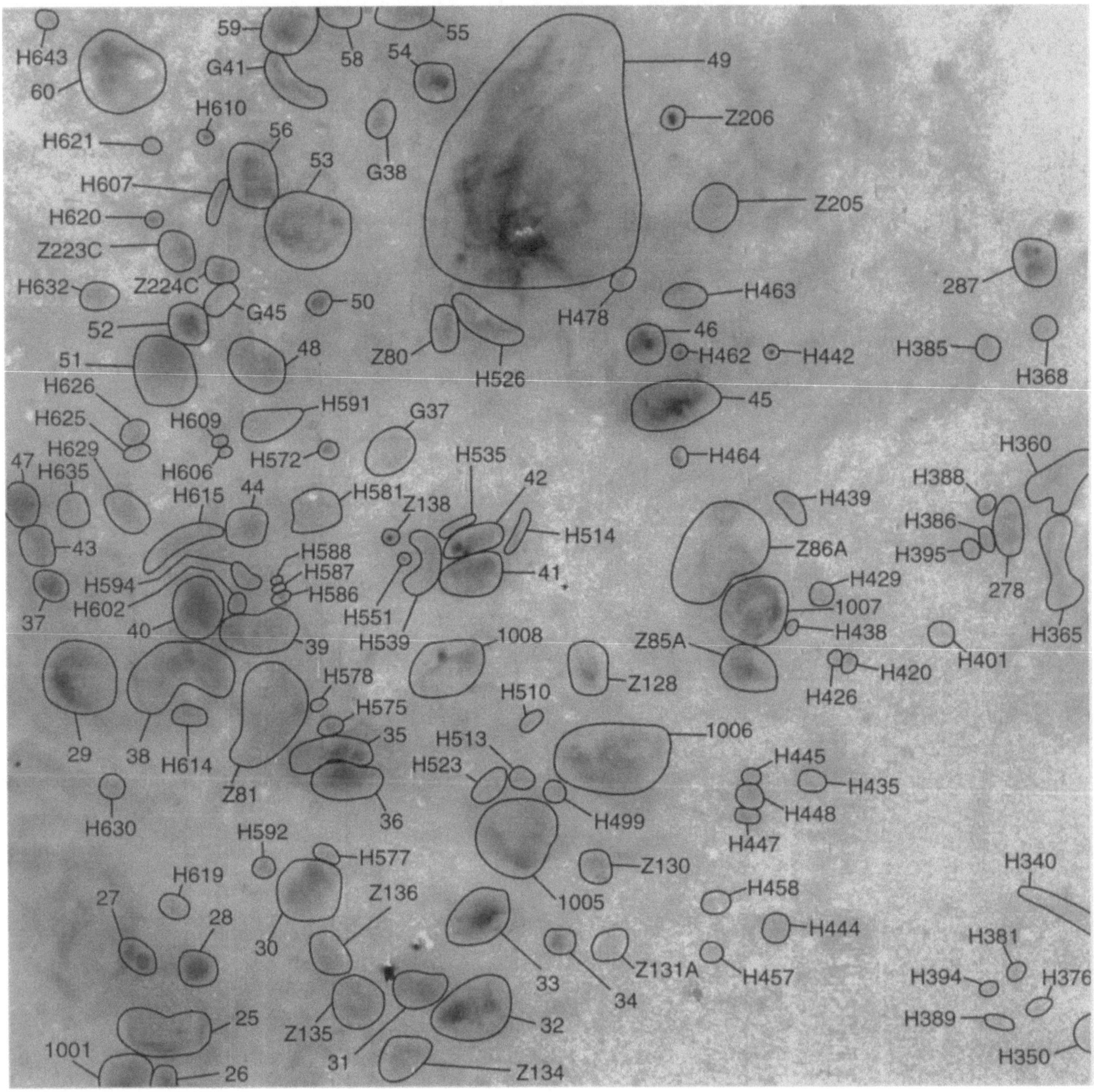
H643
60
59
G41
58
55
54
49
Z206
H610
H621
56
53
G38
H607
Z205
H620
Z223C
287
H632
Z224C
G45
50
H478
H463
52
46
51
48
Z80
H526
H462
H442
H385
H368
H626
H591
45
H625
H609
G37
H629
H606
H572
H464
H360
47
H635
H615
44
H581
Z138
H535
42
H439
H388
H386
43
H514
Z86A
H395
H588
H587
41
H429
278
H594
H586
H602
37
40
H551
H539
39
1007
H438
1008
Z85A
H365
H401
H420
Z128
H426
H578
H575
H510
29
38
H614
35
1006
H513
H445
H523
H435
Z81
36
H499
H448
H630
H447
H592
H577
Z130
H340
H619
27
Z136
1005
H458
28
30
H444
H381
33
Z131A
H457
34
H394
H376
25
Z135
32
H389
1001
31
Z134
26
H350

M33

Chart 47

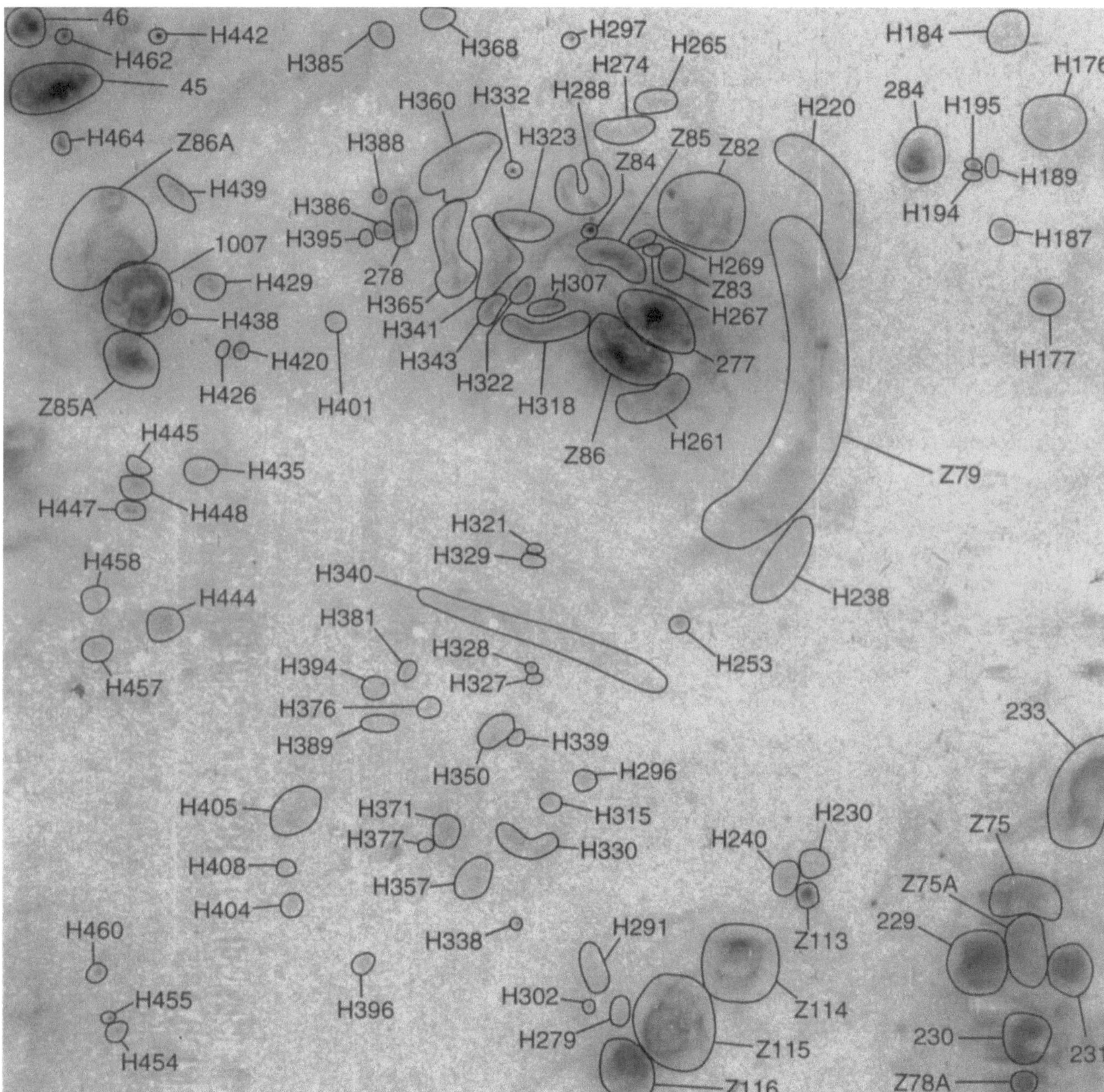
46
H442
H462
45
H385
H368
H297
H265
H184
H176
H274
H360
H332
H288
H220
284
H195
H464
Z86A
H388
H323
Z85
Z82
H439
Z84
H189
H386
H194
H395
H187
1007
H269
H429
278
Z83
H307
H365
H267
H438
H341
H177
H420
H343
277
H426
H322
H401
H318
Z85A
H445
H261
Z86
H435
Z79
H447
H448
H321
H329
H458
H340
H238
H444
H381
H253
H328
H394
H327
H457
H376
233
H389
H339
H350
H296
H405
H315
H230
H371
H240
Z75
H377
H330
H408
Z75A
H357
H404
H460
H338
H291
Z113
229
H302
Z114
H455
H396
H279
Z115
230
231
H454
Z116
Z78A

M33

Chart 48

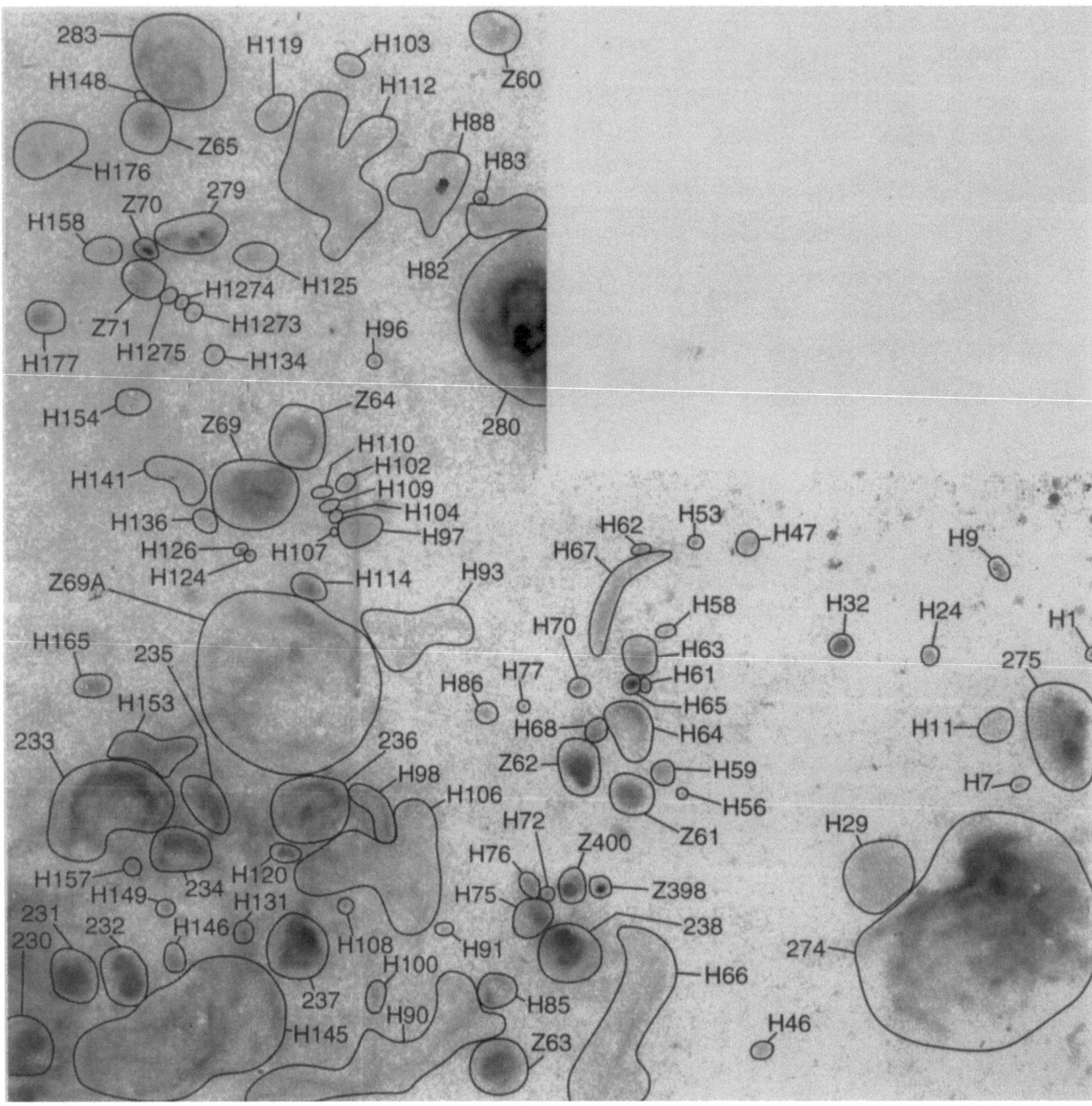
283
H119
H103
H148
H112
Z60
H88
Z65
H176
H83
279
Z70
H158
H82
H1274
H125
H1273
Z71
H96
H1275
H177
H134
H154
Z64
Z69
280
H110
H102
H141
H109
H104
H136
H97
H126
H107
H124
H114
Z69A
H93
H62
H53
H47
H9
H67
H58
H32
H24
H1
H70
H63
H165
235
H61
275
H77
H86
H65
H153
H68
H64
H11
233
236
Z62
H59
H98
H7
H106
H56
H72
Z61
H29
Z400
H76
H157
H120
234
Z398
H149
H75
H131
231
232
238
H146
230
H108
H91
274
H66
H100
237
H85
H90
H145
H46
Z63

M33

Chart 49

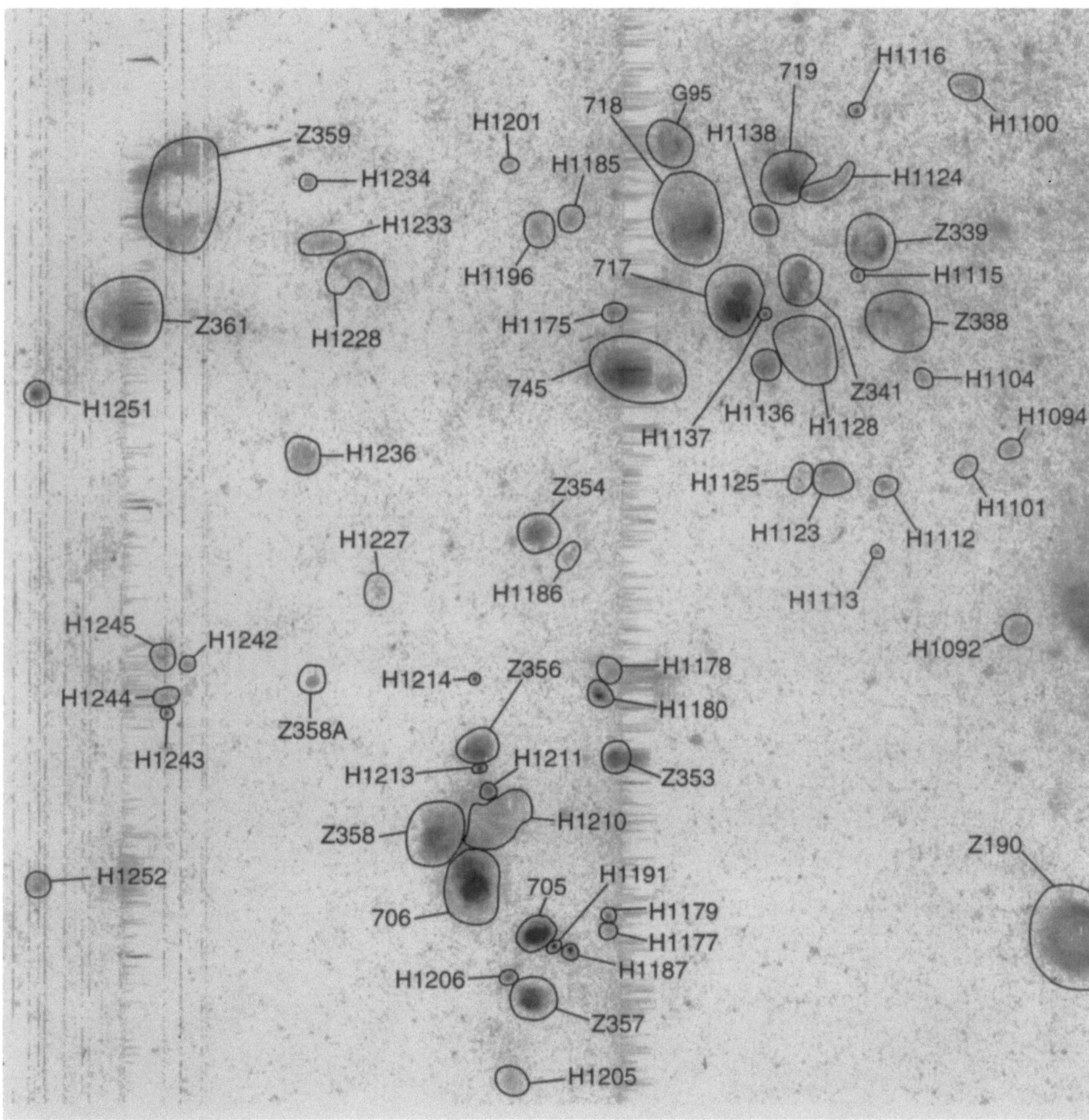
H1116
719
G95
718
H1201
Z359
H1138
H1100
H1185
H1234
H1124
H1233
Z339
H1196
717
H1115
Z361
H1228
H1175
Z338
H1104
745
Z341
H1251
H1136
H1128
H1094
H1137
H1236
H1125
Z354
H1101
H1123
H1112
H1227
H1186
H1113
H1245
H1242
H1092
H1214
Z356
H1178
H1244
H1180
Z358A
H1243
H1213
H1211
Z353
H1210
Z358
Z190
H1252
705
H1191
706
H1179
H1177
H1187
H1206
Z357
H1205

M33

Chart 50

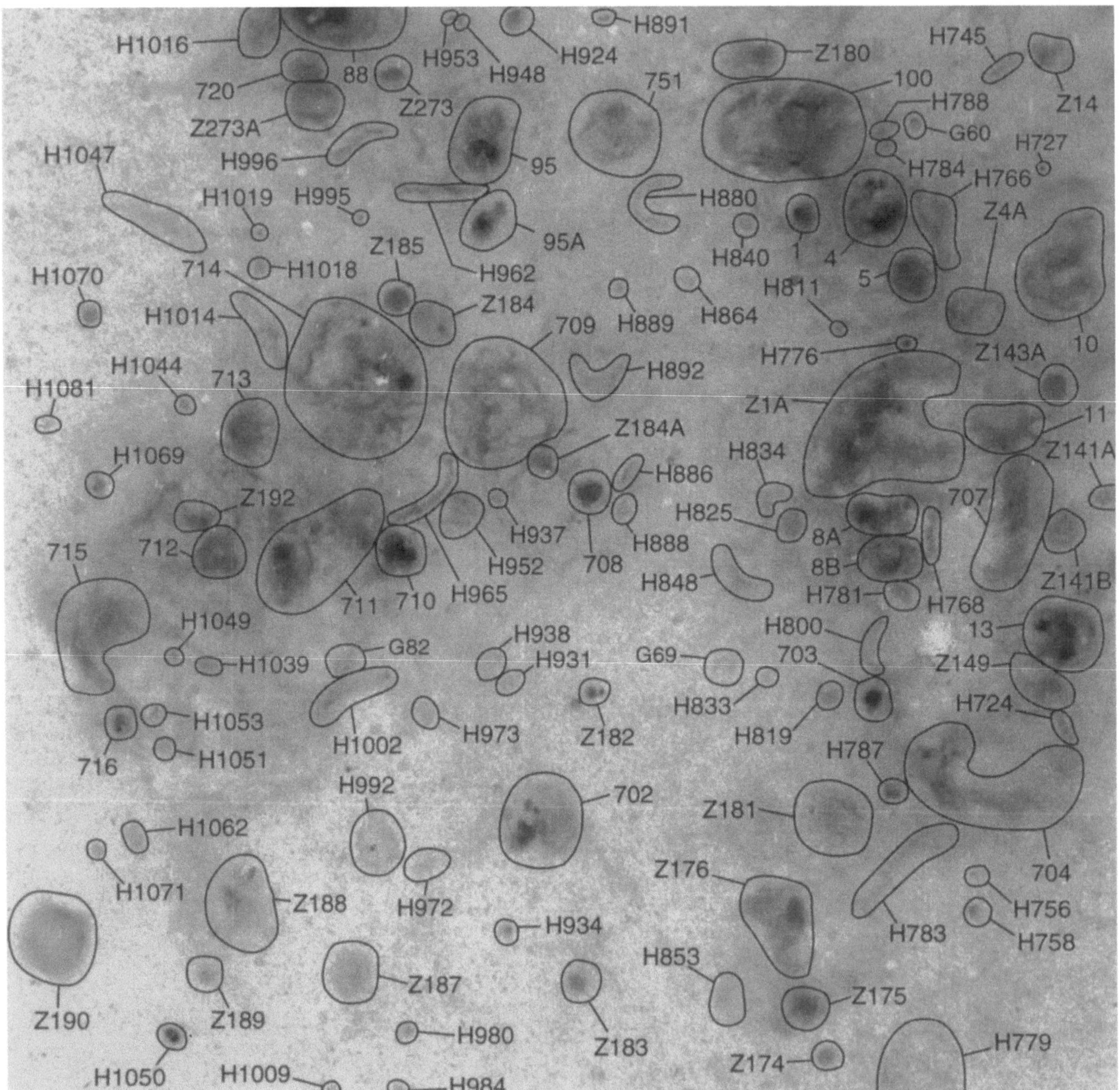
H1016
88
H953
H948
H924
H891
Z180
H745
Z14
720
Z273
Z273A
751
100
H788
G60
H727
H1047
H996
95
H784
H766
H1019
H995
H880
Z4A
95A
Z185
H840
1
4
H1070
714
H1018
H962
H811
5
Z184
H889
H864
H1014
709
H776
Z143A
10
H1044
713
H892
H1081
Z1A
11
Z184A
H834
Z141A
H1069
H886
Z192
707
H825
715
712
H937
H888
8A
708
H952
8B
Z141B
H848
H781
H768
711
710
H965
H1049
H800
13
H938
G82
H1039
H931
G69
703
Z149
H833
H1053
H973
Z182
H819
H724
H1002
H787
716
H1051
H992
702
Z181
H1062
704
Z176
H1071
Z188
H972
H756
H934
H783
H758
H853
Z187
Z175
Z190
Z189
H980
H779
Z183
Z174
H1050
H1009
H984

M33

Chart 51

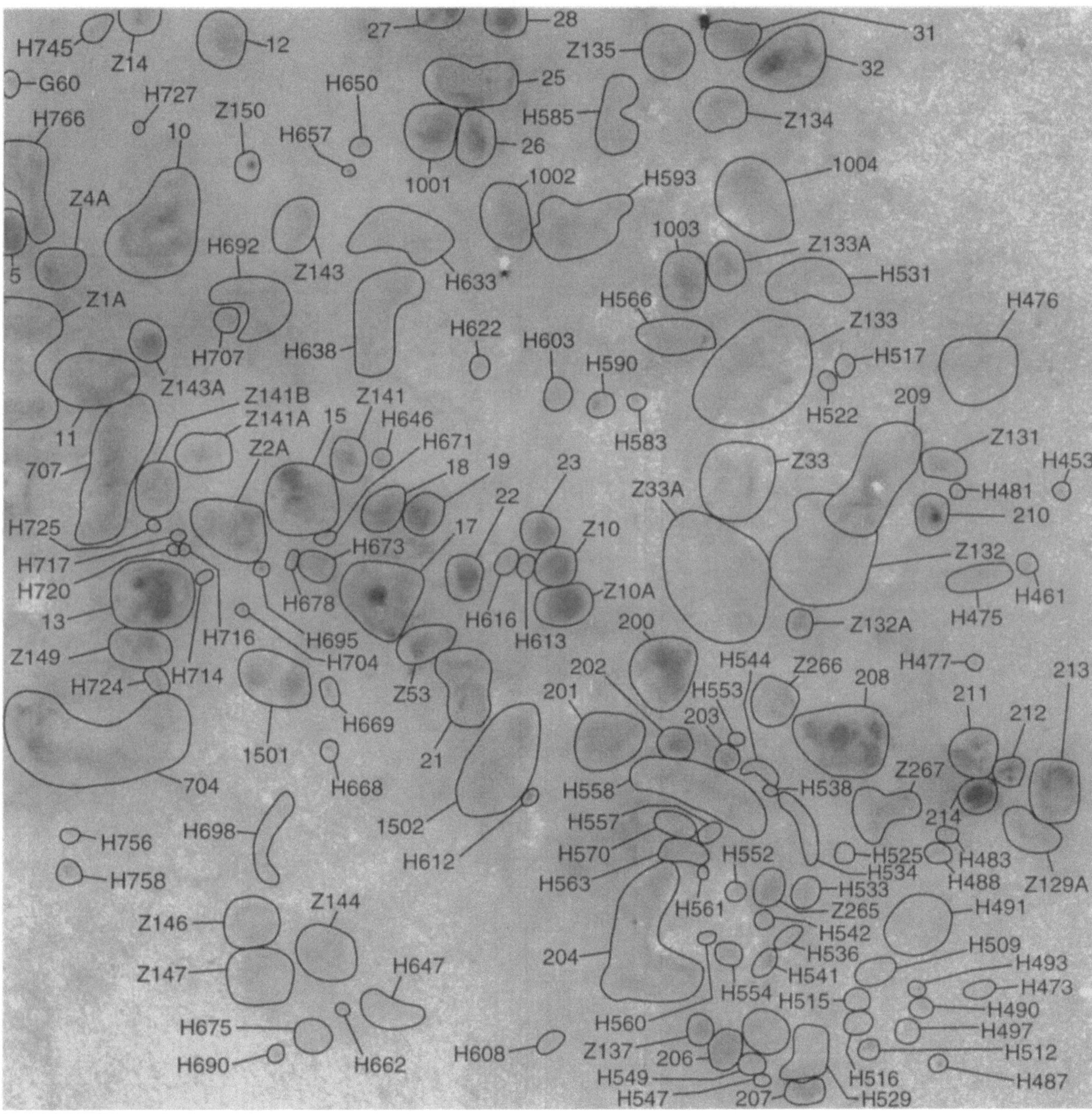
H745
Z14
12
27
28
31
Z135
32
G60
H727
Z150
H650
25
H766
10
H657
H585
Z134
26
1001
1002
H593
1004
Z4A
1003
Z133A
H692
Z143
H633
H531
Z1A
H566
H476
H622
H603
Z133
H707
H638
H590
H517
Z143A
Z141B
Z141
H522
209
11
Z141A
15
H646
H583
Z131
707
Z2A
H671
H453
18
19
23
Z33
22
Z33A
H481
210
H725
17
Z10
H717
H673
Z132
H720
H678
Z10A
H461
13
H616
H475
H716
H695
H613
200
Z132A
Z149
H704
H544
Z266
H477
H724
H714
202
H553
208
213
Z53
201
211
H669
203
212
1501
21
H538
Z267
704
H668
H558
214
H557
1502
H756
H698
H570
H552
H525
H483
H612
H534
H758
H563
H533
H488
Z129A
H561
Z265
H491
Z146
Z144
H542
H536
H509
H647
204
H541
H493
Z147
H554
H515
H473
H490
H675
H560
H497
Z137
206
H512
H690
H662
H608
H549
H516
H487
H547
207
H529

M33

Chart 52

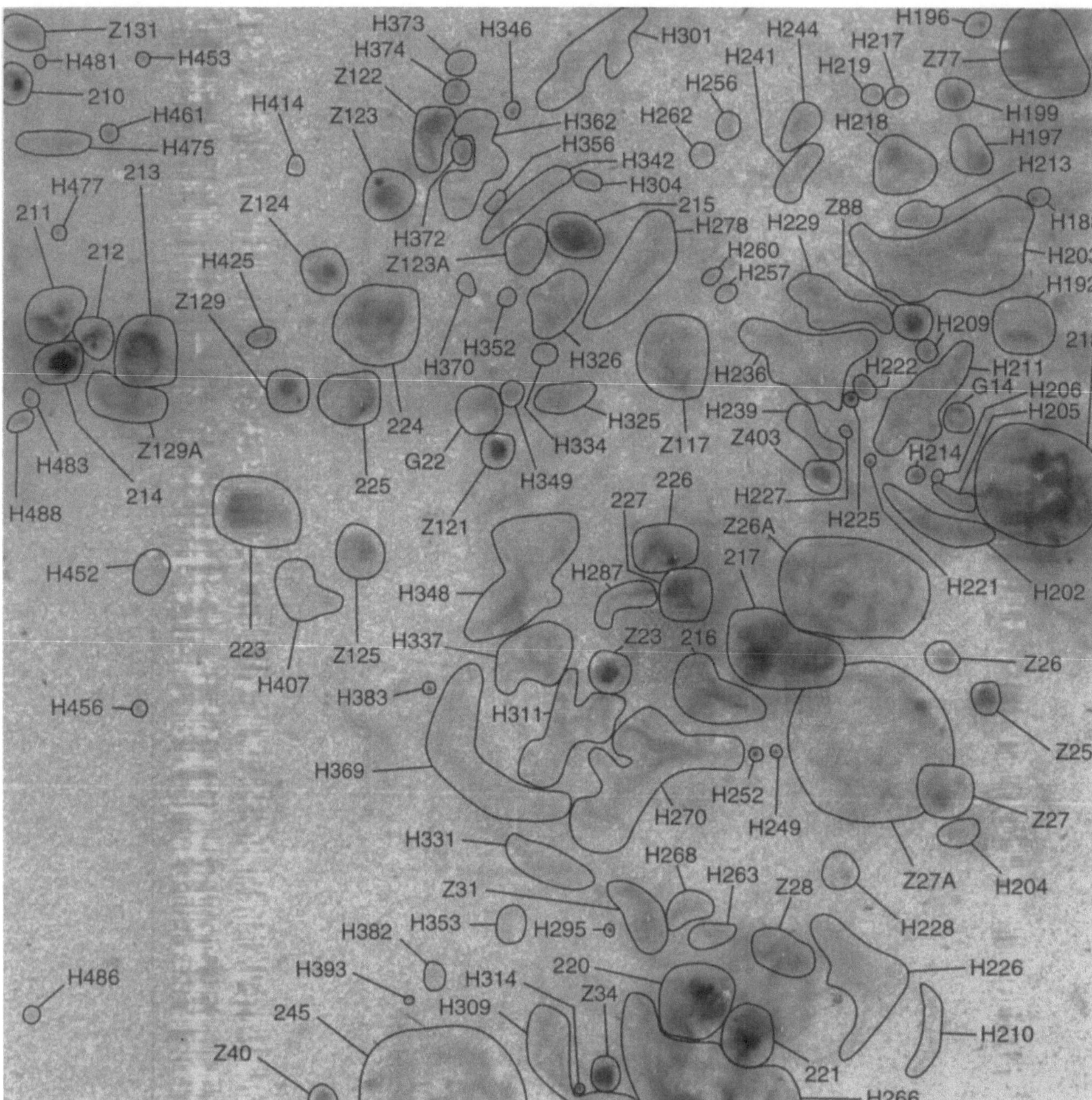
Z131
H481
H453
210
H461
H475
H477
211
213
212
H414
H373
H374
Z122
Z123
H346
H301
H362
H356
H342
H304
215
H278
H372
Z123A
Z124
H425
Z129
H370
H352
H326
H325
H334
H349
G22
Z121
224
225
Z129A
H483
H488
214
H244
H241
H256
H262
H196
H217
H219
Z77
H199
H197
H213
H218
H188
H203
H192
Z88
H229
H260
H257
H209
218
H211
G14
H206
H205
H222
H236
H239
Z403
Z117
H214
H227
H225
226
227
Z26A
H221
H202
217
H287
H348
H337
Z23
216
Z26
Z25
H383
H311
H369
H252
H249
H270
Z27
Z27A
H204
H331
H268
H263
Z28
H228
Z31
H353
H295
H382
H226
H393
H314
220
Z34
H309
245
Z40
221
H210
H266
H486
H456
H452
223
H407
Z125

M33

Chart 53

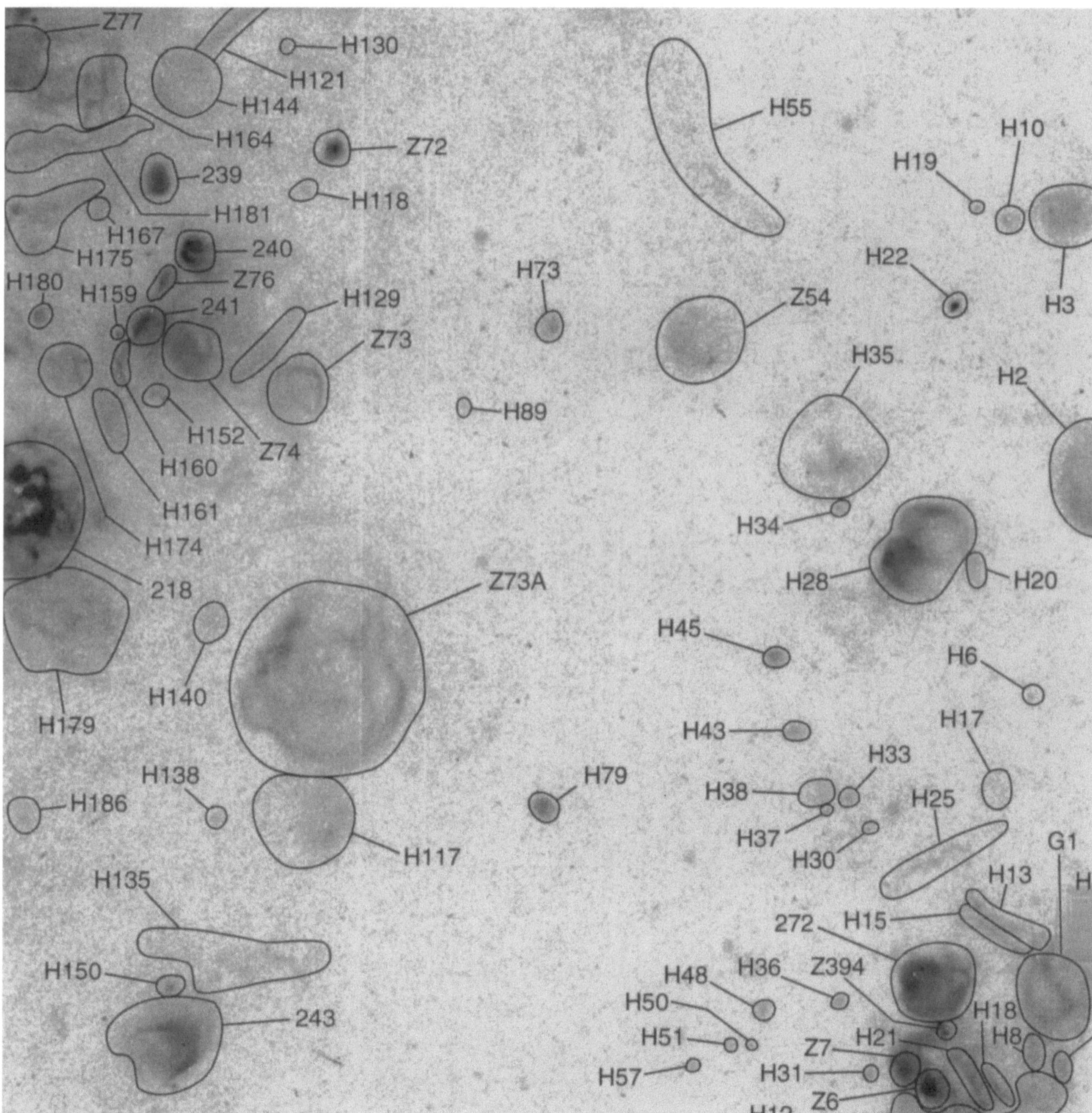
Z77
H130
H121
H144
H164
Z72
239
H118
H181
H167
H175
240
Z76
H180
H159
241
H129
Z73
H55
H10
H19
H3
H22
H73
Z54
H35
H2
H89
H152
Z74
H160
H161
H174
218
H34
H28
H20
Z73A
H45
H6
H140
H179
H43
H17
H33
H138
H79
H38
H186
H37
H30
H25
H117
G1
H13
H4
H135
272
H15
H150
H48
H36
Z394
243
H50
H18
H51
H21
H8
Z7
H57
H31
Z6
H12

M33

Chart 54

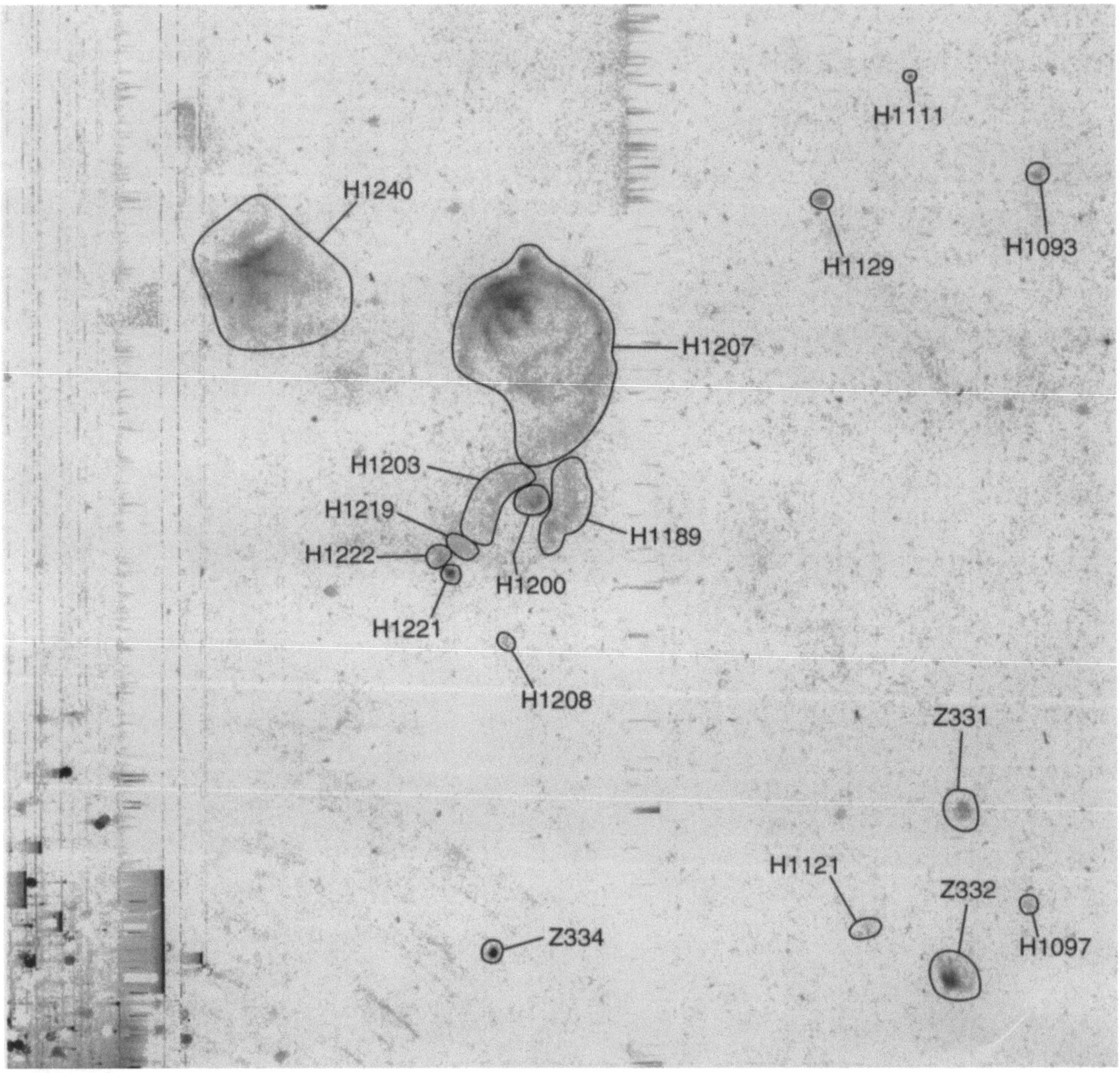

H1111
H1240
H1129
H1093
H1207
H1203
H1219
H1189
H1222
H1200
H1221
H1208
Z331
H1121
Z332
Z334
H1097

M33

Chart 55

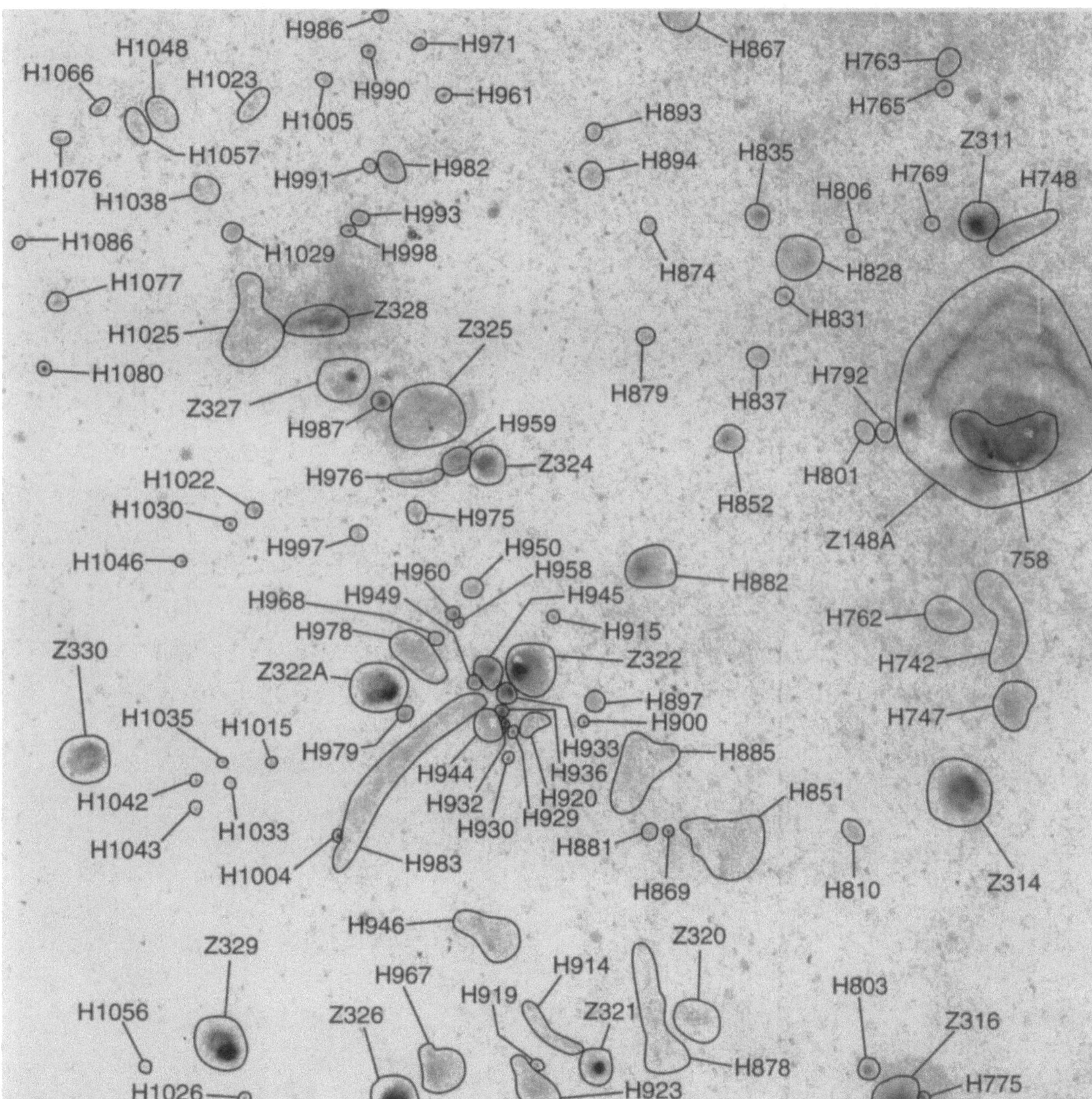
H986
H971
H867
H1048
H1066
H1023
H990
H961
H763
H765
H1005
H893
H1057
H1076
H991
H982
H894
H835
Z311
H769
H748
H806
H1038
H993
H1086
H1029
H998
H874
H828
H1077
Z328
H831
H1025
Z325
H1080
H879
H837
H792
Z327
H987
H959
H976
Z324
H801
H1022
H852
H1030
H975
Z148A
H997
758
H1046
H950
H960
H958
H882
H968
H949
H945
H762
H978
H915
Z330
Z322
H742
Z322A
H897
H900
H747
H1035
H1015
H933
H885
H979
H936
H944
H920
H851
H1042
H932
H929
H930
H1033
H881
H1043
H983
H810
Z314
H1004
H869
H946
Z320
Z329
H914
H967
H803
H919
H1056
Z326
Z321
Z316
H878
H923
H775
H1026

M33

Chart 56

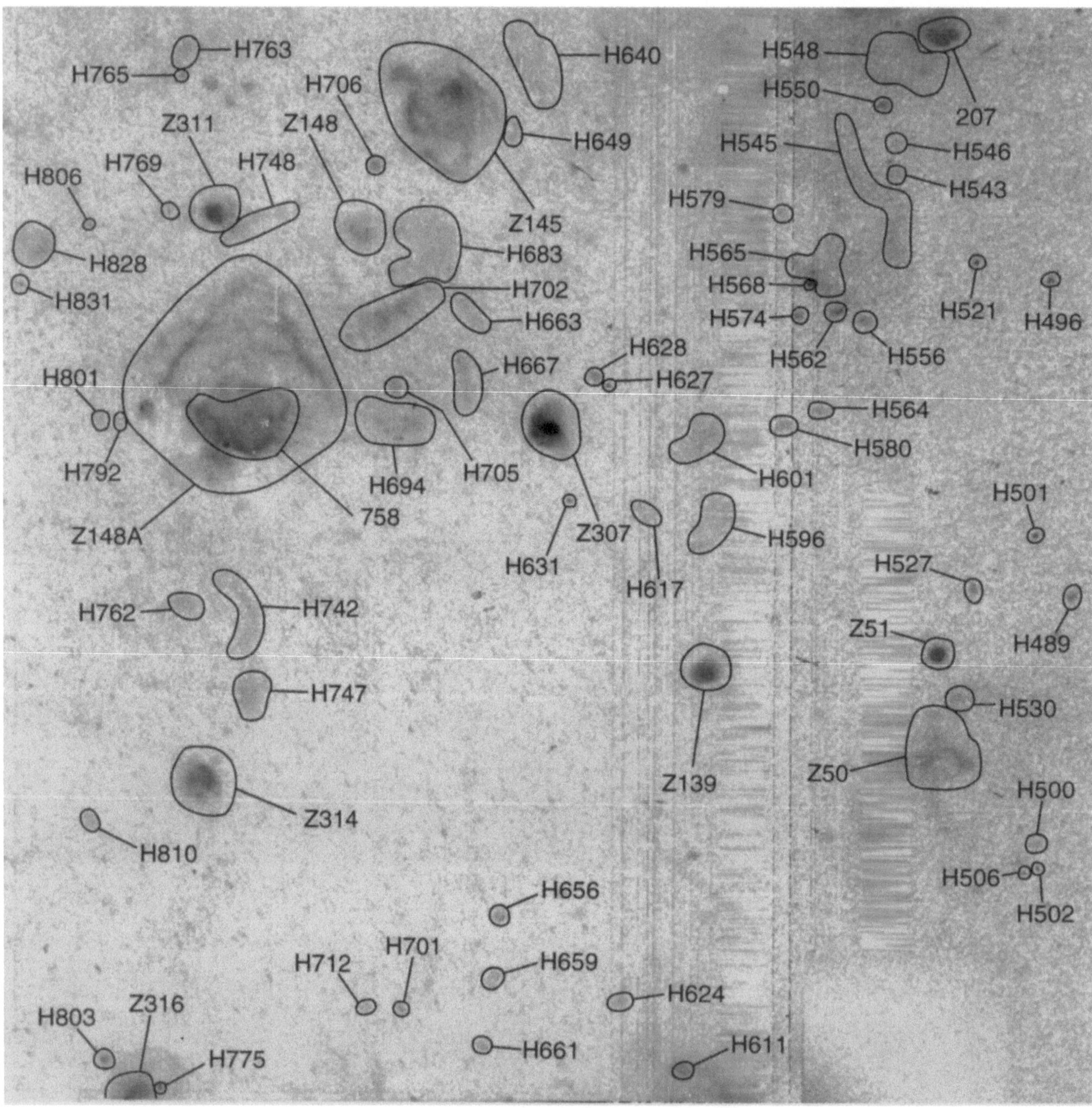
H763
H765
H706
H640
H548
H550
207
Z311
Z148
H649
H545
H546
H806
H769
H748
H543
H579
Z145
H828
H683
H565
H702
H568
H521
H831
H496
H663
H574
H562
H556
H628
H667
H627
H801
H564
H580
H792
H705
H694
H601
H501
758
Z148A
H631
Z307
H596
H617
H527
H762
H742
H489
Z51
H747
H530
Z50
Z139
H500
Z314
H810
H506
H502
H656
H701
H712
H659
H624
H803
Z316
H611
H775
H661

M33

Chart 57

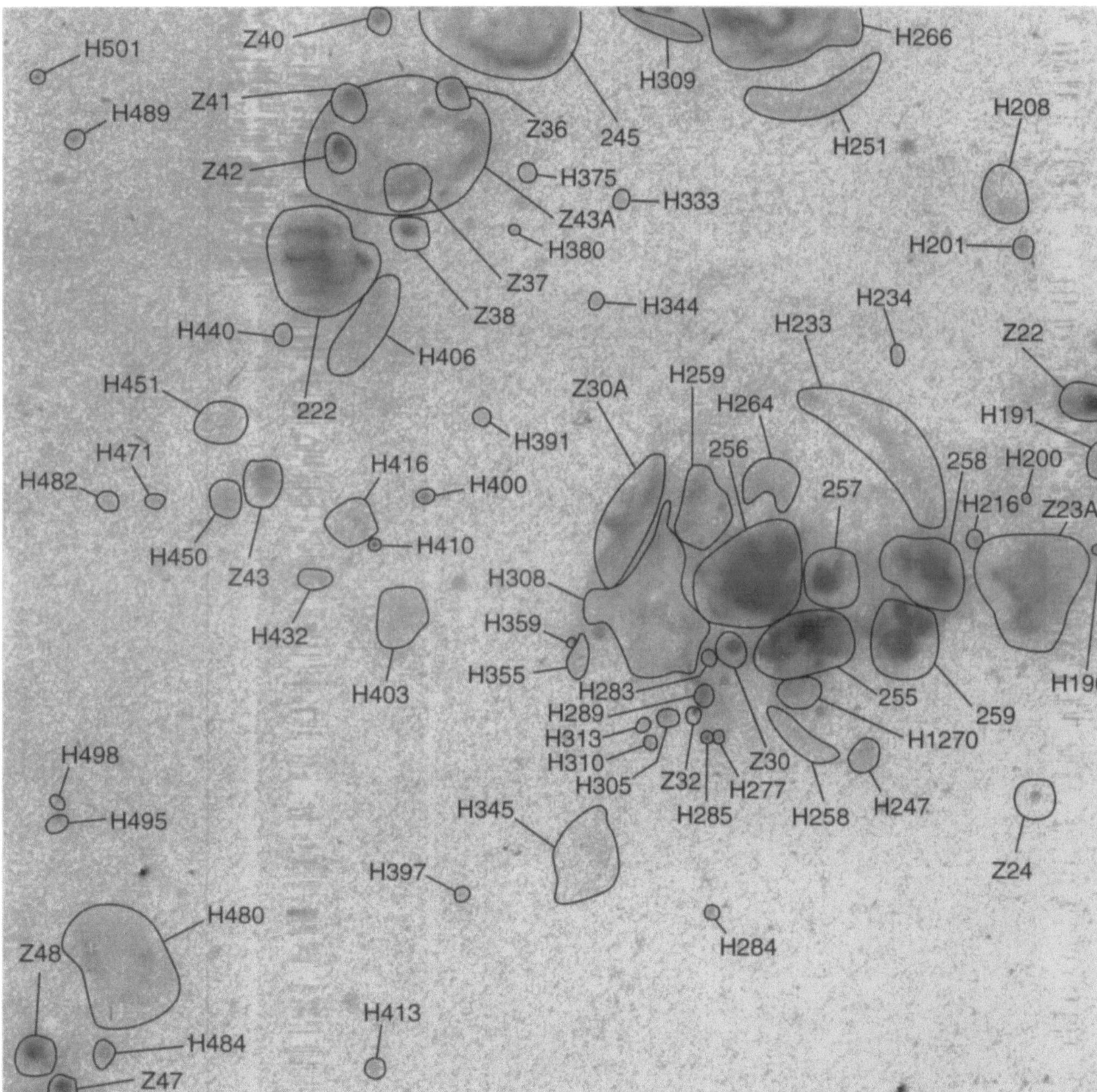
H501
Z40
H266
H309
H489
Z41
Z36
245
H208
H251
Z42
H375
H333
Z43A
H380
H201
Z37
H344
H234
Z38
H233
Z22
H440
H406
H451
222
Z30A
H259
H264
H391
H191
H471
256
258
H200
H482
H416
H400
257
H216
Z23A
H450
H410
Z43
H308
H432
H359
H403
H355
H283
255
H190
H289
259
H1270
H313
H310
Z30
H305
Z32
H277
H498
H495
H345
H285
H258
H247
Z24
H397
H480
H284
Z48
H413
H484
Z47

M33

Chart 58

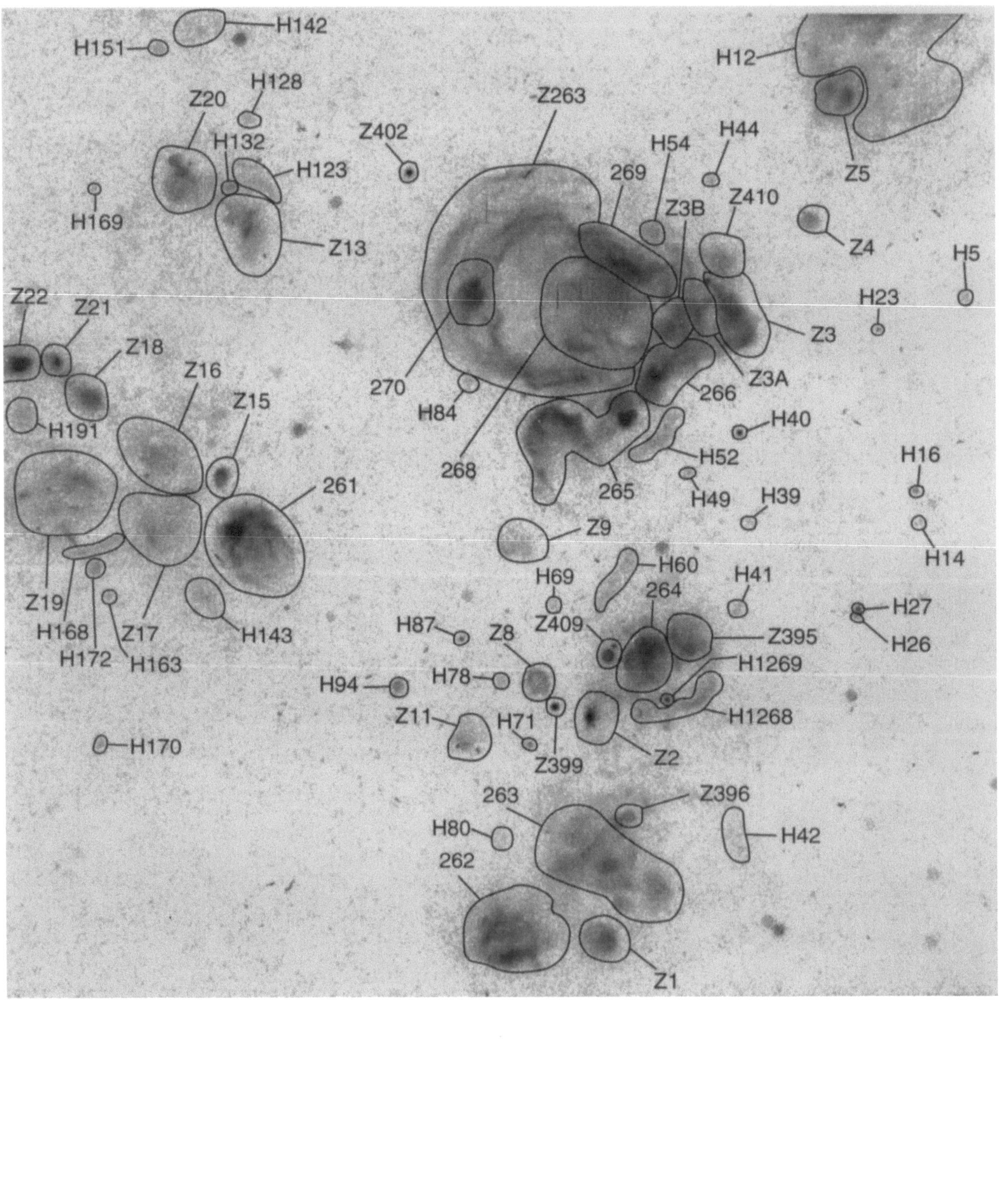
H142
H151
H128
Z20
H132
H123
H169
Z13
Z402
Z263
H12
Z5
H54
H44
269
Z3B
Z410
Z4
H5
H23
Z3
Z3A
266
270
H84
268
H40
H52
H49
265
H39
H16
H14
Z22
Z21
Z18
Z16
Z15
H191
261
Z9
H60
H69
H41
264
Z409
Z395
H27
H26
H1269
H87
Z8
H78
H94
H1268
Z11
H71
Z399
Z2
H170
Z19
H168
H172
Z17
H163
H143
263
Z396
H80
H42
262
Z1

M33

Chart 59

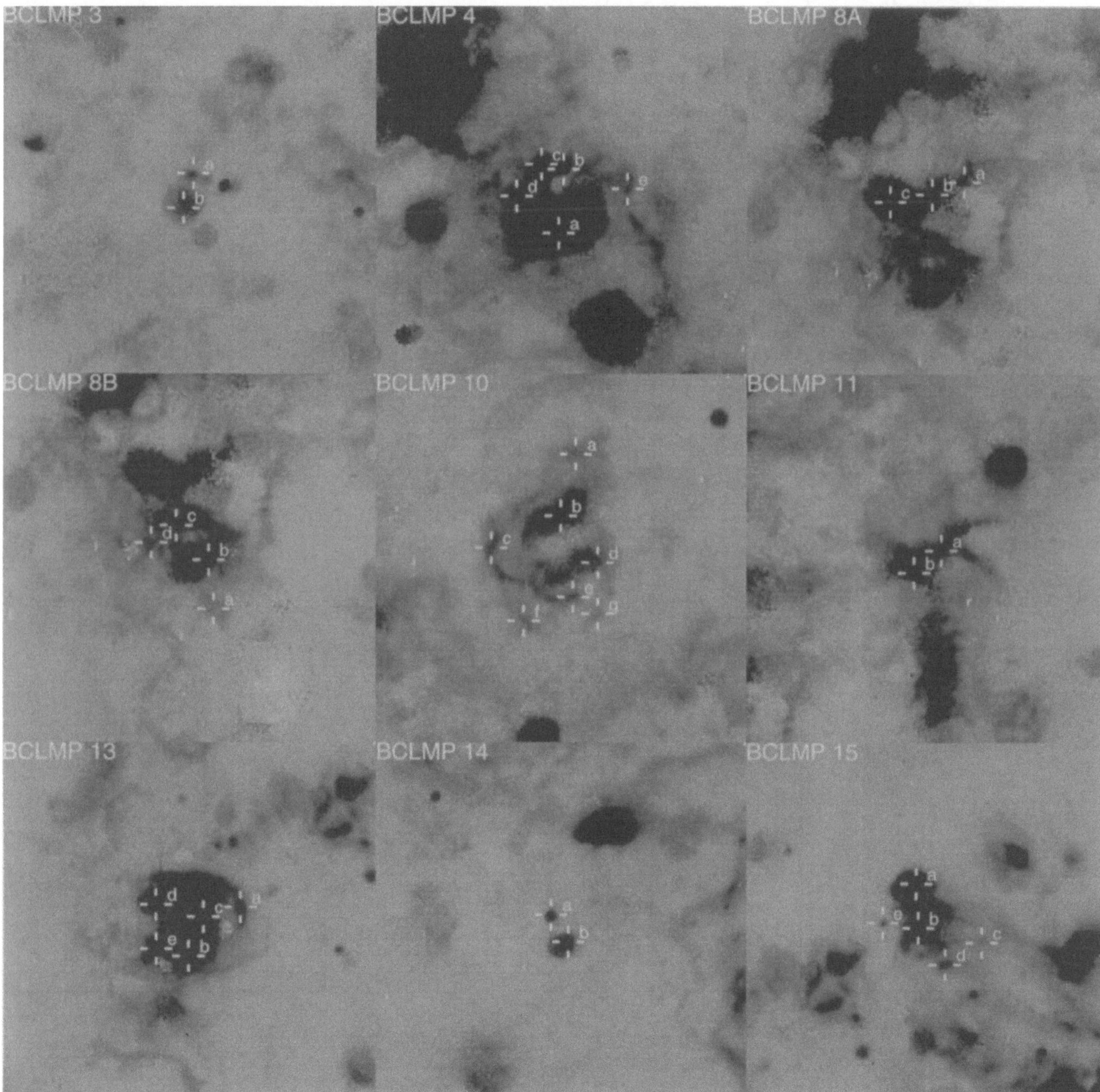
BCLMP 3
a
b
BCLMP 4
c
b
d
e
a
BCLMP 8A
c
b
a
BCLMP 8B
c
d
b
a
BCLMP 10
a
b
c
d
e
f
g
BCLMP 11
a
b
BCLMP 13
d
c
a
e
b
BCLMP 14
a
b
BCLMP 15
a
e
b
c
d

M33

Chart 60

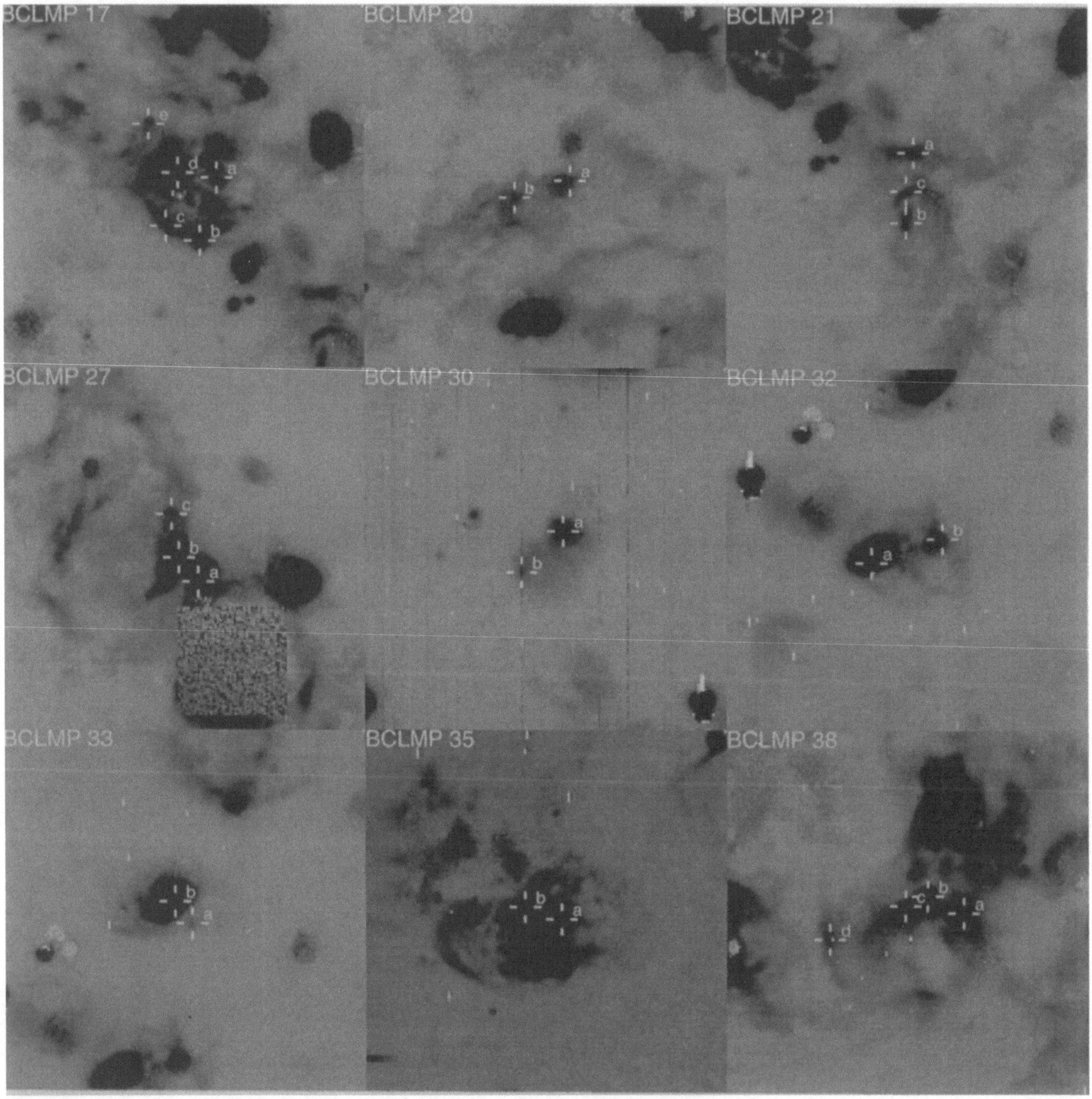
BCLMP 17
e
d
a
c
b
BCLMP 20
a
b
BCLMP 21
a
c
b
BCLMP 27
c
b
a
BCLMP 30
a
b
BCLMP 32
b
a
BCLMP 33
b
a
BCLMP 35
b
a
BCLMP 38
c
b
a
d

M33

Chart 61

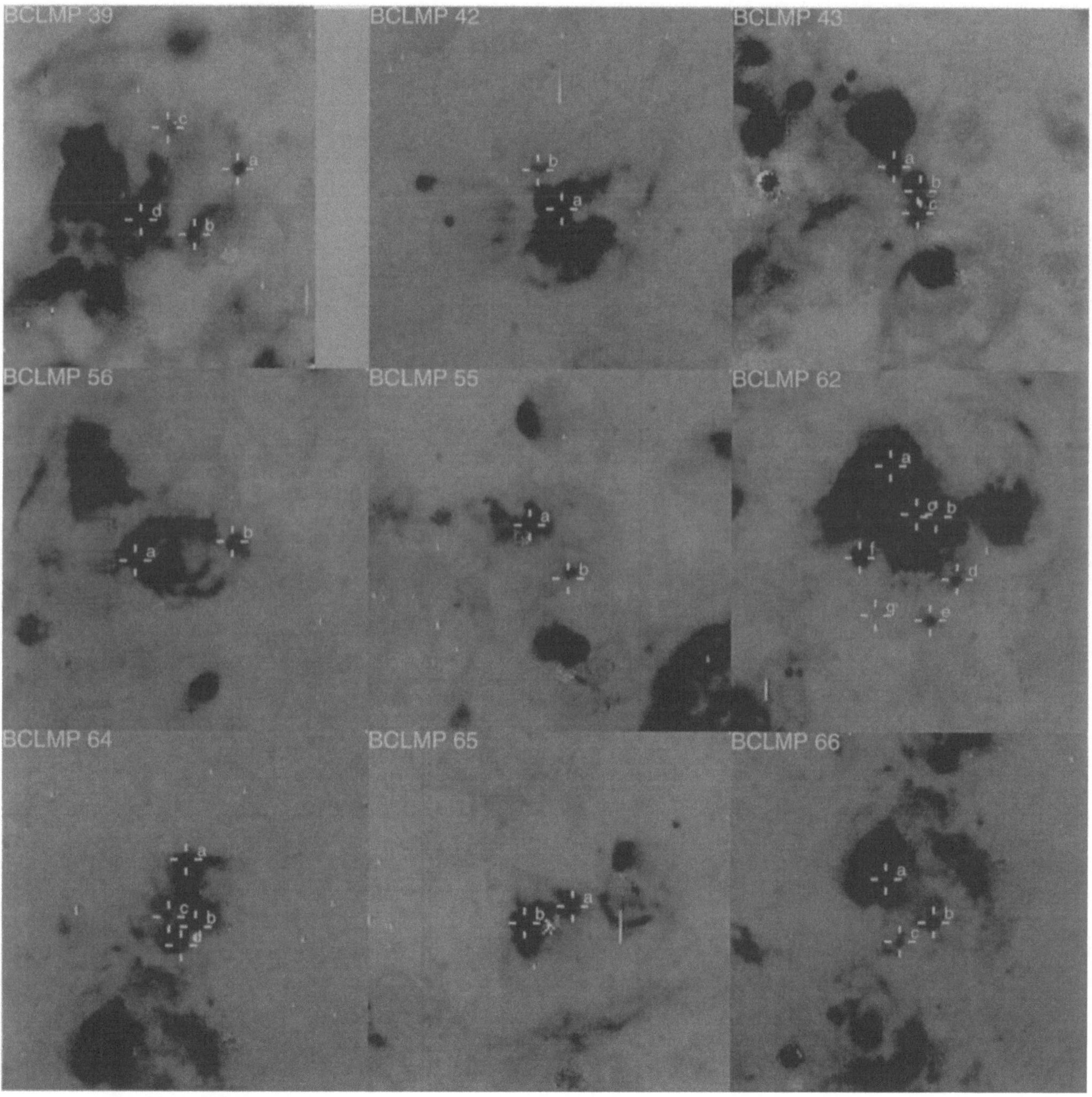

BCLMP 39
BCLMP 42
BCLMP 43
BCLMP 56
BCLMP 55
BCLMP 62
BCLMP 64
BCLMP 65
BCLMP 66

M33

Chart 62

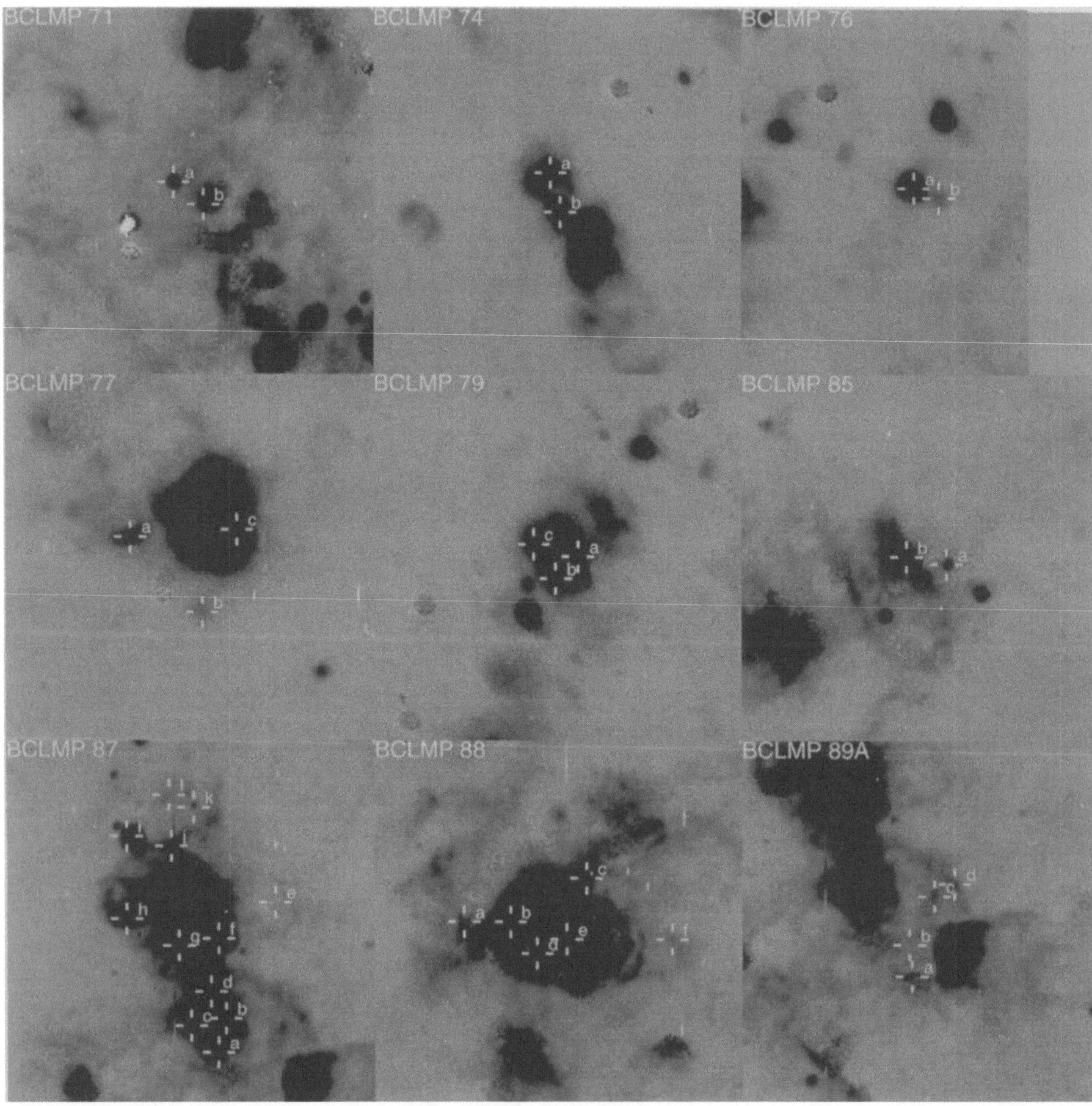
BCLMP 71
a
b
BCLMP 74
a
b
BCLMP 76
a
b
BCLMP 77
a
c
b
BCLMP 79
c
a
b
BCLMP 85
b
a
BCLMP 87
k
e
h
g
f
d
b
c
a
BCLMP 88
c
a
b
e
d
f
BCLMP 89A
d
c
b
a

M33

Chart 63

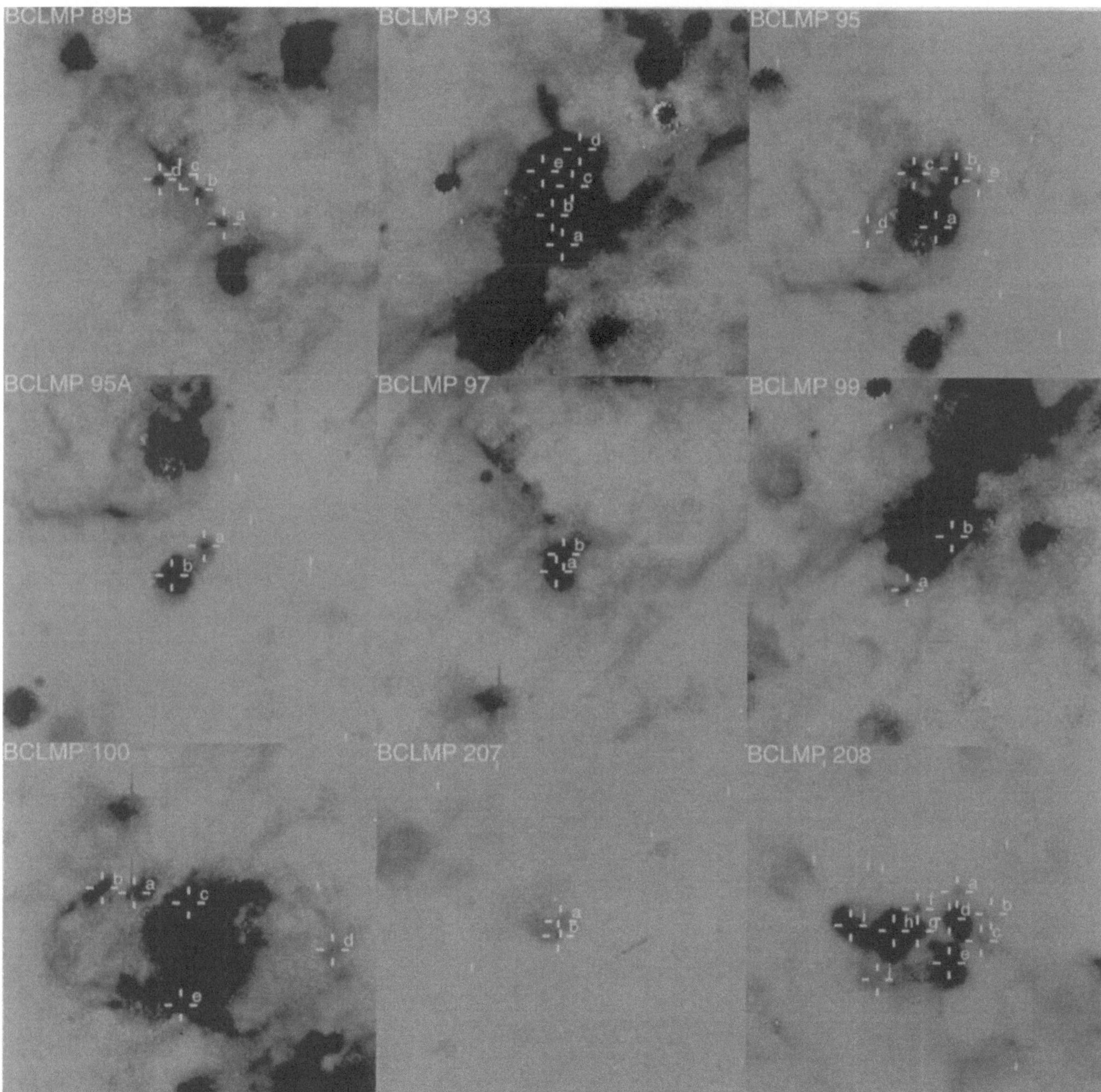
BCLMP 89B
BCLMP 93
BCLMP 95
BCLMP 95A
BCLMP 97
BCLMP 99
BCLMP 100
BCLMP 207
BCLMP 208

M33

Chart 64

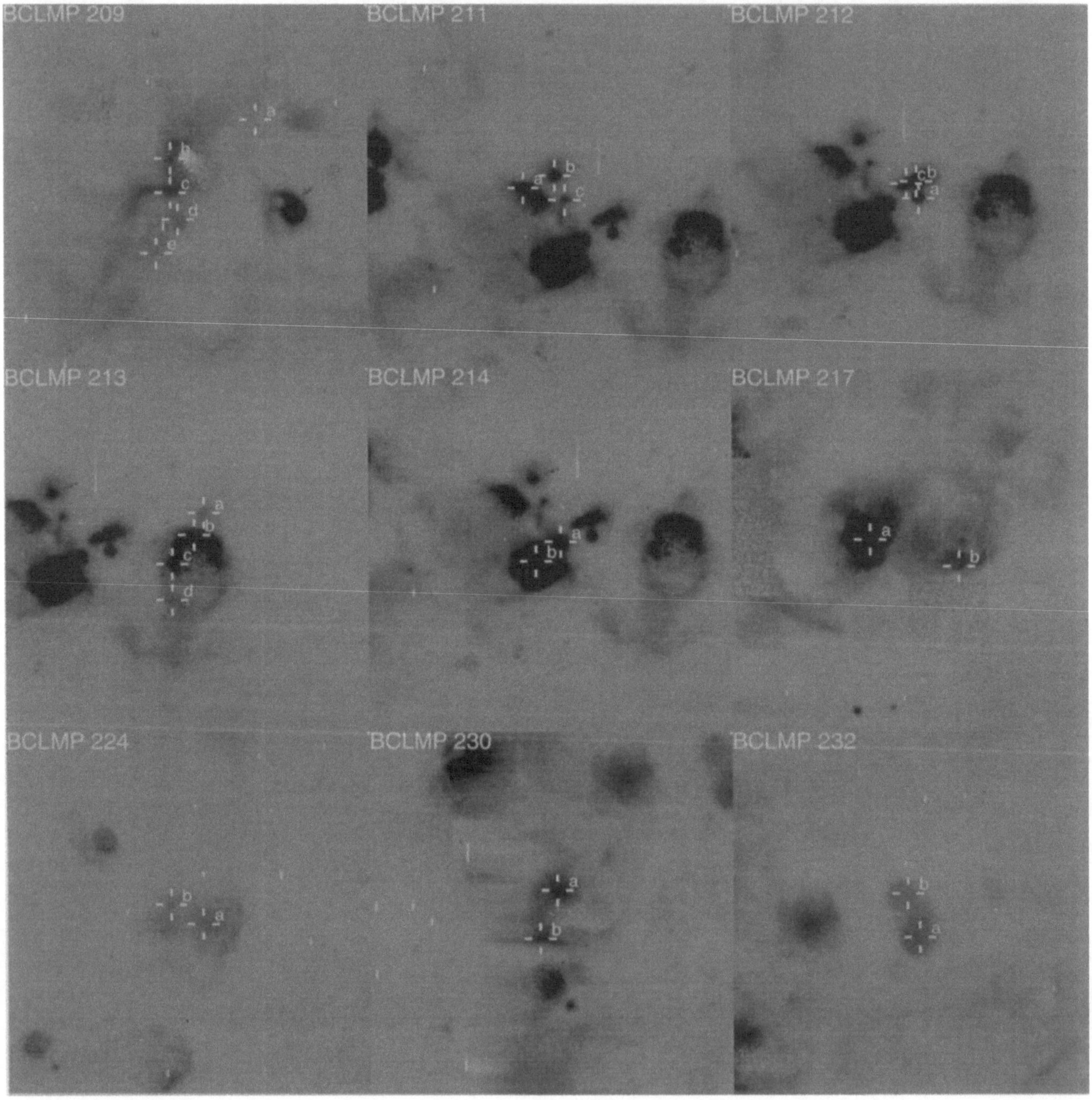
BCLMP 209
a
b
c
d
e
BCLMP 211
a
b
c
BCLMP 212
c
b
a
BCLMP 213
a
b
c
d
BCLMP 214
a
b
BCLMP 217
a
b
BCLMP 224
b
a
BCLMP 230
a
b
BCLMP 232
b
a

M33

Chart 65

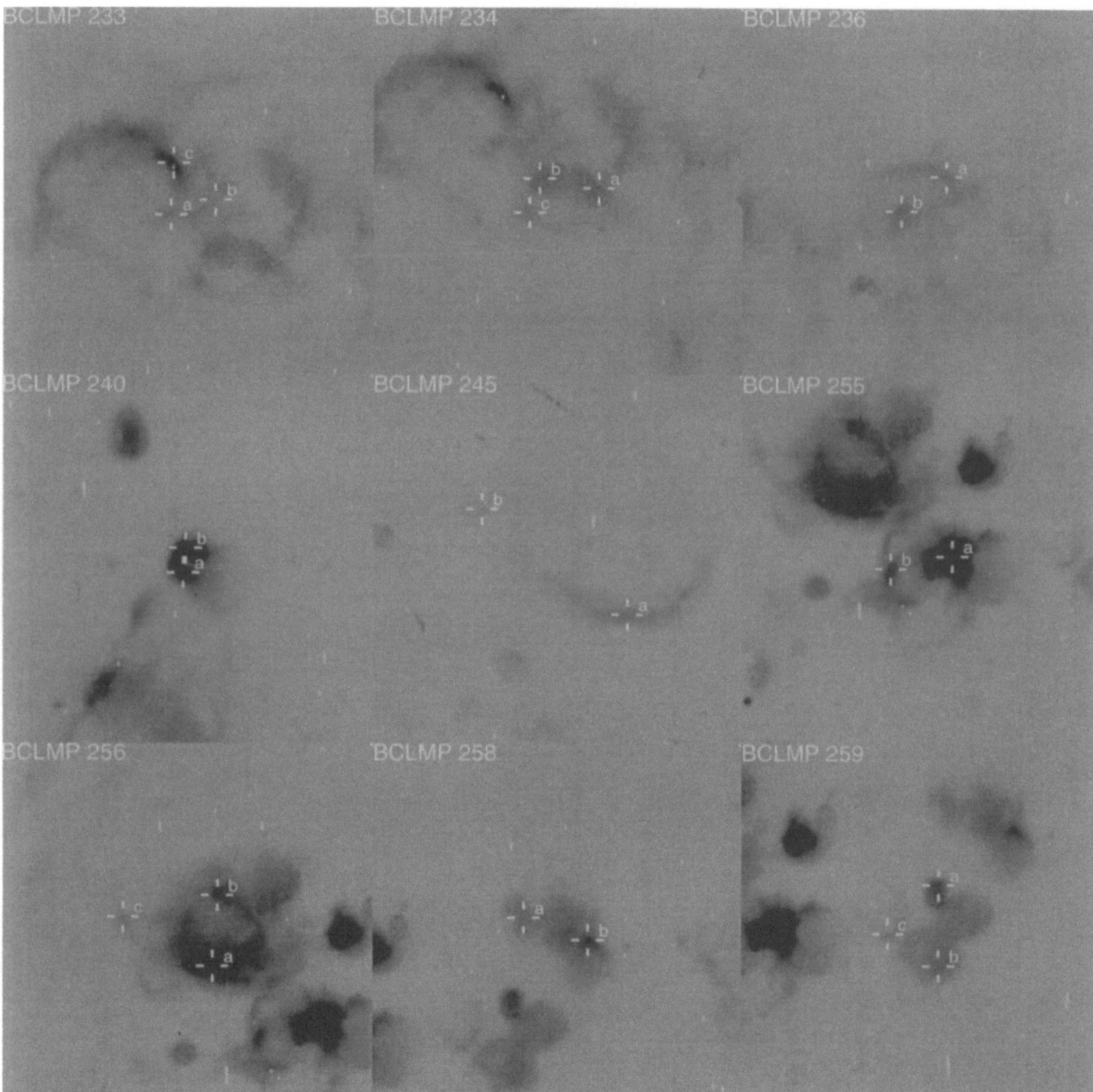
BCLMP 233
c
b
a
BCLMP 234
b
a
c
BCLMP 236
a
b
BCLMP 240
b
a
BCLMP 245
b
a
BCLMP 255
a
b
BCLMP 256
b
c
a
BCLMP 258
a
b
BCLMP 259
a
c
b

M33

Chart 66

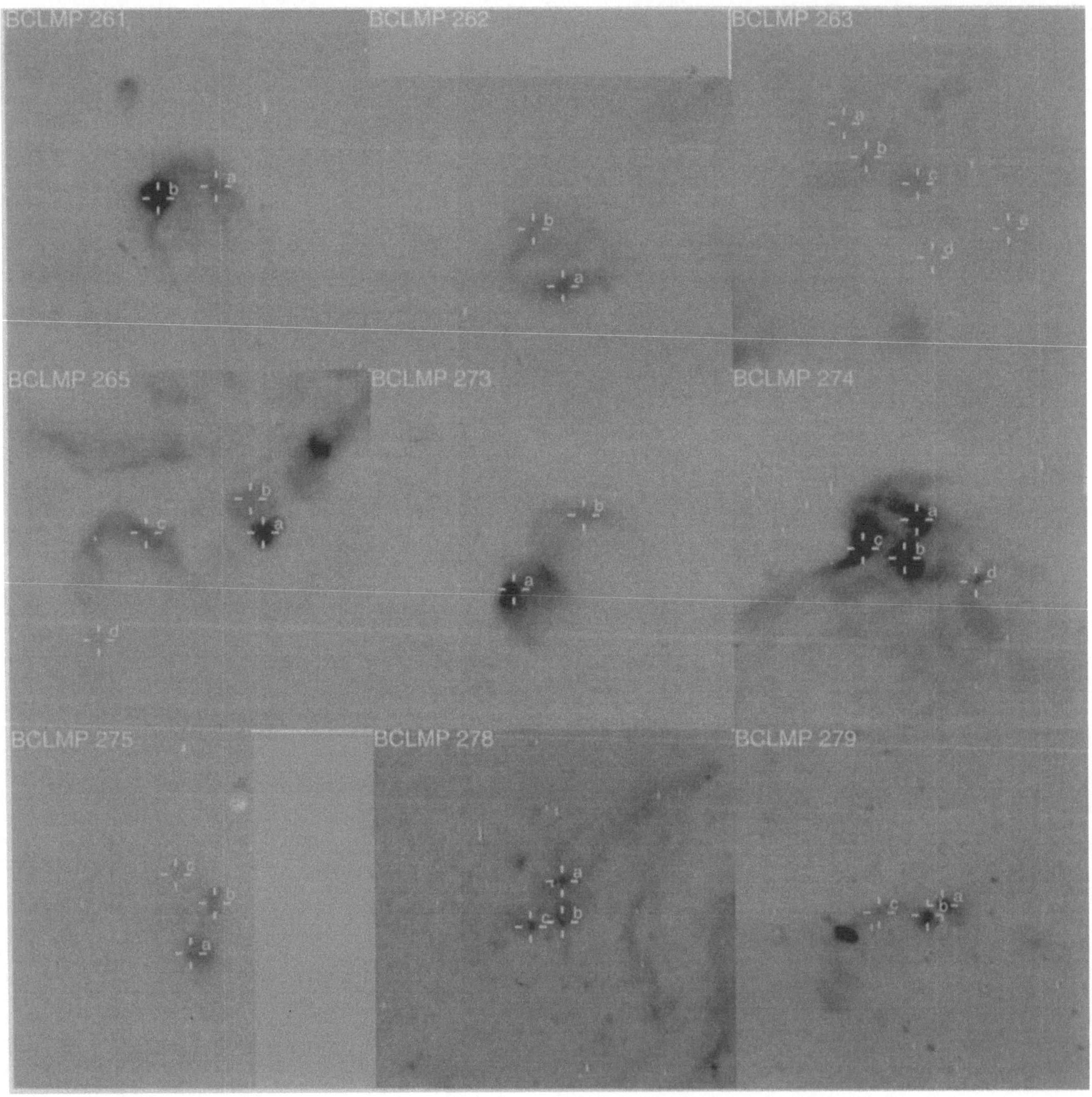
BCLMP 261
BCLMP 262
BCLMP 263
BCLMP 265
BCLMP 273
BCLMP 274
BCLMP 275
BCLMP 278
BCLMP 279

M33

Chart 67

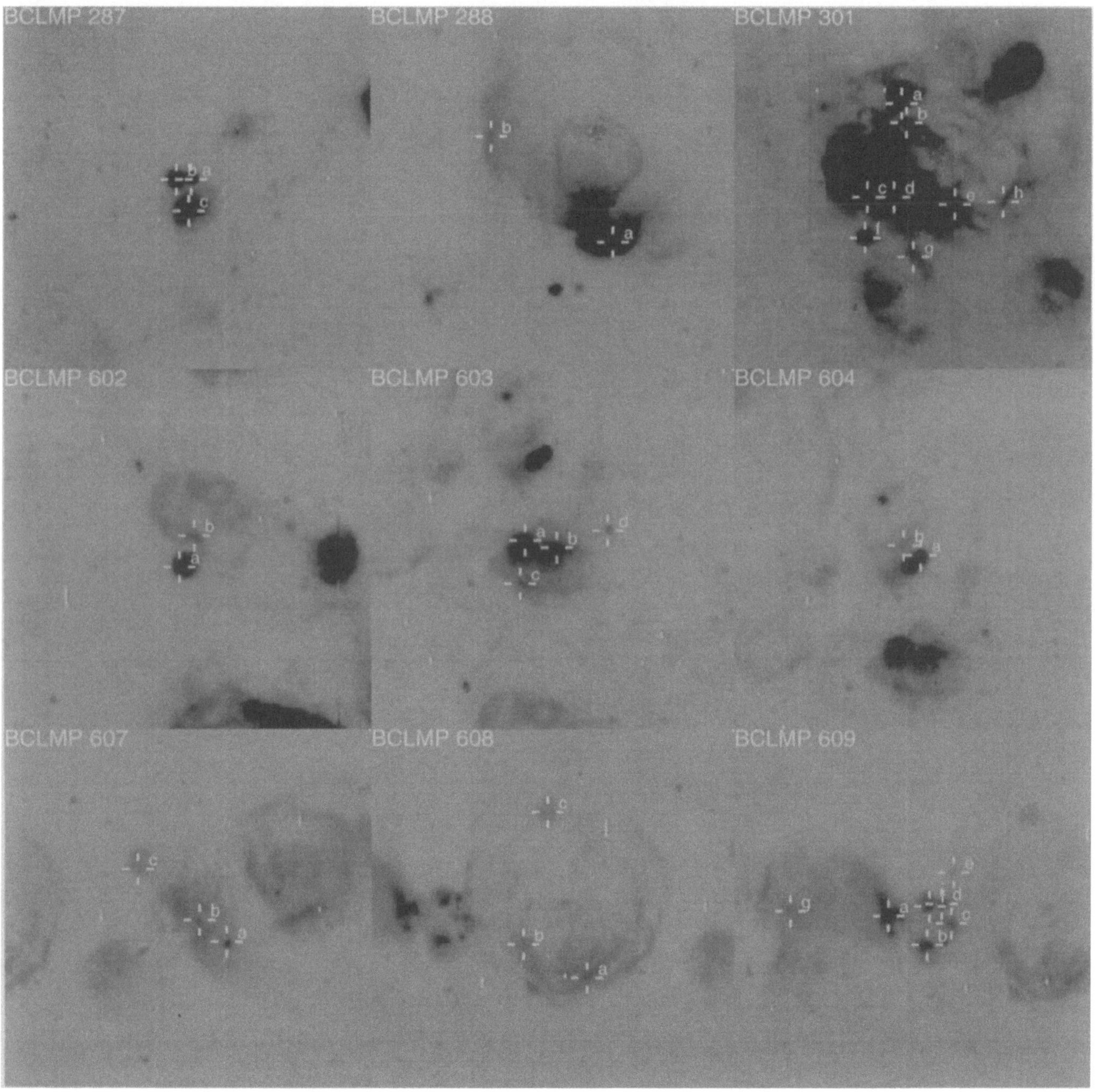
BCLMP 287
BCLMP 288
BCLMP 301
BCLMP 602
BCLMP 603
BCLMP 604
BCLMP 607
BCLMP 608
BCLMP 609

M33

Chart 68

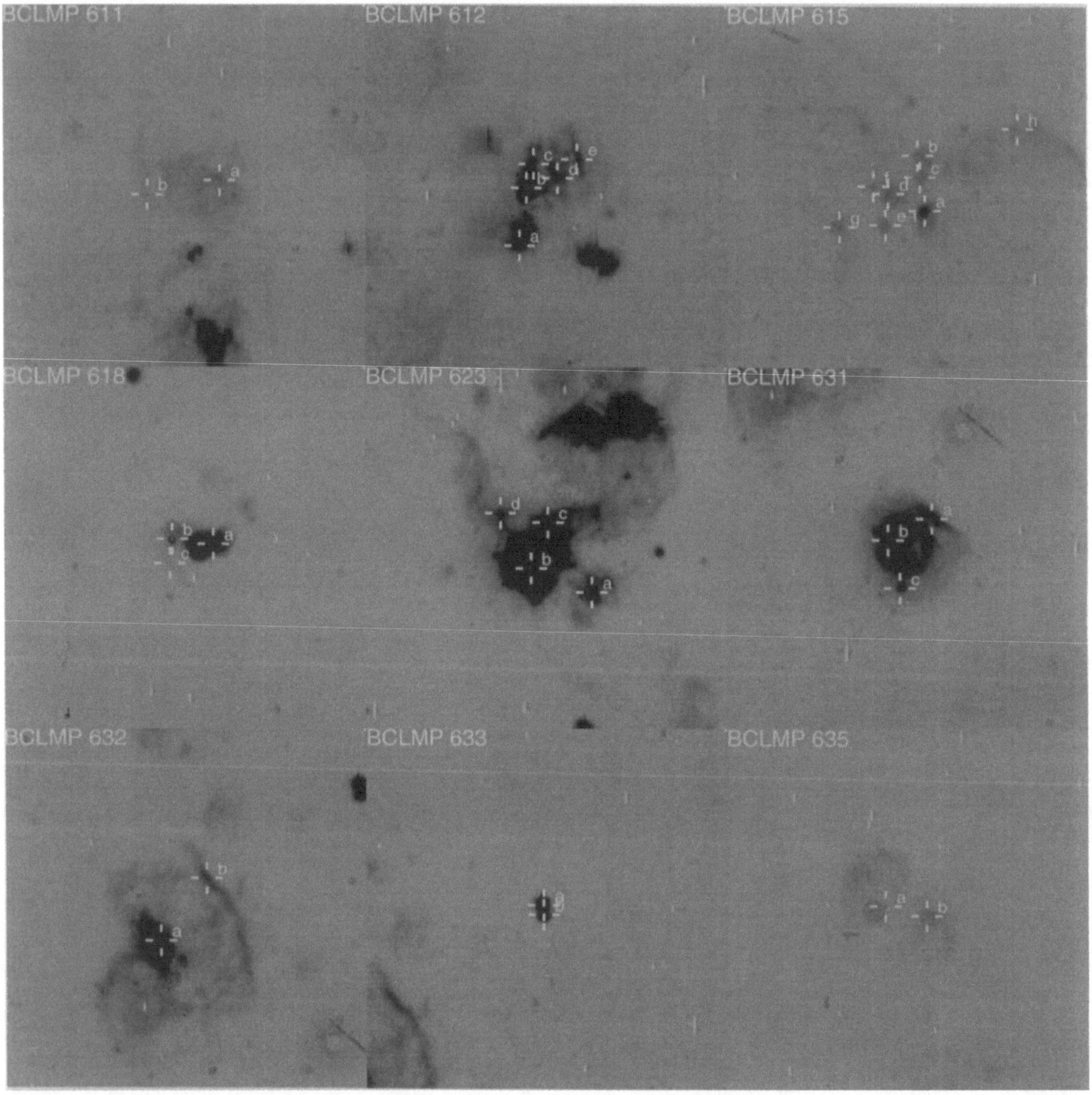
BCLMP 611
BCLMP 612
BCLMP 615
BCLMP 618
BCLMP 623
BCLMP 631
BCLMP 632
BCLMP 633
BCLMP 635

M33

Chart 69

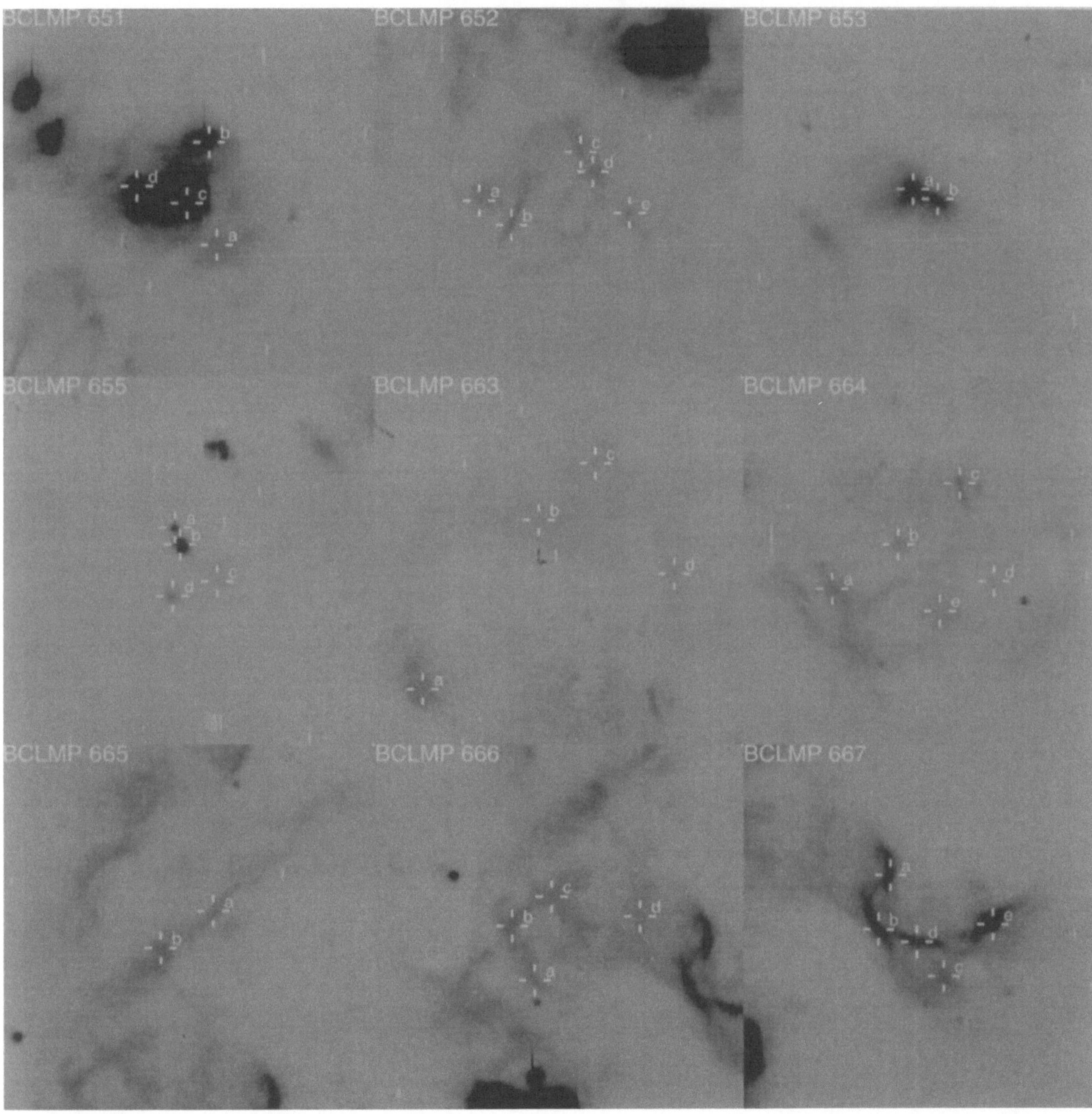
BCLMP 651
b
d
c
a
BCLMP 652
c
d
a
b
e
BCLMP 653
a
b
BCLMP 655
a
b
c
d
BCLMP 663
c
b
d
a
BCLMP 664
c
b
a
d
e
BCLMP 665
a
b
BCLMP 666
c
b
d
a
BCLMP 667
a
b
d
e
c

M33

Chart 70

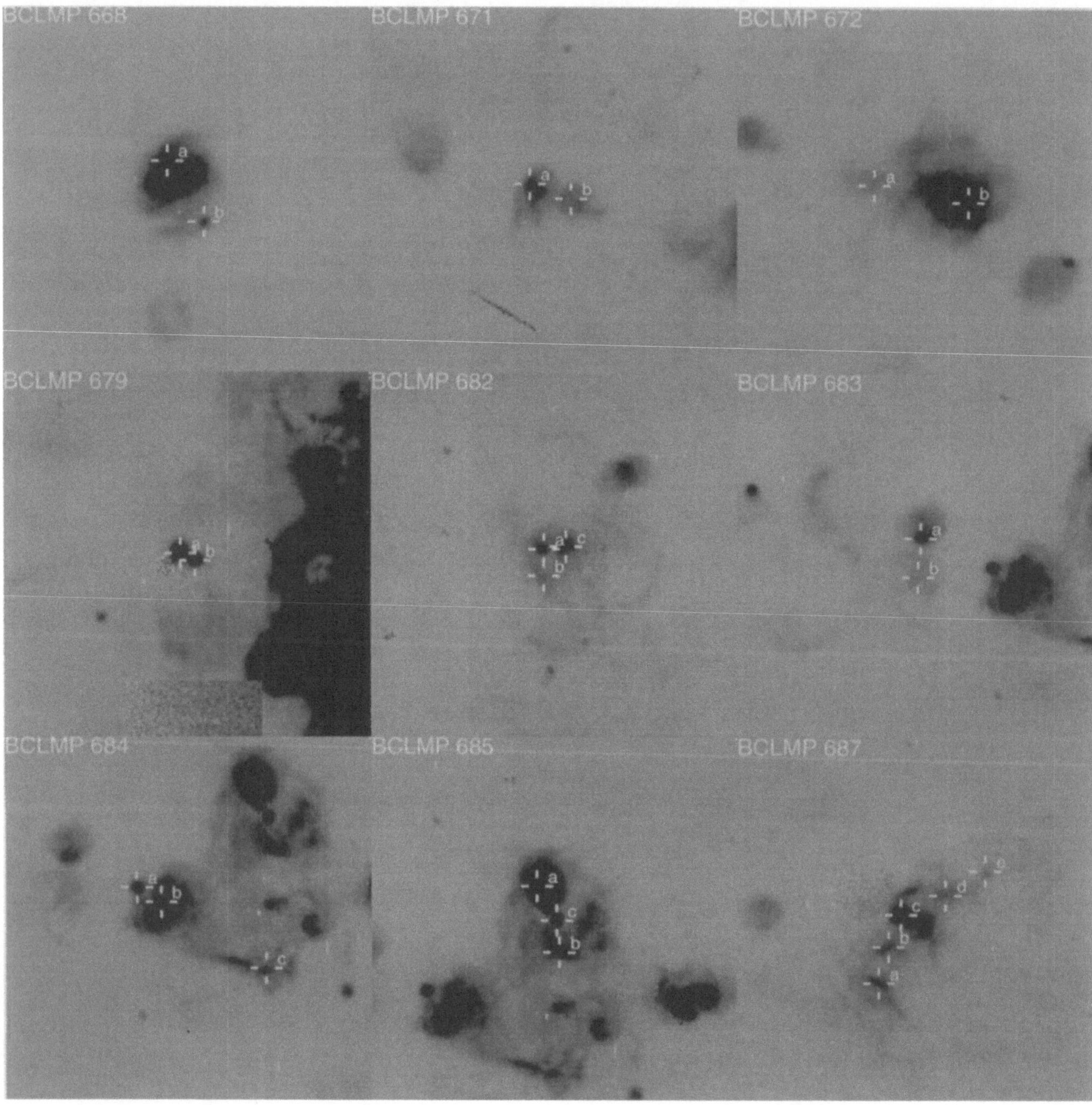
BCLMP 668
a
b
BCLMP 671
a
b
BCLMP 672
a
b
BCLMP 679
a
b
BCLMP 682
a
c
b
BCLMP 683
a
b
BCLMP 684
a
b
c
BCLMP 685
a
c
b
BCLMP 687
e
d
c
b
a

M33

Chart 71

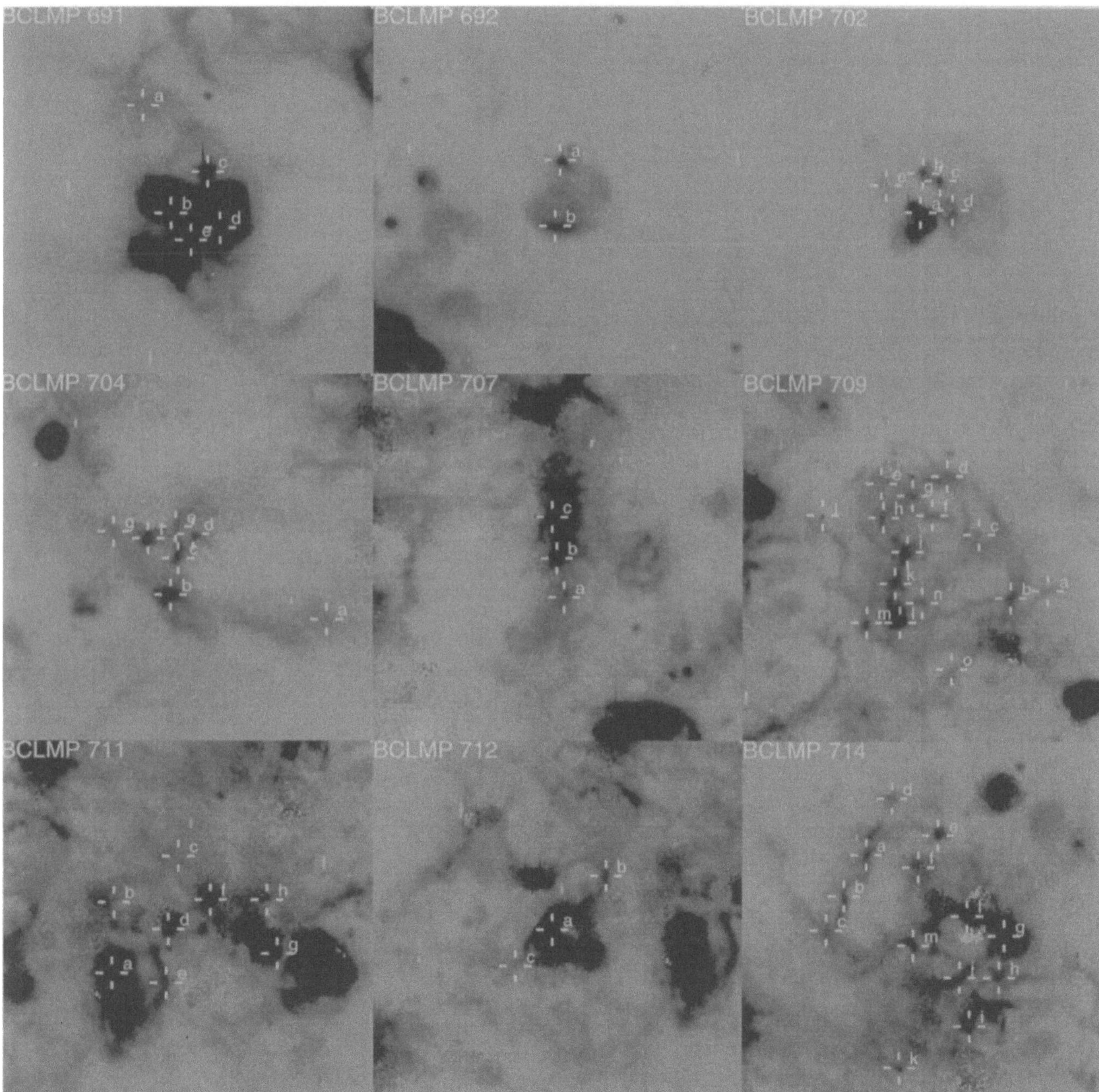
BCLMP 691
BCLMP 692
BCLMP 702
BCLMP 704
BCLMP 707
BCLMP 709
BCLMP 711
BCLMP 712
BCLMP 714

M33

Chart 72

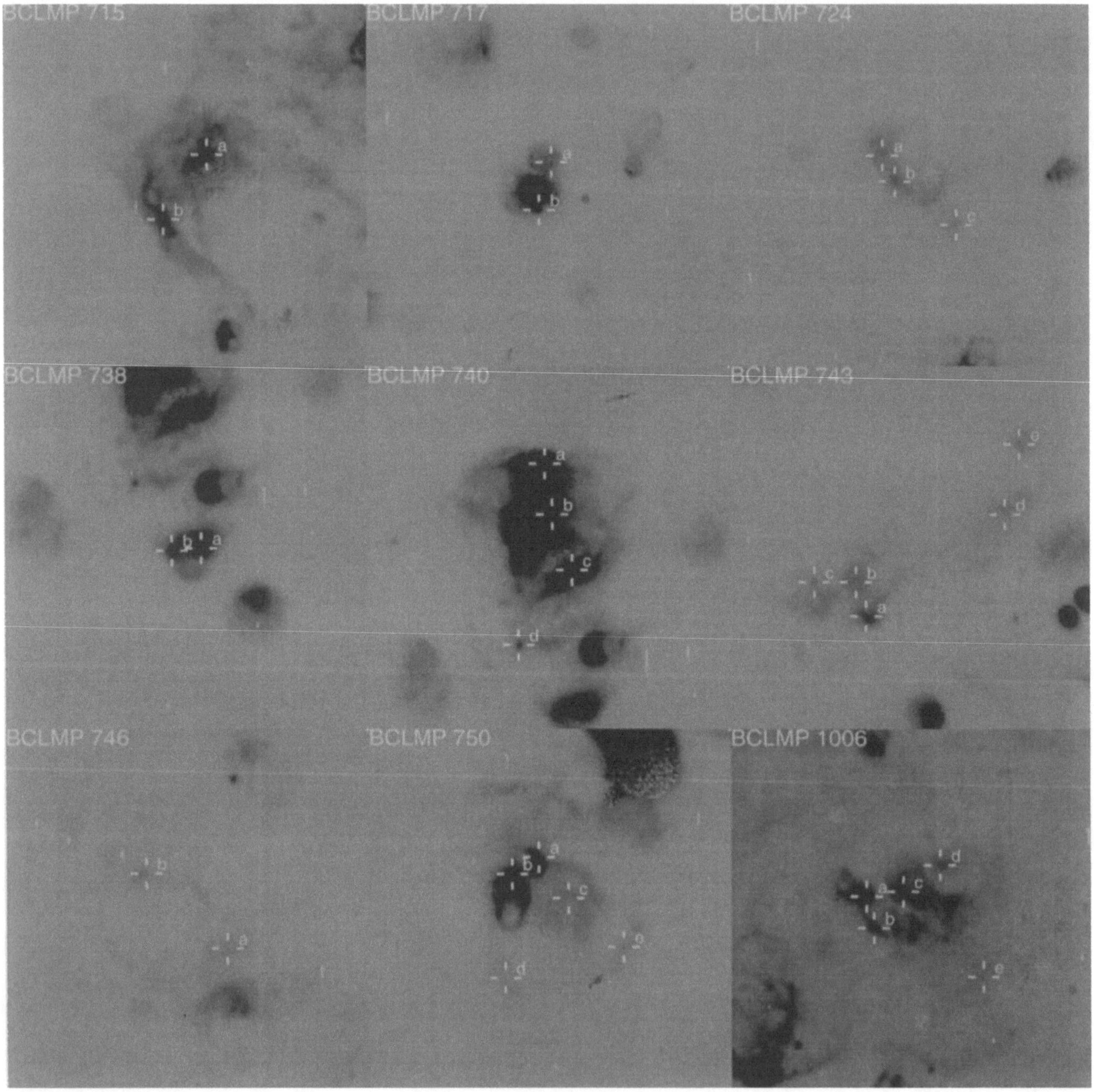
BCLMP 715
BCLMP 717
BCLMP 724
BCLMP 738
BCLMP 740
BCLMP 743
BCLMP 746
BCLMP 750
BCLMP 1006

M33

Chart 73

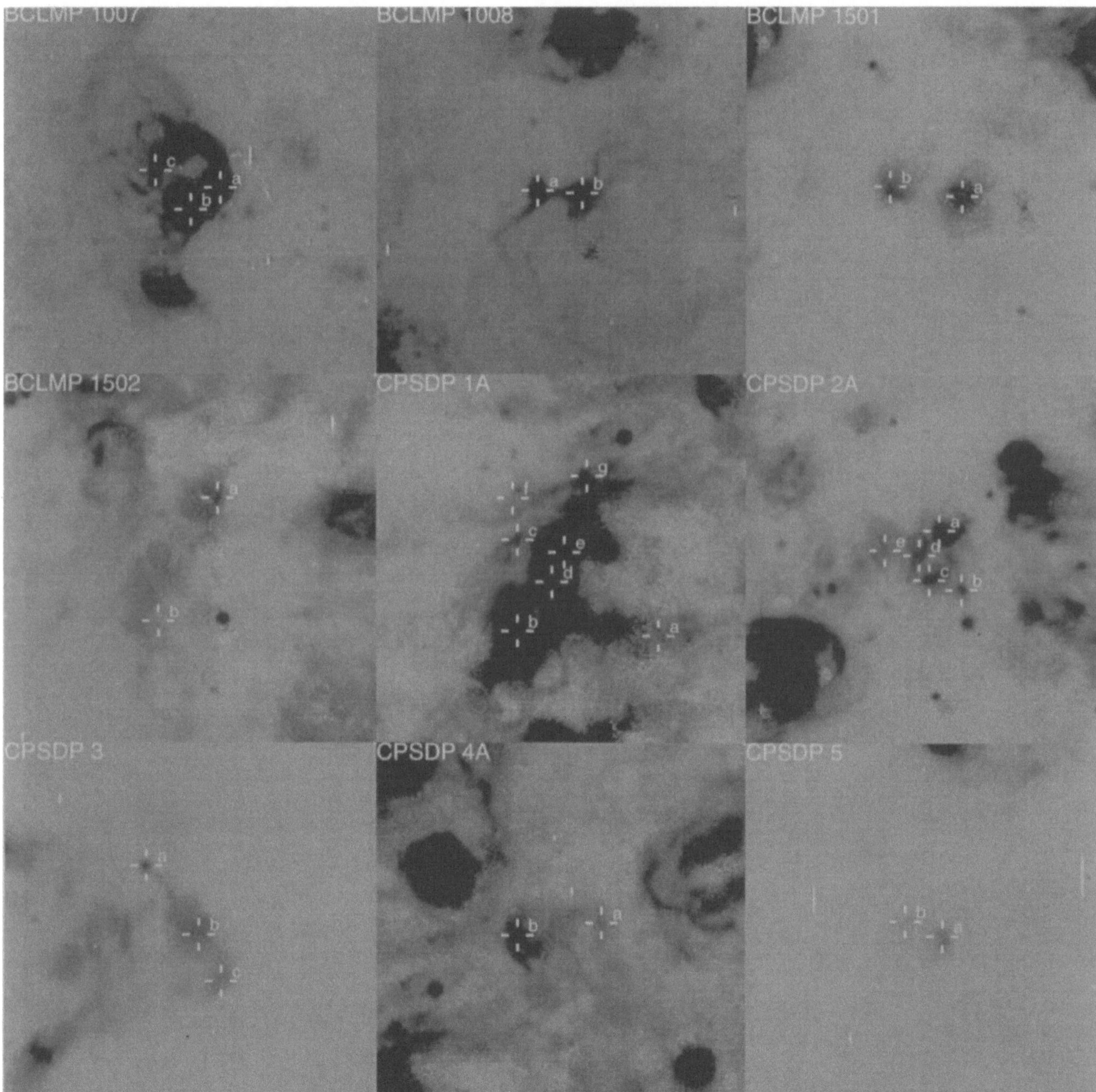
BCLMP 1007
BCLMP 1008
BCLMP 1501
BCLMP 1502
CPSDP 1A
CPSDP 2A
CPSDP 3
CPSDP 4A
CPSDP 5

M33

Chart 74

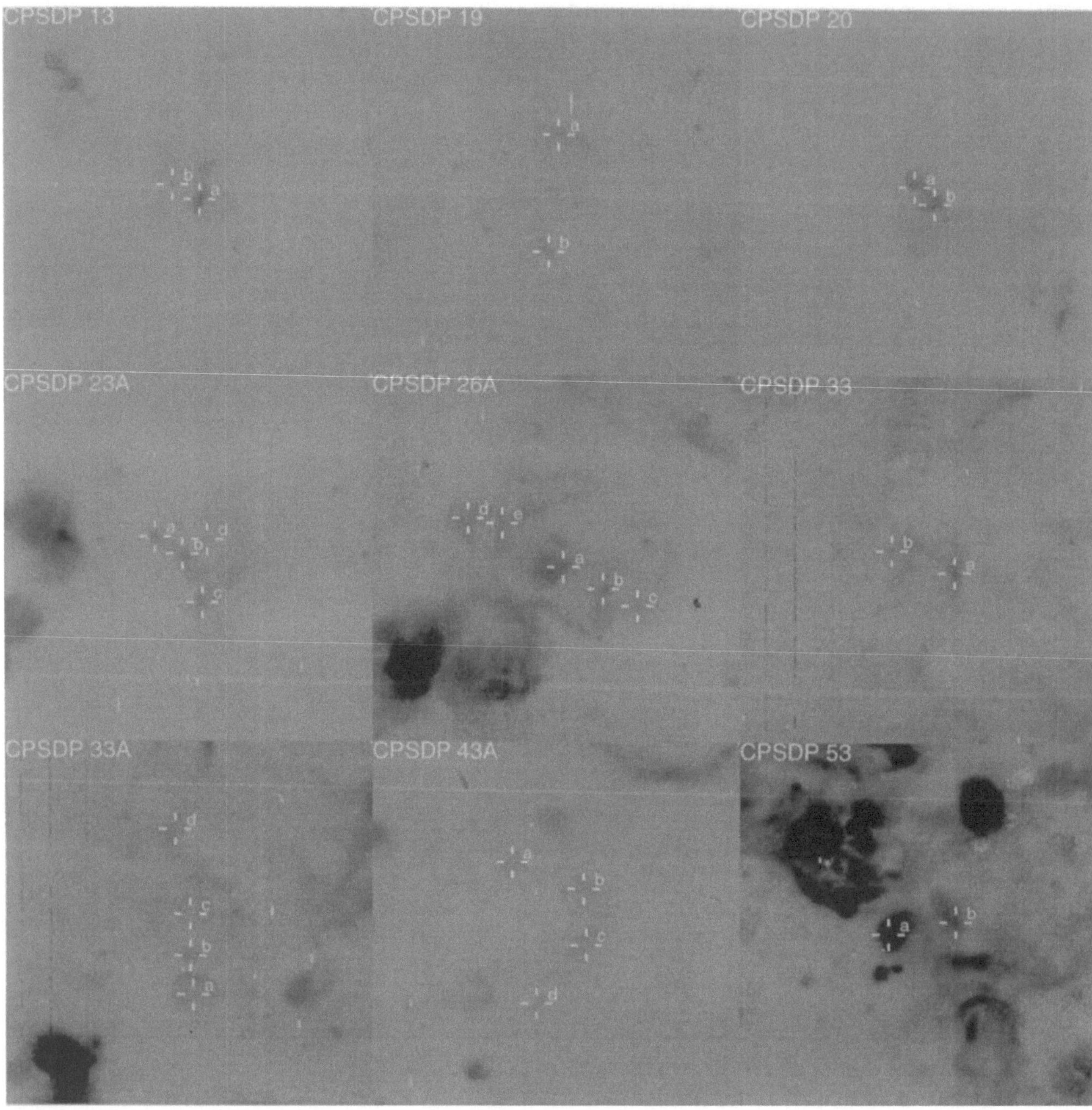
CPSDP 13
CPSDP 19
CPSDP 20
CPSDP 23A
CPSDP 26A
CPSDP 33
CPSDP 33A
CPSDP 43A
CPSDP 53

M33

Chart 75

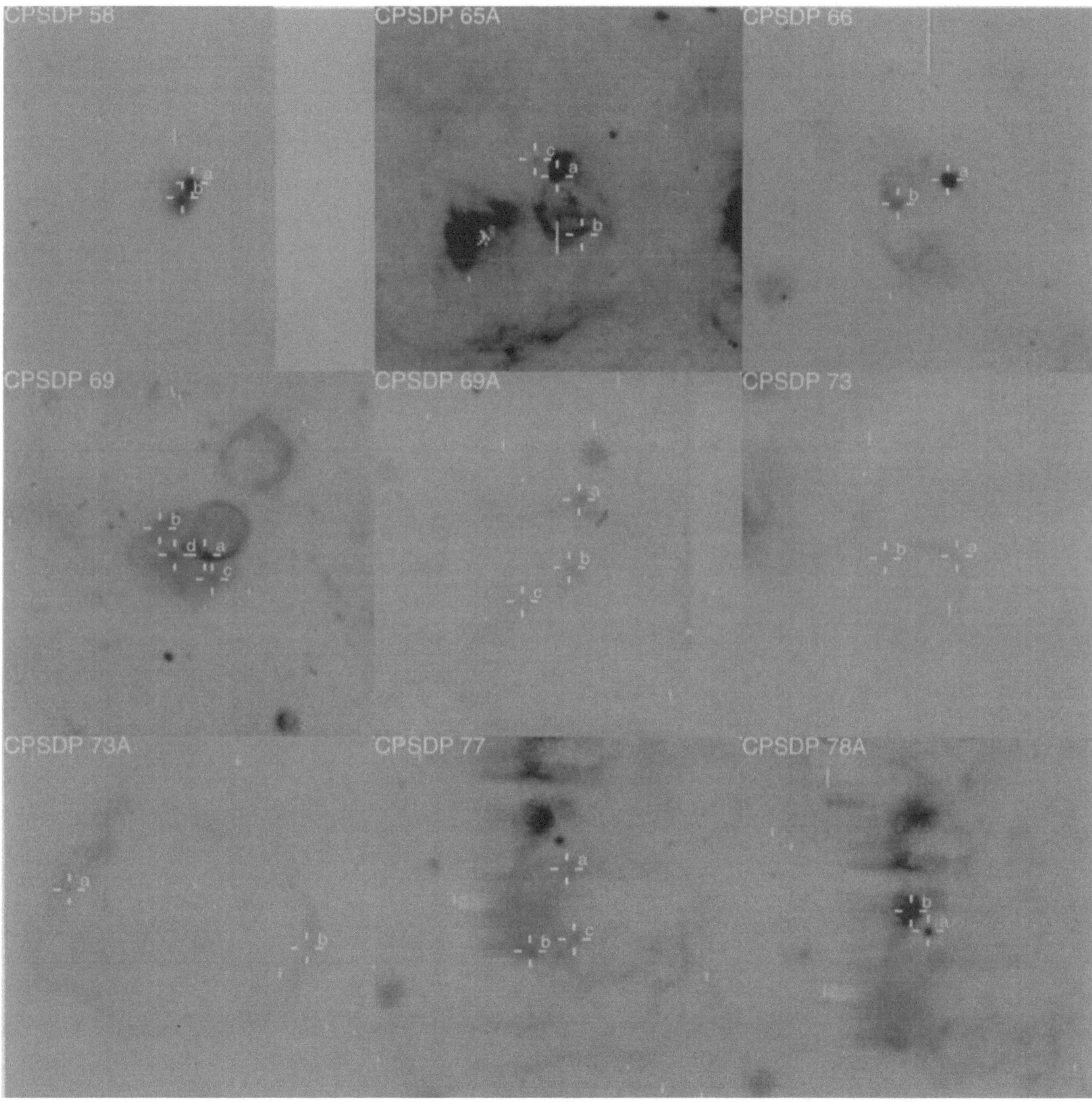
CPSDP 58
a
b
CPSDP 65A
c
a
b
CPSDP 66
a
b
CPSDP 69
b
d
a
c
CPSDP 69A
a
b
c
CPSDP 73
b
a
CPSDP 73A
a
b
CPSDP 77
a
b
c
CPSDP 78A
b
a

M33

Chart 76

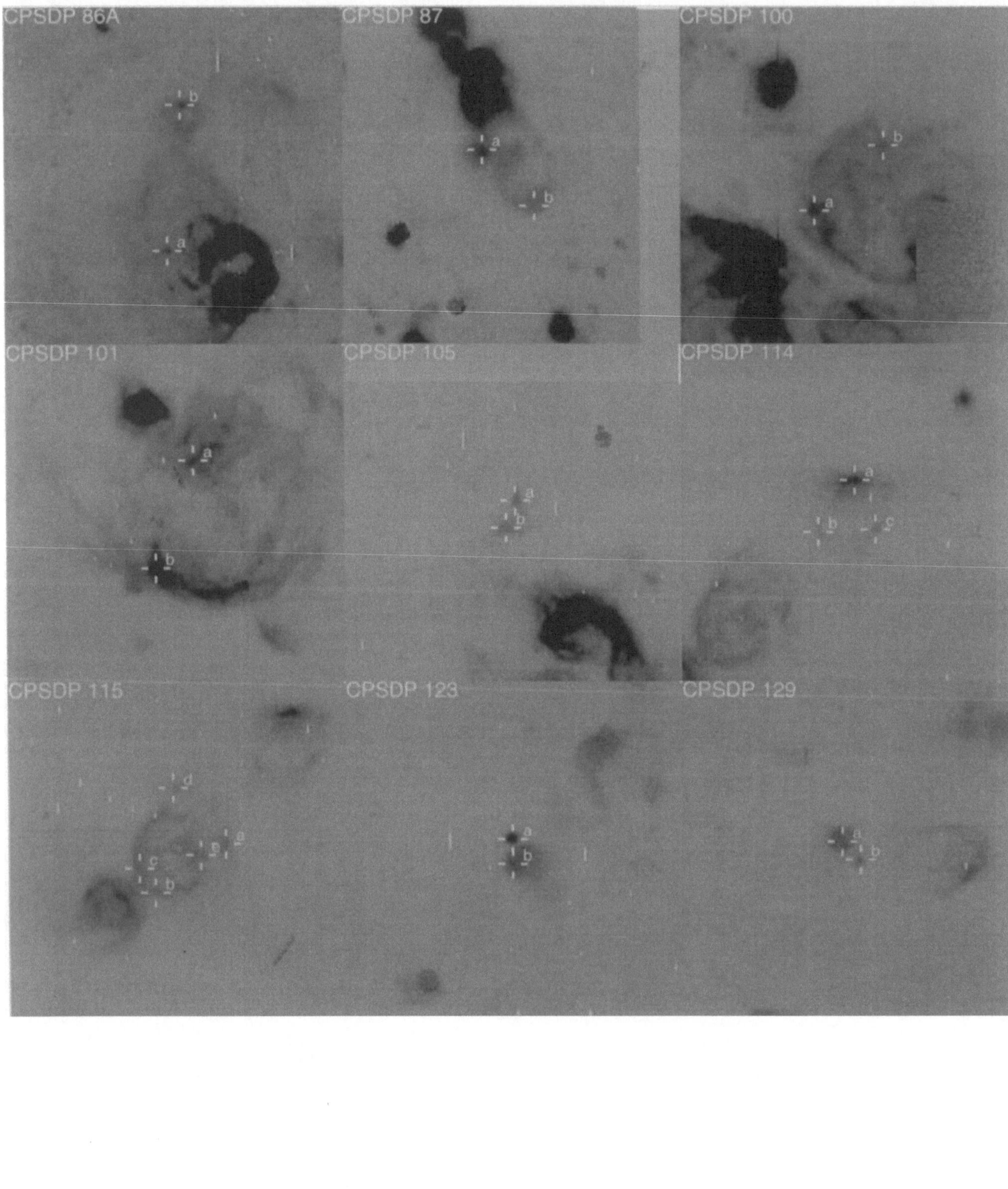
CPSDP 86A
CPSDP 87
CPSDP 100
CPSDP 101
CPSDP 105
CPSDP 114
CPSDP 115
CPSDP 123
CPSDP 129

M33

Chart 77

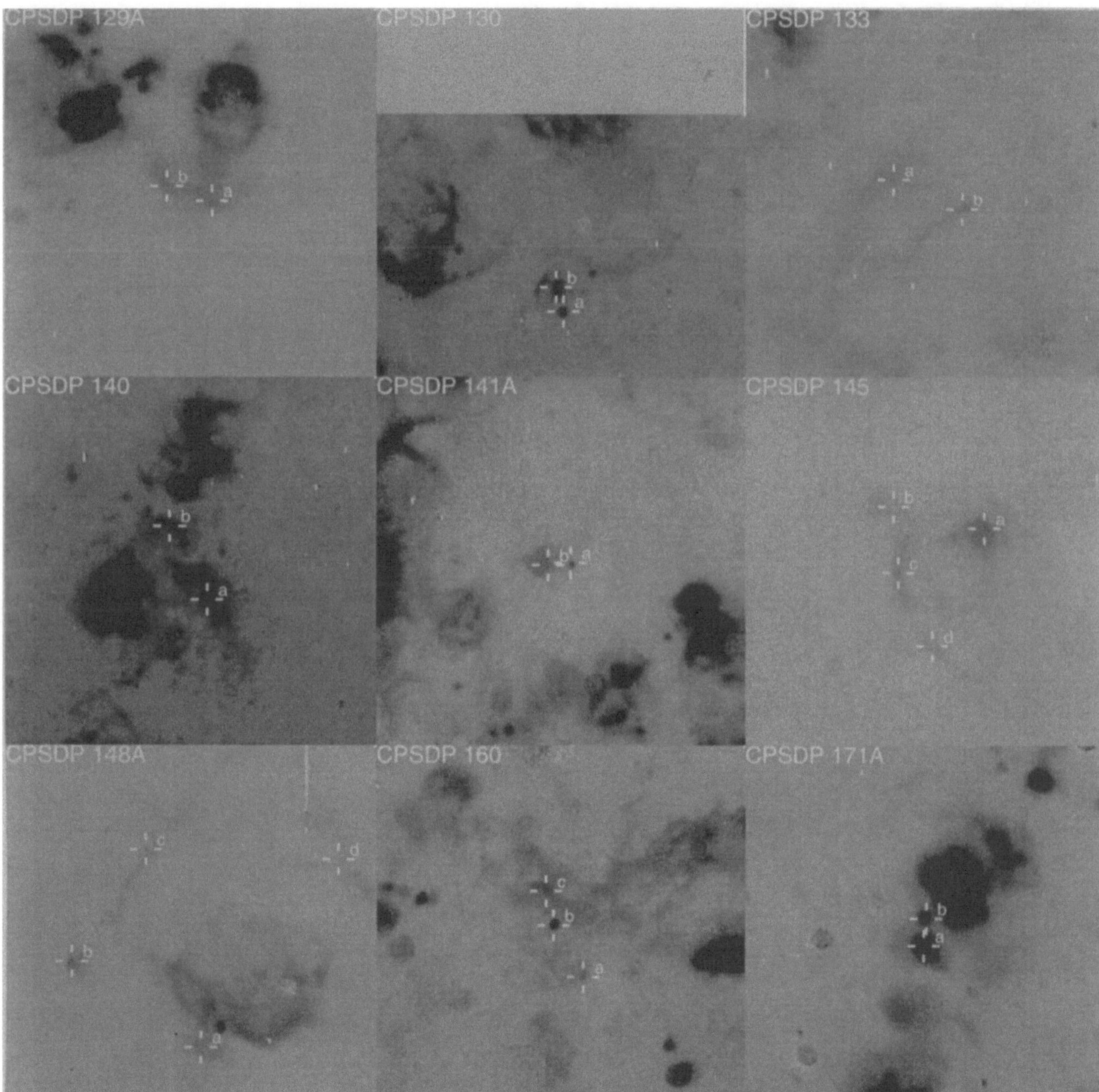
CPSDP 129A
b
a
CPSDP 130
b
a
CPSDP 133
a
b
CPSDP 140
b
a
CPSDP 141A
b
a
CPSDP 145
b
a
c
d
CPSDP 148A
c
d
b
a
CPSDP 160
c
b
a
CPSDP 171A
b
a

M33

Chart 78

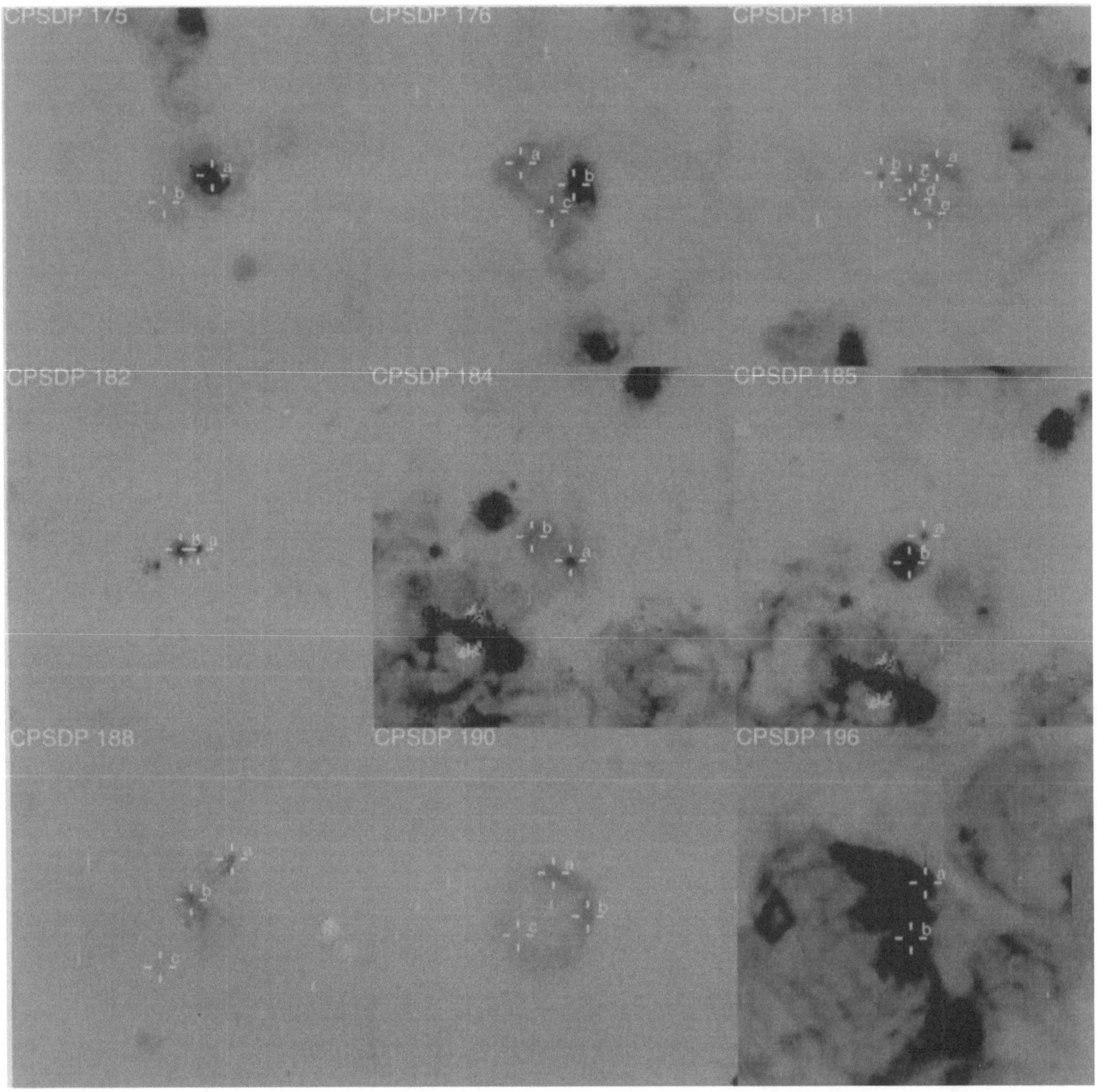
CPSDP 175
CPSDP 176
CPSDP 181
CPSDP 182
CPSDP 184
CPSDP 185
CPSDP 188
CPSDP 190
CPSDP 196

M33

Chart 79

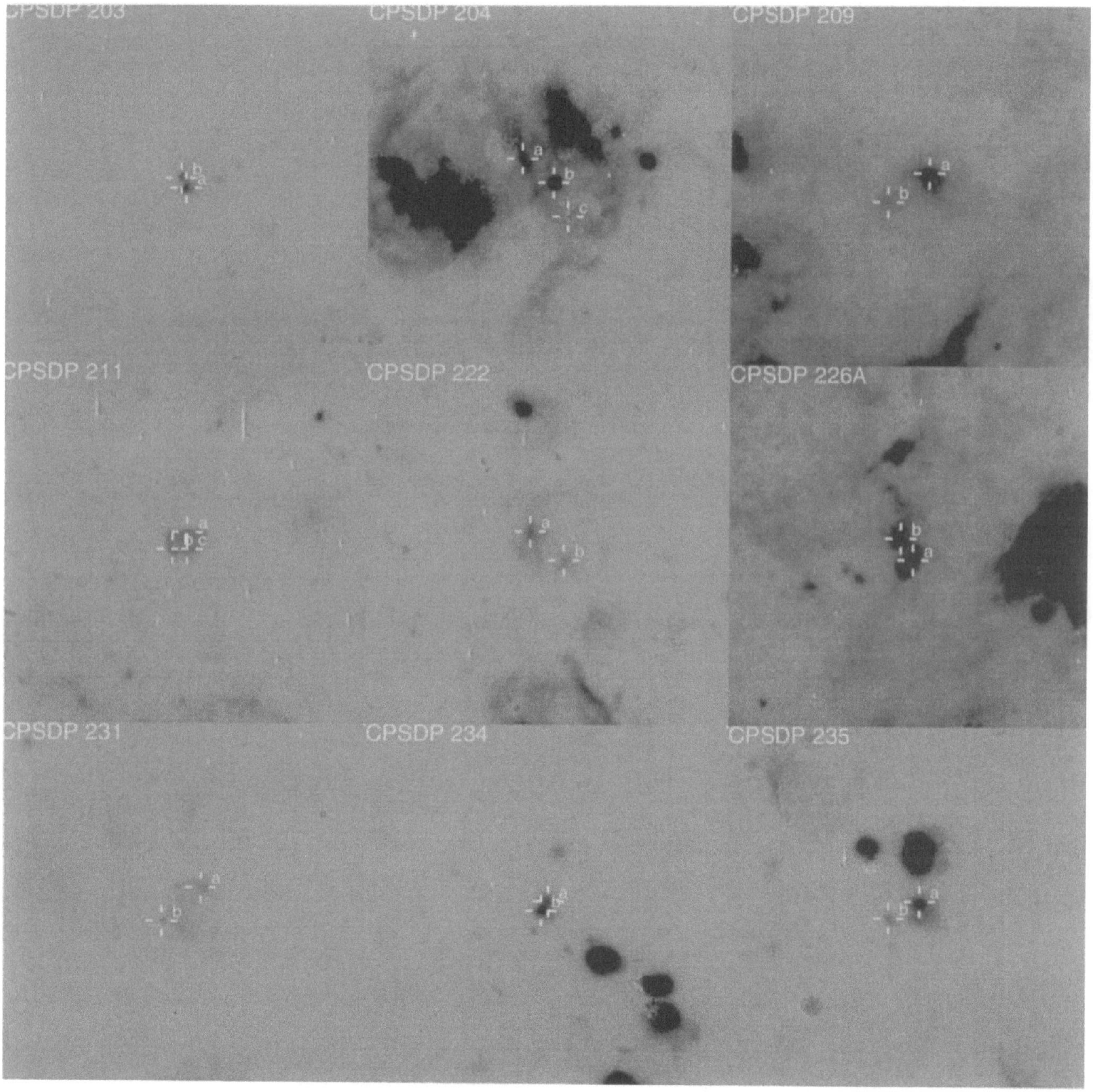
CPSDP 203
b
a
CPSDP 204
a
b
c
CPSDP 209
a
b
CPSDP 211
a
b
c
CPSDP 222
a
b
CPSDP 226A
b
a
CPSDP 231
a
b
CPSDP 234
a
CPSDP 235
a
b

M33

Chart 80

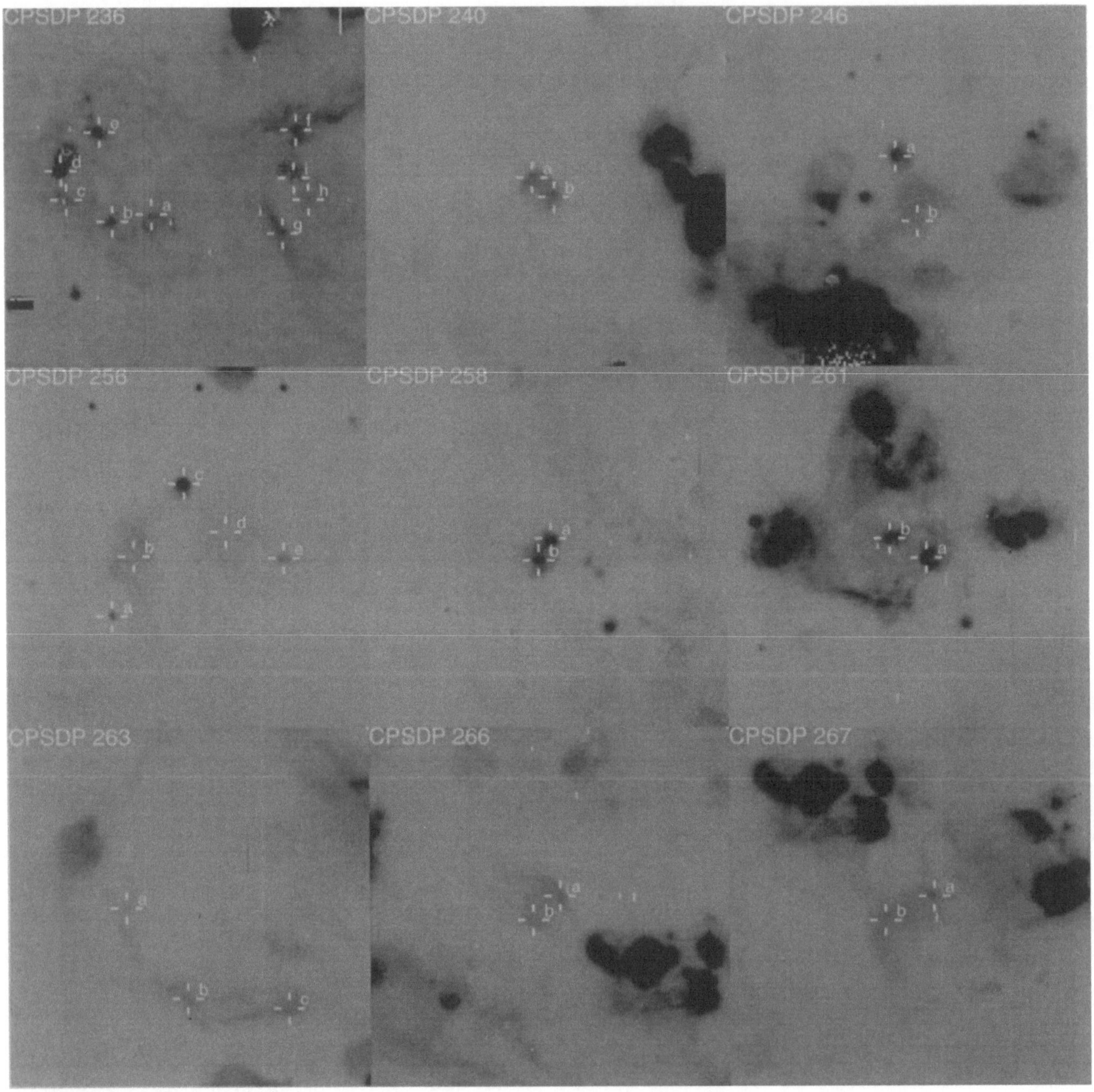

CPSDP 236
CPSDP 240
CPSDP 246
CPSDP 256
CPSDP 258
CPSDP 261
CPSDP 263
CPSDP 266
CPSDP 267

M33

Chart 81

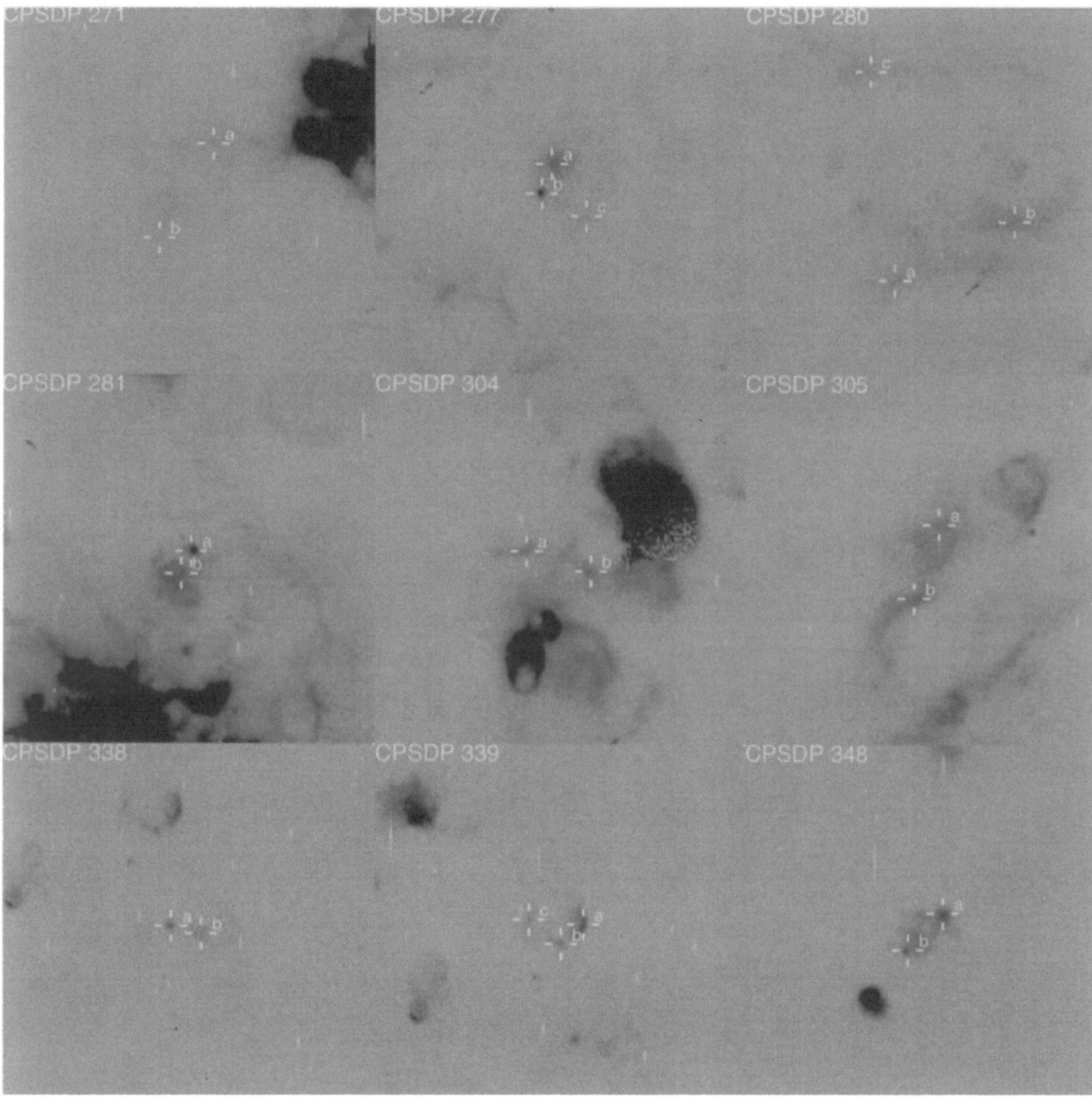
CPSDP 271
CPSDP 277
CPSDP 280
CPSDP 281
CPSDP 304
CPSDP 305
CPSDP 338
CPSDP 339
CPSDP 348

M33

Chart 82

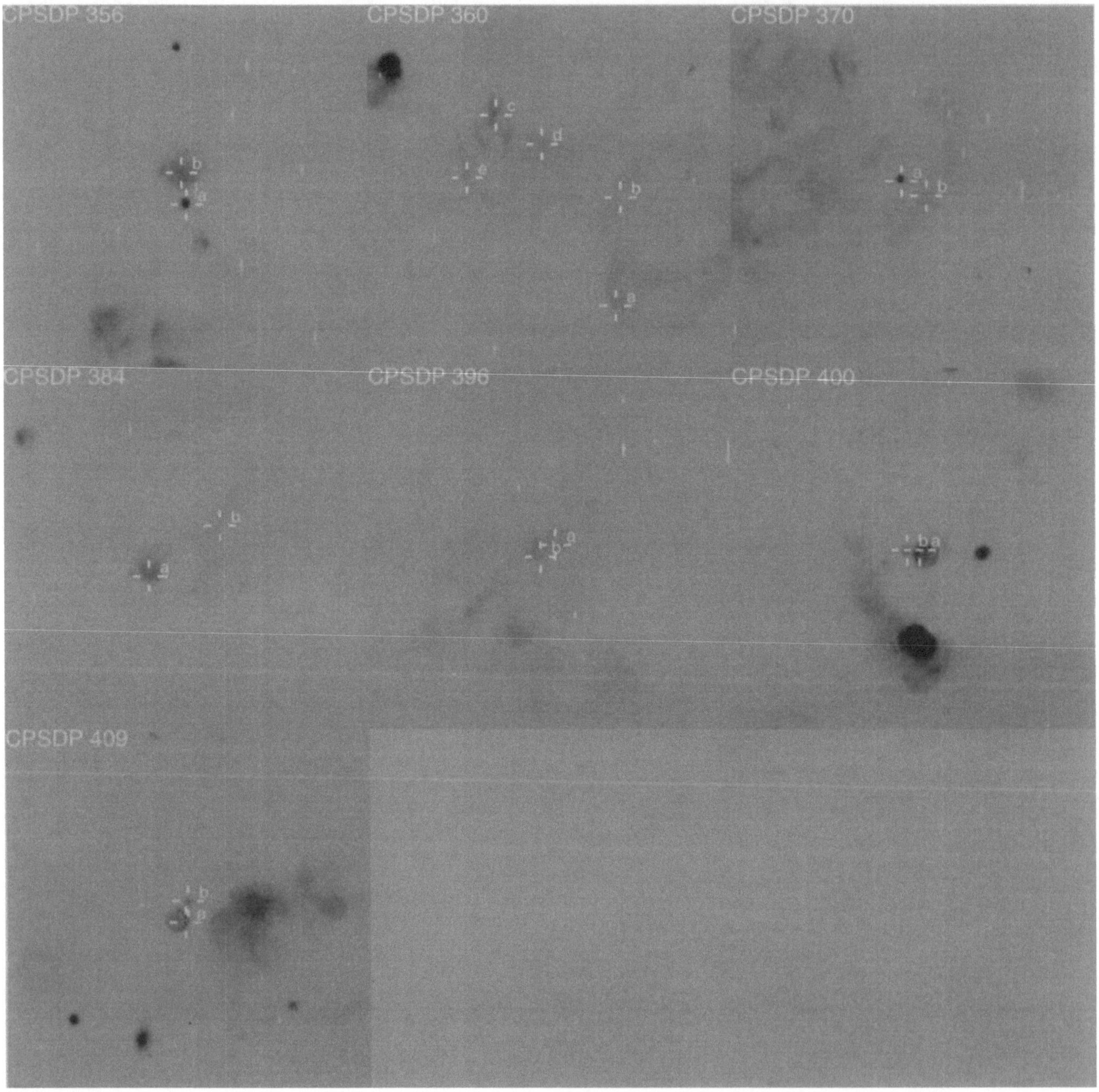
CPSDP 356
CPSDP 360
CPSDP 370
CPSDP 384
CPSDP 396
CPSDP 400
CPSDP 409

M33

Chart 83

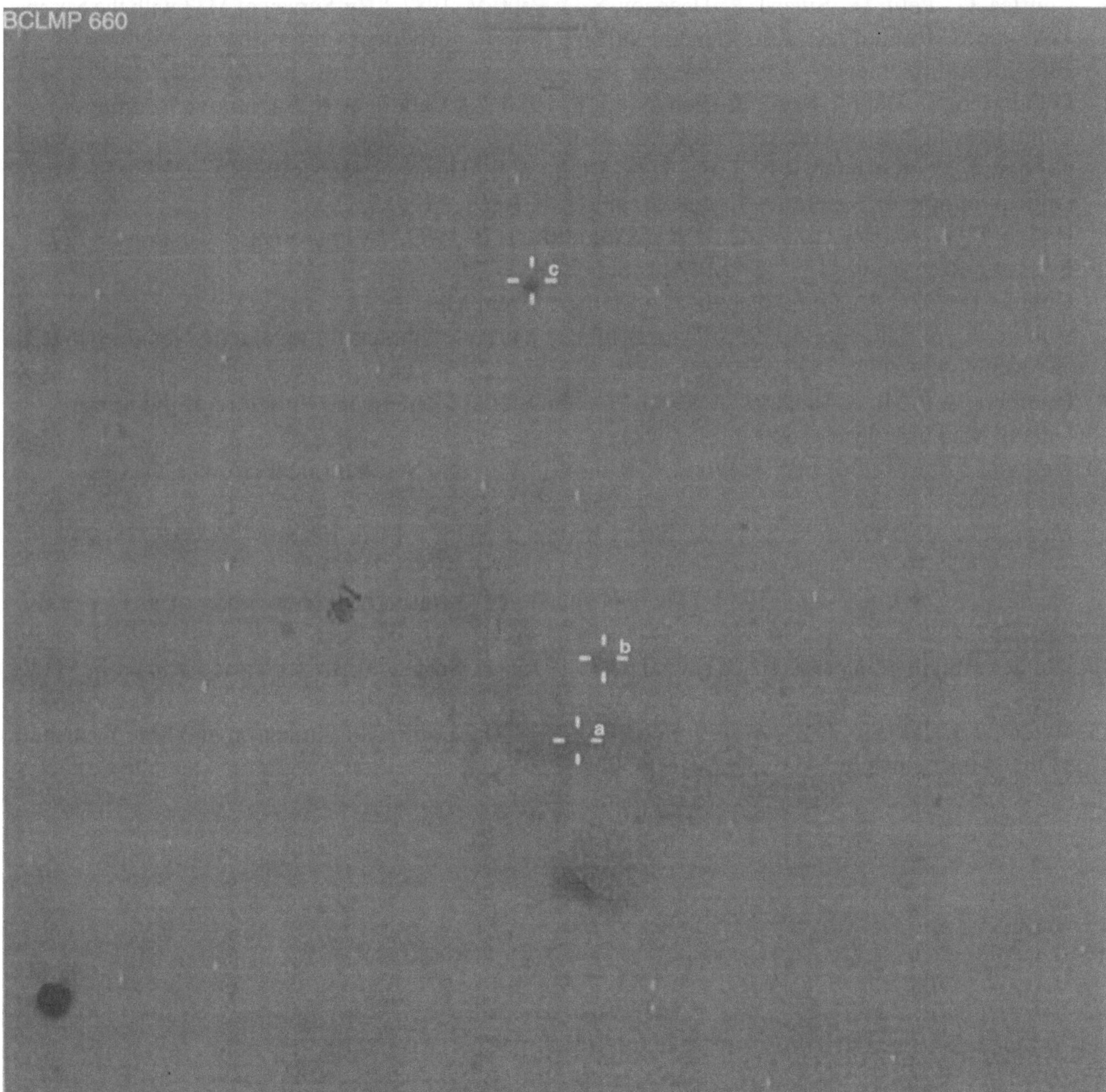

M33

References

1. Boulesteix, J., Courtès, G., Laval, A., Monnet, G., & Petit, H. 1974, "An Optical Study of M33: I. Morphology of the Gas", A&A, 37, 33.
2. Christian, C.A., & Schommer, R.A. 1982, "The Cluster System of M33", ApJS, 49, 405.
3. Courtès, G., Petit, H., SivanJ.-P., Dodonov, S., & Petit, M. 1987, "Hα Survey of M33 with the Six-meter Telescope: Morphology of the General Diffuse Emission, Evidence for a Chaotic Medium of Bubbles and Filaments", A&A, 174, 28.
4. D'Odorico, S., Dopita, M.A., & Bienvenuti, P. 1980, "A Catalogue of Supernova Remnant Candidates in Nearby Galaxies", A&AS, 40, 67.
5. Gordon, S.M., Kirshner, R.P., Long, K.S., Blair, W.P., Duric, N., & Smith, R.C. 1998, "A New Optical Sample of Supernova Remnants in M33", ApJS, 117, 89.
6. Hodge, P.W., Balsley, J., Wyder, T.K., & Skelton, B.P. 1999, "A Photometric Catalog of 1272 Emission Regions in M33", PASP, 111, 685.
7. Hubble, E. 1926, "A Spiral Nebula as a Stellar System Messier 33", ApJ, 63, 236
8. Hubble, E., & Sandage, A. 1953, "The Brightest Variable Stars in Extragalactic Nebulae. I. M31 and M33", ApJ, 118, 353.
9. Humphreys, R.M., & Sandage, A. 1980, "On the Stellar Content and Structure of the Spiral Galaxy M33", ApJS, 44, 319.
10. Kinman, T.D., Mould, J.R., & Wood, P.R. 1987, "Variable Stars in Local Group Galaxies. I. M33", AJ, 93, 833.
11. Melnick, J., & D'Odorico, S. 1978, "Star Clusters in M33. I. Identification", A&AS, 34, 249.
12. Sabbadin, F. 1979, "Supernova Remnants in M33", A&A, 80, 212.
13. Sandage, A., & Carlson, C. 1983, "The Distance to M33 Based on a New Study of its Cepheids", ApJ, 267, L25.
14. van den Bergh, S., Herbst, E., & Kowal, C.T. 1975, "A Survey of Bright Variable Stars in M33", ApJS, 29, 303.
15. Wyder, T.K., Hodge, P.W., & Skelton, B.P. 1997, "The Luminosity Function and Size Distribution of the H II Regions in M33", PASP, 109, 927.

PART II

THE ATLAS OF DWARF GALAXIES IN THE LOCAL GROUP

And I

Alternate name(s): Andromeda I
Right Ascension (2000): 00h 45m 43s
Declination (2000): +38d 00.4m
Type: dE0
Apparent magnitude (V): 12.75
Color (B-V): 0.75
(U-B): 0.29
(V-R):
Color Excess, E(B-V): 0.04
Absolute magnitude (M_V): -11.8
Distance (kpc): 805
Radial velocity (solar, km/sec):

Objects Identified On the Atlas:
Globular clusters: none
Source:
Open clusters: none
Source:
OB associations: none
Source:
Variable stars: none
Source:
Carbon stars: none
Source:
HII regions: none
Source:
Planetary nebulae: none
Source:
Dust clouds: none
Source:

Andromeda I

Chart 84

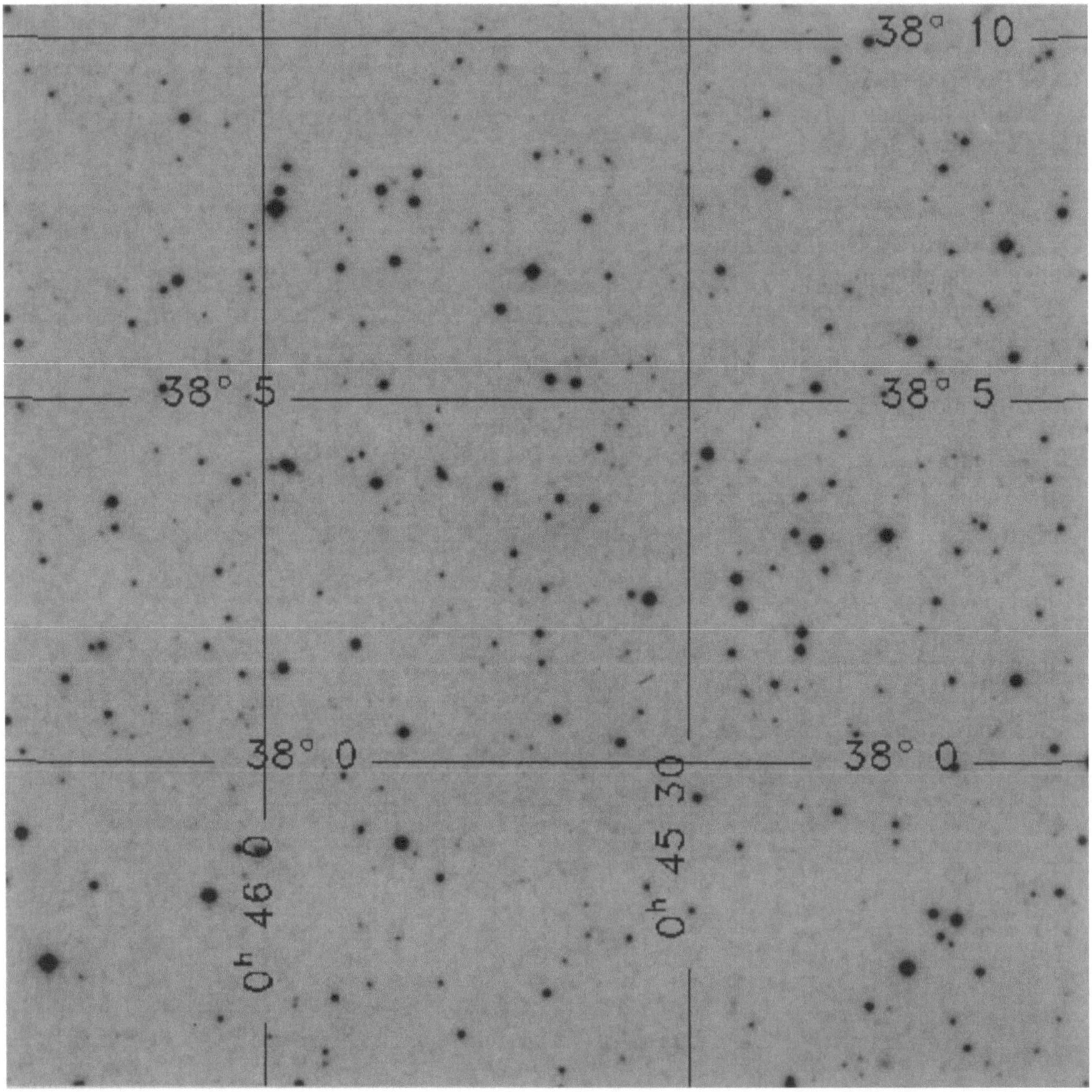

Andromeda I

Chart 85

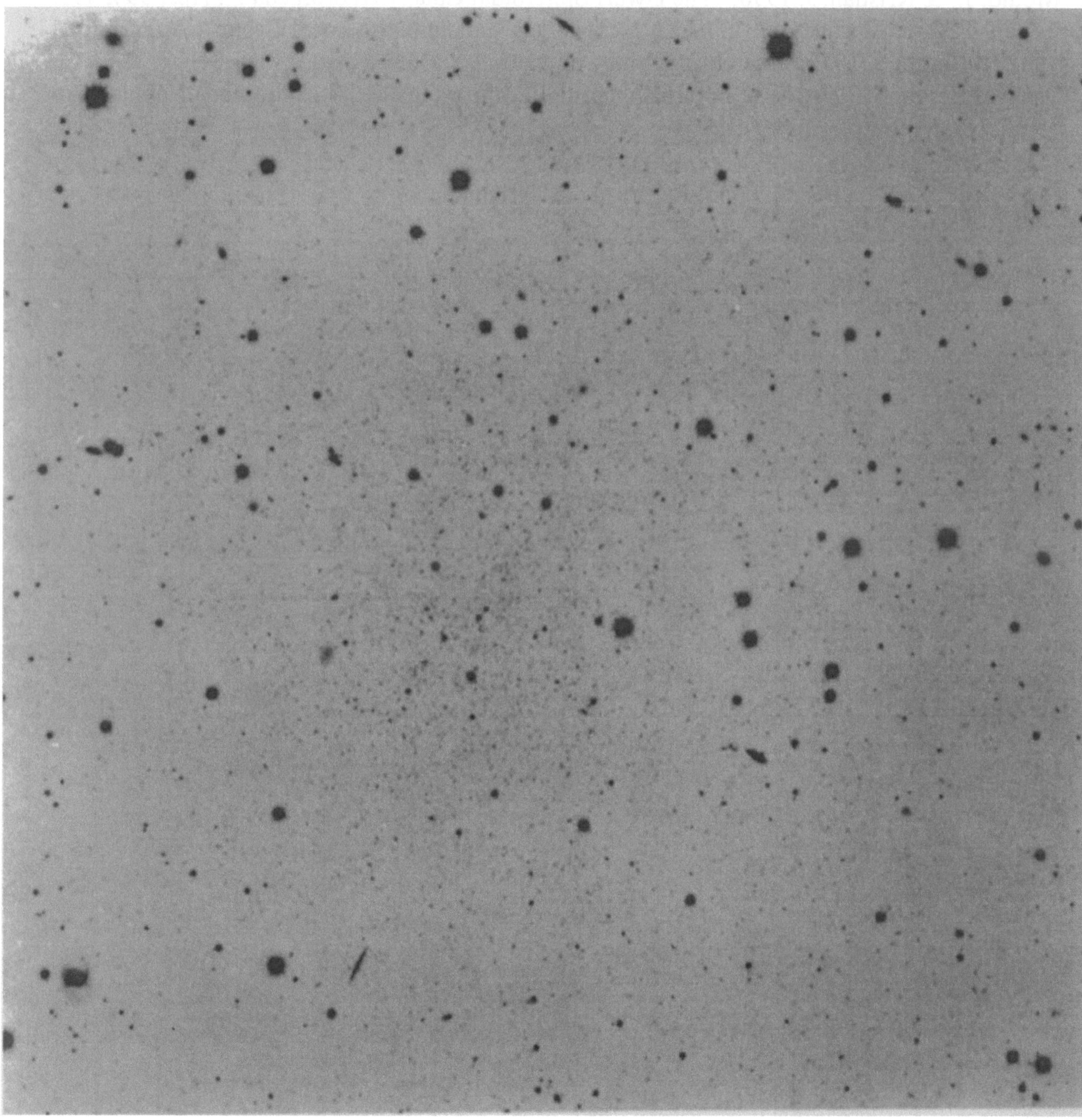

Andromeda I

References

1. Caldwell, N., Armandroff, T.E., Seitzer, P., & Da Costa, G.S. 1992, "The Dwarf Spheroidal Companions to M31: Surface Brightness Profiles", AJ, 103, 840.
2. Da Costa, G.S., Armandroff, T.E., Caldwell, N., & Seitzer, P. 1996, "The Dwarf Spheroidal Companions to M31: WFPC2 Observations of Andromeda I", AJ, 112, 2576.
3. Mould, J., & Kristian, J. 1990, "The Dwarf Spheroidal Galaxy Andromeda I", ApJ, 354, 438.
4. Thuan, T.X., & Martin, G.E. 1979, "A New Dwarf Galaxy in the Local Group", ApJ, 232, L11.
5. van den Bergh, S. 1972, "Search for Faint Companions to M31", ApJ, 171, L31.
6. van den Bergh, S. 1974, "The Dwarf Spheroidal Companions to the Andromeda Nebula", ApJ, 191, 271.

And II

Alternate name(s): Andromeda II
Right Ascension (2000): 01h 16m 11s
Declination (2000): +33d 21.7m
Type: dE3
Apparent magnitude (V): 12.7
Color (B-V):
(U-B):
(V-R):
Color Excess, E(B-V): 0.08
Absolute magnitude (M_V): -11.1
Distance (kpc): 525
Radial velocity (solar, km/sec):

Objects Identified On the Atlas:
Globular clusters: none
Source:
Open clusters: none
Source:
OB associations: none
Source:
Variable stars: none
Source:
Carbon stars: none
Source:
HII regions: none
Source:
Planetary nebulae: none
Source:
Dust clouds: none
Source:

Andromeda II

Chart 86

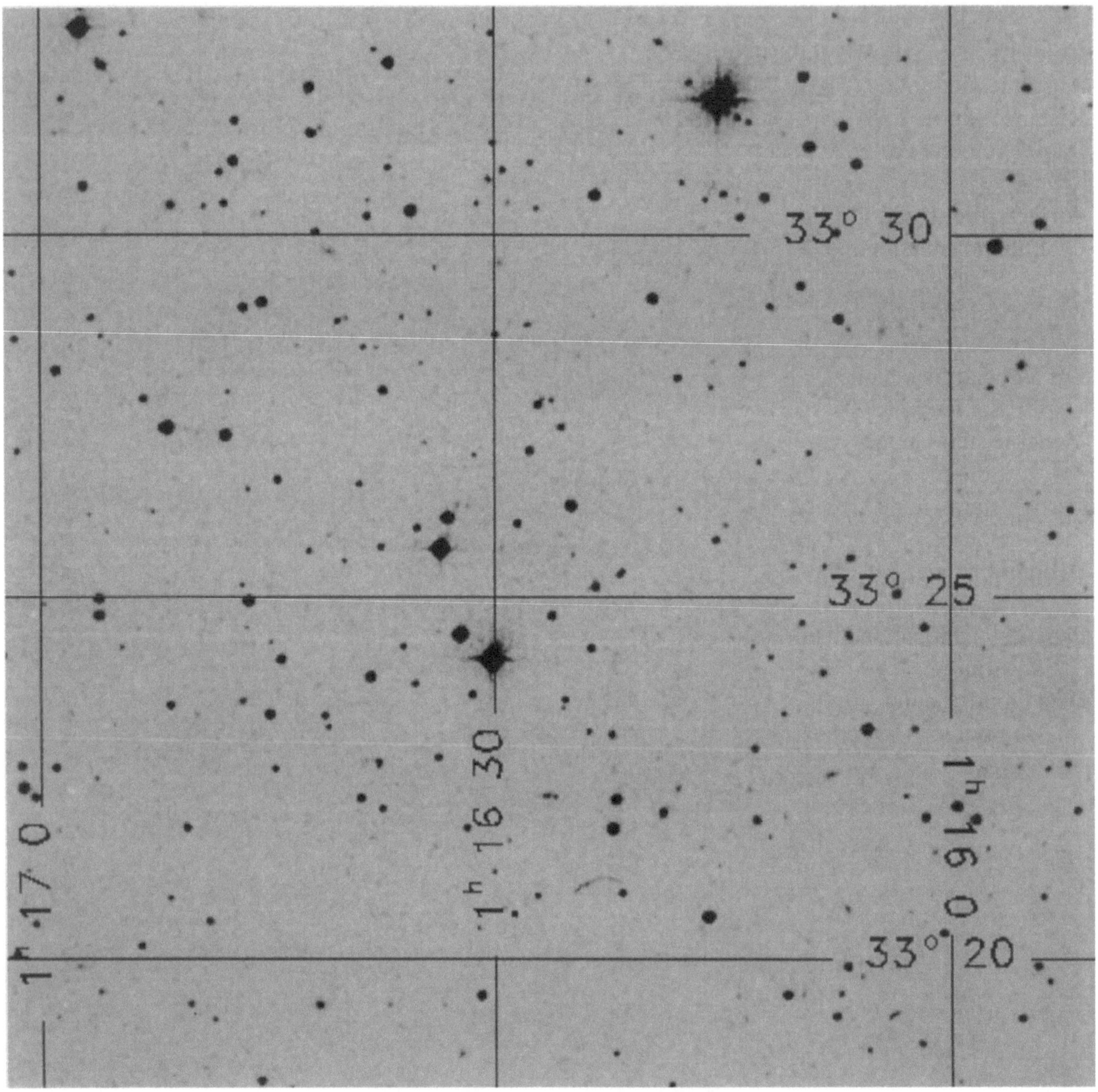

Andromeda II

Chart 87

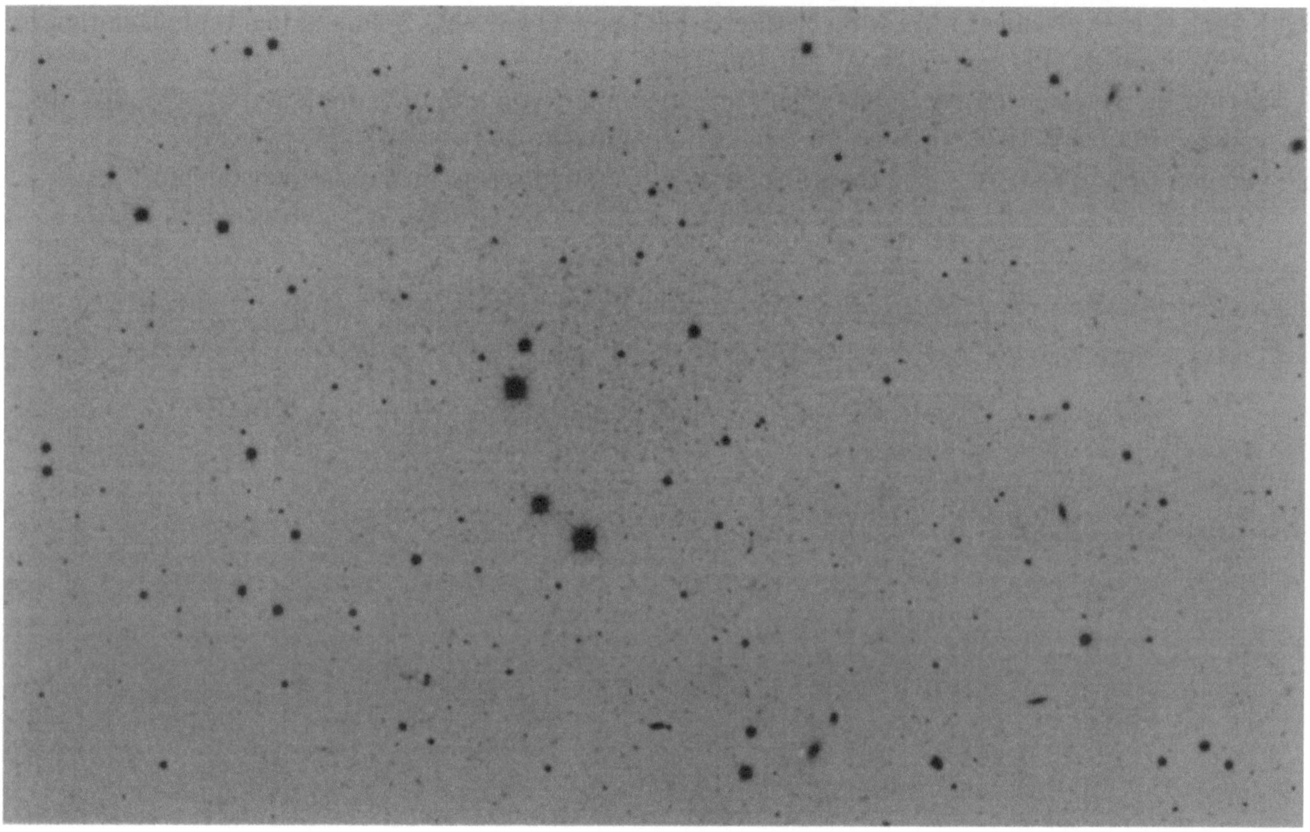

Andromeda II

References

1. Aaronson, M., Gordon, G., Mould, J., Olszewski, E., & Suntzeff, N. 1985, "The Extended Branch of the Andromeda II Dwarf Spheroidal Galaxy", ApJ, 296, L7.
2. Caldwell, N., Armandroff, T.E., Seitzer, P., & Da Costa, G.S. 1992, "The Dwarf Spheroidal Companions to M31: Surface Brightness Profiles", AJ, 103, 840.
3. König, C.H.B., Nemec, J.M., Mould, J.R., & Fahlman, G.G. 1993, "Color-Magnitude Diagram for the Andromeda II Dwarf Galaxy", AJ, 106, 1819.
4. Mould, J., & Aaronson, M. 1986, "The Formation and Evolution of Carbon Stars", ApJ, 303, 10.
5. van den Bergh, S. 1972, "Search for Faint Companions to M31", ApJ, 171, L31.
6. van den Bergh, S. 1974, "The Dwarf Spheroidal Companions to the Andromeda Nebula", ApJ, 191, 271.

And III

Alternate name(s): Andromeda III
Right Ascension (2000): 00h 35m 17s
Declination (2000): +36d 30.5m
Type: dE6
Apparent magnitude (V): 14.21
Color (B-V):
(U-B):
(V-R):
Color Excess, E(B-V): 0.05
Absolute magnitude (M_V): -10.2
Distance (kpc): 760
Radial velocity (solar, km/sec):

Objects Identified On the Atlas:
Globular clusters: none
Source:
Open clusters: none
Source:
OB associations: none
Source:
Variable stars: none
Source:
Carbon stars: none
Source:
HII regions: none
Source:
Planetary nebulae: none
Source:
Dust clouds: none
Source:

Andromeda III

Chart 88

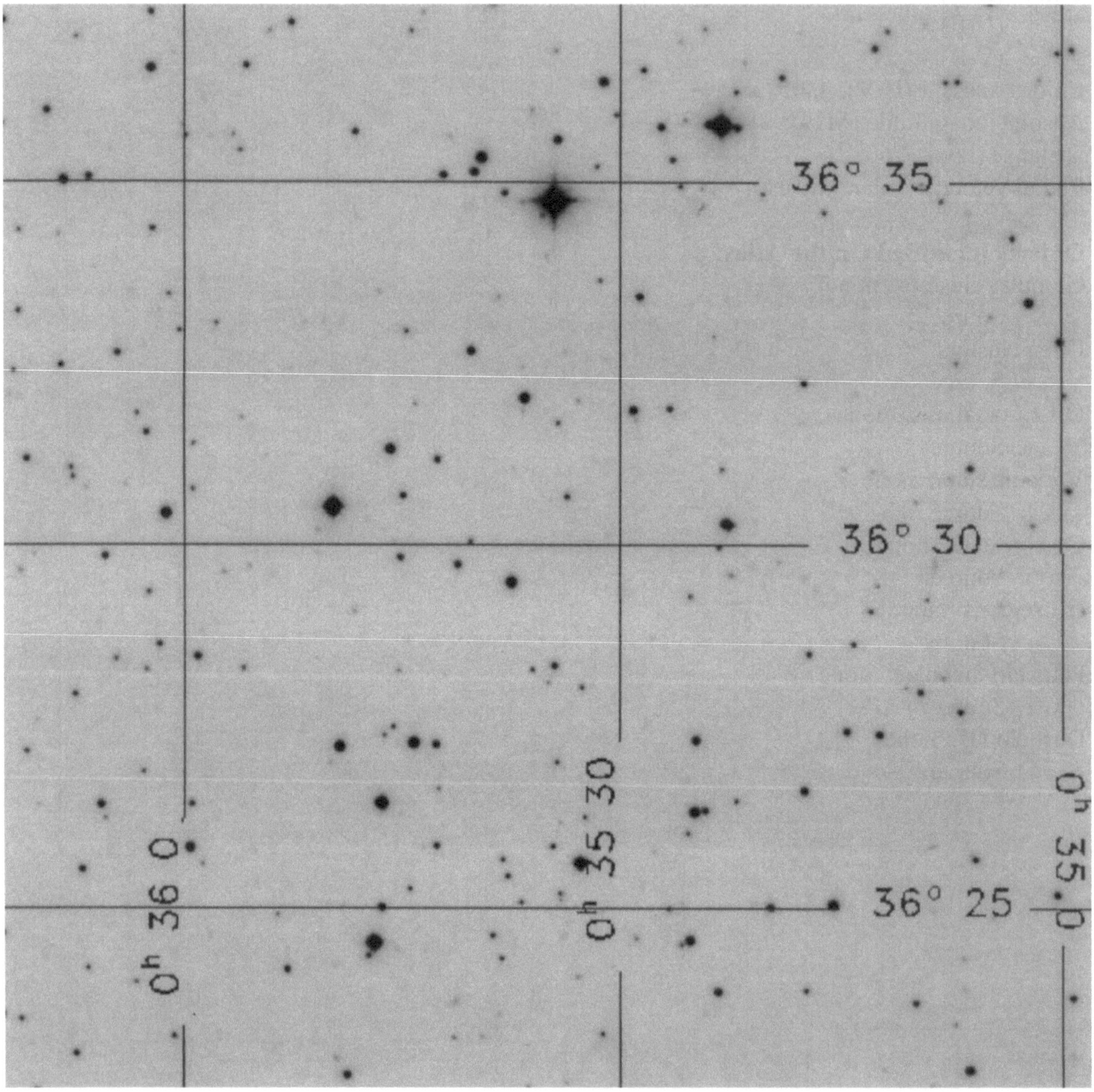

Andromeda III

Chart 89

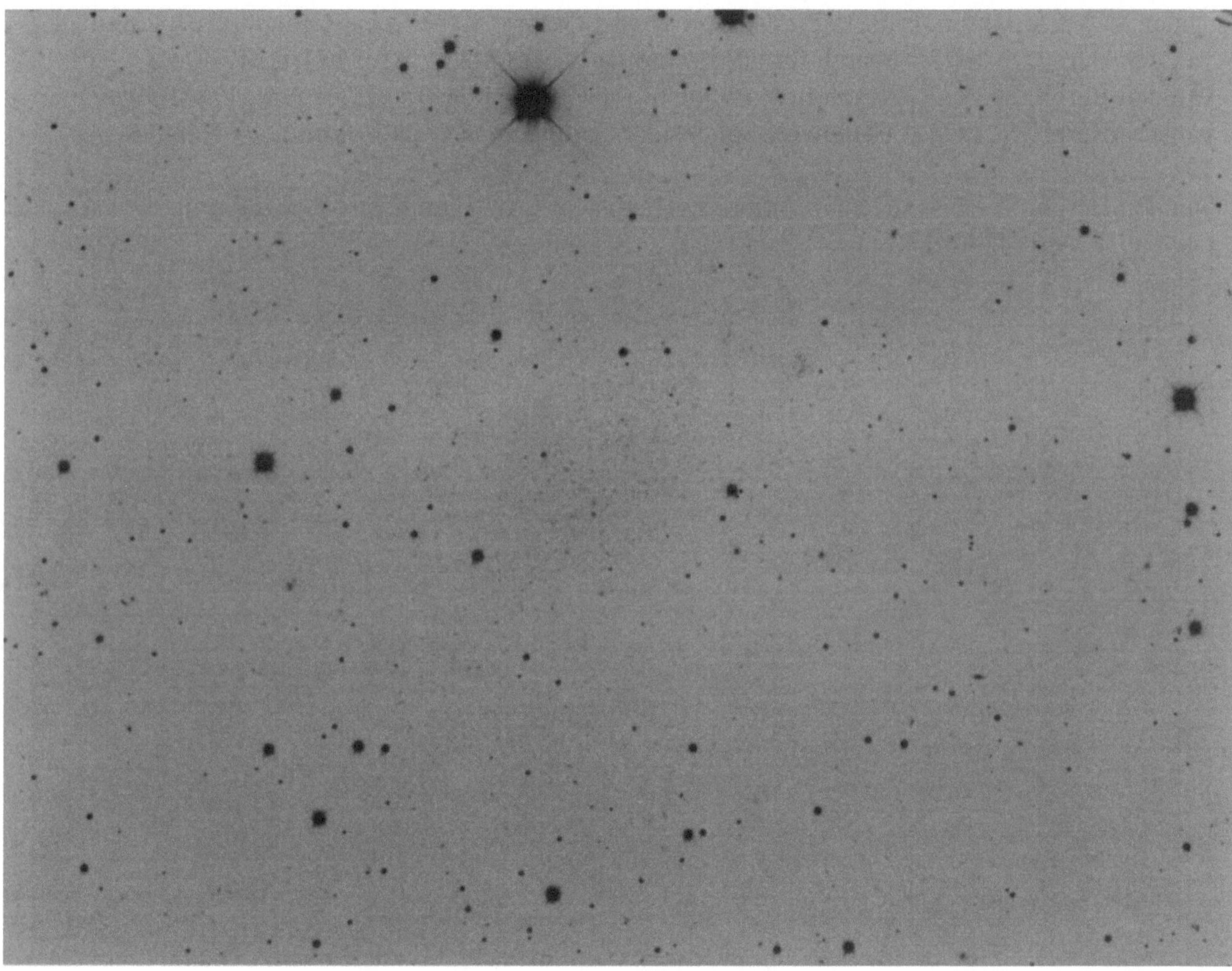

Andromeda III

References

1. Armandroff, T.E., Da Costa, G.S., Caldwell, N., & Seitzer, P. 1993, "The Dwarf Spheroidal Companions to M31: A Color-Magnitude Diagram for And III", AJ, 106, 986.
2. Caldwell, N., Armandroff, T.E., Seitzer, P., & Da Costa, G.S. 1992, "The Dwarf Spheroidal Companions to M31: Surface Brightness Profiles", AJ, 103, 840.
3. Thuan, T.X., & Martin, G.E. 1979, "A New Dwarf Galaxy in the Local Group", ApJ, 232, L11.
4. van den Bergh, S. 1972, "Search for Faint Companions to M31", ApJ, 171, L31.
5. van den Bergh, S. 1972, "Resolution of One of the Companions to M31", ApJ, 178, L99.
6. van den Bergh, S. 1974, "The Dwarf Spheroidal Companions to the Andromeda Nebula", ApJ, 191, 271.
7. van den Bergh, S. 1994, "The Evolutionary History of Low-Luminosity Local Group Dwarf Galaxies", ApJ, 428, 617.

Antlia

Alternate name(s): AM 1001-270
Right Ascension (2000): 10h 04m 04s
Declination (2000): -27d 19.8m
Type: dE/Irr
Apparent magnitude (V): 14.8
Color (B-V):
(U-B):
(V-R): 0.05
Color Excess, E(B-V): 0.05
Absolute magnitude (M_V): -10.7
Distance (kpc): 1235
Radial velocity (solar, km/sec): 361

Objects Identified On the Atlas:
Globular clusters: none
Source:
Open clusters: none
Source:
OB associations: none
Source:
Variable stars: none
Source:
Carbon stars: none
Source:
HII regions: none
Source:
Planetary nebulae: none
Source:
Dust clouds: none
Source:

Antlia

Chart 90

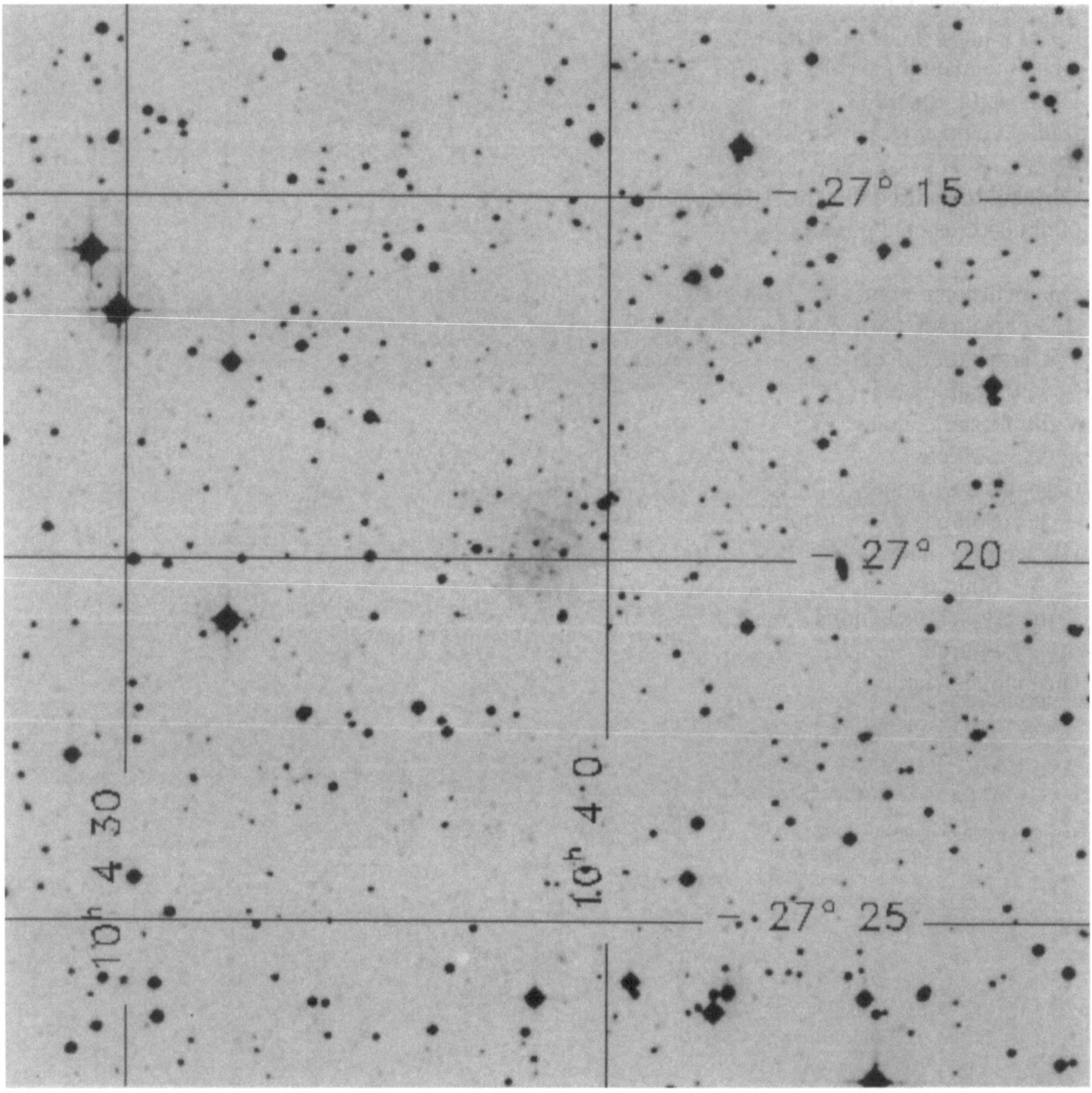

Antlia

Chart 91

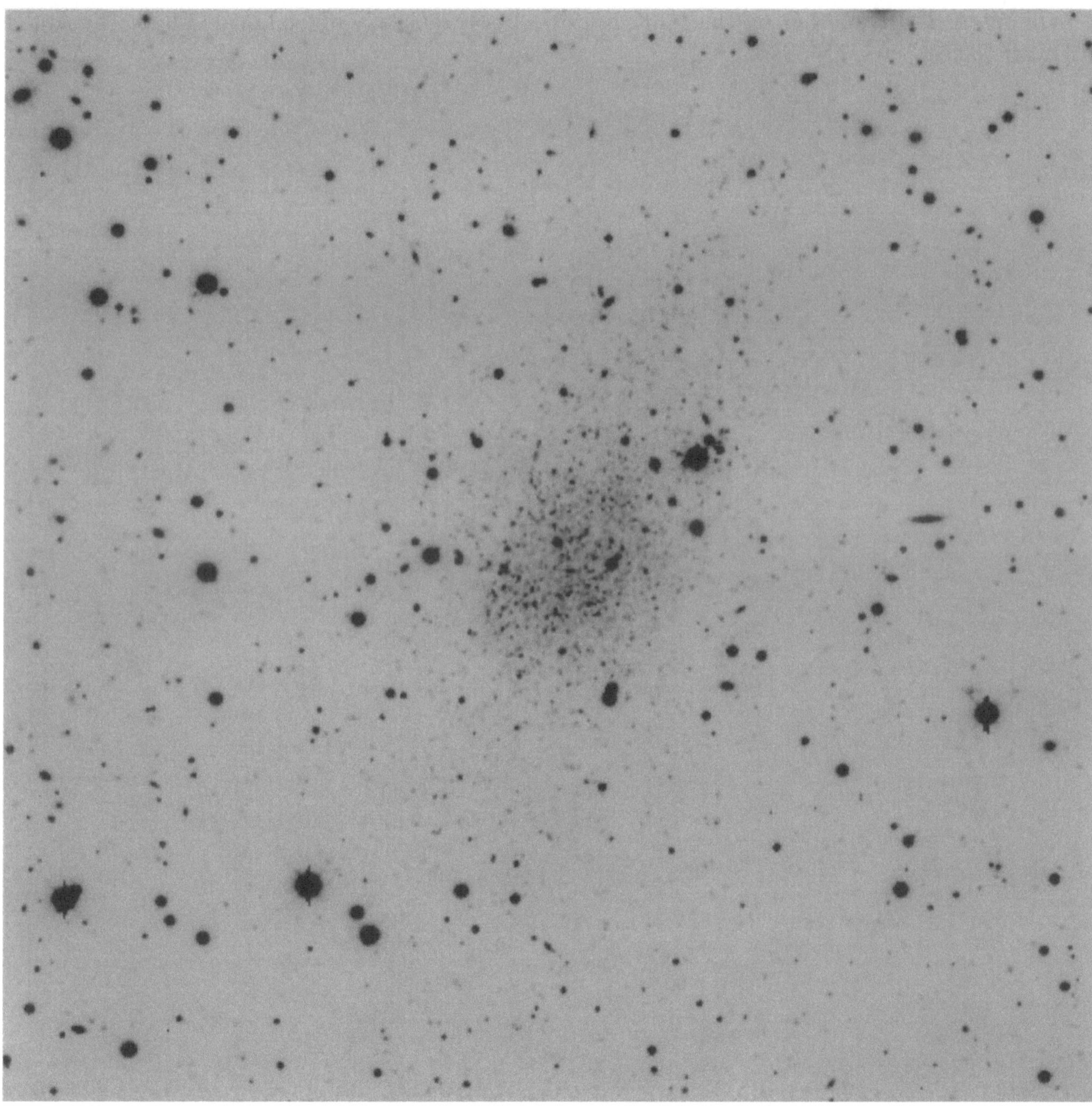

Antlia

References

1. Aparicio, A., Dalcanton, J.J., Gallart, C., Martinez-Delgado, D. 1997, "The Nature of the Antlia Galaxy: A New Dwarf Irregular in the Outskirts of the Local Group", AJ, 114, 1447.
2. Sarajedini, A., Claver, C.V., & Ostheimer, J.C., Jr. 1997, "The Properties of the Antlia Dwarf Galaxy", AJ, 114, 2505.
3. Whiting, A.B., Irwin, M.G., & Hau, G.K.T. 1997, "A New Galaxy in the Local Group: The Antlia Dwarf Galaxy", AJ, 114, 996.

Carina

Alternate name(s): AM 0640-505
Right Ascension (2000): 06h 41m 37s
Declination (2000): -50d 58.0m
Type: dE4
Apparent magnitude (V): 10.85
Color (B-V):
(U-B):
(V-R):
Color Excess, E(B-V): 0.04
Absolute magnitude (M_V): -9.2
Distance (kpc): 101
Radial velocity (solar, km/sec): 223

Objects Identified On the Atlas:
Globular clusters: none
Source:
Open clusters: none
Source:
OB associations: none
Source:
Variable stars: none
Source:
Carbon stars: 11; designation: C
Source: [(2], [(3)]
HII regions: none
Source:
Planetary nebulae: none
Source:
Dust clouds: none
Source:

Carina

Chart 92

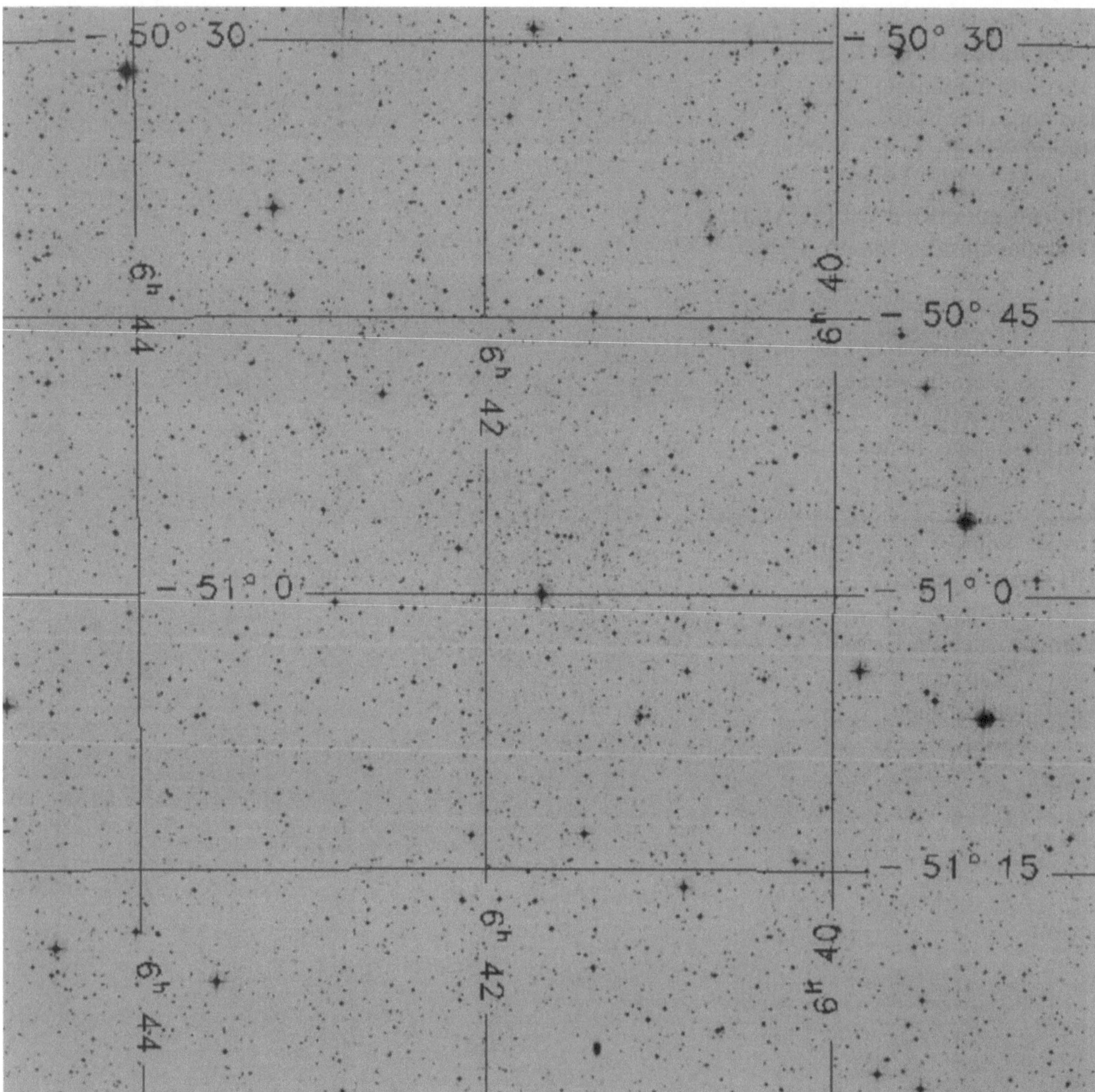

Carina

Chart 93

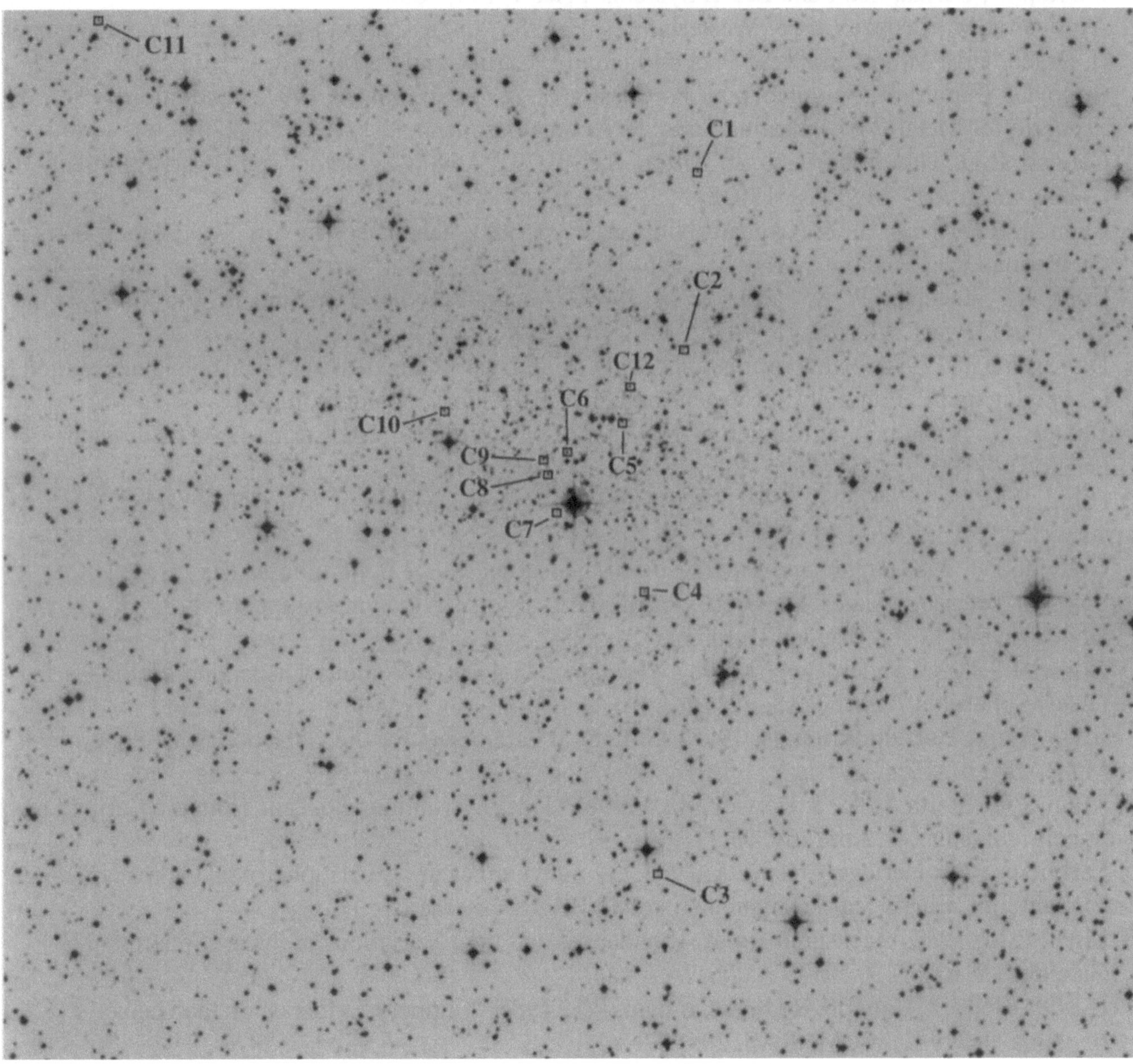

Carina

References

1. Armandroff, T.E., & Da Costa, G.S. 1991, "Metallicities for Old Stellar Systems from Ca II Triplet Strengths in Member Giants", AJ, 101, 1329.
2. Azzopardi, M., Lequeux, J., & Westerlund, B.E. 1985, "New Carbon Stars in Spheroidal Galaxies. I. Sculptor, Carina, Leo I and Leo II Systems", A&A, 144, 388.
3. Azzopardi, M., Lequeux, J., & Westerlund, B.E. 1985, "New Carbon Stars in Spheroidal Galaxies. II. Draco, Ursa Minor, and New Data on Sculptor, Carina, and Leo I", A&A, 161, 232.
4. Bono, G., Caputo, F., Santolamazza, P., Cassisi, S., & Piersimoni, A. 1997, "Evolutionary Scenario for Metal-Poor Pulsating Stars. II. Anomalous Cepheids", AJ, 113, 2209.
5. Cannon, R.D., Hawarden, T.G., & Tritton, S.B. 1977, "A New Sculptor-type Dwarf Elliptical Galaxy in Carina", MNRAS, 180, 81P
6. Cannon, R.D., Niss, B., & Nørgaard-Nielson, H.U. 1980, "The Discovery of Two Carbon Stars in the Carina Dwarf Galaxy", MNRAS, 196, 1P
7. Cook, K., Schechter, P., & Aaronson, M. 1983, "Accurate Radial Velocities for Carbon Stars in the Carina Dwarf", BAAS, 15, 907.
8. Da Costa, G.S. 1994, "New results for Local Group Dwarf Galaxies", in Dwarf Galaxies, edited by G. Meylan and P. Prugniel (European Southern Observatory, Munich), p. 221.
9. Demers, S., Beland, S., & Kunkel, W.E. 1983, "The Structure of the Carina Dwarf Elliptical Galaxy", PASP, 95, 354.
10. Faber, S.M., & Lin, D.N.C. 1983, "Is there Non-Luminous Matter in Dwarf Spheroidal Galaxies?", ApJ, 266, L17.
11. Godwin, P.J. 1985, "Core and Tidal Radii of the Carina Dwarf Spheroidal Galaxy from UK Schmidt Telescope Plates", in Dynamics of Star Clusters, IAU Symposium No. 113, edited by J. Goodman and P. Hut (Reidel, Dordrecht), p. 77.
12. Godwin, P.J., & Lynden-Bell, D. 1987, "Evidence of a Small Velocity Dispersion for the Carina Dwarf Spheroidal?", MNRAS, 22, 7P
13. Irwin, M., & Hatzidimitriou, D. 1995, "Structural Parameters for the Galactic Dwarf Spheroidals", MNRAS, 277, 1354.
14. Kuhn, J.R., Smith, H.A., & Hawley, S.L. 1996, "Tidal Disruption and Tails from the Carina Dwarf Spheroidal Galaxy", ApJ, 469, 93.
15. Lariviere, P., & Noriega-Crespo, A. 1993, "Gas Dynamics in Dwarf-Spheroidal Galaxies: Explaining Carina's Star Formation History", BAAS, 25, 1451.
16. Light, R.M. 1987, "Measurement of Abundances and their Variations in Dwarf Spheroidal Galaxies", BAAS, 19, 1033.
17. Lynden-Bell, D., Cannon, R.D., & Godwin, P.J.. 1983, "Slippery Evidence on the Galaxy's Invisible Heavy Halo", MNRAS, 204, 87P
18. Mateo, M., Olszewski, E.W., Pryor, C., Welch, D.L., & Fischer, P. 1993, "The Carina Dwarf Spheroidal Galaxy: How Dark Is It?", AJ, 105, 510.
19. McNamara, D.H., 1995, "Period-Luminosity Relations of SX Phoenicis Stars", AJ, 109, 1751.
20. Mighell, K.J. 1990, "The Carina Dwarf Spheroidal – An Old Galaxy with a Large Intermediate-Age Population", A&AS, 82, 1.

21. Mighell, K.J. 1990, "The Stellar Luminosity Function for the Carina Dwarf Spheroidal Galaxy", A&AS, 82, 207.
22. Mighell, K.J. 1992, "The Star-formation History of the Carina Dwarf Spheroidal Galaxy", Proc. Ast. Soc. Aust., 10, 83.
23. Mighell, K.J. 1997, "WFPC2 Observations of the Carina Dwarf Spheroidal Galaxy", AJ, 114, 1458.
24. Mighell, K.J., & Butcher, H.R. 1992, "Star Formation Histories of Galaxies from their Stellar Luminosity Functions", A&A, 255, 26.
25. Mould, J., & Aaronson, M. 1983, "The Carina Dwarf Spheroidal – An Intermediate Age Galaxy", ApJ, 273, 530.
26. Mould, J.R., Bothun, G.D., Hall. P.J., Staveley-Smith, L., & Wright, A.E. 1990, "An Upper Limit on the H I Content of the Carina Dwarf Spheroidal Galaxy", ApJ, 362, L55.
27. Mould, J.R., Cannon, R.D., Aaronson, M., & Frogel, J.A. 1982, "Carbon Stars in the Carina Dwarf Spheroidal Galaxy", ApJ, 254, 500.
28. Nemec, J., & Mateo, M. 1990, "Dwarf Cepheids in the Carina Dwarf Galaxy", in Evolution of the Universe of Galaxies, ASP Conf. Ser. 10, edited by R.C. Kron (Astronomical Society of the Pacific, San Francisco), p. 134.
29. Nemec, J.M., Nemec, A.F.L., & Lutz, T.E. 1994, "Period-Luminosity-Metallicity Relations, Pulsation Modes, Absolute Magnitudes and Distances for Population II Variable Stars", AJ, 108, 222.
30. Olszewski, E.W., Peterson, R.C., & Aaronson, M. 1986, "High Precision Radial Velocities for Faint Giants. Radial Velocities of Extreme Halo Systems and the Mass of the Galaxy", ApJ, 302, L45.
31. Saha, A., Monet, D.G., & Seitzer, P. 1986, "RR Lyrae Stars in the Carina Dwarf Galaxy", AJ, 92, 302.
32. Seitzer, P., Da Costa, G.S., Armandroff, T., Frogel, J., & Light, R. 1990, "The Mass of the Carina Dwarf Galaxy", BAAS, 22, 755.
33. Seitzer, P., & Frogel. J.A. 1985, "Radial Velocities of Carbon Stars in Three Dwarf Spheroidal Galaxies", AJ, 90, 1796.
34. Smecker-Hane, T.A., Stetson, P.B., Hesser, J.E., & Lehnert, M.D. 1994, "The Stellar Populations of the Carina Dwarf Spheroidal Galaxy. I. A New Color-Magnitude Diagram for the Giant and Horizontal Branches", AJ, 108, 507.
35. van den Bergh, S. 1994, "The Evolutionary History of Low-Luminosity Local Group Dwarf Galaxies", ApJ 428, 617.
36. Zinnecker, H. 1994, "ROSAT Observations Towards the Carina and Sculptor Dwarf Galaxies", in Dwarf Galaxies, edited by G. Meylan and P. Prugniel (European Southern Observatory, Munich), p. 231.

Draco

Alternate name(s): DDO 208
Right Ascension (2000): 17h 20 m 19s
Declination (2000): +57d 54.8m
Type: dE3
Apparent magnitude (V): 10.9
Color (B-V): 0.87
(U-B): -0.11
(V-R):
Color Excess, E(B-V): 0.03
Absolute magnitude (M_V): -8.8
Distance (kpc): 86
Radial velocity (solar, km/sec): -296

Objects Identified On the Atlas:
Globular clusters: none
Source:
Open clusters: none
Source:
OB associations: none
Source:
Variable stars: 3 anomalous Cepheids; designation: V (a 4th is beyond chart edge)
Source: [8], [54]
Carbon stars: 4; designation: C
Source: [3], [7]
HII regions: none
Source:
Planetary nebulae: none
Source:
Dust clouds: none
Source:

Draco

Chart 94

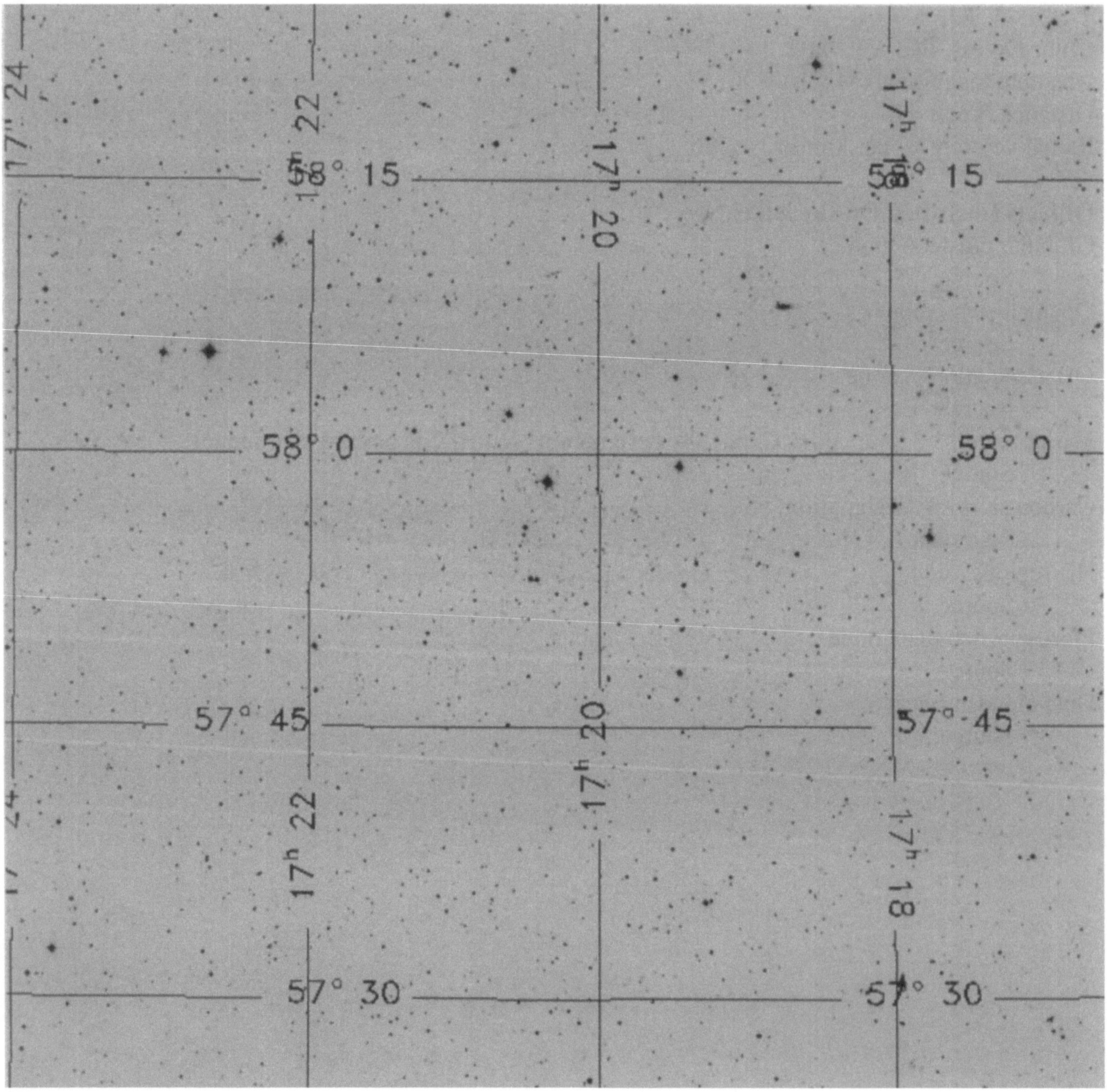

Draco

Chart 95

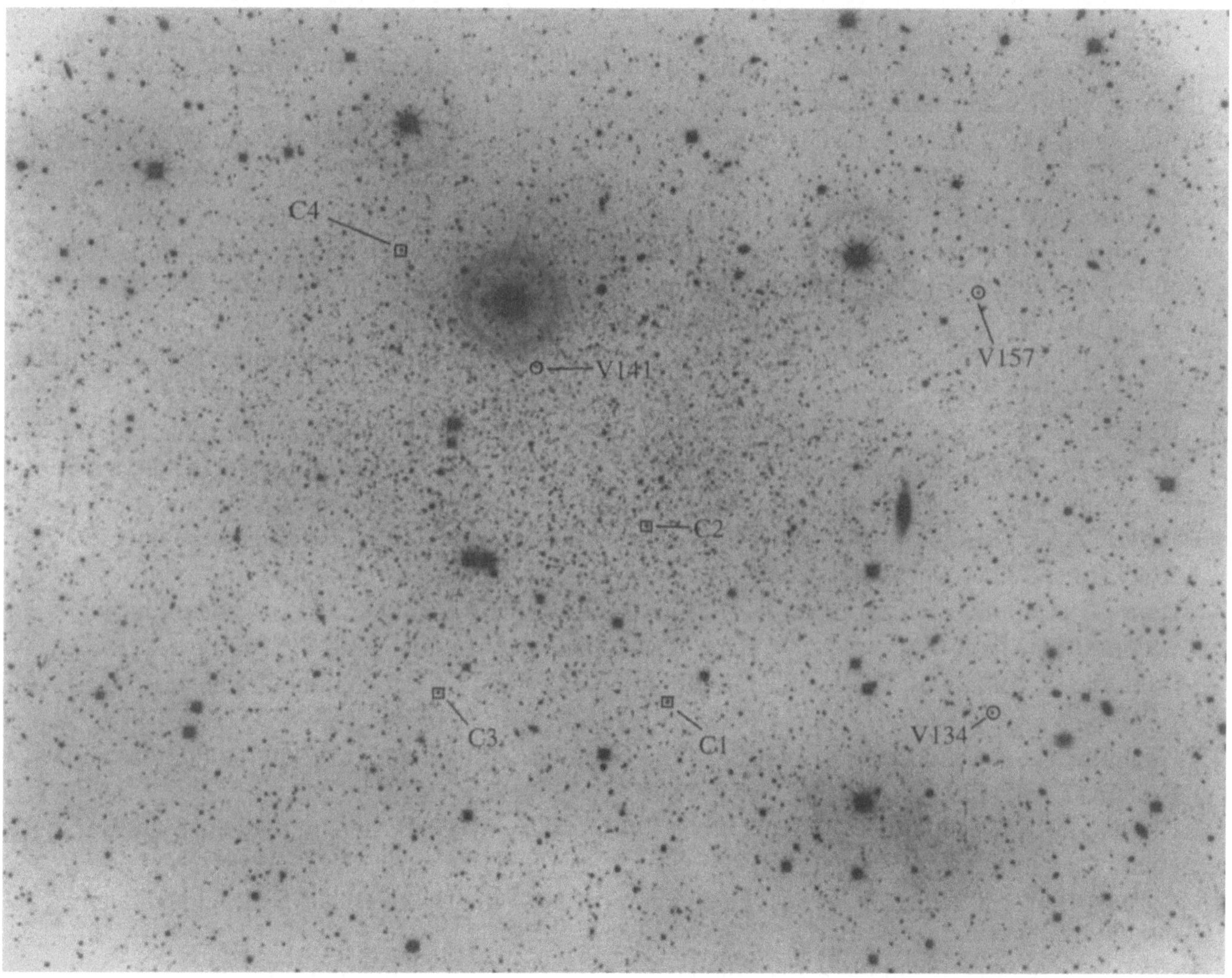

Draco

References

1. Aaronson, M. 1983, "Accurate Radial Velocities for Carbon Stars in Draco and Ursa Minor: The First Hint of a Dwarf Spheroidal Mass-to-Light Ratio", ApJ, 266, L11.
2. Aaronson, M. & Cook, K. 1983, "Weighing Draco – A Progress Report", BAAS, 15, 907.
3. Aaronson, M., Liebert, J., & Stocke, J. 1982, "Discovery of Carbon Stars in the Draco Dwarf Spheroidal Galaxy", ApJ, 254, 507.
4. Aaronson, M., & Mould, J. 1985, "Infrared Photometry and the Comparative Stellar Content of Dwarf Spheroidals in the Galactic Halo", ApJ, 290, 191.
5. Aaronson, M., & Olszewski, E. 1987, "The Search for Dark Matter in Draco and Ursa Minor: A Three-Year Progress Report", in Dark Matter in the Universe, IAU Symposium No. 117, edited by J. Kormendy and G.R. Knapp (Reidel, Dordrecht), p. 153.
6. Armandroff, T.E., Olszewski, E.W., & Pryor, C. 1995, "The Mass-to-Light Ratios of the Draco and Ursa Minor Dwarf Spheroidal Galaxies. I. Radial Velocities from Multifiber Spectroscopy", AJ, 110, 2131.
7. Azzopardi, M., Lequeux, J., & Westerlund, B.E. 1986, "New Carbon Stars in Spheroidal Galaxies. II. Draco, Ursa Minor, and New Data on Sculptor, Carina, and Leo I", A&A, 161, 232.
8. Baade, W., & Swope, H.H. 1961, "The Draco System, a Dwarf Galaxy", AJ, 66, 300.
9. Bell, R.A. 1985, "Abundances of Giant Stars in the Draco and Ursa Minor Dwarf Galaxies", PASP, 97, 219.
10. Bell, R.A., & Gustafsson, B. 1976, "Stellar Abundances in the Draco Dwarf Spheroidal Galaxy", RGO Bull., 182, 109.
11. Bono, G., Caputo, F., Santolamazza, P., Cassisi, S., & Piersimoni, A. 1997, "Evolutionary Scenario for Metal-Poor Pulsating Stars. II. Anomalous Cepheids", AJ, 113, 2209.
12. Canterna, R. 1975, "Metal Abundances of Draco and Eight Galactic Globular Clusters", ApJ, 200, L63.
13. Canterna, R., & Schommer, R.A. 1978, "The Chemical Abundances of Extreme Population II Systems", ApJ, 219, L119.
14. Caputo, F., & Castellani, V. 1971, "On the Evolutionary Stage of Stars in the Draco Dwarf Galaxy", Astrophys. & Sp. Sci., 14, 323.
15. Carney, B.W., & Seitzer, P. 1986, "Deep Photometry of the Draco Dwarf Spheroidal Galaxy", AJ, 92, 23.
16. Castellani, V. 1975, "On the Age of the Draco Dwarf Galaxy", MNRAS, 172, 59P
17. Cowley, A.P., Hartwick, F.D.A., Sargent, W.L.W. 1978, "Spectra of Giant Stars in Distant Satellites of the Galaxy", ApJ, 220, 453.
18. Demers, S., Battinelli, P., Irwin, M.J., & Kunkel, W.E. 1995, "A Search for Stellar Aggregates in Dwarf Spheroidal Galaxies", MNRAS, 274, 491.
19. Duepree, R.G., & Hodson, S.W. 1977, "The Instability Strip of the Anomalous Cepheids in the Draco Dwarf Spheroidal Galaxy", ApJ, 218, 654.
20. Faber, S.M., & Lin, D.N.C. 1983, "Is there Nonluminous Matter in Dwarf Spheroidal Galaxies?", ApJ, 266, L17.

21. Geisler, D., & Sarajedini, A. 1996, "Standards Giant Branches in the Washington System: Calibration and Application to the Draco, Ursa Minor, and Sextans Dwarf Spheroidal Galaxies", in Formation of the Galactic Halo…Inside and Out, edited by H. Morrison and A. Sarajedini (Astronomical Society of the Pacific, San Francisco), p. 524.
22. Gerhard, O.E., & Spergel, D.N. 1992, "Dwarf Spheroidal Galaxies and the Mass of the Neutrino", ApJ, 389, L9.
23. Goranskij, V.P. 1982, "Double Mode RR Lyrae Stars in the Draco Dwarf Galaxy", Astron. Tsirkulyar, 1216, 5.
24. Hargreaves, J.C., Gilmore, G., Irwin, M.J., & Carter, D. 1996, "A Dynamical Study of the Draco Dwarf Spheroidal Galaxy", MNRAS, 282, 305.
25. Hartwick, F.D.A., & McClure, R.D. 1974, "Physical Characteristics of Giant Stars in the Draco Dwarf Spheroidal Galaxy", ApJ, 193, 321.
26. Hill, J.M., Olszewski, E.W., Lesser, M.P., Aaronson, M., & Eisenhamer, J.D. 1986, "New Member Giants in Dwarf Spheroidal Galaxies Using the MX Spectrometer", BAAS, 18, 996.
27. Hodge, P.W. 1964, "Distribution of Stars in the Draco Dwarf Galaxy", AJ, 69, 853.
28. Irwin, M., & Hatzidimitriou, D. 1995, "Structural Parameters for the Galactic Dwarf Spheroidals", MNRAS, 277, 1354.
29. Kinman, T.D., Carbon, D.F., Suntzeff, N.B., & Kraft, R.P. 1981, "Carbon Abundances in Giant Stars of the Draco Dwarf Galaxy", in Astrophysical Parameters for Globular Clusters, edited by A.G.D. Philip and D.S. Hayes (L. Davis Press, Schenectady), p. 451.
30. Kinman, T.D., & Kraft, R.P. 1980, "The Giant Stars of the Draco Dwarf Galaxy. I. Photometry and Exploratory Spectroscopy", AJ, 85, 415.
31. Kinman, T.D., Kraft, R.P., & Suntzeff, N.B. 1980, "On the Metal Abundance of Giants in the Draco Dwarf Galaxy – Preliminary Results of a Spectroscopic Survey", in Physical Processes in Red Giants, edited by A. Renzini and I. Iben, Jr. (Reidel, Dordrecht), p.71.
32. Knapp, G.R., Kerr, F.J., & Bowers, P.F. 1978, "Upper Limits to the H I Content of the Dwarf Spheroidal Galaxies", AJ, 83, 360.
33. Lake, G. 1990, "The Distribution of Dark Matter in Draco and Ursa Minor", MNRAS, 244, 701.
34. Lehnert, M.D., Bell, R.A., Hesser, J.E., & Oke, J.B. 1992, "Abundances for Giant Stars in the Draco Dwarf Galaxy", ApJ, 395, 466.
35. McClure, R.D., & Hesser, J.E. 1981, "Strengths of Spectral Features of Giant Stars in Outlying Halo Clusters", ApJ, 246, 136.
36. Munari, U. 1991, "BVI Photometry of the Symbiotic Star C-1 in the Draco Dwarf Galaxy", Inf. Bull. Var. Stars 3605.
37. Munari, U. 1991, "Spectroscopy of the Symbiotic Carbon Star C-1 in the Draco Dwarf Galaxy", A&A, 251, 103.
38. Munari, U. 1991, "Infrared Properties of the Carbon Symbiotic Star C-1 in the Draco Dwarf Galaxy", Inf. Bull. Var. Stars 3648.
39. Nemec, J.M. 1985, "Double-Mode RR Lyrae Stars in the Draco Dwarf Galaxy", AJ, 90, 204.
40. Nemec, J.M., Nemec, A.F.L., & Lutz, T.E. 1994, "Period-Luminosity-Metallicity Relations, Pulsation Modes, Absolute Magnitudes, and Distances for Population II Variable Stars", AJ, 108, 222.
41. Norris, J., & Zinn, R. 1975, "The Cepheid Variables and the Stellar Populations of the Dwarf Spheroidal Galaxies", ApJ, 202, 335.

42. Olszewski, E.W., Aaronson, M., & Hill, J.M. 1995, "Nine Seasons of Velocity Measurements in the Draco and Ursa Minor Dwarf Spheroidal Galaxies with the MMT Echelle", AJ, 110, 2120.
43. Olszewski, E.W., Pryor, C., & Armandroff, T.E. 1996, "The Mass-to-Light Ratios of the Draco and Ursa Minor Dwarf Spheroidal Galaxies. II. The Binary Population and its Effect on the Measured Velocity Dispersions of Dwarf Spheroidals", AJ, 111, 750.
44. Pryor, C., & Kormendy, J. 1990, "The Dark Matter Halos of Draco and Ursa Minor", AJ, 100, 127.
45. Silbermann, N.A., Harris, H.C., & Smith, H.A. 1996, "Variable Stars in the Draco Dwarf Spheroidal Galaxy", in Formation of the Galactic Halo…Inside and Out, edited by H. Morrison and A. Sarajedini (Astronomical Society of the Pacific, San Francisco), p. 536.
46. Smith, G.H. 1984, "The Chemical Inhomogeneity of the Draco Dwarf Spheroidal Galaxy", AJ, 89, 801.
47. Stetson, P.B. 1979, "The Dwarf Spheroidal Galaxy in Draco. I. New *BV* Photometry", AJ, 84, 1149.
48. Stetson, P.B. 1979, "The Dwarf Spheroidal Galaxy n Draco. II. Galactic Foreground Reddening", AJ, 84, 1167.
49. Stetson, P.B. 1980, "The Dwarf Spheroidal Galaxy in Draco. III. Proper Motion Membership Probabilities", AJ, 85, 387.
50. Stetson, P.B. 1980, "The Dwarf Spheroidal Galaxy in Draco. IV. On the Mean Metal Abundance and Metal-Abundance Range of the Red Giant Stars", AJ, 85, 398.
51. Stetson, P.B., "Spectroscopy of Giant Stars in the Draco and Ursa Minor Dwarf Galaxies", PASP, 96, 128.
52. Stetson, P.B., VandenBergh, D.A., & McClure, R.D. 1985, "The Main-Sequence Turnoff of the Draco Dwarf Galaxy", PASP, 97, 908.
53. Strobel, N.V., & Lake, G. 1994, "Black Hole Models for the Dark Matter in Draco and Ursa Minor", ApJ, 424, L83.
54. Swope, H.H. 1968, "Thirteen Periodic Variable Stars Brighter than the Normal RR Lyrae-Type Variables in Four Dwarf Galaxies", AJ, 73, S204.
55. van den Bergh, S. 1994, "The Evolutionary History of Low-Luminosity Local Group Dwarf Galaxies", ApJ, 428, 617.
56. Wehlau, A., & Demers, S. 1986, "Magnitudes of Bright Variables in the Draco Dwarf Galaxy", Inf. Bull. Var. Stars 2914.
57. Wilson, A.G. 1955, "Sculptor-Type Systems in the Local Group of Galaxies", PASP, 67, 27.
58. Zinn, R. 1978, "The Metal Abundance Range in the Draco Dwarf Galaxy", ApJ, 225, 790.
59. Zinn, R. 1980, "On the Metal Abundance Range in the Draco Dwarf Galaxy", AJ, 85, 1468.
60. Zinn, R., & Searle, L. 1976, "The Masses of the Anomalous Cepheids in the Draco System", ApJ, 209, 734.

Fornax

Alternate name(s): A 0237-34
Right Ascension (2000): 02h 39m 59s
Declination (2000): -34d 27.0m
Type: dE3
Apparent magnitude (V): 7:6
Color (B-V): 0.63
(U-B): 0.08
(V-R): 0.45
Color Excess, E(B-V): 0.03
Absolute magnitude (M_V): -13.1
Distance (kpc): 138
Radial velocity (solar, km/sec): 55

Objects Identified On the Atlas:
Globular clusters: 5; designation: GC
Source:[34]
Open clusters: none
Source:
OB associations: none
Source:
Variable stars: 30 long period variables, 2 Cepheids (not shown)
Source:[16], [47]
Carbon stars: 25 (not shown)
Source:[1], [5], [26]
HII regions: none
Source: none
Planetary nebulae: 1; designation: PN
Source: [11]
Dust clouds: none
Source:

Fornax

Chart 96

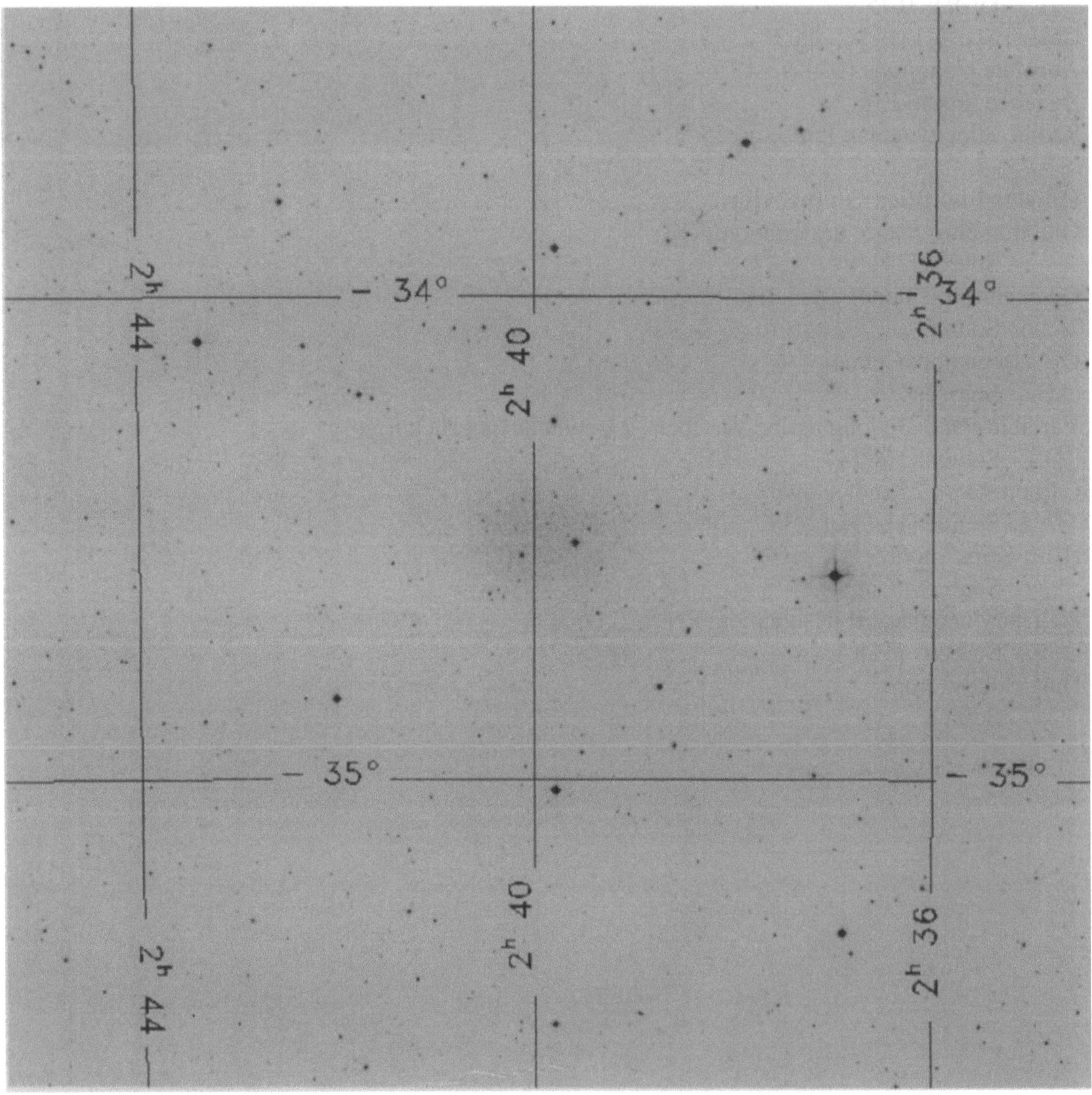

Fornax

Chart 97

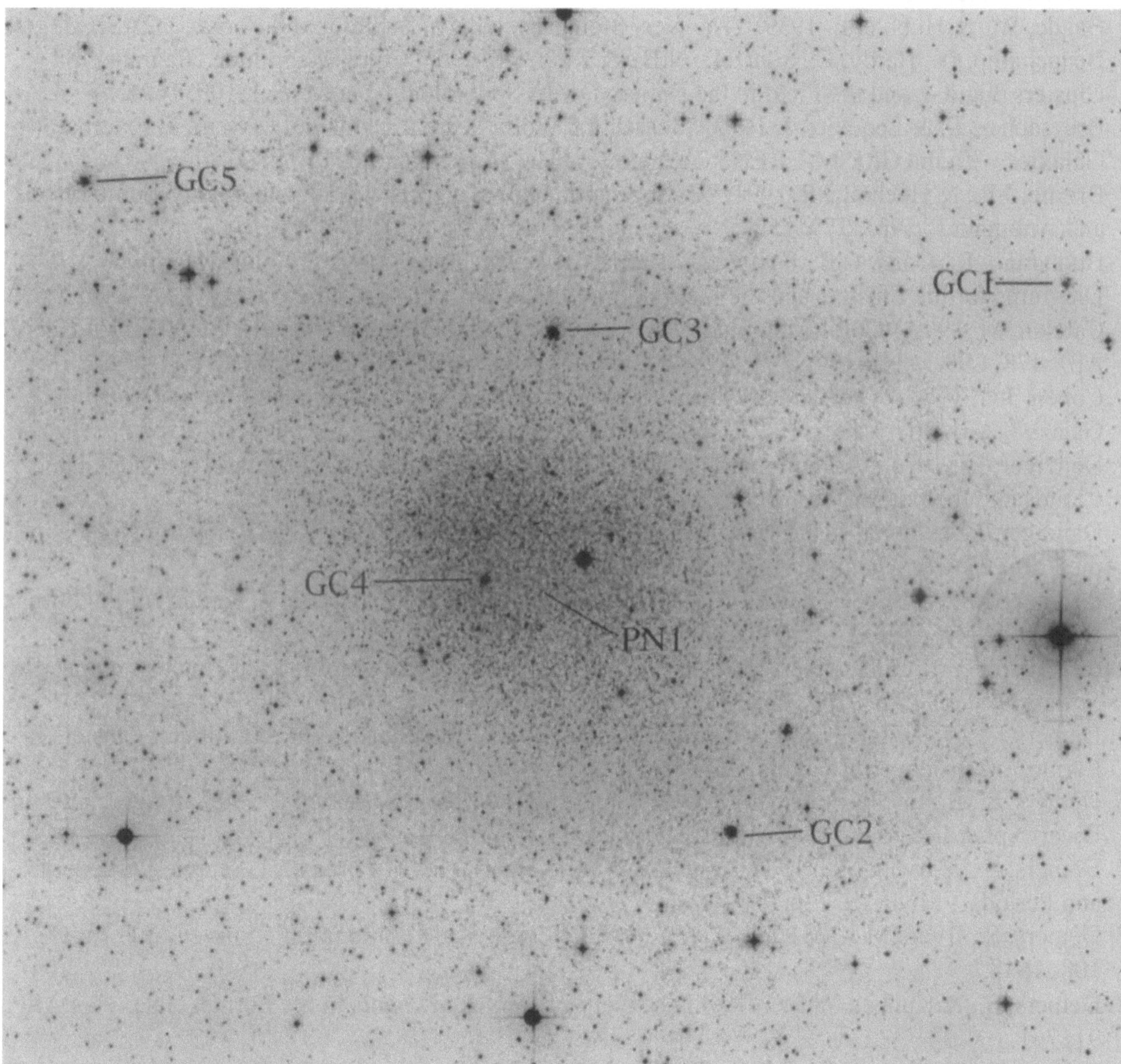

Fornax

References

1. Aaronson, M., & Mould, J. 1980, "Carbon Stars in the Fornax Dwarf Spheroidal Galaxy", ApJ, 240, 804.
2. Aaronson, M., & Olszewski, E.W. 1986, "Accurate Radial Velocities for Three Fornax Globular Clusters", AJ, 92, 580.
3. Baade, W., & Hubble, E. 1939, "The New Stellar Systems in Sculptor and Fornax", PASP, 51, 40.
4. Beauchamp, D., Hardy, E., Suntzeff, N.B., & Zinn, R. 1995, "Color-Magnitude Diagrams for Clusters 2 and 4, and the Field in the Fornax Dwarf Spheroidal Galaxy", AJ, 109, 1628.
5. Breysacher, J., & Lequeux, J. 1983, "Search for Wolf-Rayet, Carbon and M Stars in External Galaxies with the GRISM/GRENS Technique", The Messenger, 33, 21.
6. Brodie, J.P., & Huchra, J.P. 1991, "Extragalactic Globular Clusters. III. Metallicity Comparisons and Anomalies", ApJ, 379, 157.
7. Buonanno, R., Corsi, C.E., Fusi Pecci, F., Hardy, E., & Zinn, R. 1985, "Color-Magnitude Diagrams for the Clusters and the Field of the Fornax Dwarf Spheroidal Galaxy", A&A, 152, 65.
8. Catelan, M. 1997, "Color-Magnitude Diagrams of Merged Globular Clusters: Metallicity Effects", ApJ, 478, L99.
9. Cohen, J.G. 1983, "The Velocity Dispersion of the Globular Clusters in the Fornax Dwarf Galaxy", ApJ, 207, L41.
10. Danziger, I.J. 1973, "A Photometric Study of the Integrated Light of Clusters in the Magellanic Clouds and the Fornax Dwarf Galaxy", ApJ, 181, 641.
11. Danziger, I.J., Dopita, M.A., Hawarden, T.G., & Webster, B.L. 1978, "The Planetary Nebula in the Fornax Dwarf Galaxy", ApJ, 220, 458.
12. de Vaucouleurs, G., & Ables, H.D. 1967, "Integrated Magnitudes and Color Indices of the Fornax Dwarf Galaxy", ApJ, 151, 105.
13. de Vaucouleurs, G., & Ables, H.D. 1970, "Magnitude and Colors of Globular Clusters in the Fornax System", ApJ, 159, 425.
14. Demers, S. 1969, "Integrated Magnitudes and Colors of the Four Brightest Globular Clusters in Fornax", Astrophysical Letters, 3, 175.
15. Demers, S., Battinelli, P., Irwin, M.J., & Kunkel, W.E. 1995, "A Search for Stellar Aggregates in Dwarf Spheroidal Galaxies", MNRAS, 274, 491.
16. Demers, S., & Irwin, M.J. 1987, "The Long-Period Variables of Fornax – I. Search, Discovery, and Periods", MNRAS, 226, 943.
17. Demers, S., Irwin, M.J., & Kunkel, W.E. 1994, "Large Scale Photographic Survey of Fornax", AJ, 108, 1648.
18. Demers, S., & Kunkel, W.E. 1979, "Discovery of Very Red Giants in the Fornax Galaxy", PASP, 91, 761.
19. Demers, S., Kunkel, W.E., & Grondin, L. 1990, "CCD Photometry of Cluster 1 of Fornax", PASP, 102, 632.
20. Demers, S., Kunkel, W.E., & Hardy, E. 1979, "The Giant Branch of Fornax", ApJ, 232, 84.
21. Dubath, P., Meylan, G., & Mayor, M. 1992, "Core Velocity Dispersions and Metallicities of Three Globular Clusters Belonging to the Fornax Dwarf Spheroidal Galaxy", ApJ, 400, 510.

22. Eskridge, P.B. 1987, "The Global Luminosity Function of the Fornax Dwarf Elliptical Galaxy", AJ, 94, 1564.
23. Eskridge, P.B. 1988, "The Structure of the Fornax Dwarf Elliptical Galaxy. I. The Radial Profile", AJ, 96, 1352.
24. Eskridge, P.B. 1988, "The Structure of the Fornax Dwarf Elliptical Galaxy. II. Two-Dimensional Structure", AJ, 96, 1614.
25. Faber, S.M., & Lin, D.N.C. 1983, "Is There Nonluminous Matter in Dwarf Spheroidal Galaxies?", ApJ, 266, L17.
26. Frogel, J.A., Blanco, V.M., & McCarthy, M.F. 1982, "The Late-Type Stellar Content of the Fornax and Sculptor Dwarf Galaxies", ApJ, 252, 133.
27. Gizis, J.E., Mould, J.R., & Djorgovski, S. 1993, "An X-ray Image of The Fornax Dwarf Spheroidal Galaxy", PASP, 105, 871.
28. Gordon, K.C., & Kron, G.E. 1983, "Integrated Magnitudes and Colors of Clusters in the Magellanic Clouds and Fornax System", PASP, 95, 461.
29. Gratton, R.G., Ortolani, S., & Richter, O. 1986, "The Dwarf Galaxies in Fornax and in Phoenix", MSAIt, 57, 561,
30. Grebel, E.K., Roberts, W.J., & van de Rydt, F. 1994, "Population Studies in Dwarf Spheroidal Galaxies: Washington Photometry of Sculptor and Fornax", in The Local Group: Comparative and Global Properties, edited by A. Layden, R.C. Smith, and J. Storm (European Southern Observatory, Munich), p. 148.
31. Grillmair, C.J., et al. 1997, "WFPC2 Observations of the Fornax Dwarf Spheroidal", BAAS, 29, 1341.
32. Harris, H.C., & Canterna, R. 1977, "Integrated Colors of Globular Clusters in the Galaxy, Fornax, and M31", AJ, 82, 798.
33. Harris, W.E. 1983, "A Comment on the Colors of Globular Clusters in Elliptical Galaxies", PASP, 95, 21.
34. Hodge, P.W. 1961, "The Fornax Dwarf Galaxy. I. The Globular Clusters", AJ, 66, 83.
35. Hodge, P.W. 1961, "The Fornax Dwarf Galaxy. II. The Distribution of Stars", AJ, 66, 249.
36. Hodge, P.W. 1965, "Integrated Photometry of the Clusters of the Fornax Dwarf Galaxy", ApJ, 141, 308.
37. Hodge, P.W. 1969, "The Color Indices of the Globular Clusters of the Fornax Dwarf Galaxy", PASP, 81, 875.
38. Hodge, P.W., & Smith, D.W. 1974, "The Structure of the Fornax Dwarf Galaxy", ApJ, 188, 19.
39. Huchra, J.P., Brodie, J.P., Caldwell, N., Christian, C., & Schommer, R. 1996, "Extragalactic Globular Clusters. IV. The Data", ApJS, 102, 29.
40. Irwin, M., & Hatzidimitriou, D. 1995, "Structural Parameters for the Galactic Dwarf Spheroidals", MNRAS, 277, 1354.
41. Jarvis, B., & Seitzer, P. 1990, "Resolving the Fornax Globular Clusters with the NTT", Messenger, 59, 52.
42. Jones, T.J., & Hyland, A.R. 1982, "Multiaperture *JHK* Photometry of the Globular Clusters of the Fornax Dwarf Spheroidal Galaxy", MNRAS, 200, 509.
43. Jørgensen, U.G., & Jimenez, R. 1997, "Globular Clusters 1 and 3 in the Fornax Dwarf Galaxy", A&A, 317, 54.
44. Knapp, G.R., Kerr, F.J., & Bowers, P.F. 1978, "Upper Limits to the H I Content of the Dwarf Spheroidal Galaxies", AJ, 83, 360.

45. Kron, G.E., & Mayall, N.U. 1960, “Photoelectric Photometry of Galactic and Extragalactic Star Clusters”, AJ, 65, 581.
46. Light, R.M. 1987, “Measurement of Abundances and their Variations in Dwarf Spheroidal Galaxies”, BAAS, 19, 1033.
47. Light, R.M., Armandroff, T.E., & Zinn, R. 1986, “Two Cepheid Variables in the Fornax Dwarf Galaxy”, AJ, 92, 43.
48. Lundgren, K. 1990, “Red Giants in the Fornax Dwarf Elliptical Galaxy”, A&A, 233, 21.
49. Lynden-Bell, D., Cannon, R.D., & Godwin, P.J. 1983, “Slippery Evidence on the Galaxy’s Invisible Heavy Halo”, MNRAS, 204, 87P.
50. Mateo, M. 1992, “Hunting for Dark Matter”, in Carnegie Institution of Washington Yearbook 91 (Port City Press, Baltimore), p. 34.
51. Mateo, M., Olszewski, E., Welch, D.L., Fischer, P., & Kunkel, W. 1991, “A Kinematic Study of the Fornax Dwarf Spheroidal Galaxy”, AJ, 102, 914.
52. Mateo, M., Olszewski, E., Welch, D.L., Fischer, P. & Kunkel, W. 1991, “The Dark Matter Content of the Fornax and Carina Dwarf galaxies”, BAAS, 23, 919.
53. McDowell, J.C., & Godwin, P.J. 1986, “An Accurate Position and Radial Velocity for the Planetary Nebula in the Fornax Dwarf Galaxy”, Observatory, 106, 19.
54. Paltoglou, G. 1989, “The Nature of the Fornax Dwarf Spheroidal Galaxy”, BAAS, 21, 1130.
55. Paltoglou, G., & Freeman, K.C. 1987, “Dynamics of the Fornax Dwarf Spheroidal Galaxy”, in Structure and Dynamics of Elliptical Galaxies, IAU Symposium No. 127, edited by T. de Zeeuw (Reidel, Dordrecht), p. 447.
56. Rodgers, A.W., & Harding, P. 1989, “Spectroscopy of the Globular Clusters in the Fornax Dwarf Elliptical Galaxy”, PASP, 101, 563.
57. Rodgers, A.W., & Roberts, W.H. 1994, “Structure of the Globular Clusters in the Fornax Dwarf Elliptical Galaxy”, AJ, 107, 1737.
58. Sagar, R., Hawkins, M.R.S., & Cannon, R.D. 1990, “The Giant Branch of the Fornax Spheroidal Galaxy”, MNRAS, 242, 167.
59. Seitzer, P., & Frogel, J.A. 1985, “Radial Velocities of Carbon Stars in Three Dwarf Spheroidal Galaxies”, AJ, 90, 1796.
60. Shapley, H. 1938, “Two Stellar Systems of a New Kind”, Nature, 142, 715.
61. Shapley, H. 1939, “Galactic and Extragalactic Studies. II. Notes on the Peculiar Stellar Systems in Sculptor and Fornax”, Proceedings of the NAS, 25, 565.
62. Smith, E.O., Neill, J.D., Mighell, K.J., & Rich, R.M. 1996, “Fornax Globular Clusters 1 and 5: A Confirmed Extragalactic Second Parameter Pair”, AJ, 111, 1596.
63. Tinney, C.G. 1995, “A Gravitational Lens Candidate behind the Fornax Dwarf Spheroidal Galaxy”, MNRAS, 277, 609.
64. van den Bergh, S. 1994, “The Evolutionary History of Low-Luminosity Local Group Dwarf Galaxies”, ApJ, 428, 617.
65. van den Bergh, S. 1996, “Mergers of Globular Clusters”, ApJ, 471, L31.
66. Verner, G., Demers, S., Hardy, E., & Kunkel, W.E. 1981, “The Globular Clusters of Fornax”, AJ, 86, 357
67. Webbink, R.F. 1985, “Structure Parameters of Galactic Globular Clusters”, in Dynamics of Star Clusters, IAU Symposium No. 113, edited by J. Goodman and P. Hut (Reidel, Dordrecht), p. 541.
68. Westerlund, B.E. 1979, “Red Stars in Nearby Galaxies”, ESO Messenger, 19, 7.

69. Westerlund, B.E. 1983, "Discovery of an S Star in the Fornax Dwarf Elliptical Galaxy", A&A, 118, L5.
70. Westerlund, B.E., Edvardsson, B., & Lundgren, K. 1987, "Red Stars in the Fornax Dwarf Galaxy", A&A, 178, 41.
71. Zinn, R. 1992, "The Dwarf Spheroidal Satellite Galaxies of the Milky Way", in The Globular Cluster-Galaxy Connection: Globular Clusters in the Context of Their Parent Galaxies, edited by G.H. Smith and J.P. Brodie.
72. Zinn, R., & Persson, S.E. 1981, "The Ages and Metallicities of the Globular Clusters in the Fornax Dwarf Spheroidal Galaxy", ApJ, 247, 849.

IC 10

Alternate name(s): UGC 192
Right Ascension (2000): 00h 20m 25s
Declination (2000): 59d 17.5m
Type: dIrr
Apparent magnitude (V): 11.55
Color (B-V): 1.37
(U-B): 0.28
(V-R):
Color Excess, E(B-V): 0.87
Absolute magnitude (M_V): -15.6
Distance (kpc): 825
Radial velocity (solar, km/sec): -344

Objects Identified On the Atlas:
Globular clusters: none
Source:
Open clusters: none
Source:
OB associations: 13; designation: A
Source: this atlas
Variable stars: 13; designation: Arabic numerals
Source: [36]
Carbon stars: none
Source:
HII regions: 144; designation: Arabic numerals
Source: [17]
Planetary nebulae: none
Source:
Dust clouds: 4; designation: D
Source: [this atlas]

IC 10

Chart 98

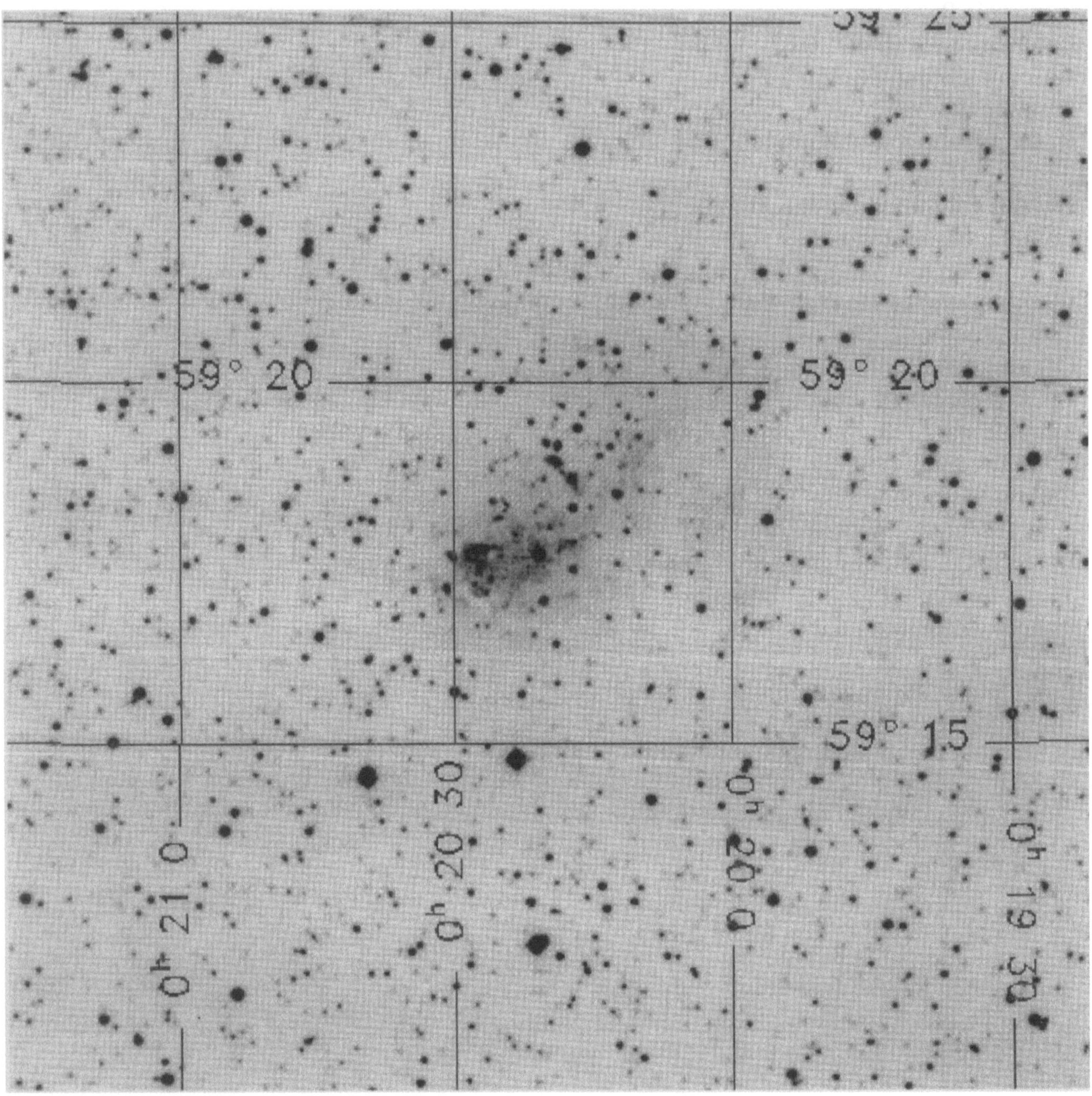

IC 10

Chart 99

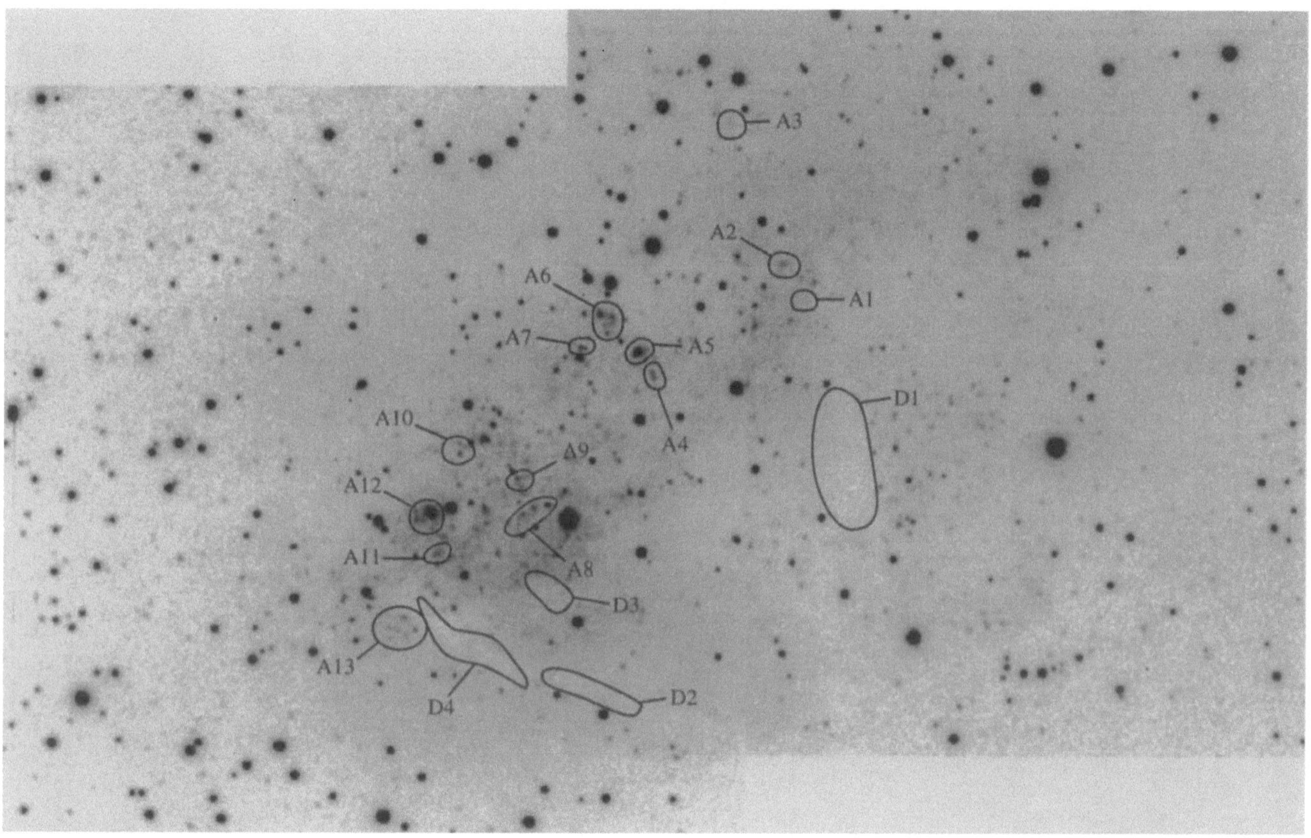

IC 10

Chart 100

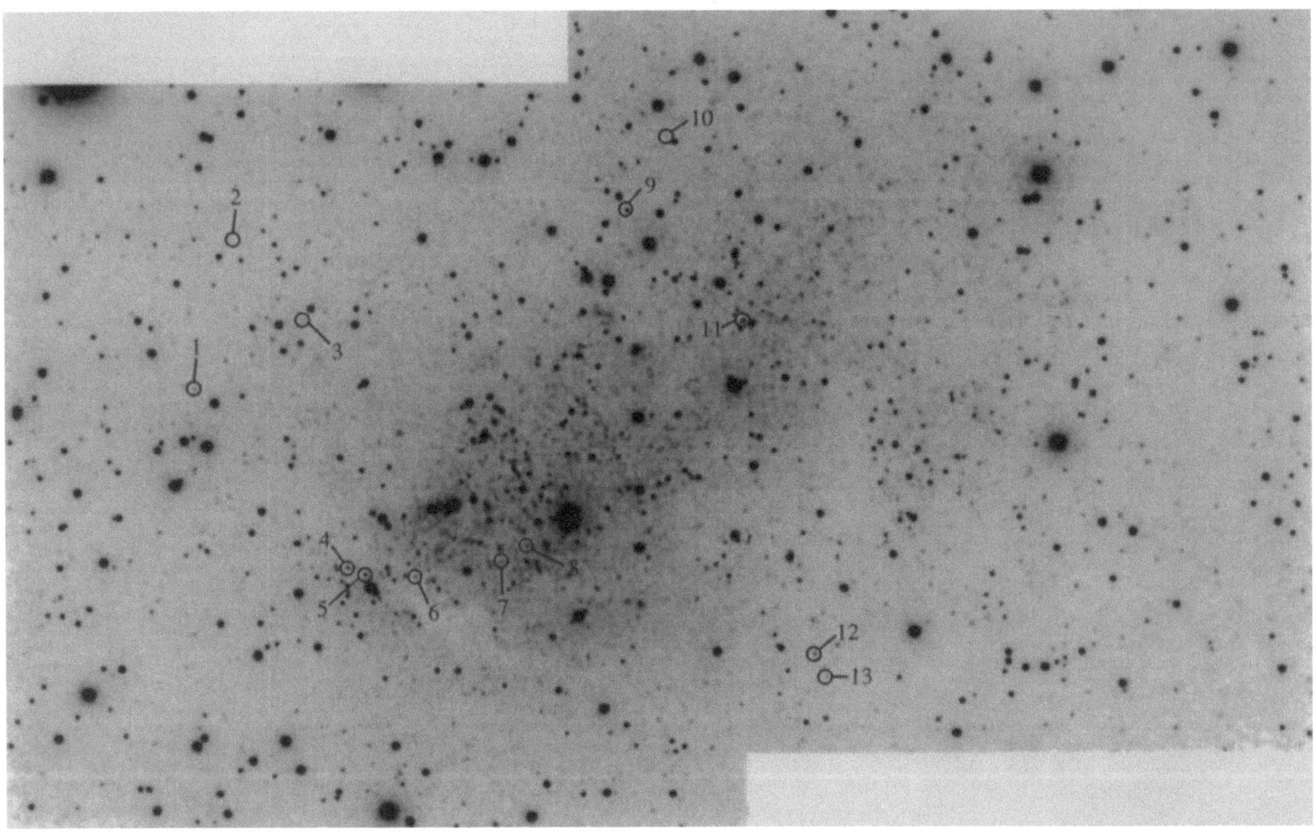

IC 10

Chart 101

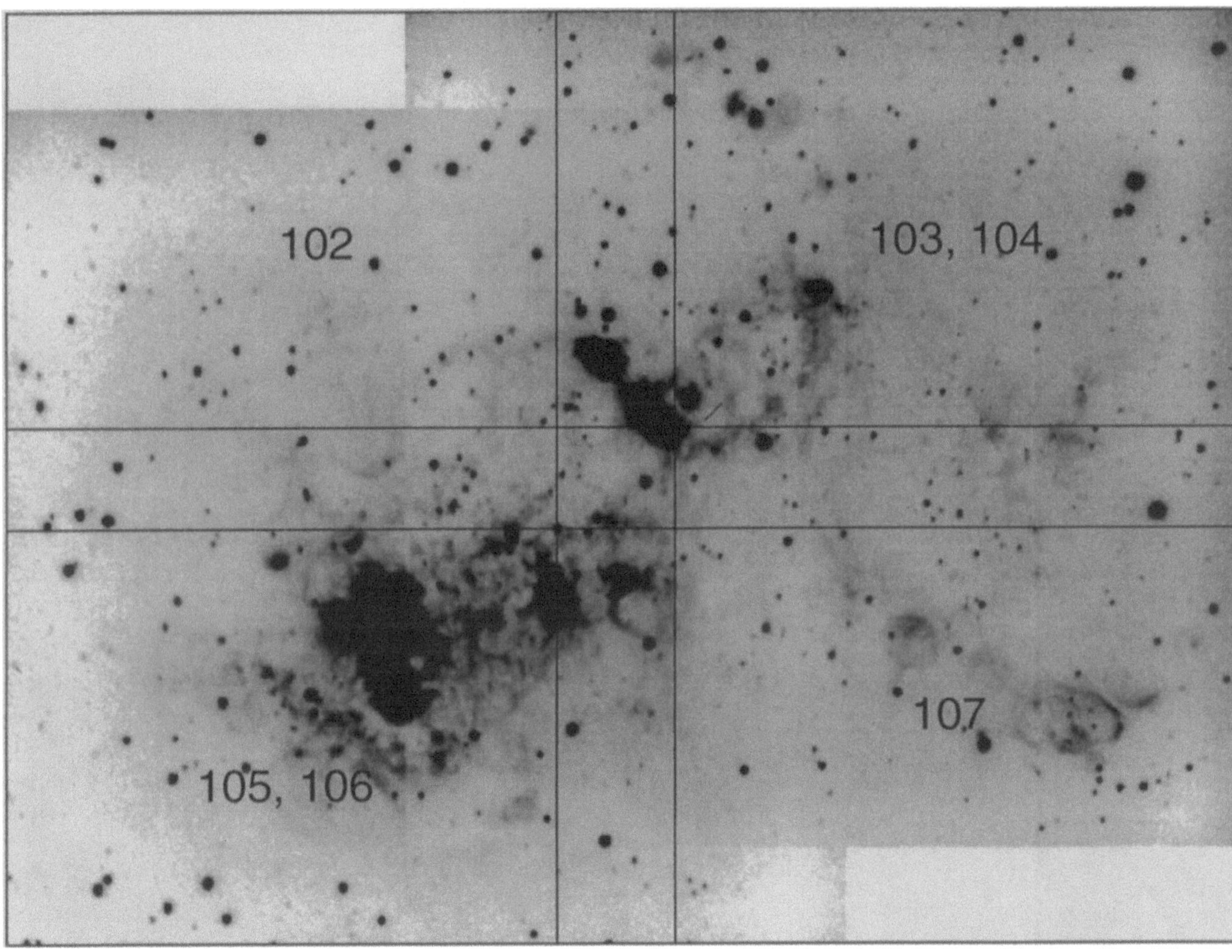

IC 10

Chart 102

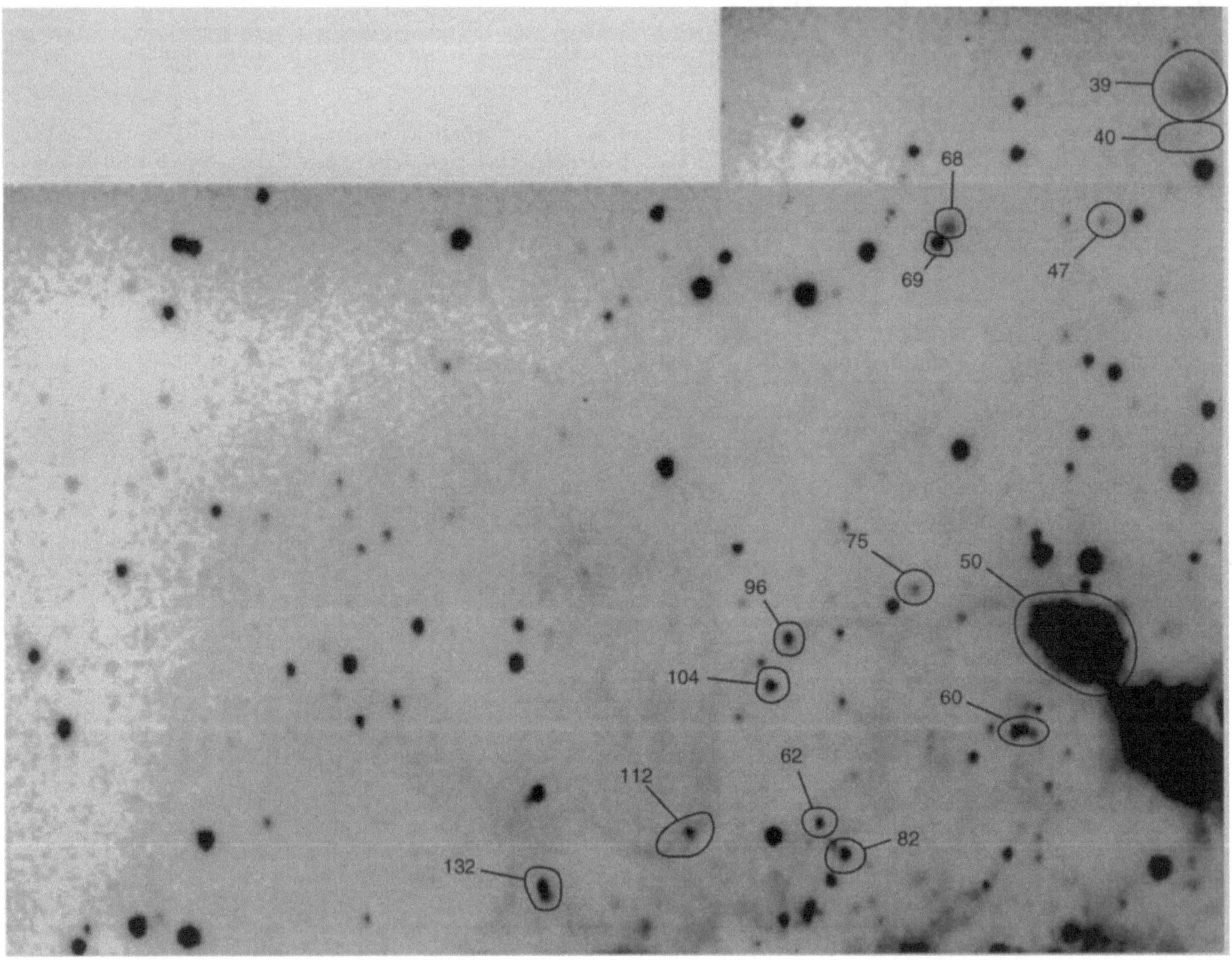

IC 10

Chart 103

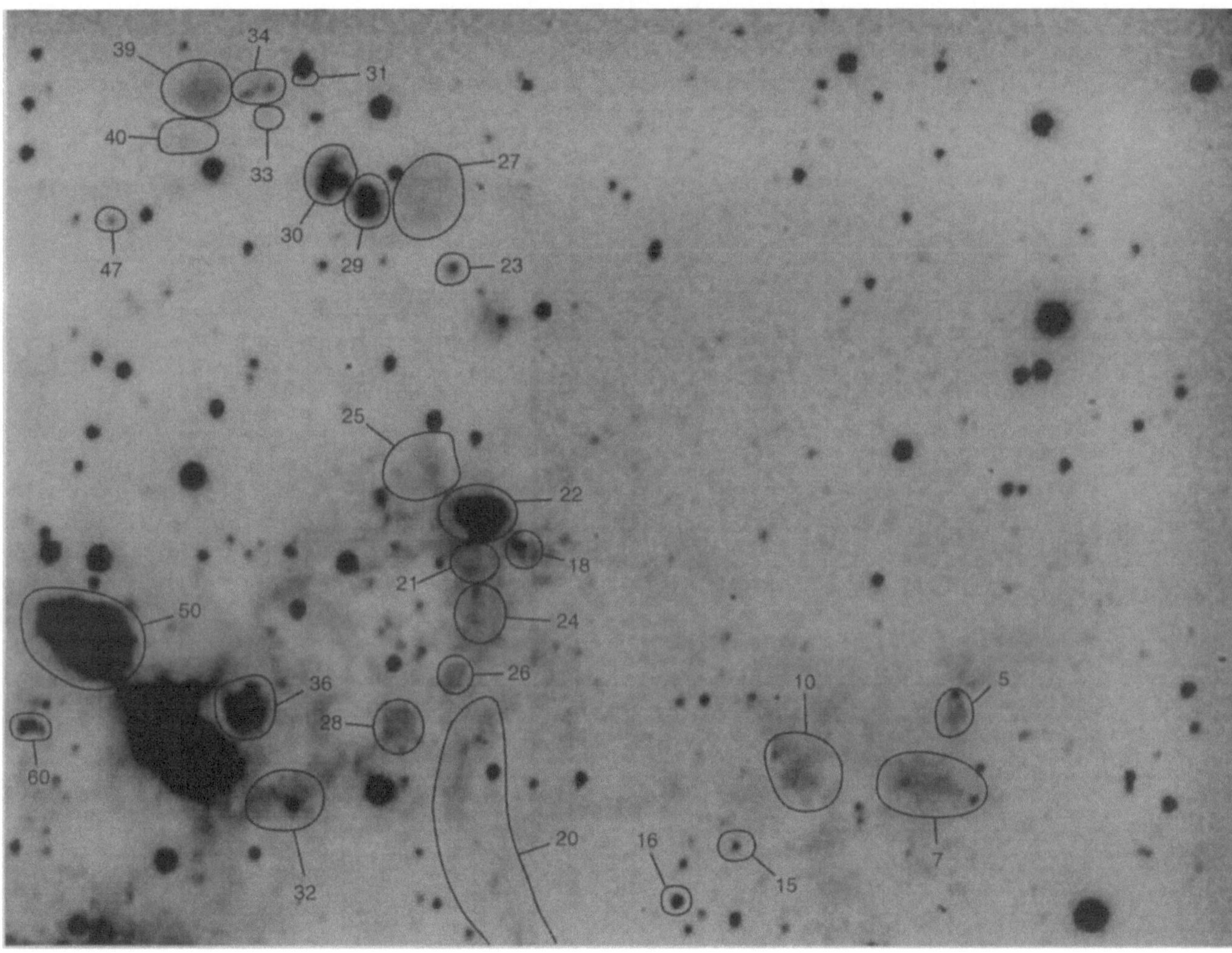

IC 10

Chart 104

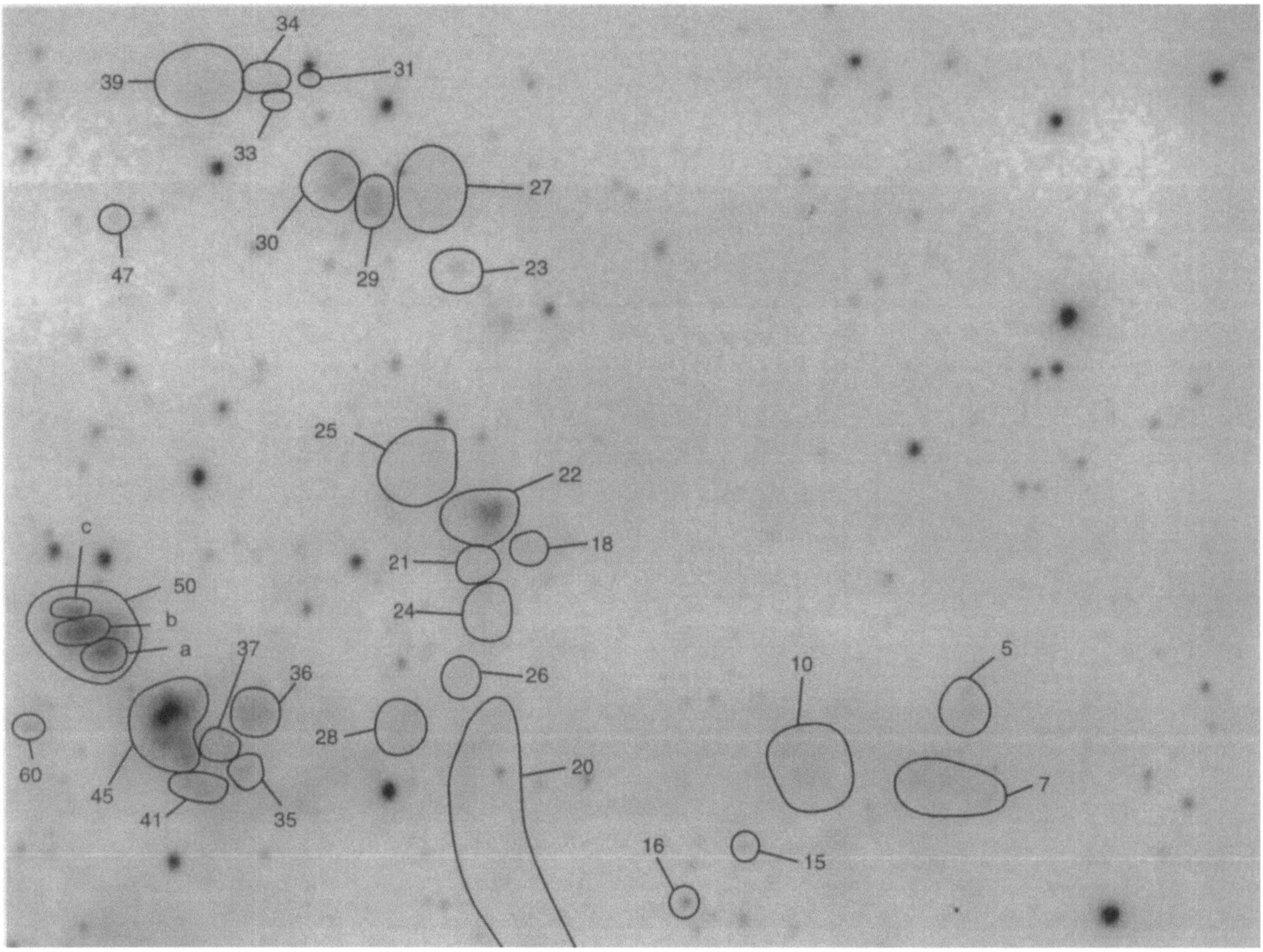

IC 10

Chart 105

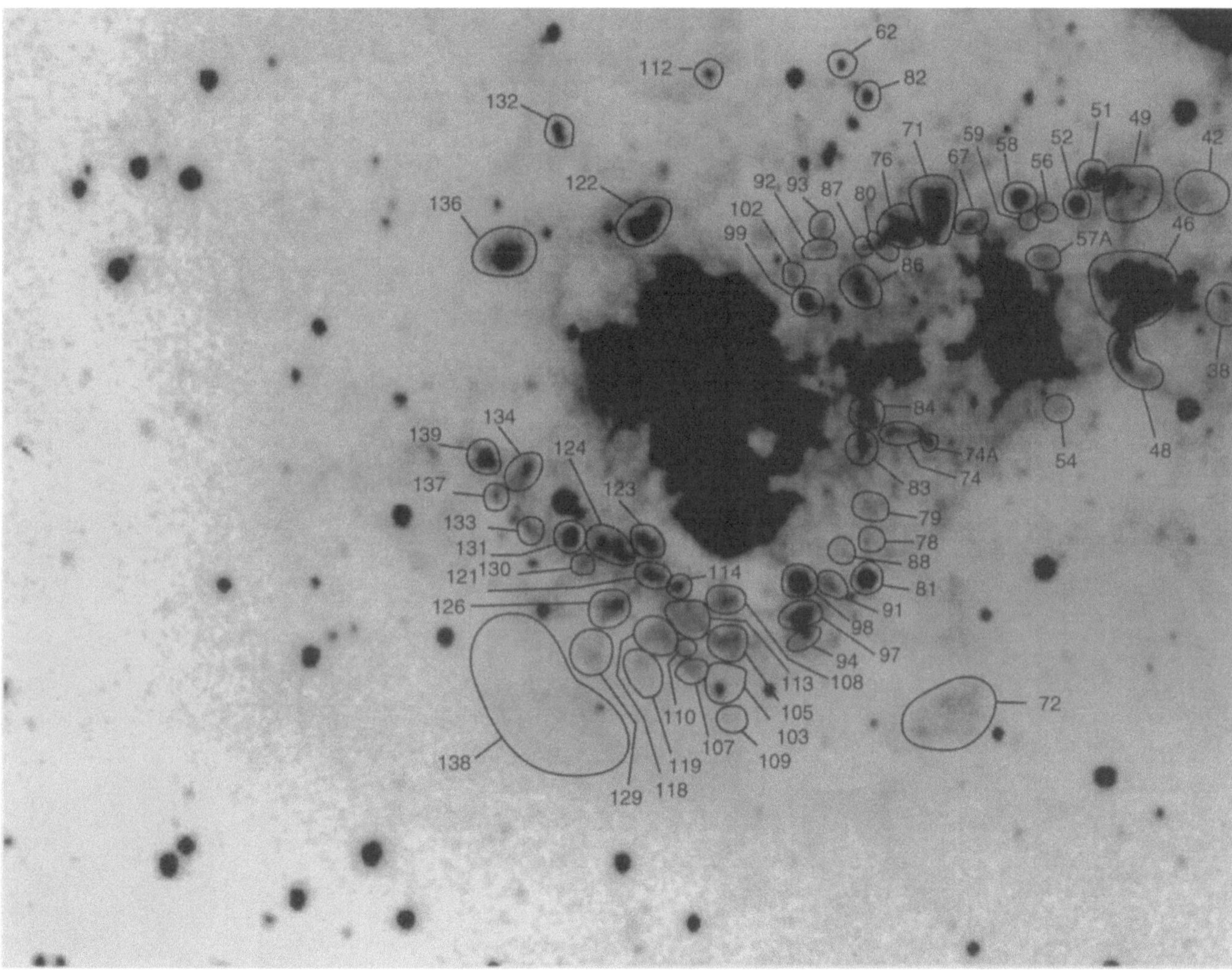

IC 10

Chart 106

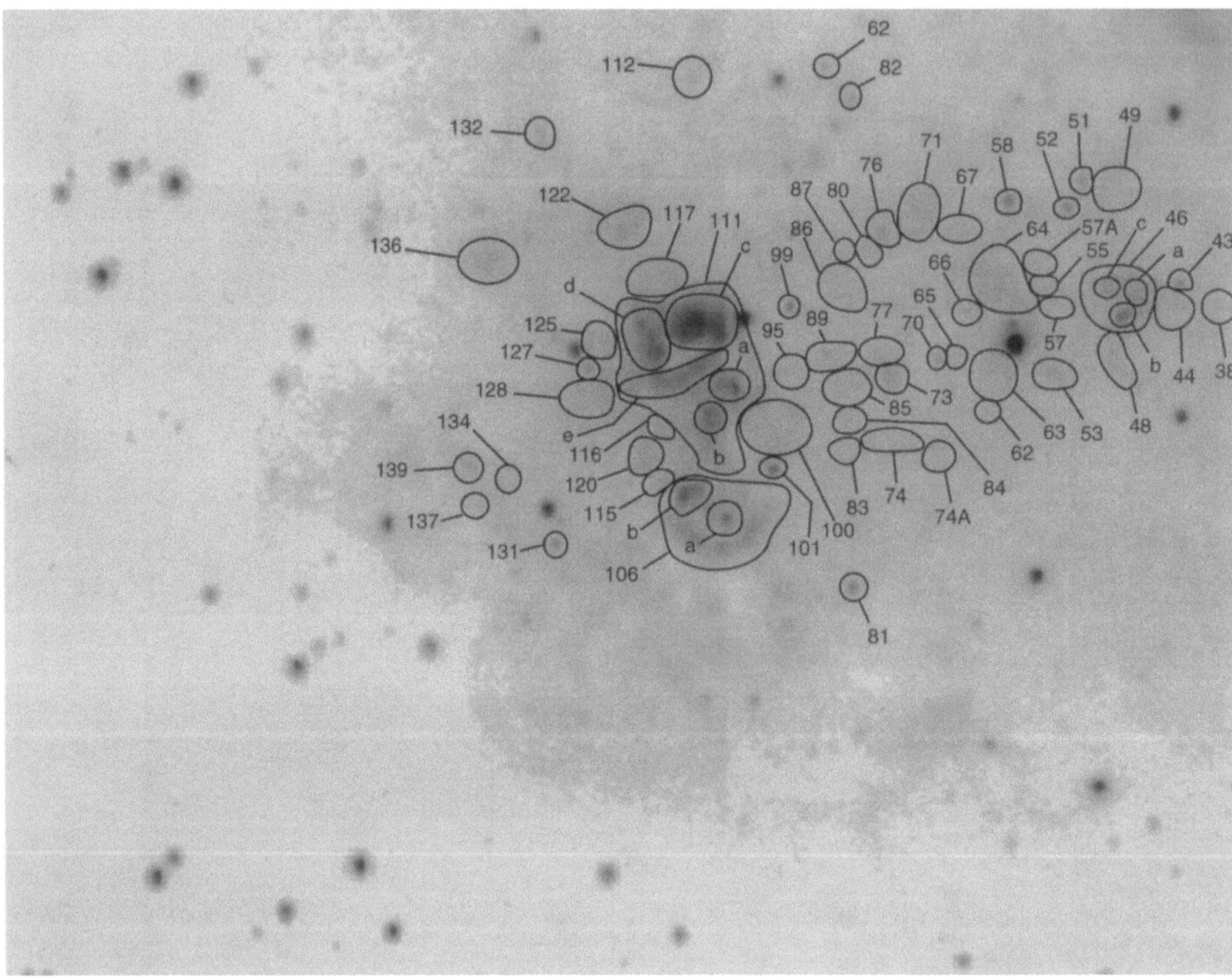

IC 10

Chart 107

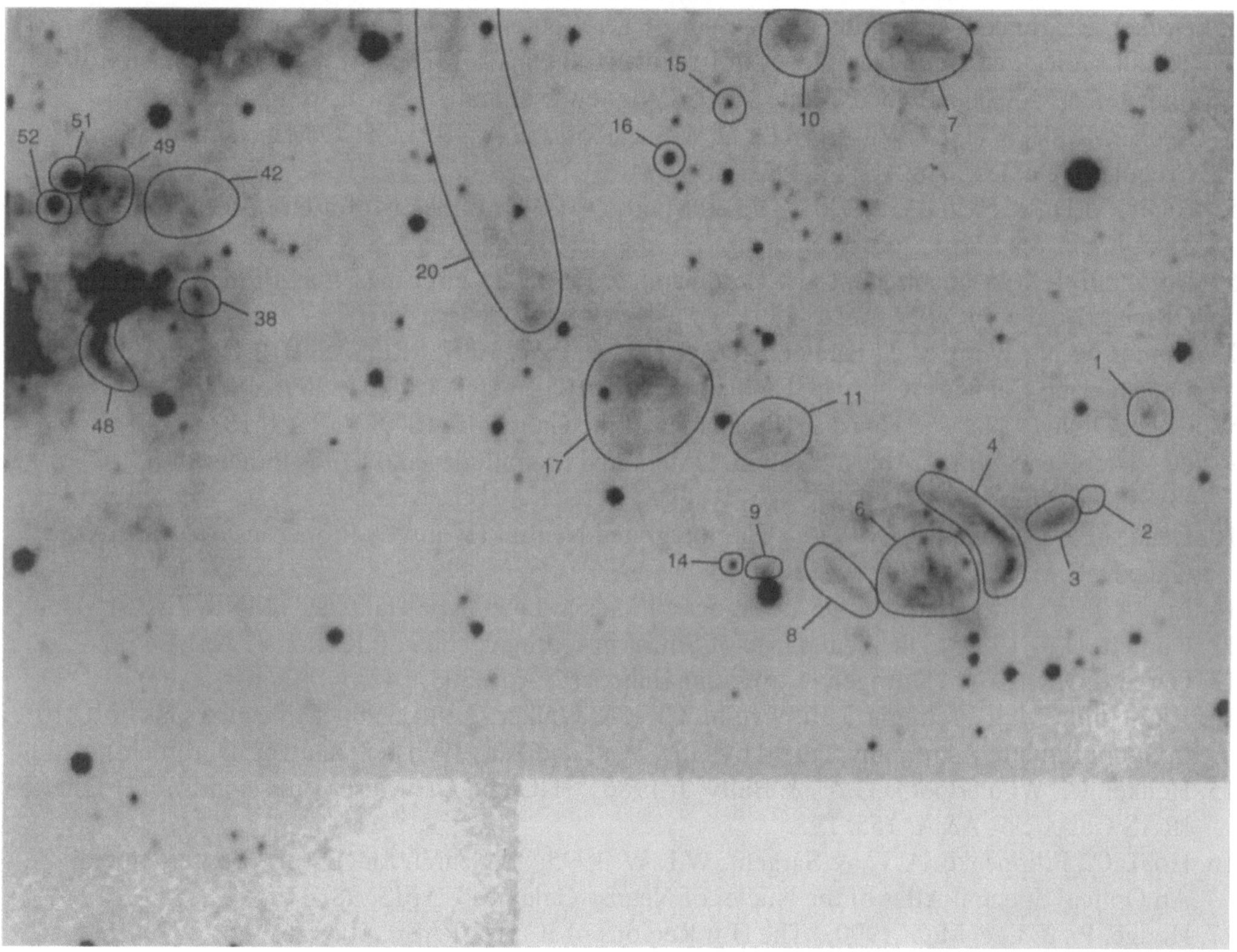

IC 10

References

1. Argon, A.L., Greenhill, L.J., Moran, J.M., Reid, M.J., Menten, K.M., Henkel, C., & Inoue, M. 1994, "The Angular Structure and Intraday Variability of a Water Vapor Maser in IC 10", ApJ, 422, 586.
2. Baan, W.I., & Haschick, A. 1994, "The H_2O Masers in IC 10 SE", ApJ, 424, L33.
3. Becker, R., Henkel, C., Schilke, P., Wilson, T.L., Wouterloot, J.G.A., & Appenzeller, I. 1989, "Giant Molecular Cloud Complexes in Irregular Galaxies", in The Physics and Chemistry of Interstellar Molecular Clouds, edited by G. Winnewisser and J.T. Armstrong, p. 276.
4. Becker, R., Henkel, C., Wilson, T.L., & Wouterloot, J.G.A. 1993, "H_2O Masers in Nearby Irregular Galaxies", A&A, 268, 483.
5. Bicay, M.D., & Helou, G. 1990, "The 60 Micron to 20 Centimeter Infrared-to-Radio Ratio Within Spiral Galaxies", ApJ, 362, 59.
6. Bottinelli, L., Gouguenheim, L., & Heidmann, J. 1972, "21-cm Line Study of the Heavily Obscured Galaxy IC 10", A&A, 18, 121.
7. Brandt, W.N., Ward, M.J., Fabian, A.C., & Hodge, P.W. 1997, "ROSAT HRI Observations of the Local Group Galaxies IC 10, NGC 147 and NGC 185", MNRAS, 291, 709.
8. Cohen, R.J. 1979, "The Unusual Kinematics of the Galaxy IC 10", MNRAS, 187, 839.
9. de Vaucouleurs, G., & Ables, H. 1965, "Integrated Magnitudes and Color Indices of IC 10", PASP, 77, 272.
10. Dean, J.F., & Davies, R.D. 1975, "The Integrated Neutral Hydrogen Properties of Nearby Galaxies", MNRAS, 170, 503.
11. Fitt, A.J., & Alexander, P. 1993, "Magnetic Fields in Late-Type Galaxies", MNRAS, 261, 445.
12. Garnett, D.R. 1989, "The Abundance of Sulfur in Extragalactic H II Regions", ApJ, 345, 282.
13. Garnett, D.R. 1990, "Nitrogen in Irregular Galaxies", ApJ, 363, 142.
14. Hauschildt, M., Fairclough, J.H., Wright, G.S., & Walker, T.M. 1990, "Molecular Gas in IC 10", in Submillimetre Astronomy, edited by G.D. Watt and A.S. Webster (Kluwer, Dordrecht), p. 233.
15. Henkel, C., Wouterloot, J.G.A., & Bally, J. 1986, "H_2O and OH Maser Emission from Bright IRAS Galaxies", A&A, 155, 193.
16. Ho, L.C., Filippenko, A.V., & Sargent, W.L.W. 1995, "A Search for 'Dwarf' Seyfert Nuclei. II. An Optical Spectral Atlas of the Nuclei of Nearby Galaxies", ApJS, 98, 477.
17. Hodge, P., & Lee, M.G. 1990, "The H II Regions of IC 10", PASP, 102, 26.
18. Hodge, P., Wilcots, E., & Miller, B. 1994, "High Resolution H I Study of IC 10", BAAS, 26, 1436.
19. Huchtmeier, W. 1979, "The Giant H I Envelope of the Irregular Galaxy IC 10", A&A, 75, 170.
20. Hunter, D.A., & Gallagher, III, J.S. 1985, "Star-Forming Properties and Histories of Dwarf Irregular Galaxies: Down but not Out", ApJS, 58, 533.
21. Hunter, D.A., Hawley, W.N., and Gallagher, III, J.S. 1993, "A Survey for Extra-H II Region Ionized Gas Structures in Irregular Galaxies", AJ, 106, 1797.
22. Jacoby, G.H., & Lesser, M.P. 1981, "Distances to Nearby Galaxies Derived from Planetary Nebulae", AJ, 86, 185.

23. Karachentsev, I.D., & Tikhonov, N.A. 1993, "Photometric Distances to the Nearby Galaxies IC 10, IC 342, and UGCA 86, Visible Through the Milky Way", A&AS, 100, 227.
24. Klein, U., & Gräve, R. 1986, "Multi-frequency Radio Continuum Observations of Dwarf Irregular Galaxies", A&A, 161, 155.
25. Klein, U., Gräve, R., & Wielebinski, R. 1983, "A Survey of the Distribution of λ2.8 cm Radio Continuum in Nearby Galaxies", A&A, 117, 332.
26. Lequeux, J., Peimbert, M., Rayo, J.F., Serrano, A., & Torres-Peimbert, S. 1979, "Chemical Composition and Evolution of Irregular and Blue Compact Galaxies", A&A, 80, 155.
27. Madden, S.C., Poglitsch, A., Geis, N., Stacey, G.J., & Townes, C.H. 1997, "[C II] 158 Micron Observations of IC 10: Evidence for Hidden Molecular Hydrogen in Irregular Galaxies", ApJ, 483, 200.
28. Massey, P., & Armandroff, T.E. 1992, "IC 10: A 'Poor Cousin' Rich in Wolf-Rayet Stars", AJ, 103, 1159.
29. Massey, P., & Armandroff, T.E. 1995, "The Massive Star Content, Reddening, and Distance of the Nearby Irregular Galaxy IC 10", AJ, 109, 2470.
30. Mayall, N.U. 1935, "An Extra-Galactic Object 3° from the Plane of the Galaxy", PASP, 47, 317.
31. Mayall, N.U. 1941, "The Radial Velocity of IC 10", PASP 53, 122.
32. Ohta, K., Sasaki. M., & Saitō, M. 1988, "CO Clouds in the Dwarf Irregular Galaxy IC 10", PASJ, 40, 653.
33. Ohta, K., Sasaki, M., Yamada, T., Saitō, M., & Nakai, N. 1992, "Aperture Synthesis Observations of Molecular Clouds in the Dwarf Irregular Galaxy IC 10", PASJ, 44, 585.
34. Roberts, M.S. 1962, "A Study of Neutral Hydrogen in IC 10", AJ 67, 431.
35. Rowan-Robinson, M., Phillips, T.G., & White, G. 1980, "A Survey of Nearby Galaxies for CO", A&A, 82, 381.
36. Saha, A., Hoessel, J.J., Krist, J., & Danielson, G.E. 1996, "Variable Stars in the Dwarf Galaxy IC 10", AJ, 111, 197.
37. Saitō, M., Sasaki, M., Ohta, K., & Yamada, T. 1992, "An H-Alpha Velocity Field in the Irregular Galaxy IC 10", PASJ, 44, 593.
38. Shostak, G.S. 1974, "Aperture Synthesis Study of H I in the Galaxy IC 10", A&A, 31, 97.
39. Shostak, G.S., & Allen, R.J. 1980, "Accurate H I Profiles of Several Nearby Galaxies", A&A, 81, 167.
40. Shostak, G.S., & Skillman, E.D. 1989, "Neutral Hydrogen Observations of the Irregular Galaxy IC 10", A&A, 214, 33.
41. Shostak, G.S., & van Woerden, H. 1983, "H I Observations of the Irregular Galaxy IC 10", in Internal Kinematics and Dynamics of Galaxies, IAU Symposium No. 100, edited by E. Athanassoula (Reidel, Dordrecht), p. 33.
42. Skillman, E.D. 1988, "A Nonthermal Superbubble in the Irregular Galaxy IC 10", in Supernova Remnants and the Interstellar Medium, IAU Colloquium No. 101, edited by R.S. Roger and T.L. Landecker (Cambridge University Press, Cambridge), p. 465.
43. Thronson, Jr., H.A., Hunter, D.A., Casey, S., & Harper, D.A. 1990, "Sub-Millimeter Continuum Emission from Galaxies: Star Formation and the Interstellar Medium in the Local Group Dwarf IC 10", ApJ, 355, 94.
44. Wilson, C.D. 1995, "The Metallicity Dependence of the CO-to-H_2 Conversion Factor from Observations of Local Group Galaxies", ApJ, 448, L97.

45. Wilson, C.D., & Reid, I.N. 1991, "Giant Molecular Clouds and the CO –to-H_2 Conversion Factor in IC 10", ApJ, 366, L11.
46. Wilson, C.D., Welch, D.L., Reid, I.N., Saha, A., & Hoessel, J. 1996, "The Distance to IC 10 from Near-Infrared Observations of Cepheids", AJ, 111, 1106.
47. Yang, H., & Skillman, E.D. 1993, "A Nonthermal Superbubble in the Irregular Galaxy IC 10", AJ 106, 1448.

IC 1613

Alternate name(s): DDO 8
Right Ascension (2000): 01h 04m 54s
Declination (2000): 02d 08.0m
Type: dIrr
Apparent magnitude (V): 9.59
Color (B-V): 0.55
(U-B): -0.25
(V-R):
Color Excess, E(B-V): 0.03
Absolute magnitude (M_V): -14.7
Distance (kpc): 700
Radial velocity (solar, km/sec): -234

Objects Identified On the Atlas:
Globular clusters: none
Source
Open clusters: 43 (candidates, mostly doubtful); designation: C
Source: [25]
OB associations: 20; designation: A
Source: [25]
Variable stars: 59; designation: V
Source: [56]
Carbon stars: none
Source:
HII regions: 77; designation: Arabic numerals
Source: [28]
Dust clouds: 11; designation: D
Source: [25]
Wolf-Rayet Stars: 8; designation: WR
Source: [3]

IC 1613

Chart 108

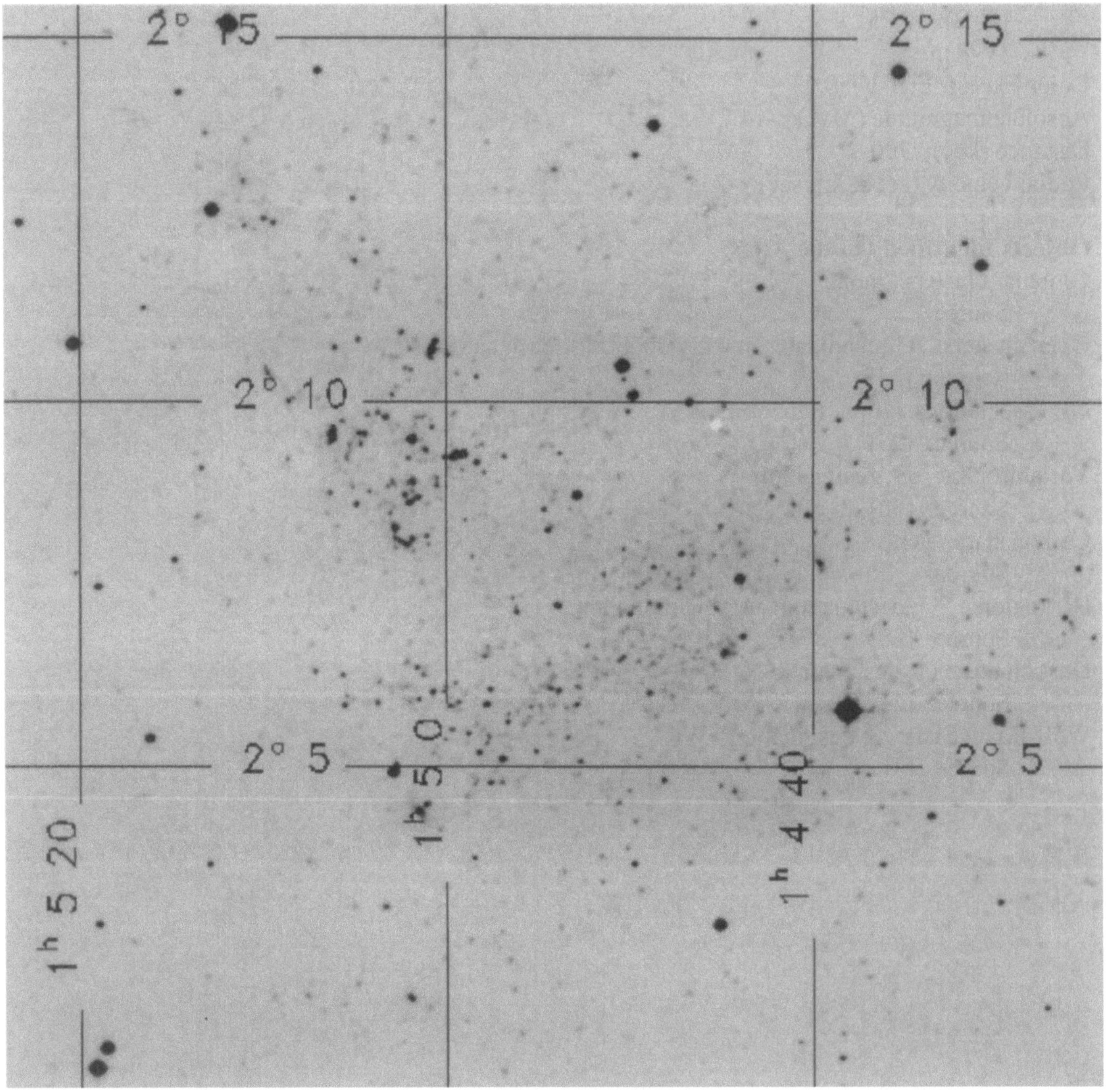

IC 1613

Chart 109

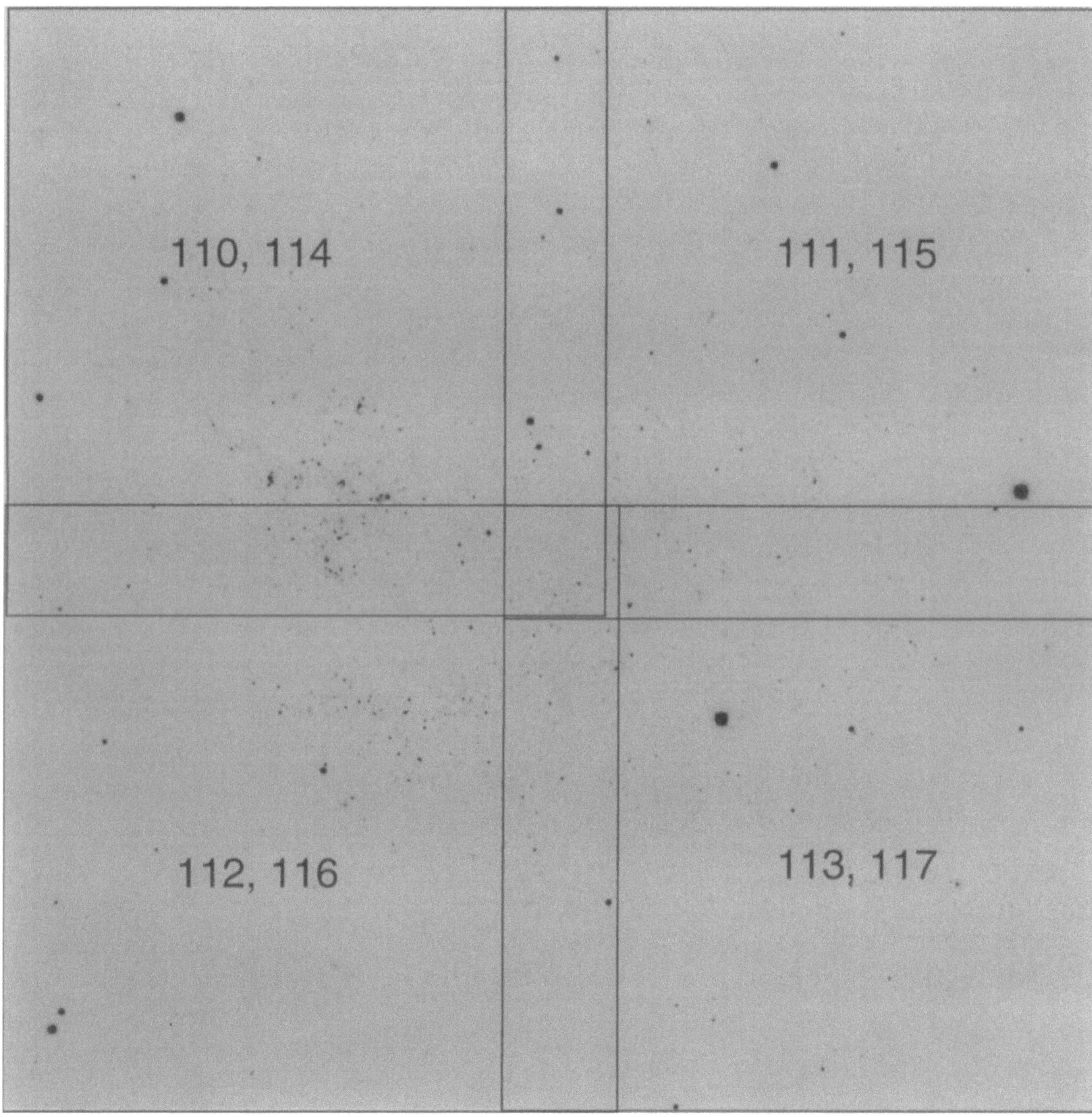

IC 1613

Chart 110

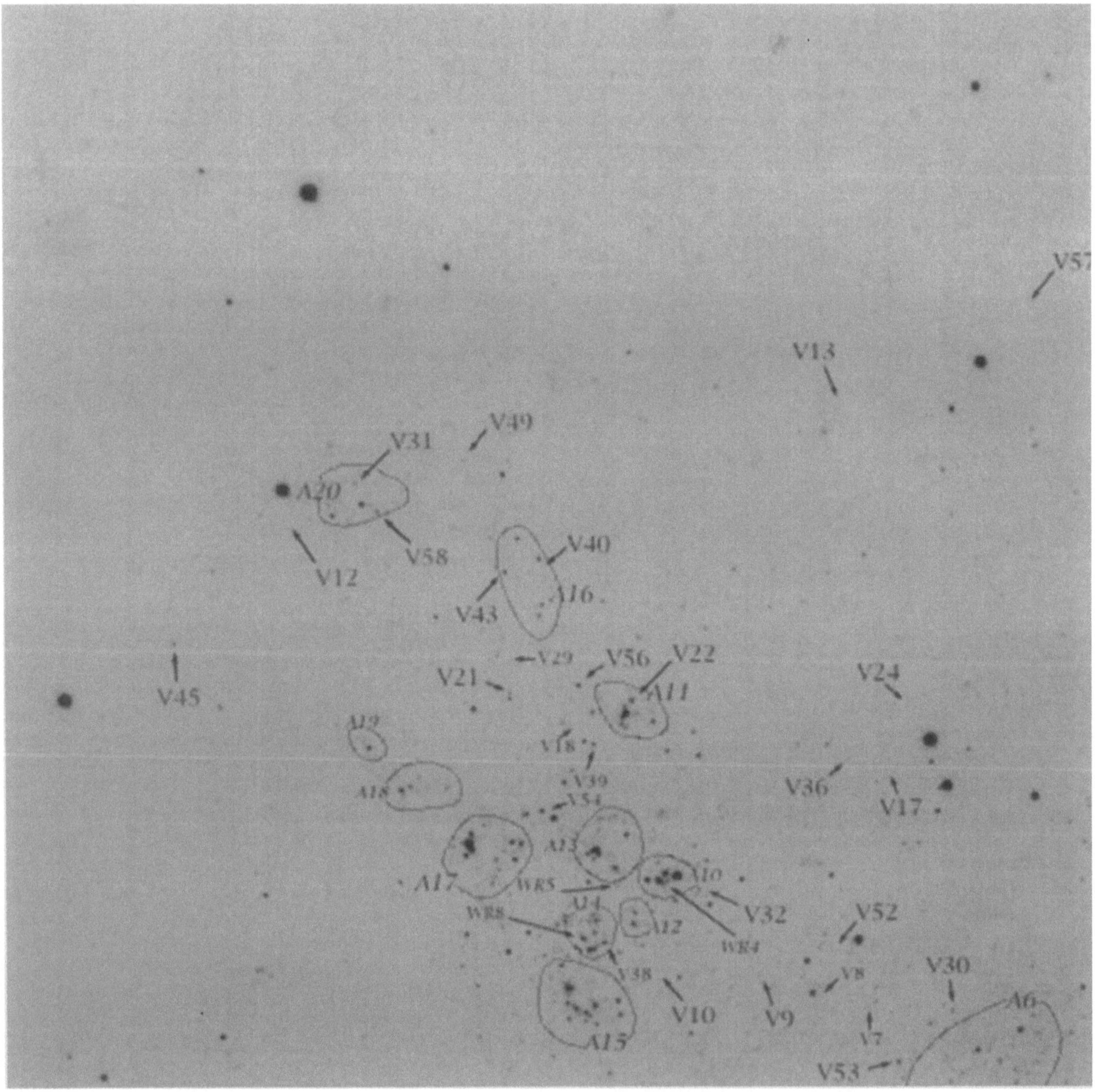

IC 1613

Chart 111

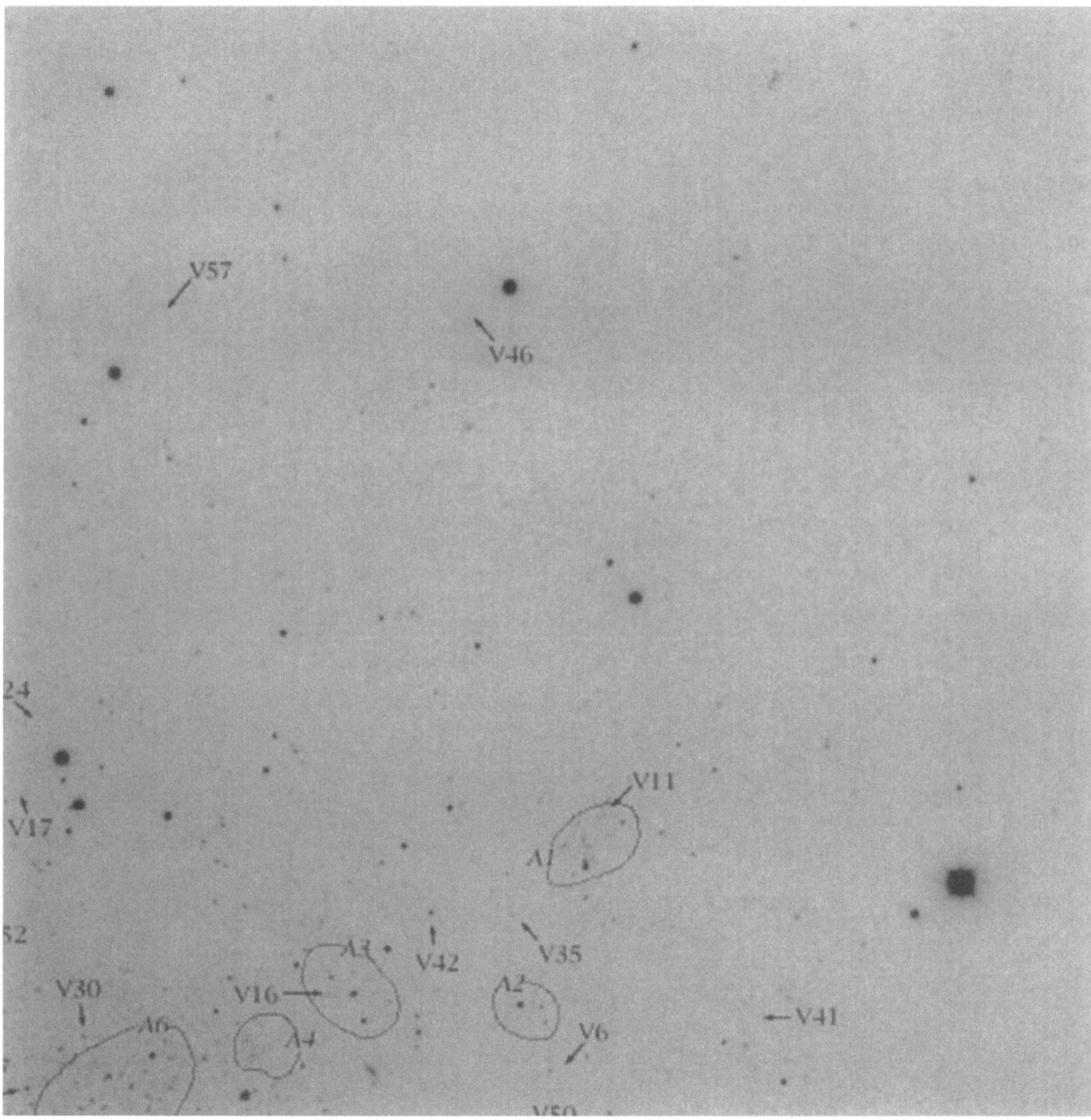

IC 1613

Chart 112

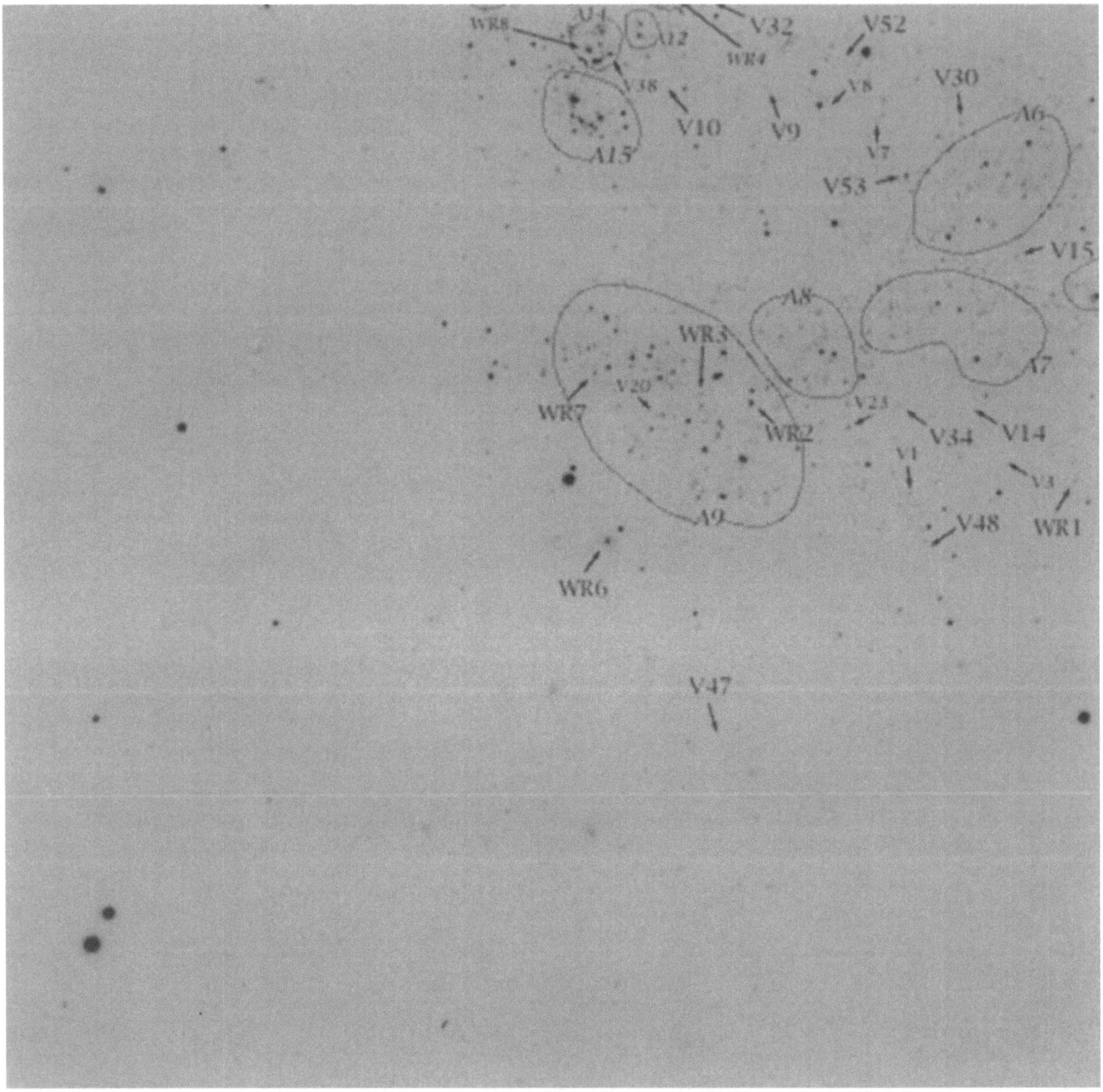

IC 1613

Chart 113

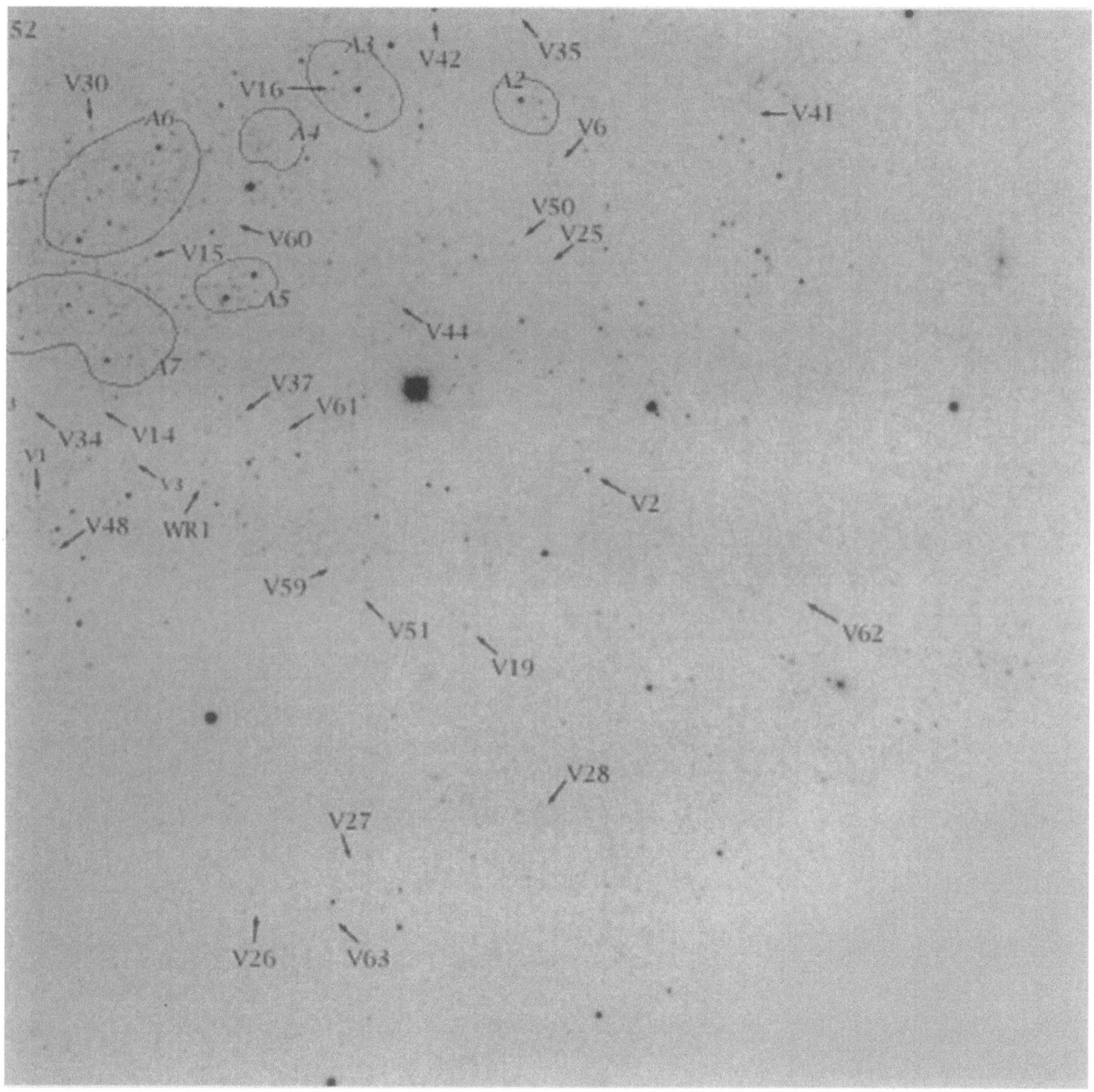

IC 1613

Chart 114

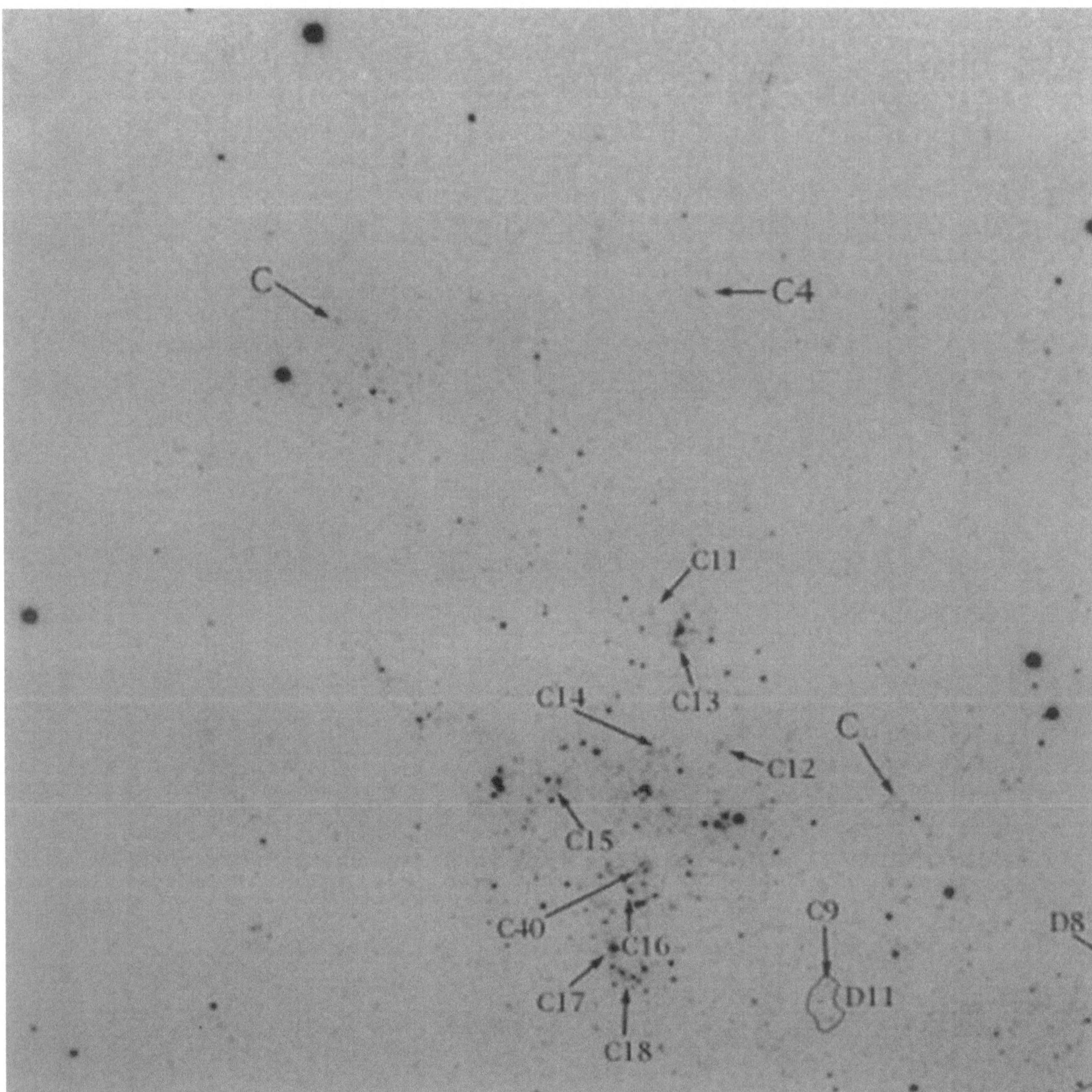

IC 1613

Chart 115

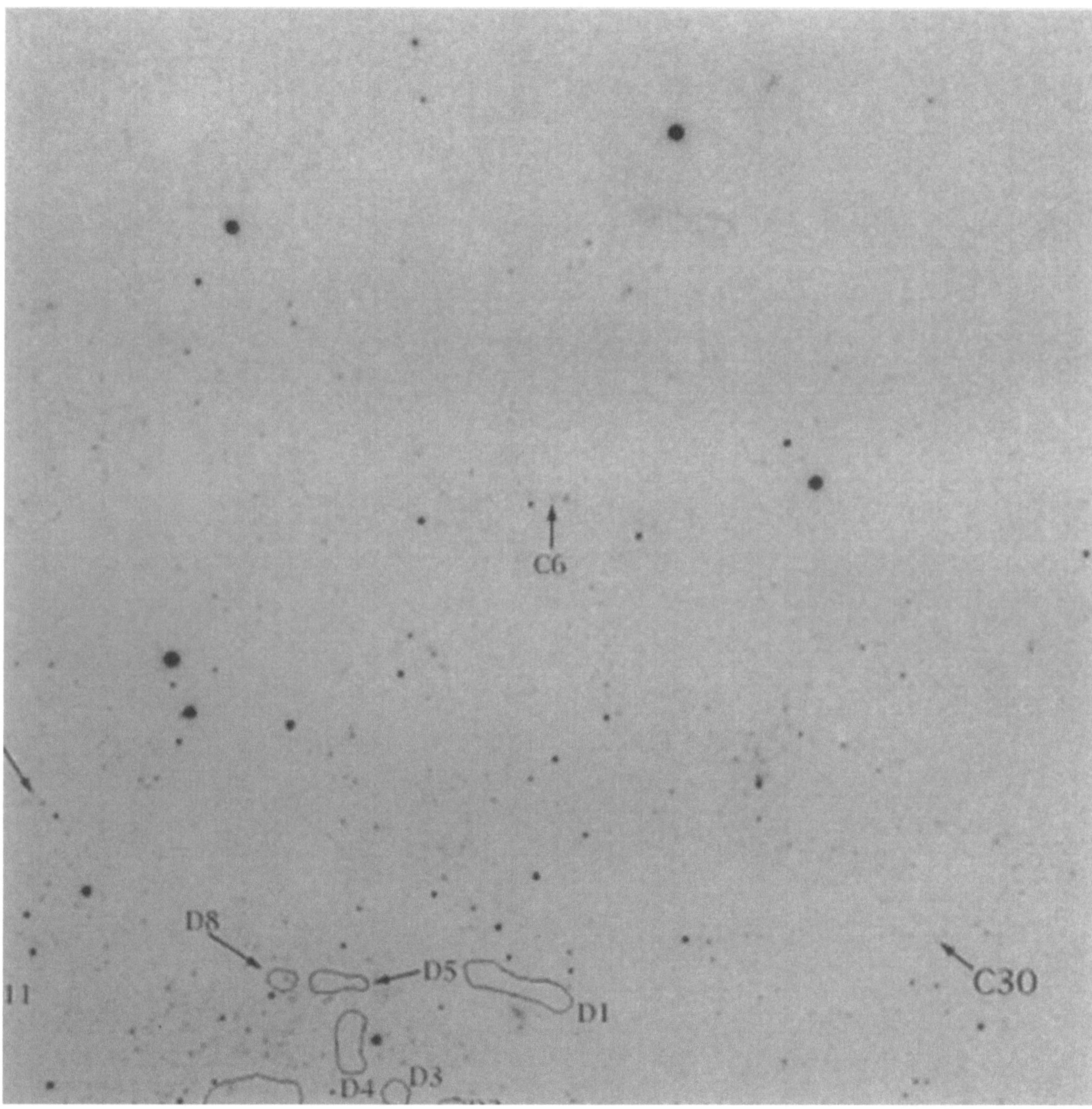

IC 1613

Chart 116

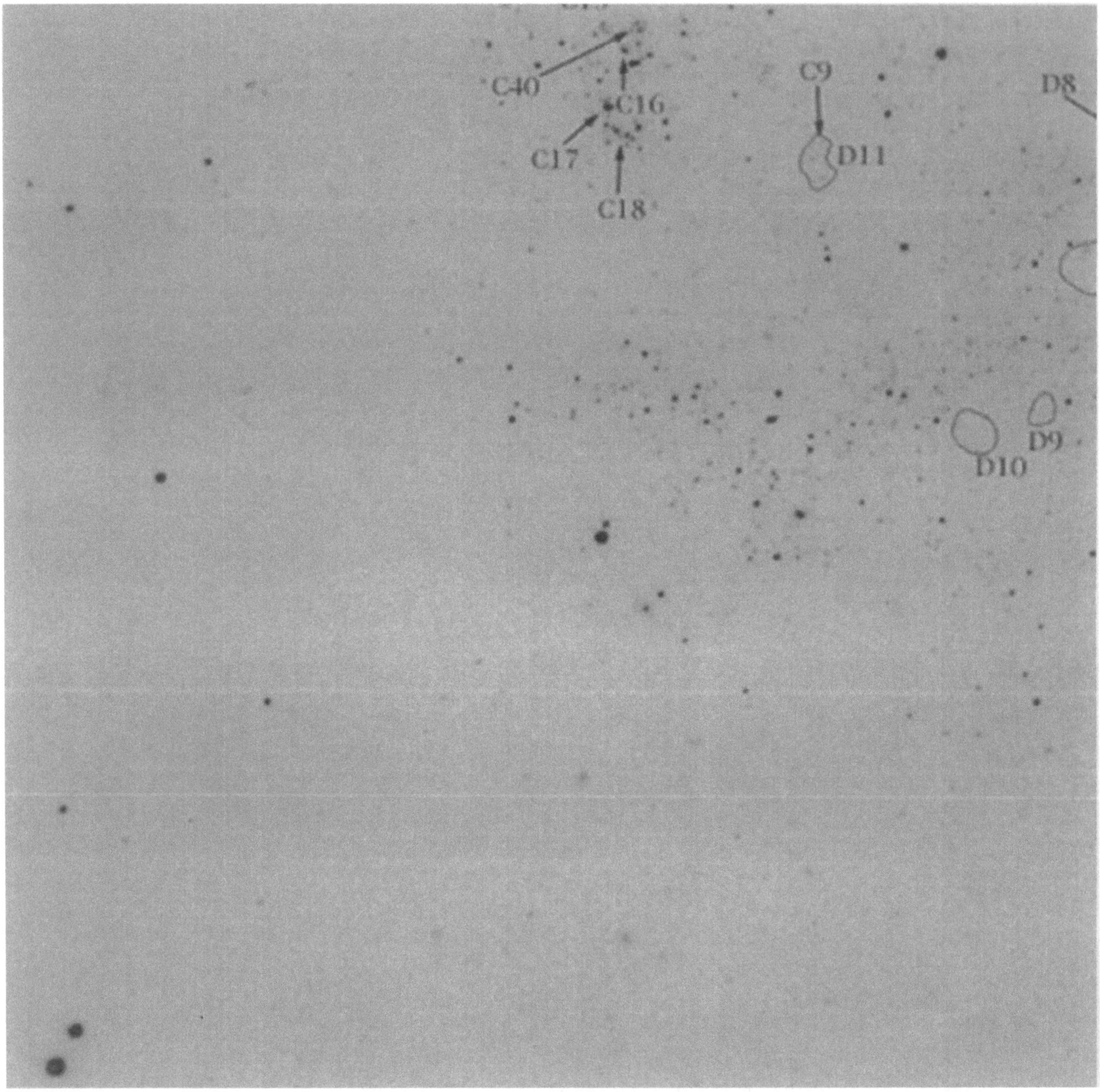

IC 1613

Chart 117

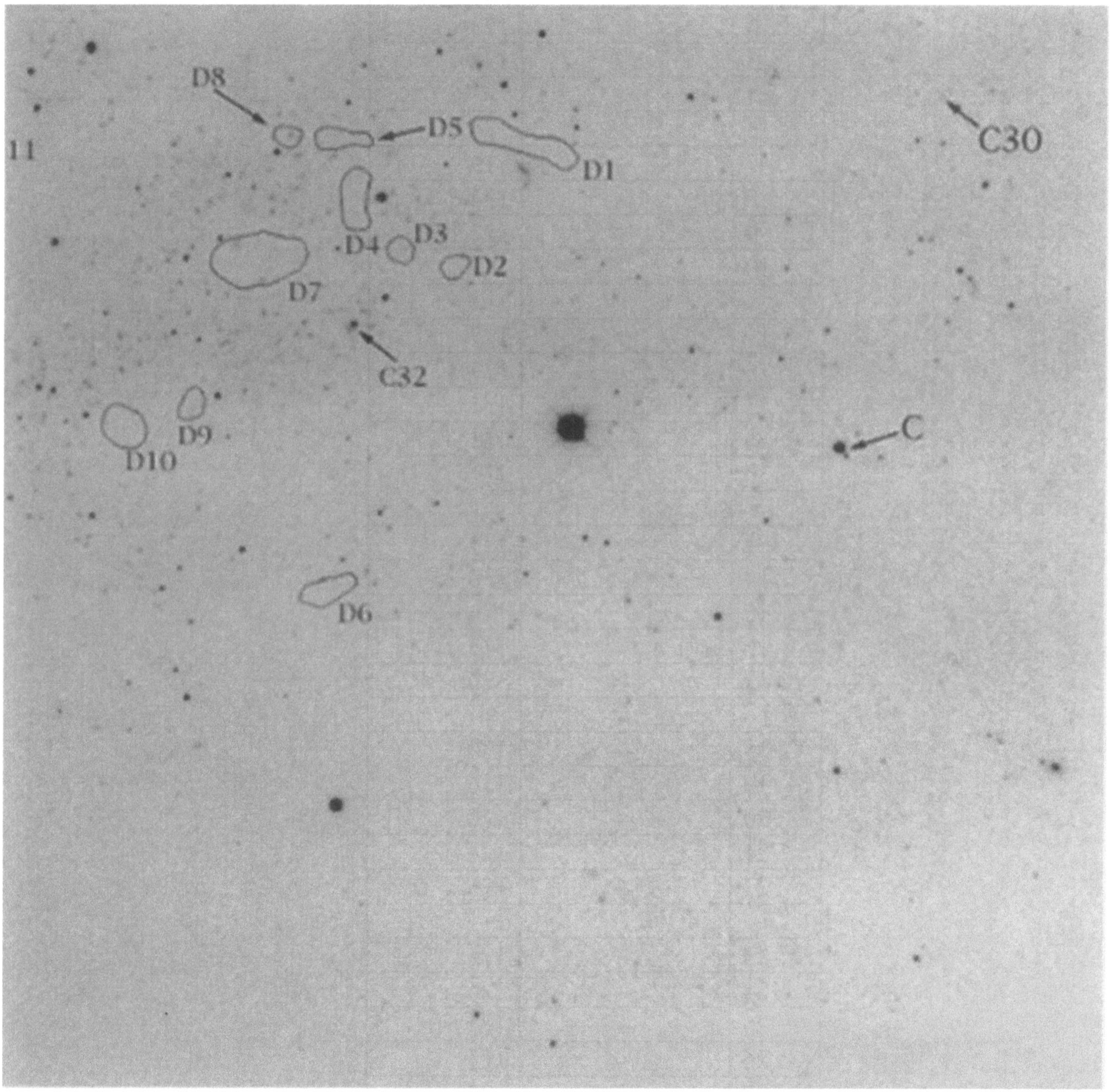

Table 5. IC 1613 Emission Regions[a]

ID	Sandage no.	Chart
1	--	128
2	--	134
3	--	128
4	--	128
5	S 1	128
6	--	134
7	--	119
8	--	134
9	--	122,128
10	--	131
11	--	127
12	--	127
13	S 2	131
13a	--	131
13b	--	131
13c	--	131
14	--	130
15	--	126
16	--	133
17	--	130,133
18	--	130
19	--	126
20	--	126
21	--	126
22	--	126
23	--	125
24	--	125
25	--	125
26	--	121
27	--	125
28		125
29	S 12	125
30		125

31	S 18	121
32	S 19	121,126
33	--	125
34	--	125
35	--	125
36	S 6	125
37	S 3	132
37a	--	132
37b	--	132
37c	--	132
37d	--	132
37e	--	132
37f	--	132
37g	--	132
38	--	125
39	--	125
40	S 7	125
40a	--	124
40b	--	124
41	--	125
42	--	129
43	--	129
44	--	121
45	S 17	125
46	--	125
46a	--	124
46b	--	124
46c	--	124
46d	S 15	124
46e	S 15	124
46f	S 16	124
46g	--	124
47	--	125
48	--	125
49	S 8	124,125
50	--	125

51	--	125
52	S 5	125
53	S 5	125
54	--	132
55	S 14	125
56	--	125
57	--	125
58	--	125
59	S 4	129
60	--	125
61	--	125
62	--	121
63	--	125
64	--	125
65	--	129
66	--	121
67	S 10	125
67a	--	124
67b	--	124
67c	--	124
67d	S 13	124
68	--	125
69	--	125
70	S 11	123
71	--	120
72	--	123
73	--	123
74	--	120
75	--	120
76	--	120
77	--	133

[a]Designations from reference 28; Sandage numbers from reference 56.

IC 1613

Chart 118

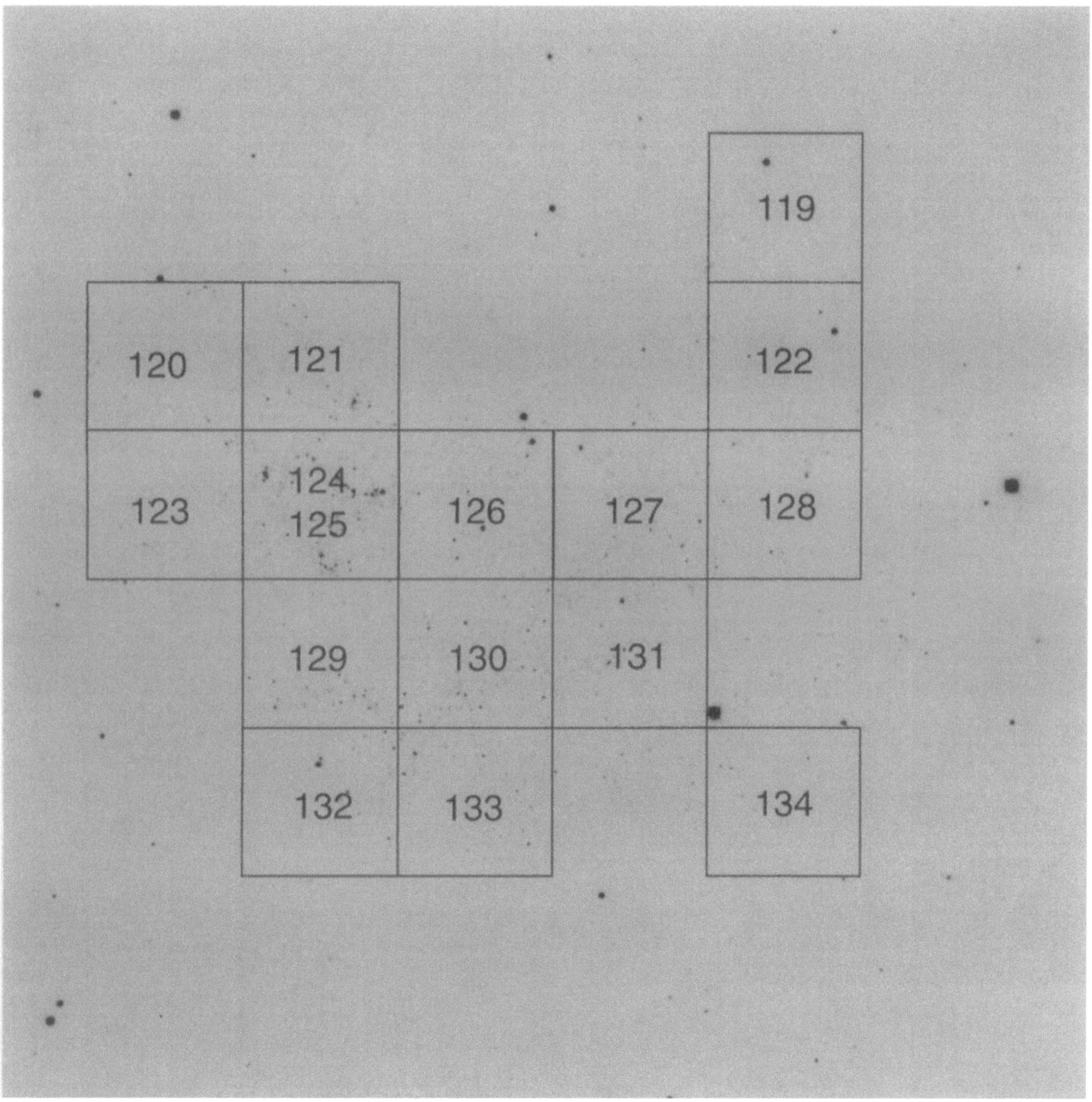

IC 1613

Chart 119

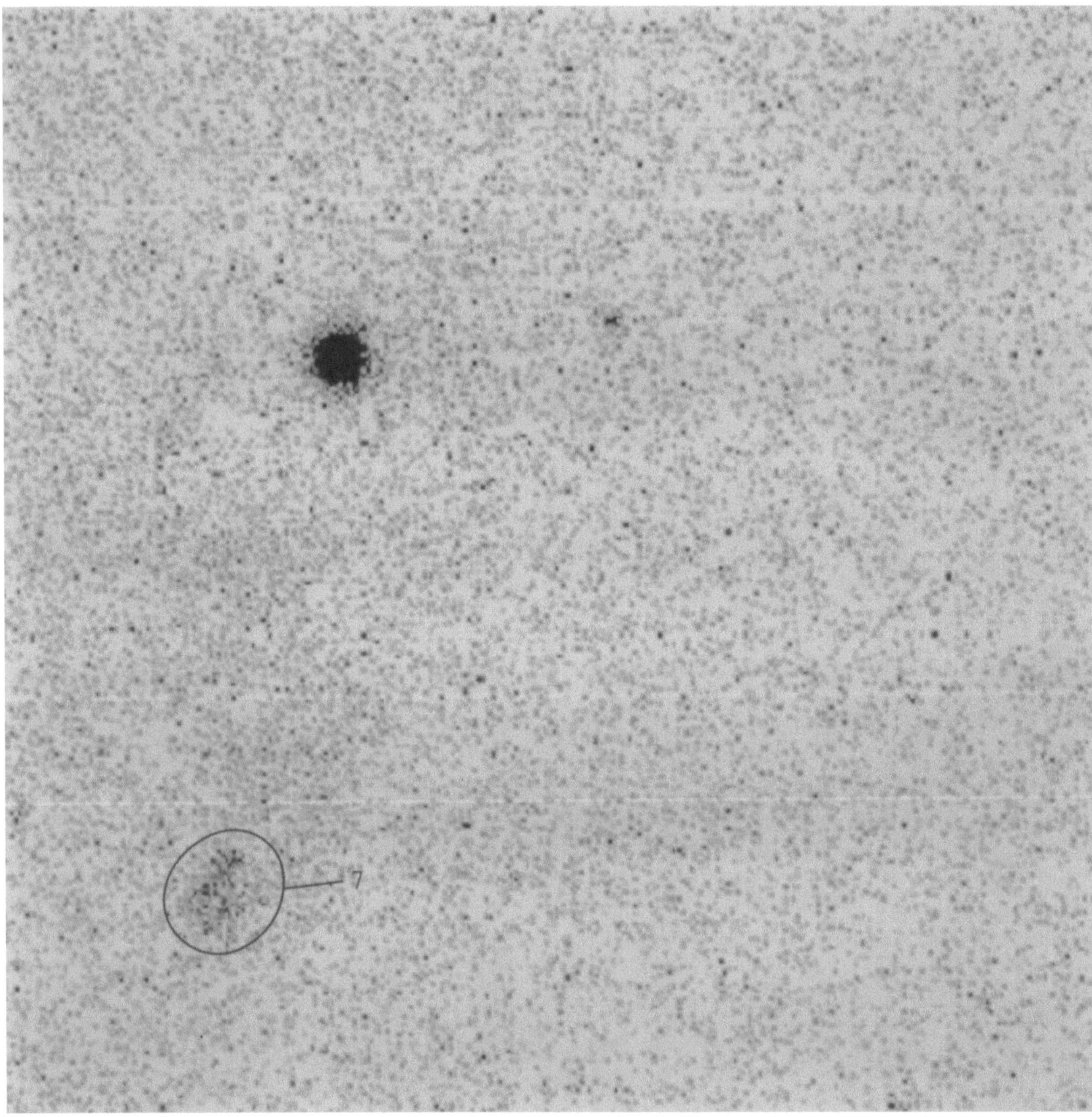

IC 1613

Chart 120

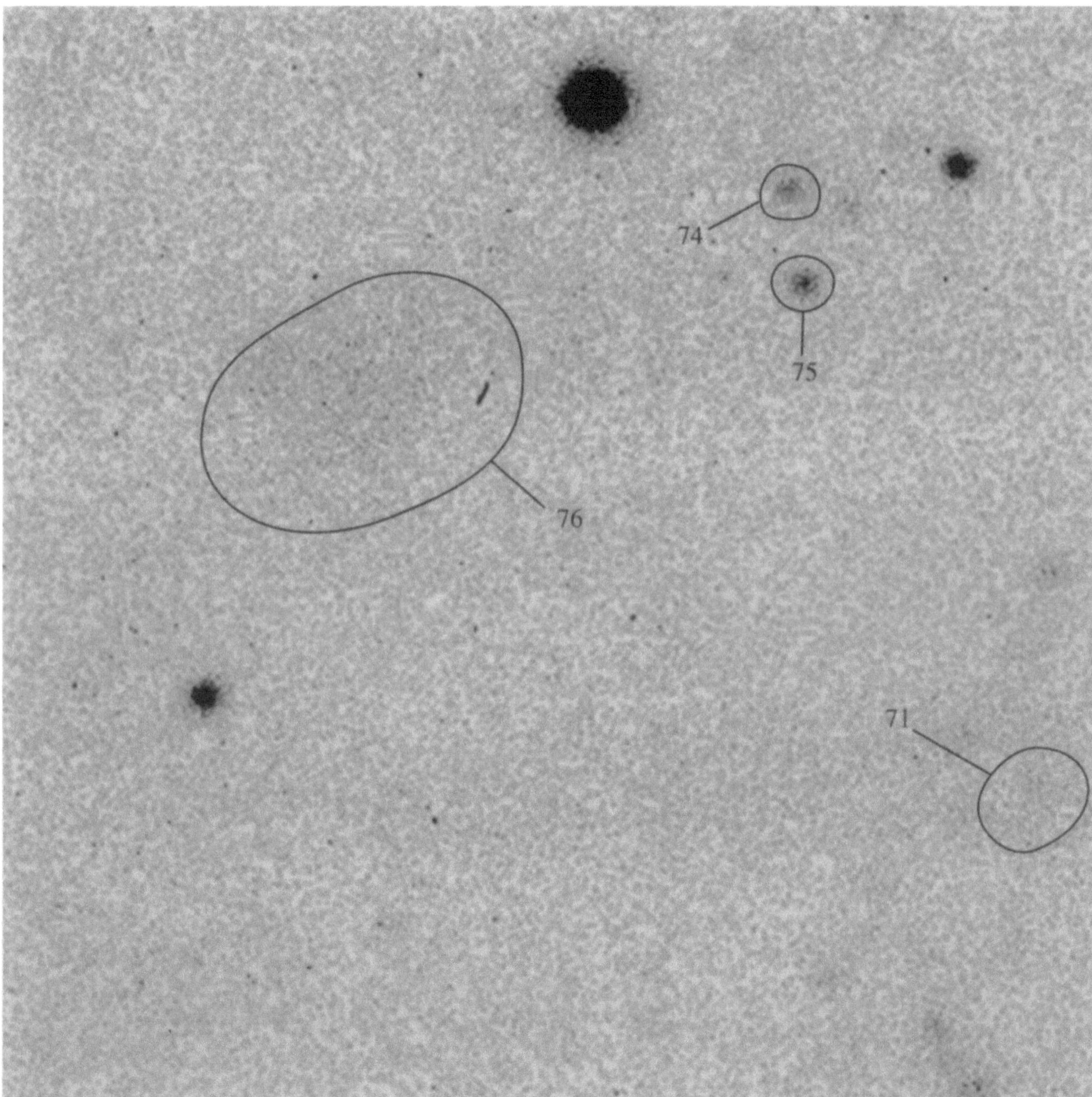

IC 1613

Chart 121

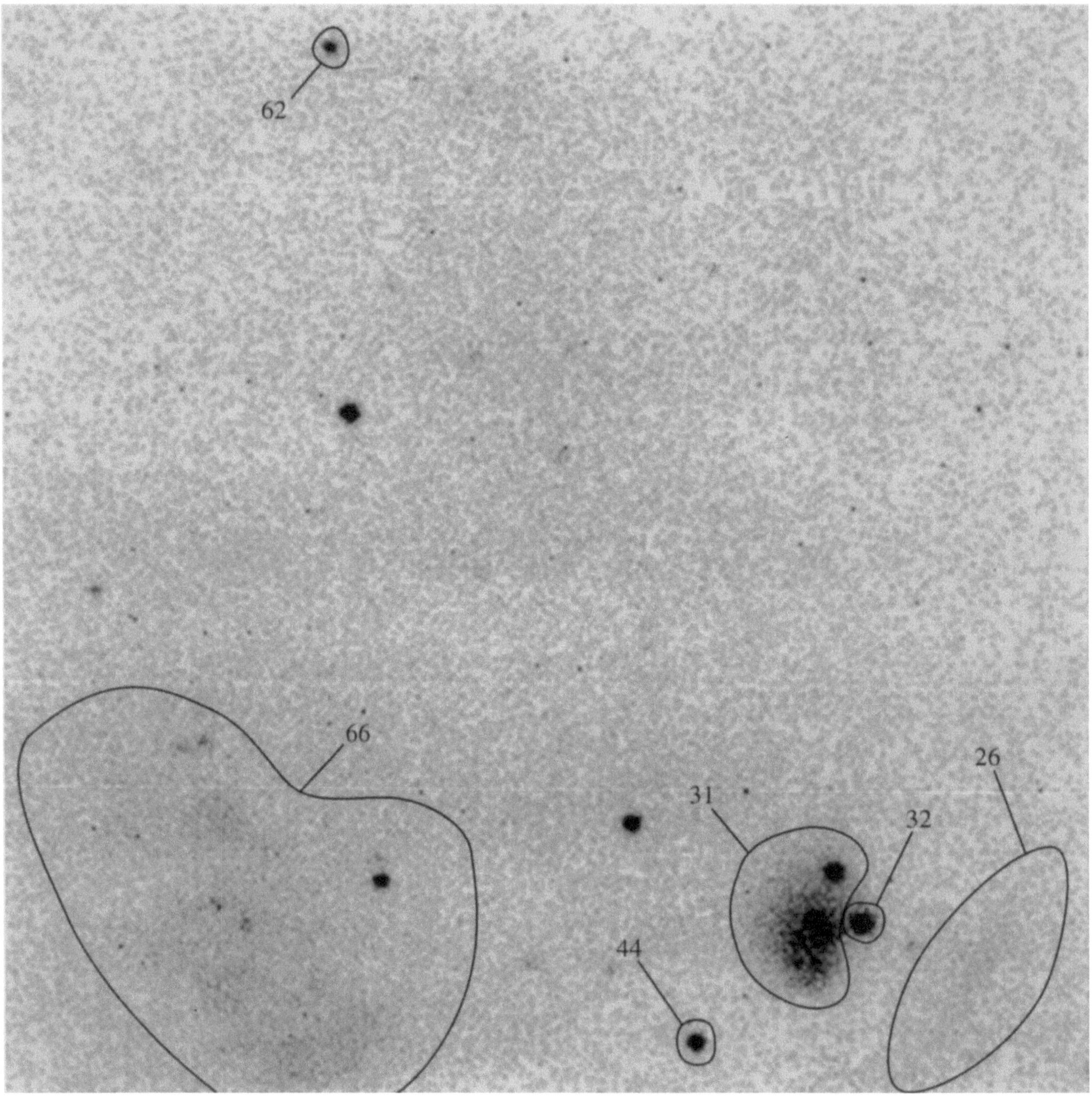

IC 1613

Chart 122

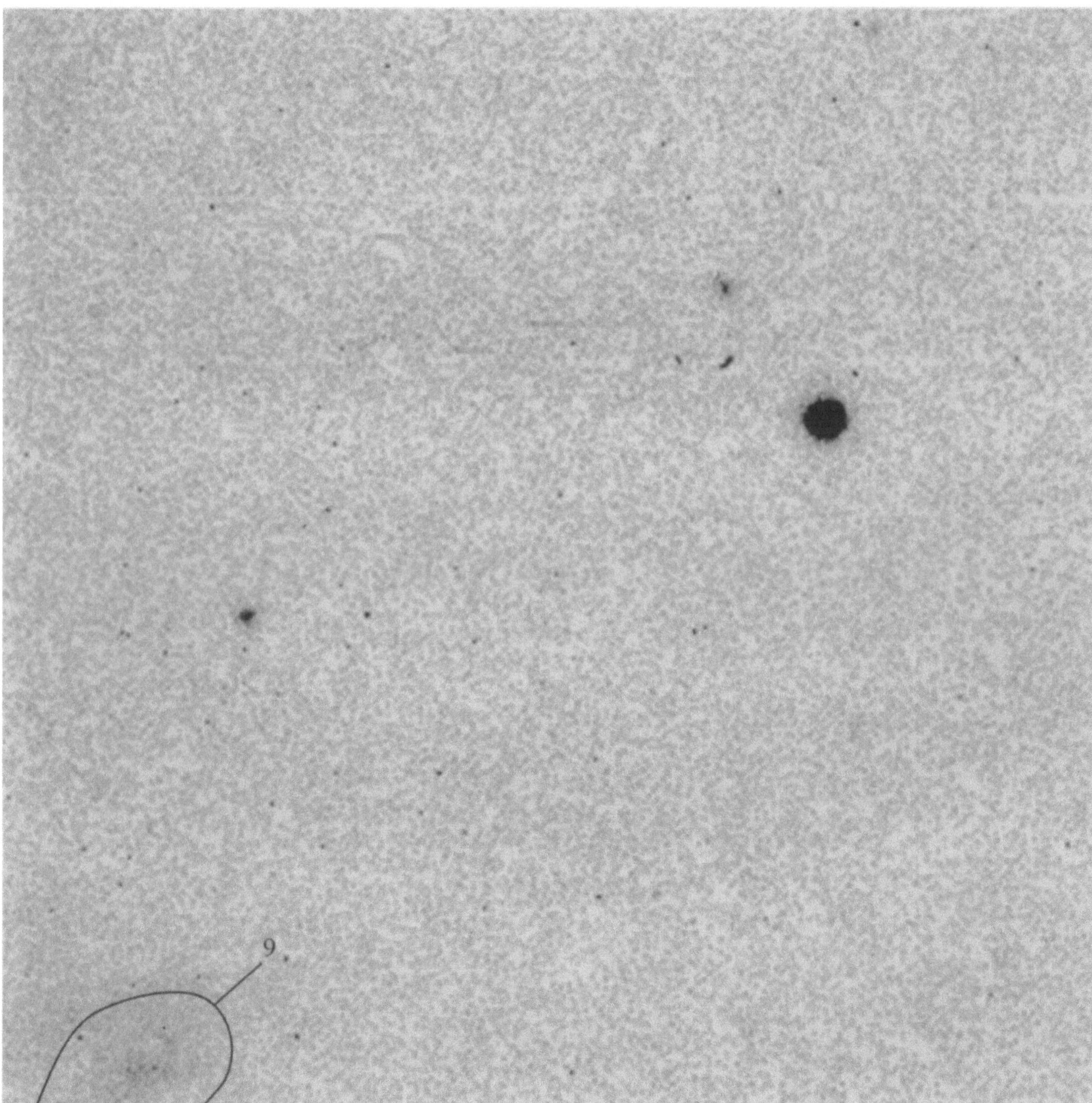

IC 1613

Chart 123

IC 1613

Chart 124

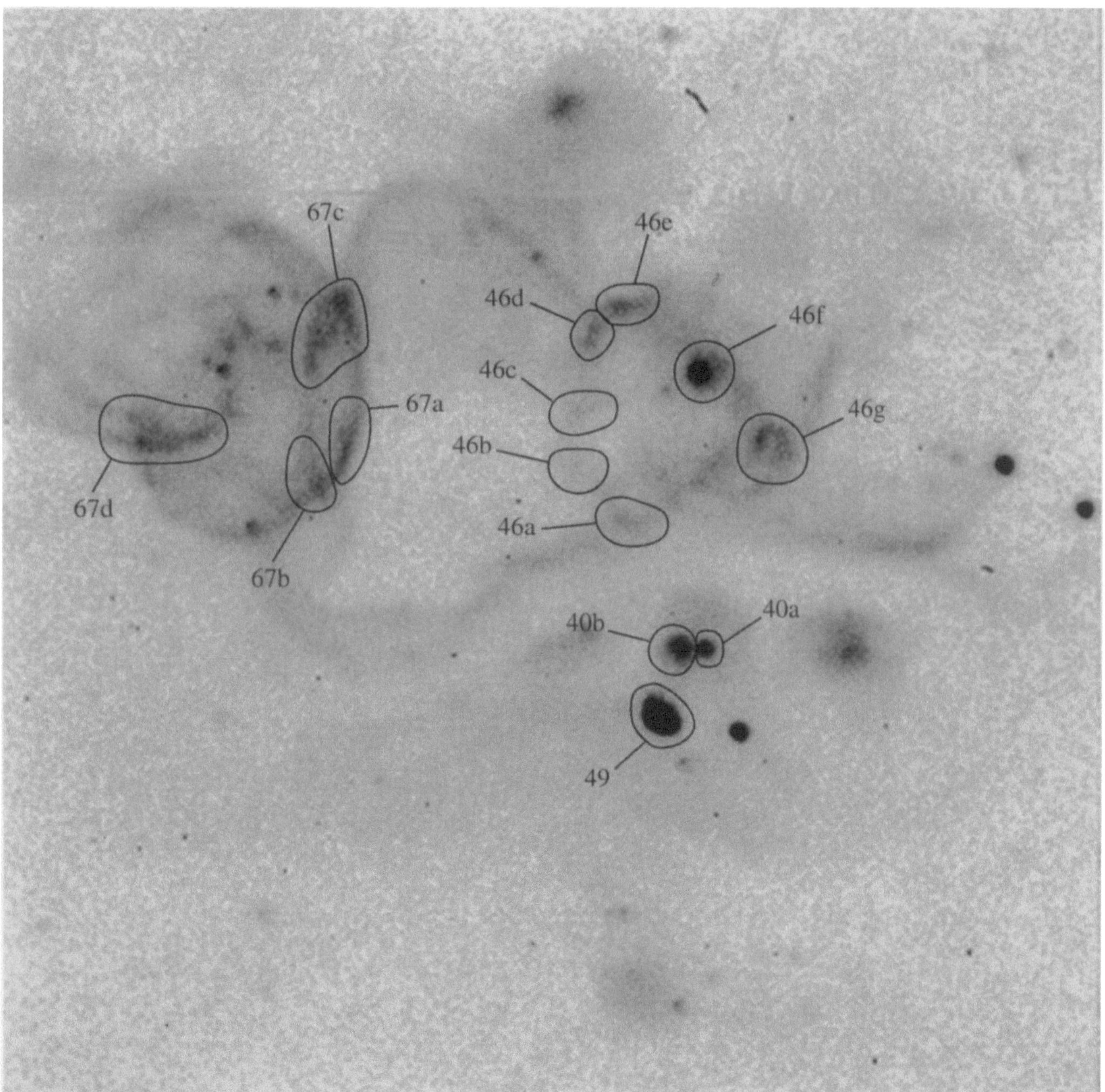

IC 1613

Chart 125

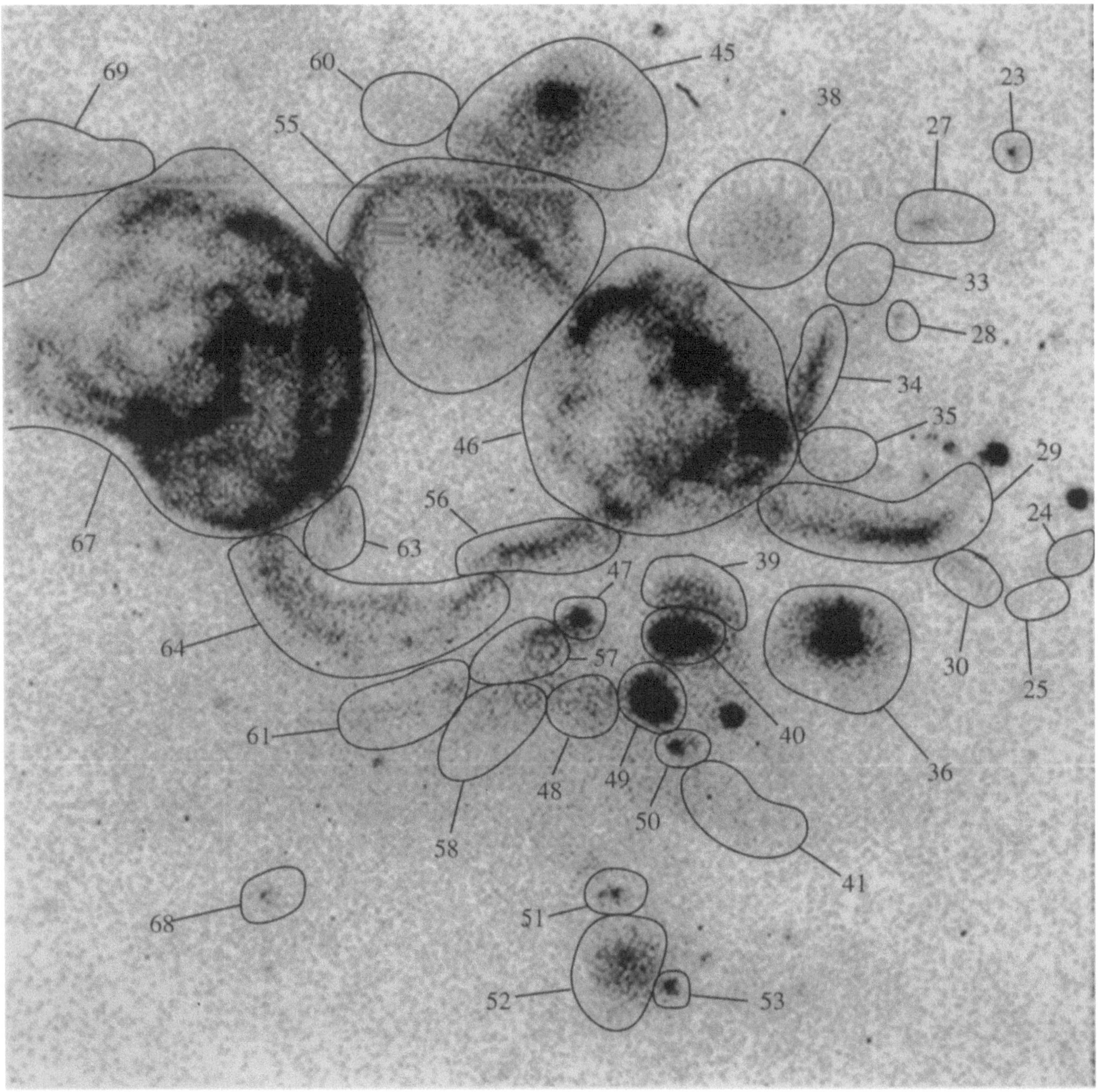

IC 1613

Chart 126

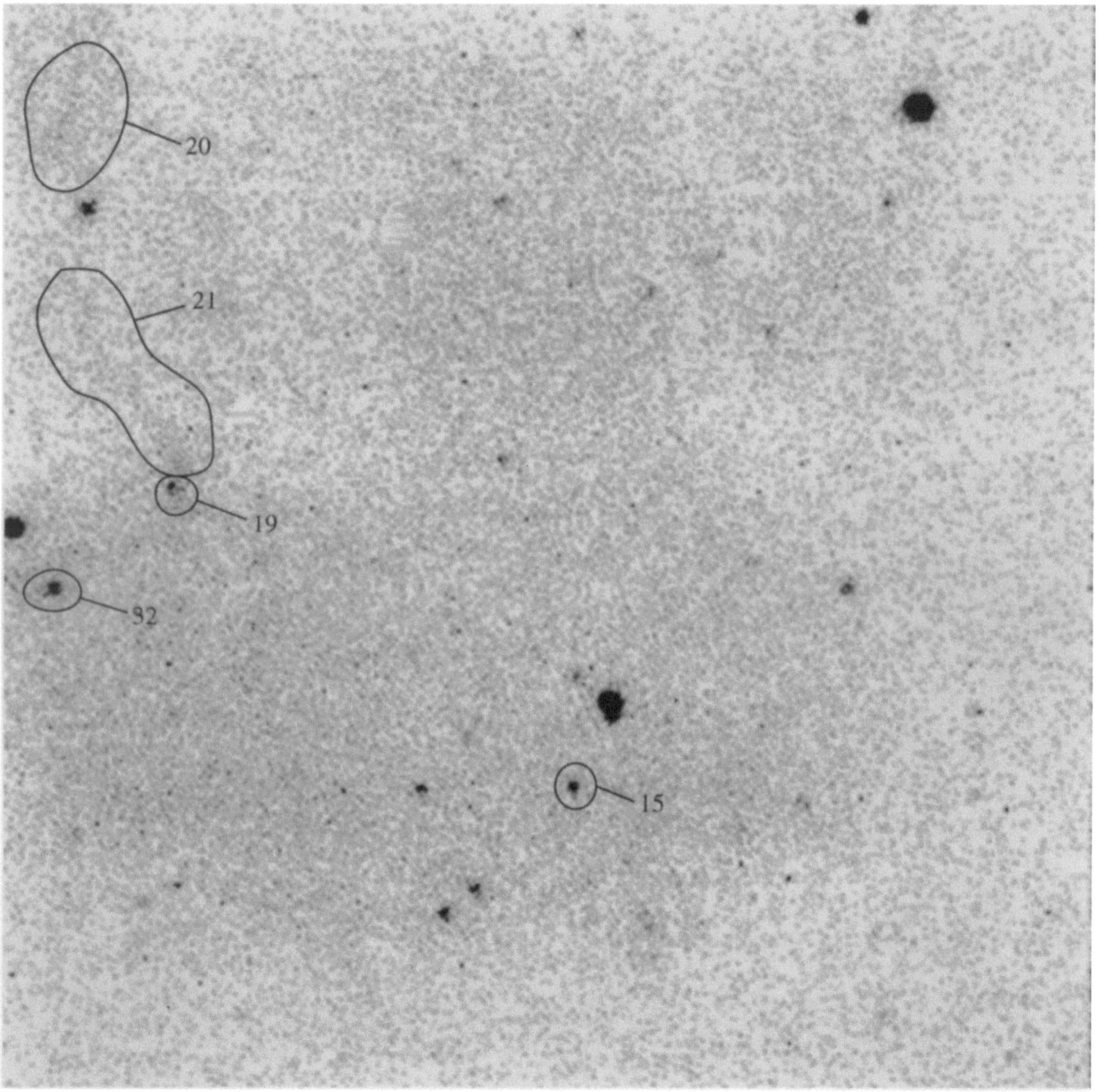

IC 1613

Chart 127

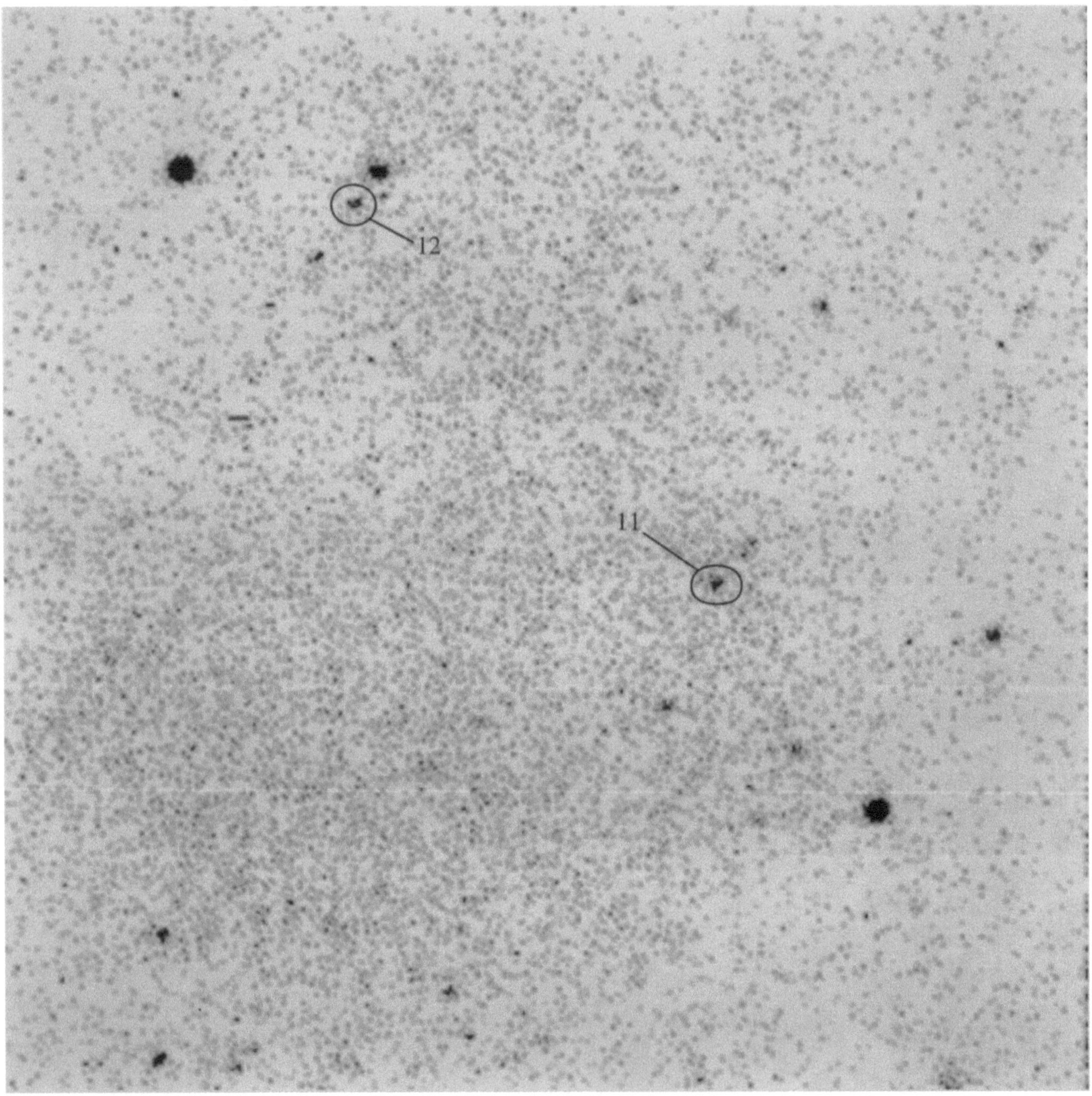

IC 1613

Chart 128

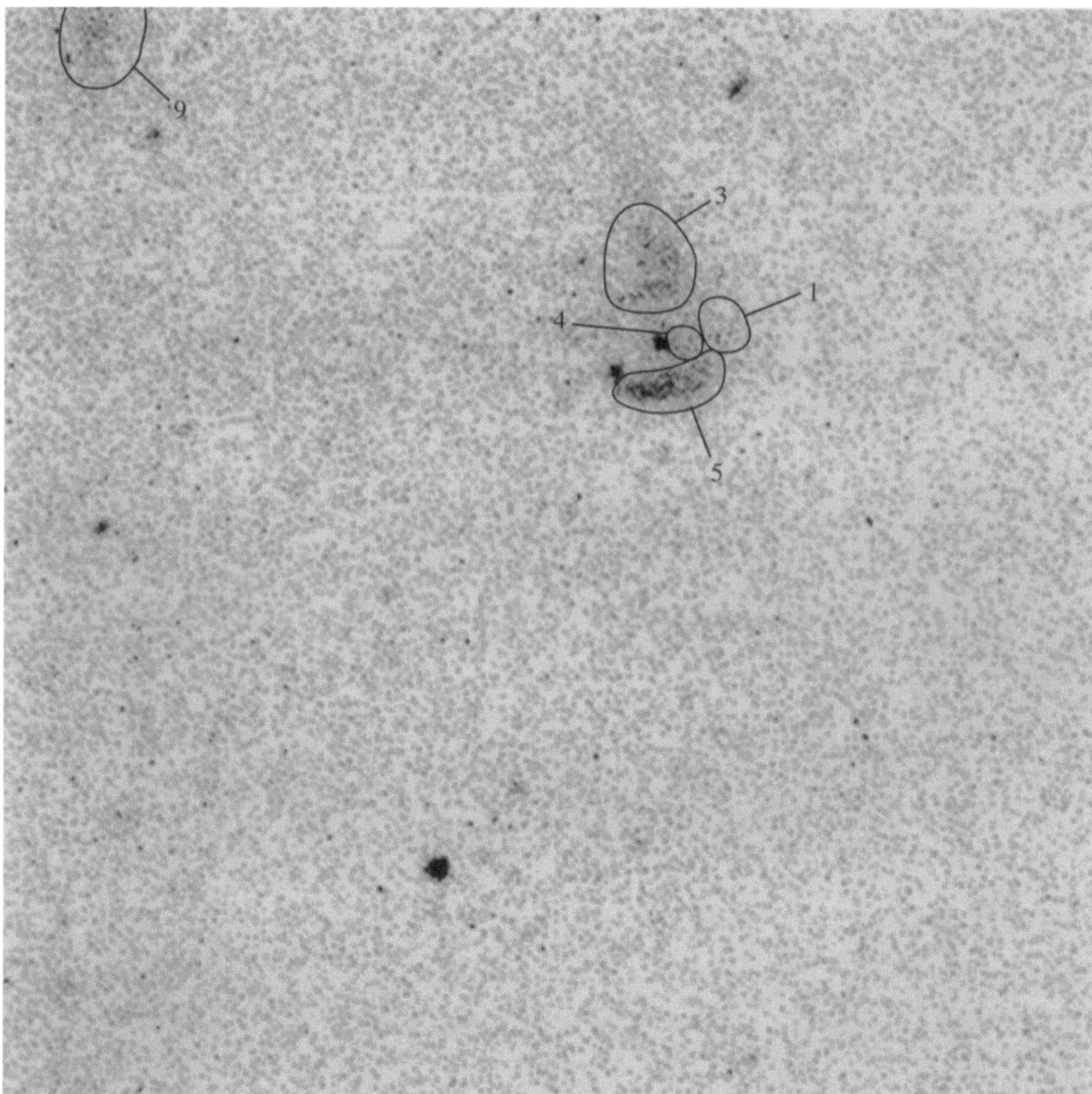

IC 1613

Chart 129

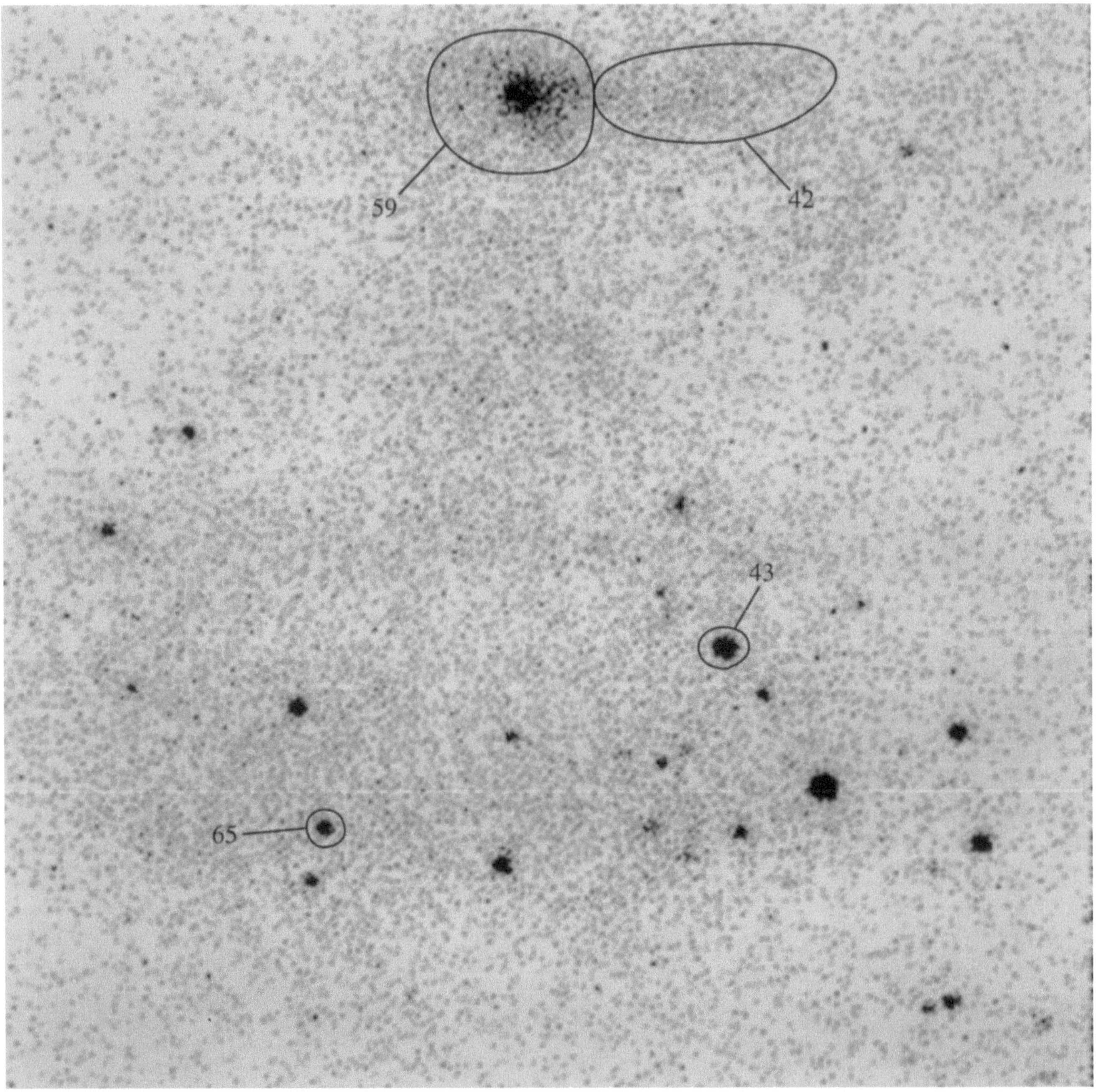

IC 1613

Chart 130

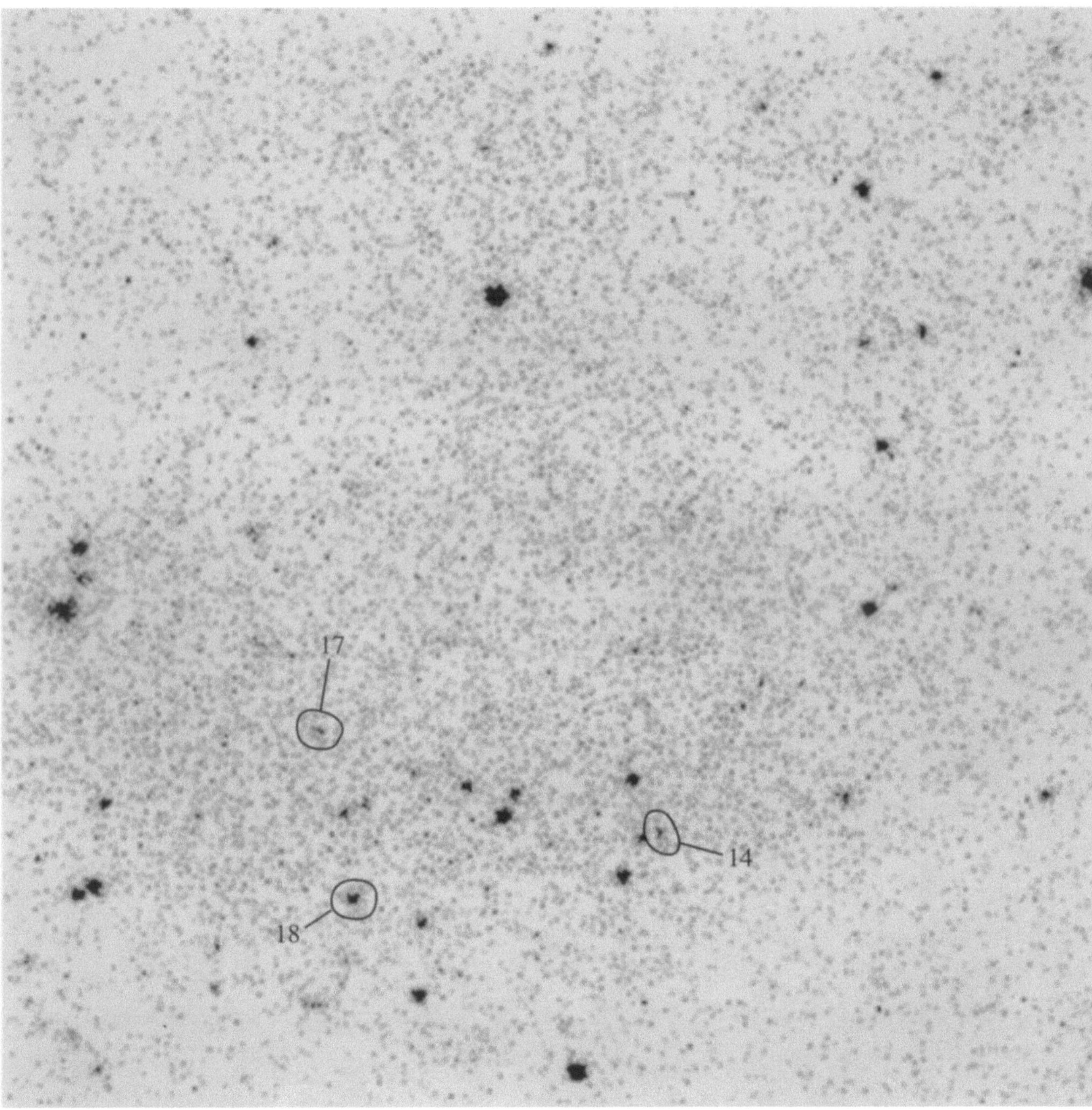

IC 1613

Chart 131

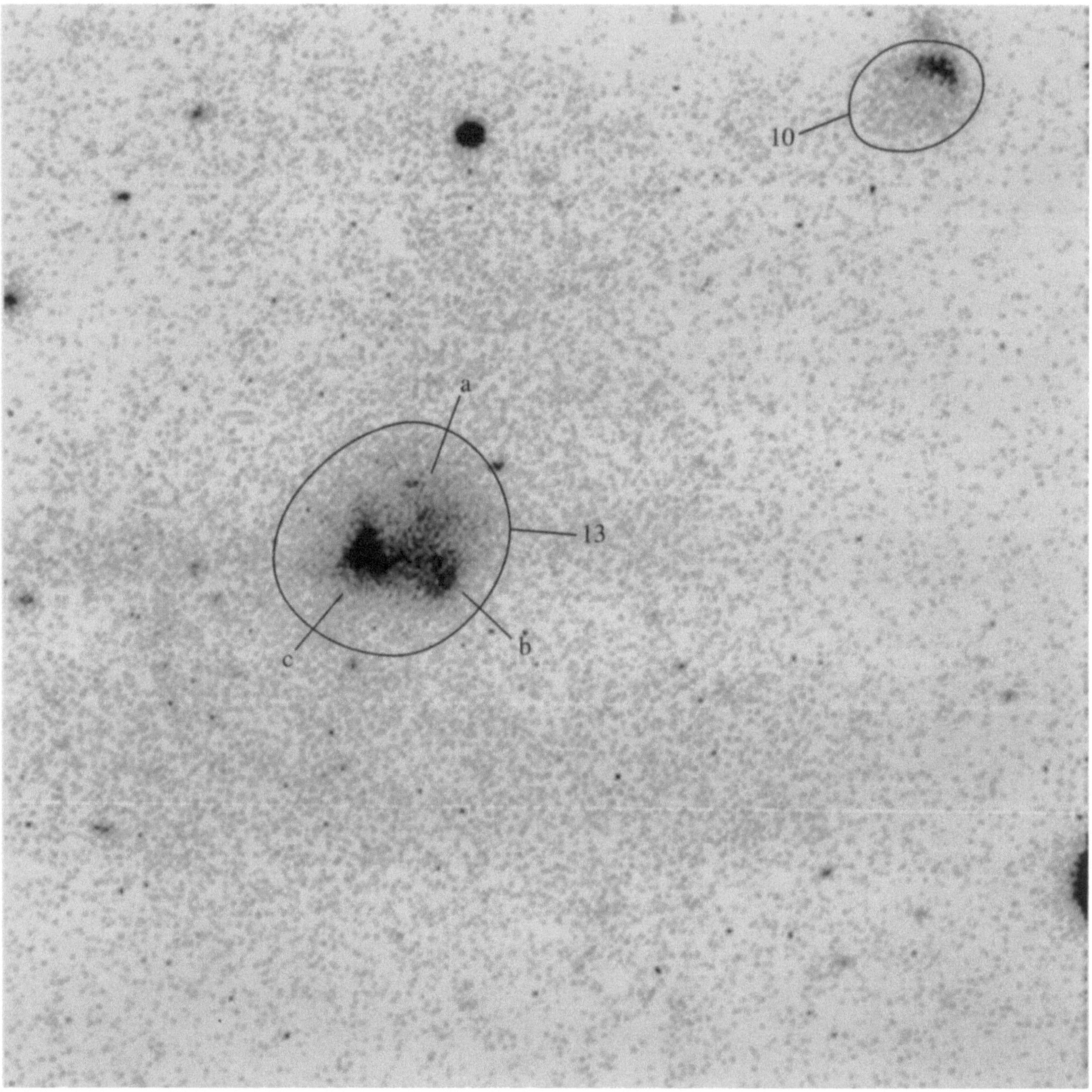

IC 1613

Chart 132

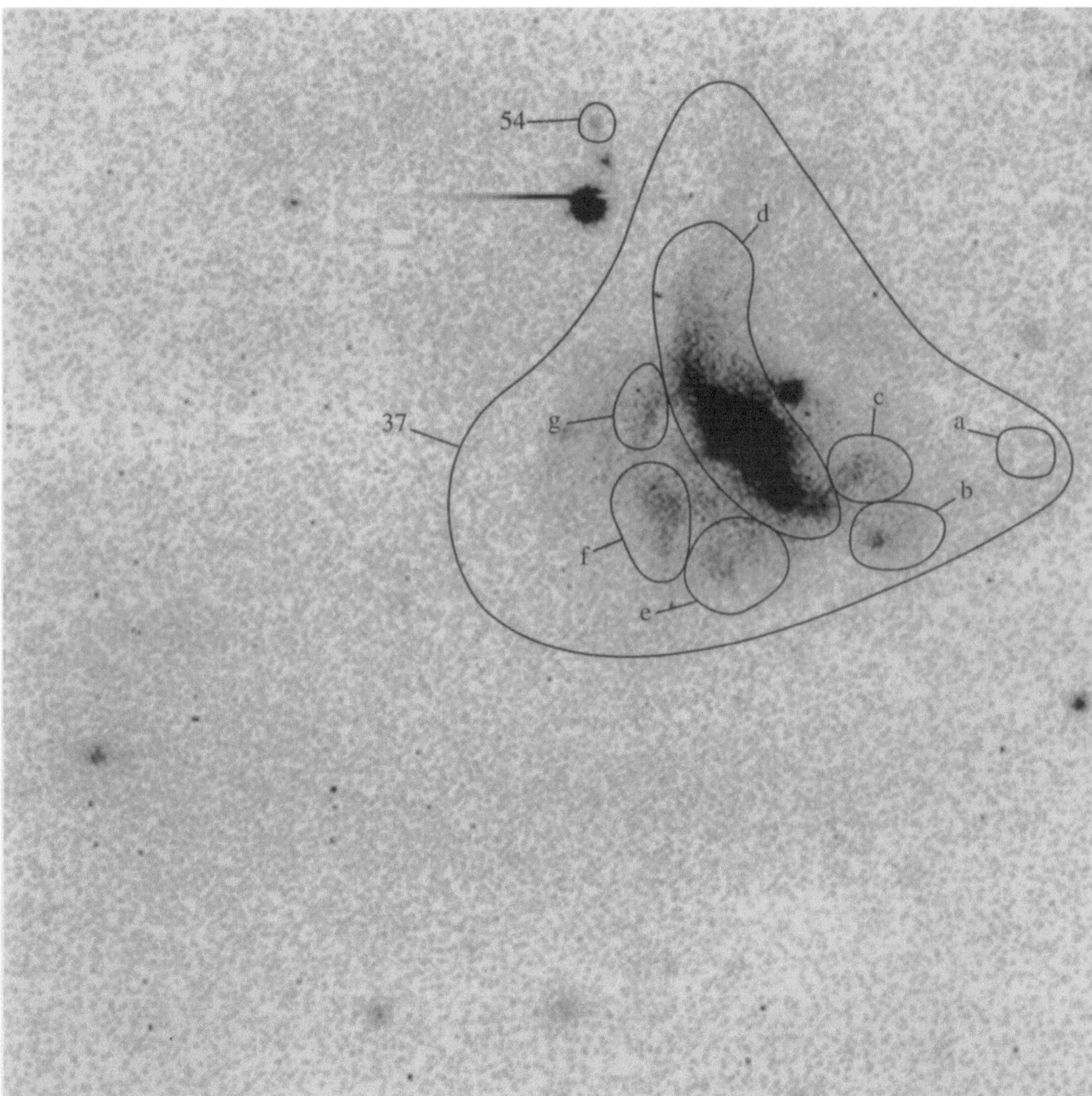

IC 1613

Chart 133

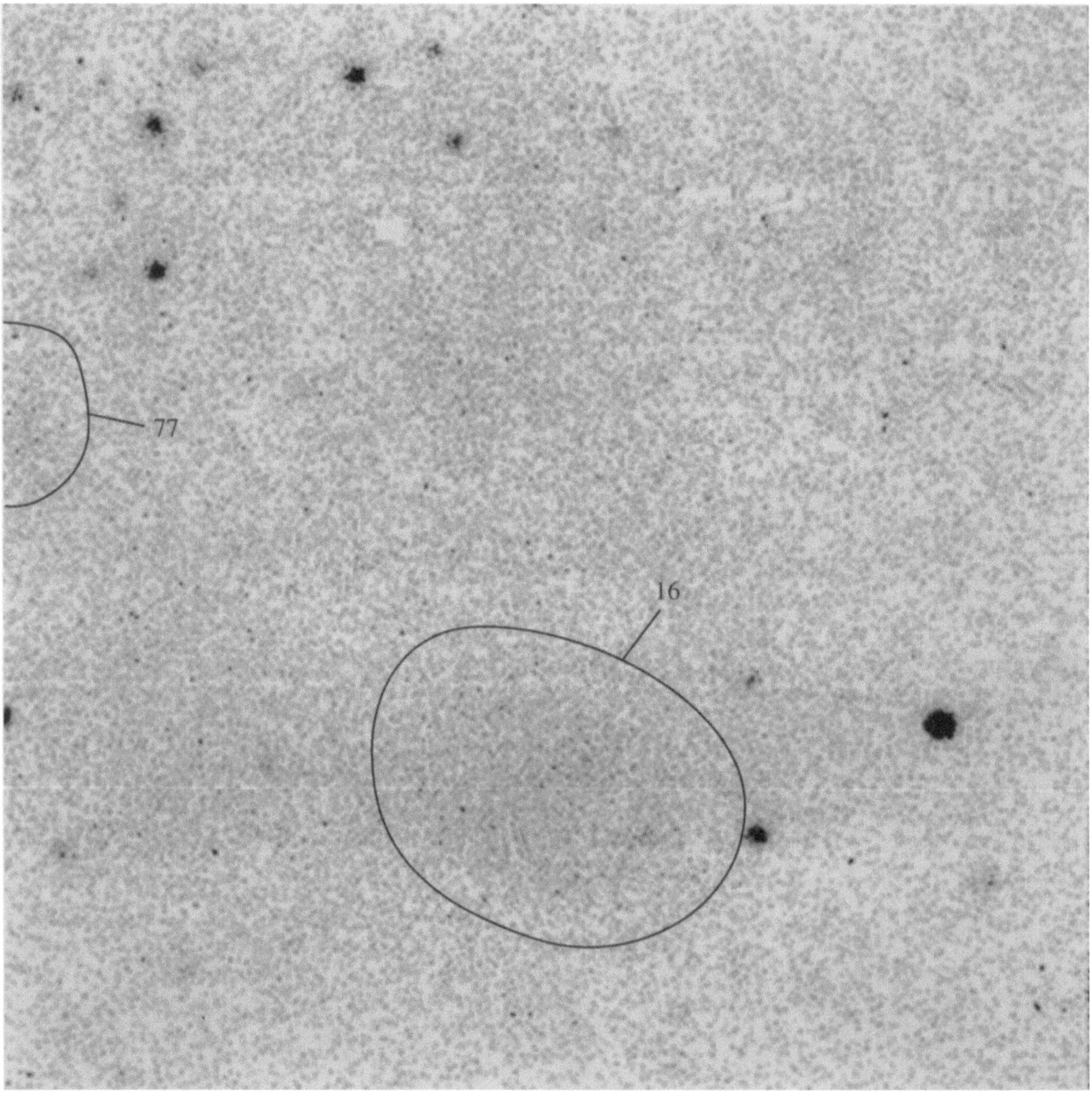

IC 1613

Chart 134

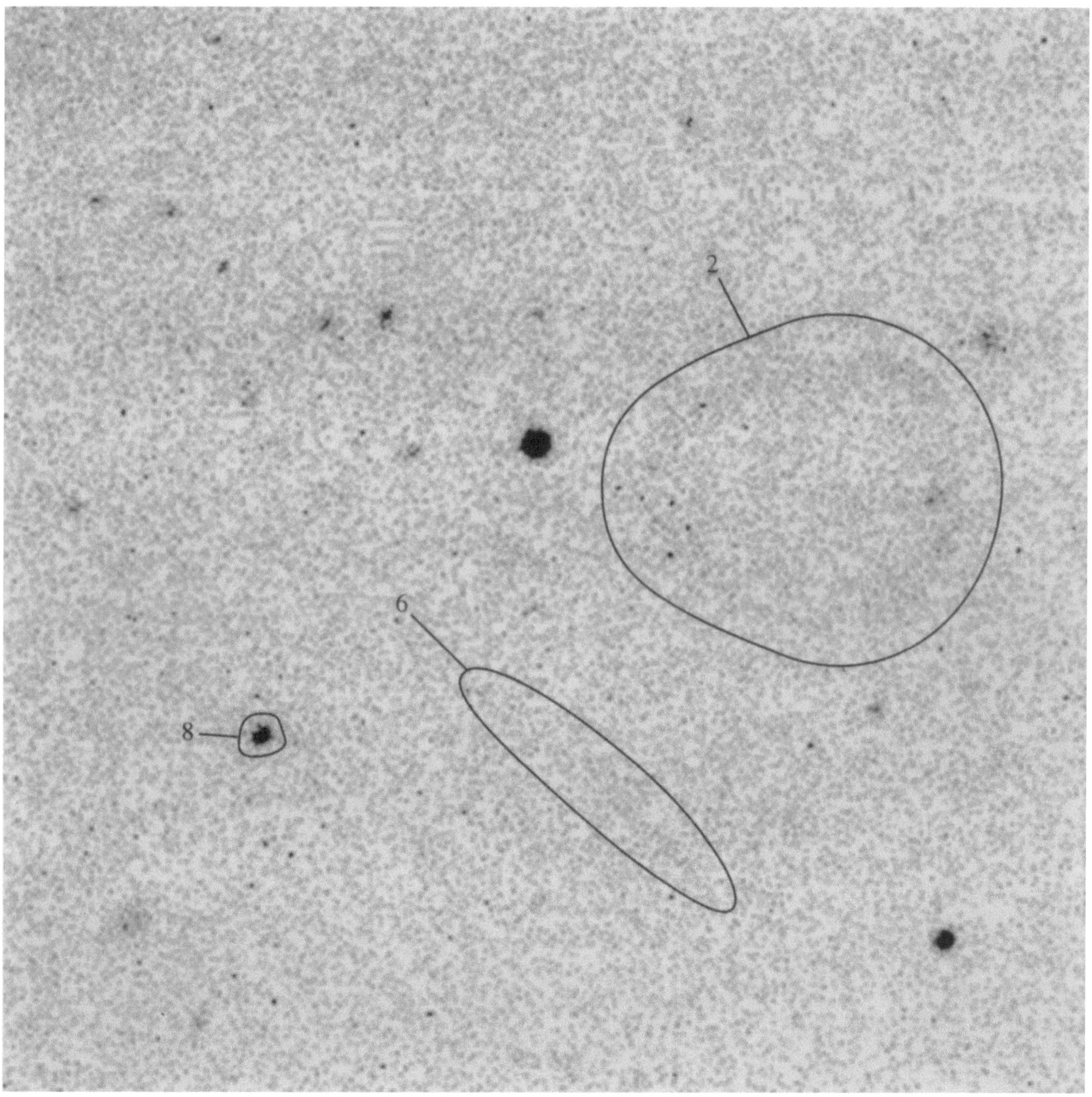

IC 1613

References

1. Aaronson, M., Cook, K.H., Norris, J. 1985, "The AGB Population of Nearby Galaxies", in Spectral Evolution of Galaxies, edited by C. Chiosi and A. Renzini (Reidel, Dordrecht), p. 171.
2. Ables, H.D. 1971, "Optical Study of Nearby Galaxies", Pub. Of the U.S. Naval Observatory, Vol. 20, Part 4, 7.
3. Armandroff, T.E., & Massey, P. 1985, "Wolf-Rayet Stars in NGC 6822 and IC 1613", ApJ, 291, 685.
4. Armandroff, T.E., & Massey, P. 1991, "Wolf-Rayet Stars in Local Group Galaxies: Numbers and Spectral Properties", AJ, 102, 927.
5. Azzopardi, M., Lequeux, J., & Maeder, A. 1988, "The Number of Wolf-Rayet Stars in Local Group Galaxies", A&A, 189, 34.
6. Baade, W. 1929, "Der Nebel NGC II 1613", Astr. Nach., 234, 407.
7. Breysacher, J., & Lequeux, J. 1983, "Search for Wolf-Rayet, Carbon and M Stars in External Galaxies with the GRISM/GRENS Technique", The Messenger, 33, 21.
8. Carlson, G., & Sandage, A. 1990, "Periods and Light Curves of 16 Cepheid Variables in IC 1613 Not Completed by Baade", ApJ, 352, 587.
9. Cook, K.H., Aaronson, M., Norris, J. 1986, "Carbon and M Stars in Nearby Galaxies: A Preliminary Survey Using a Photometric Technique", ApJ, 305, 634.
10. D'Odorico, S., & Dopita, M. 1983, "Chemical Abundances in the Interstellar Medium of Galaxies from Spectrophotometry of Supernova Remnants", in Supernova Remnants and their X-Ray Emission, IAU Symposium No. 101, edited by J. Danziger and P. Gorenstein (Reidel, Dordrecht), p. 517.
11. D'Odorico, S., Dopita, M.A., & Benvenuti, P. 1980, "A Catalog of Supernova Remnant Candidates in Nearby Galaxies", A&AS, 40, 67.
12. D'Odorico, S., & Rosa, M. 1982, "Wolf-Rayet Stars in Extragalactic H II Regions: Discovery of a Peculiar WR in IC 1613/#3", A&A, 105, 410.
13. Davidson, K., & Kinman, T.D. 1982, "Data on an Unusual Wolf-Rayet Star in the Nearby Galaxy IC 1613", PASP, 94, 634.
14. de Vaucouleurs, G. 1978, "The Extragalactic Distance Scale. II. Distances of the Nearest Galaxies from Primary Indicators", ApJ, 223, 730.
15. Dickel, J.R., D'Odorico, S., & Silverman, A. 1985, "VLA Observations of Three Extragalactic SNR at 20 and 6 cm", AJ, 90, 414.
16. Elias, J.H., & Frogel, J.A. 1985, "M Supergiants in Local Group Irregular Galaxies: Metallicities and Distances", ApJ, 289, 141.
17. Elmegreen, B.G., Elmegreen, D.M., & Morris, M. 1980, "On the Abundance of Carbon Monoxide in Galaxies: A Comparison of Spiral and Magellanic Irregular Galaxies", ApJ, 240, 455.
18. Eskridge, P.B. 1995, "IC 1613 X-1: A Galaxy Cluster at $z \sim 0.2$ Behind a Local Group Dwarf Galaxy", PASP, 107, 561.
19. Fabbiano, G., Kim. D.-W., & Trinchieri, G. 1992, "An X-Ray Catalog and Atlas of Galaxies", ApJS, 80, 531.

20. Freedman, W.L. 1986, "Recent CCD Observations of Cepheids in Nearby Galaxies: A Progress Report", in Galaxy Distances and Deviations from Universal Expansion, edited by B.F. Madore and R.B. Tully (Reidel, Dordrecht), p. 21.
21. Freedman, W.L. 1988, "New Cepheid Distances to Nearby Galaxies Based on *BVRI* CCD Photometry. I. IC 1613", ApJ, 326, 691.
22. Freedman, W.L. 1988, "Stellar Content of Nearby Galaxies. I. *BVRI* CCD Photometry for IC 1613", AJ, 96, 1248.
23. Garnett, D.R., Kennicutt, Jr., R.C., Chu, Y.-H., & Skillman, E.D. 1991, "He II Emission in Extragalactic H II Regions", ApJ, 373, 458.
24. Goss, W.M., & Lozinskaya, T.A. 1995, "Radio and IRAS Observations of Two Nebulae around WO Stars: G2.4+1.4 in the Galaxy and Sandage H II Region 3 in IC 1613", ApJ, 439, 637.
25. Hodge, P.W. 1978, "The Structure and Content of IC 1613", ApJS, 37, 145.
26. Hodge, P.W. 1980, "The Recent Evolutionary History of the Galaxies NGC 6822 and IC 1613", ApJ, 241, 125.
27. Hodge, P.W. 1986, "The Ages of Local Group Galaxies", MSAIt, 57, 553.
28. Hodge, P.W., Lee, M.G., & Gurwell, M. 1990, "The H II Regions of IC 1613", PASP, 102, 1245.
29. Hodge, P.W., Smith, T.R., Eskridge, P.B., MacGillivray, H.T., & Beard, S.M. 1991, "A Cosmos Study of IC 1613", ApJ, 369, 372.
30. Hoessel, J.G. 1986, "Luminosity Functions in Dwarf Irregulars", in Luminous Stars and Associations in Galaxies, IAU Symposium No. 116, edited by C.W.H. de Loore, A.J. Willis and P. Laskarides (Reidel, Dordrecht), p. 439.
31. Hoffman, G.L., Salpeter, E.E., Farhat, B., Roos, T., Williams, H., & Helou, G. 1996, "Arecibo H I Mapping of a Large Sample of Dwarf Irregular Galaxies", ApJS, 105, 269.
32. Huchtmeier, W.K., Seiradakis, J.H., & Materne, J. 1981, "The Neutral Hydrogen Distribution of Irregular Galaxies", A&A, 102, 134.
33. Humphreys, R.M. 1980, "Studies of Luminous Stars in Nearby Galaxies. V. The Local Group Irregulars NGC 6822 and IC 1613", ApJ, 238, 65.
34. Hunter, D.A., & Gallagher, III, J.S. 1985, "Star-Forming Properties and Histories of Dwarf Irregular Galaxies: Down but not Out", ApJS, 58, 533.
35. Hunter, D.A., Hawley, W.N., and Gallagher, III, J.S. 1993, "A Survey for Extra-H II Region Ionized Gas Structures in Irregular Galaxies", AJ, 106, 1797.
36. Kennicutt, Jr., R.C. 1984, "Structural Properties of Giant H II Regions in Nearby Galaxies", 287, 116.
37. Kingsburgh, R.L., & Barlow, M.J. 1995, "DR 1: A WO3 Star in IC 1613 and Its Surrounding Nebula, S3", A&A, 295, 171.
38. Klein, U., & Gräve, R. 1986, "Multi-Frequency Radio Continuum Observations of Dwarf Irregular Galaxies", A&A, 161, 155.
39. Lake, G. 1989, "Testing Modifications of Gravity", ApJ, 345, L17.
40. Lake, G., & Skillman, E.D. 1989, "The Mass Distribution and the Law of Gravity in the Local Group Dwarf Irregular Galaxy IC 1613", AJ 98, 1274.
41. Lequeux, J. 1979, "Comparison of the Rates of Formation of Massive Stars and of the Initial Mass Functions in Galaxies of the Local Group", A&A, 71, 1.
42. Lequeux, J. 1980, "Stellar Studies in the Large Magellanic Cloud", A&A, 85, 305.
43. Lequeux, J., Meyssonnier, N., & Azzopardi, M. 1987, "An Objective-Prism Survey of Emission-Line objects in M33 and IC 1613", A&AS, 67, 169.

44. Madore, B.F., & Freedman, W.L. 1991, "The Cepheid Distance Scale", PASP, 103, 933.
45. Massey, P., Conti. P.S., & Armandroff, T.E. 1987, "The Spectra of Extragalactic Wolf-Rayet Stars", AJ, 94, 1538.
46. McAlary, C.W., Madore, B.F., & Davis, L.E. 1984, "The Distance to IC 1613 from Infrared Photometry of Cepheids", ApJ, 276, 487.
47. Meaburn, J., Clayton, C.A., & Whitehead, M.J. 1988, "The Dynamics of a Complex of Interlocking Giant Shells in the Local Group Dwarf Galaxy IC 1613", MNRAS, 235, 479.
48. Milgrom, M. 1991, "Criticism of Lake's Analysis of the Rotation Curves of Dwarf Spirals", ApJ, 367, 490.
49. Mould, J., & Aaronson, M. 1986, "The Formation and Evolution of Carbon stars", ApJ, 303, 10.
50. Ohta, K., Tomita, A., & Saitō, M. 1993, "CO-to-H_2 Conversion Factor in Dwarf Irregular Galaxies", PASJ, 45, L21.
51. Peimbert, M., Bohigas, J., & Torres-Peimbert, S. 1988, "On the Supernova Remnant S8, and Other Gaseous Nebulae, in IC 1613", Rev. Mex. Astron. Astrof., 16, 45.
52. Price, J.S., Mason, S.F., & Gullixson, C.A. 1988, "CCD Surface Photometry of NGC 147, IC 1613, and NGC 6822", BAAS, 20, 701.
53. Price, J.S., Mason, S.F., & Gullixson, C.A. 1990, "The H II Regions of IC 1613", AJ, 100, 420.
54. Rice, W., Lonsdale, C.J., Soifer, B.T., Neugebauer, G., Kopan, E.L., Lloyd, L.A., de Jong, T., & Habing, H.J. 1988, "A Catalog of IRAS Observations of Large Optical Galaxies", ApJS, 68, 91.
55. Saha, A., Freedman, W.L., Hoessel, J.G., & Mossman, A.E. 1992, "RR Lyrae Stars in Local Group Galaxies. IV. IC 1613", AJ, 104, 1072.
56. Sandage, A. 1971, "The Distance of the Local Group Galaxy IC 1613 Obtained from Baade's Work on its Stellar Continent", ApJ, 166, 13.
57. Sandage, A. 1988, "Cepheids as Distance Indicators When Used Near their Detection Limit", PASP, 100, 935.
58. Sandage, A., & Katem, B. 1976, "The Brightest Stars in Nearby Galaxies. I. The Color-Magnitude Diagram and Luminosity Function for IC 1613", AJ, 81, 743.
59. Sandage, A., & Tammann, G.A. 1974, "Steps Toward the Hubble Constant. I. Calibration of the Linear Sizes of Extragalactic H II Regions", ApJ, 190, 525.
60. Smith, H.E. 1975, "Spectrophotometric Observations of Ionized Hydrogen Regions in Nearby Spiral and Irregular Galaxies", ApJ, 199, 591.
61. Tarrab, I. 1983, "Estimated Energy and Momentum Input to the Interstellar Medium for Several External Galaxies", A&A, 125, 308.
62. Tomita, A., Ohta, K., & Saitō, M. 1993, "H-Alpha Velocity Fields of Giant H II Regions in Four Nearby Dwarf Irregular Galaxies", PASJ, 45, 693.
63. van den Bergh, S. 1979, "Star Clusters in IC 1613 and in the Small Magellanic Cloud", ApJ, 230, 95.
64. Vangioni-Flam, E., Lequeux, J., Maucherat-Joubert, M., & Rocca-Volmerange, B. 1980, "Ultraviolet Studies of the Magellanic Clouds. II. Internal Extinction, Formation of Massive Stars, Comparison With Other Galaxies", A&A, 90, 73.
65. Volders, L., & Högbom, J.A. 1961, "Observations of Neutral Hydrogen in IC 1613, NGC 6822 and M82", Bull. Ast. Inst. Neth., 15, 307.
66. Walker, A.R. 1987, "Notes on Cepheid Variables in Galaxies More Distant Than the Magellanic Clouds", S. African Astron. Obs. Circ., 11, 125.
67. Wolf, M. 1906, "Extended Nebula Near 26 Ceti", MNRAS, 67, 91.

Leo I

Alternate name(s): DDO 74
Right Ascension (2000): 10h 08m 27s
Declination (2000): +12d 18.5m
Type: dE3
Apparent magnitude (V): 10.2
Color (B-V): 0.75
(U-B): 0.29
(V-R):
Color Excess, E(B-V): 0.01
Absolute magnitude (M_V): -11.8
Distance (kpc): 250
Radial velocity (solar, km/sec): 285

Objects Identified On the Atlas:
Globular clusters: none
Source:
Open clusters: none
Source:
OB associations: none
Source:
Variable stars: 69; designations: HWV, V
Source: [15], [19]
Carbon stars: 15; designation: ALW
Source: [2], [3], [4]
HII regions: none
Source:
Planetary nebulae: none
Source:
Dust clouds: none
Source:

Leo I

Chart 135

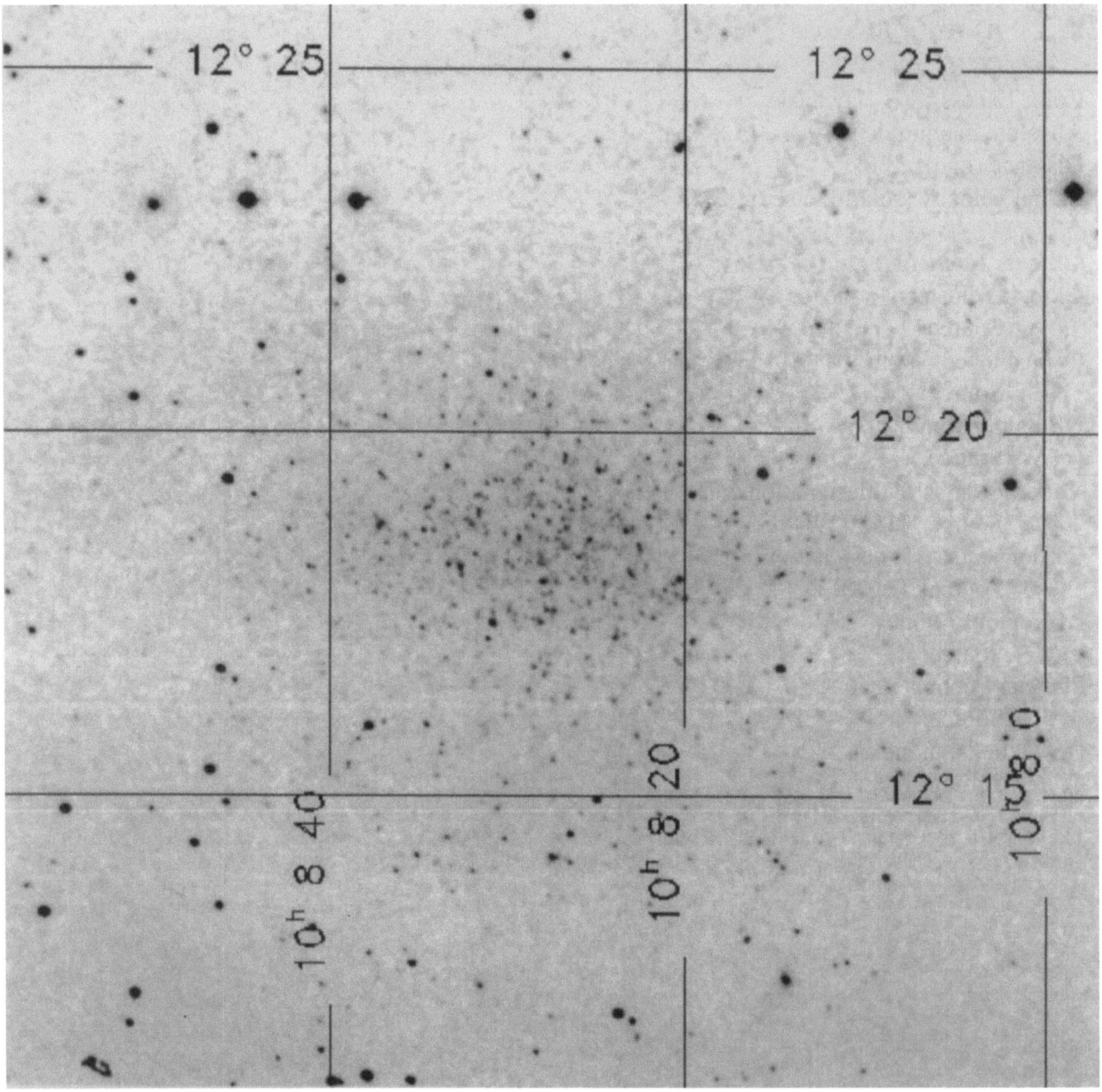

Leo I

Chart 136

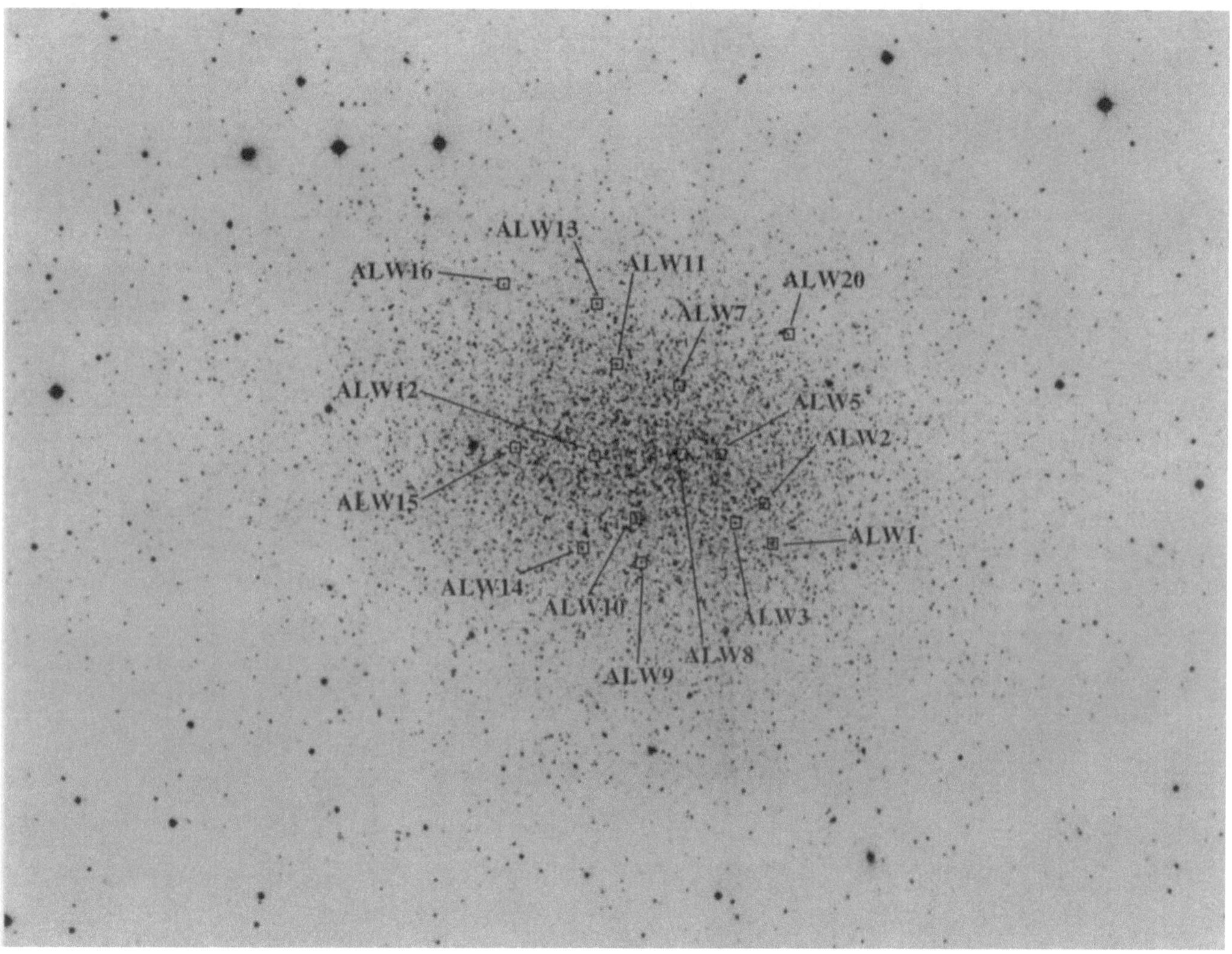

Leo I

Chart 137

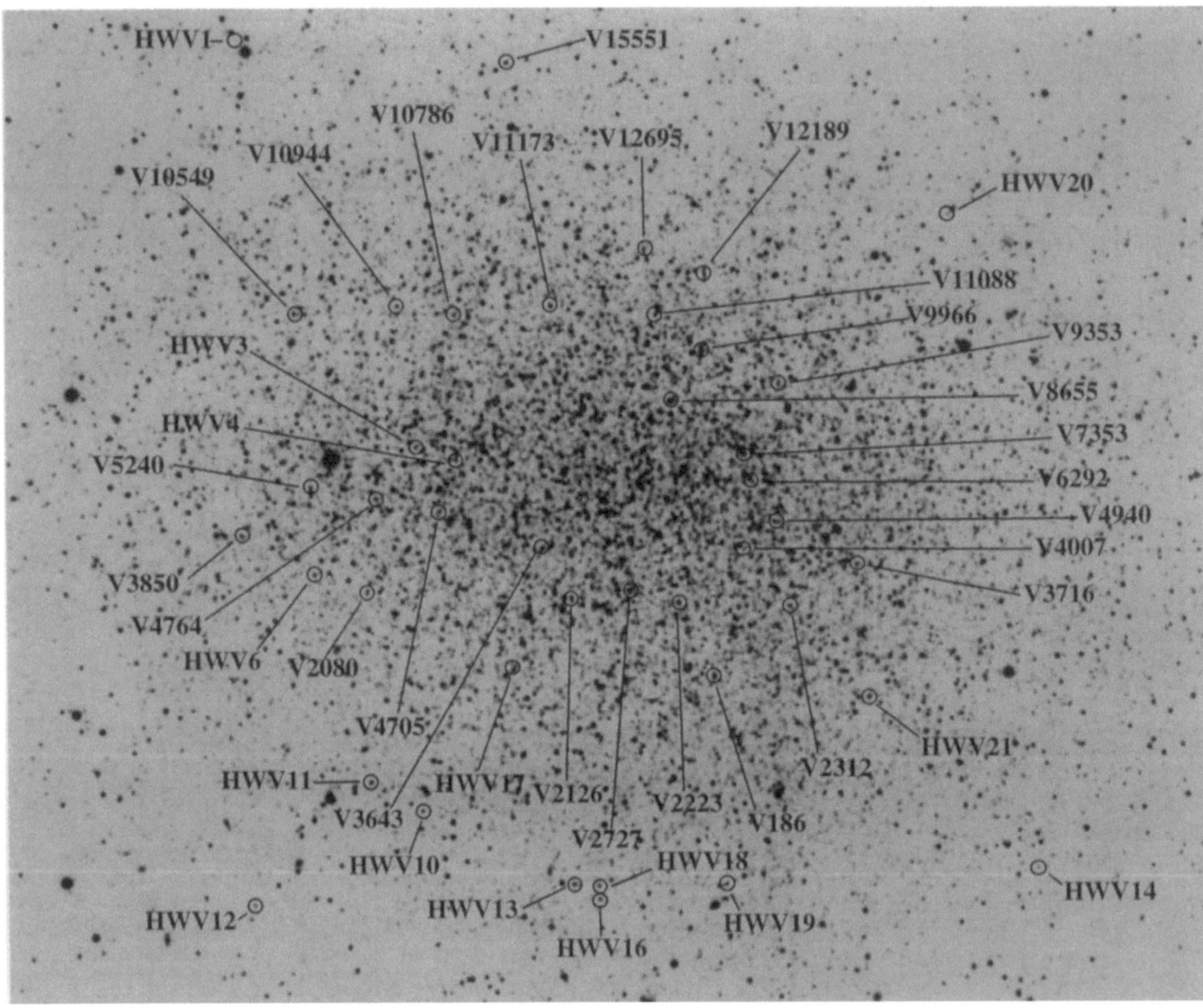

Leo I

Chart 138

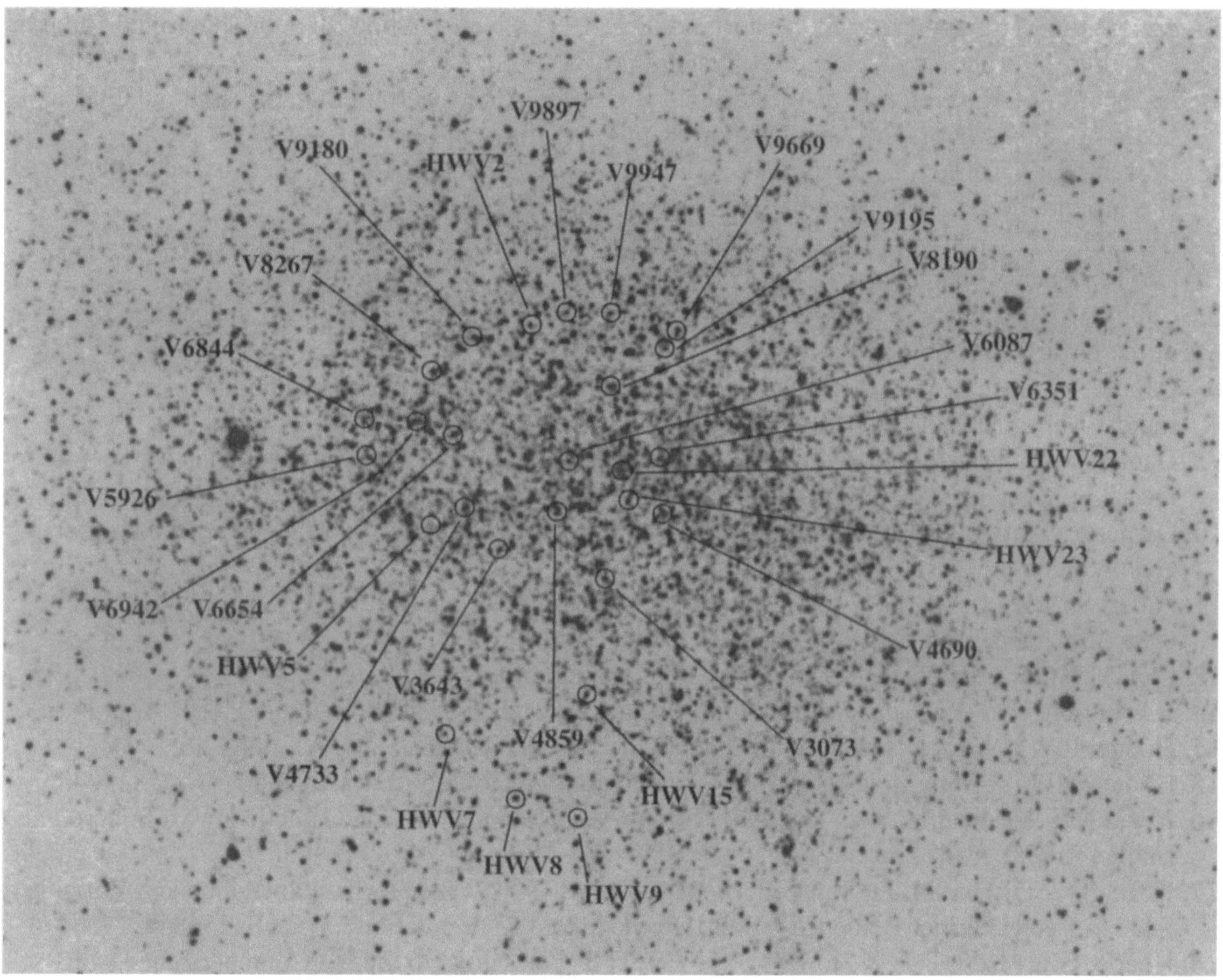

Leo I

References

1. Aaronson, M., & Mould, J. 1985, "Infrared Photometry and the Comparative Stellar Content of Dwarf Spheroidals in the Galactic Halo", ApJ, 290, 191.
2. Aaronson, M., Olszewski, E.W., & Hodge, P.W. 1983, "Carbon Stars and the Seven Dwarfs", ApJ, 267, 274.
3. Azzopardi, M., Lequeux, J., & Westerlund, B.E. 1985, "New Carbon Stars in Spheroidal Galaxies. I. Sculptor, Carina, Leo I and Leo II Systems", A&A, 144, 388.
4. Azzopardi, M., Lequeux, J., & Westerlund, B.E. 1986, "New Carbon Stars in Spheroidal Galaxies. II. Draco, Ursa Minor, and New Data on Sculptor, Carina, and Leo I", A&A, 161, 232.
5. Bowen, D.V., Tolstoy, E., Ferrara, A., Blades, J.C., & Brinks, E. 1997, "The Absence of Diffuse Gas around the Dwarf Spheroidal Leo I", ApJ, 478, 530.
6. Byrd, G., Valtonen, M., McCall, M., & Innanen, K. 1994, "Orbits of the Magellanic Clouds and Leo I in Local Group History", AJ, 107, 2055.
7. Caputo, F., Castellani, V., Degl'Innocenti, S. 1995, "'Vertical' Horizontal Branches and the Age of the Dwarf Spheroidal Galaxy Leo I", A&A, 304, 365.
8. Caputo, F., Castellani, V., Degl'Innocenti, S. 1996, "Stellar Luminosity Function as an Age Indicator for the Dwarf Spheroidal Leo I", A&A, 309, 413.
9. Demers, S., Battinelli, P., Irwin, M.J., & Kunkel, W.E. 1995, "A Search for Stellar Aggregates in Dwarf Spheroidal Galaxies", MNRAS, 274, 491.
10. Demers, S., Irwin, M.J., & Gambu, I. 1994, "Deep CCD Photometry of the Dwarf Spheroidal Galaxy Leo I", MNRAS, 266, 7.
11. Faber, S.M., & Lin, D.N.C. 1983, "Is there Non-Luminous Matter in Dwarf Spheroidal Galaxies?", ApJ, 266, L17.
12. Fox, M.F., & Pritchet, C.J. 1987, "The Giant Branch of Leo I", AJ, 93, 1381.
13. Harrington, R.G., & Wilson, A.G. 1950, "Two New Stellar Systems in Leo", PASP, 62, 118.
14. Hodge, P.W. 1963, "Distribution of Stars in the Leo I Dwarf Galaxy", AJ, 68, 470.
15. Hodge, P.W., & Wright, F.W. 1978, "Variable Stars in the Leo I Dwarf Galaxy", AJ, 83, 228.
16. Irwin, M., & Hatzidimitriou, D. 1995, "Structural Parameters for the Galactic Dwarf Spheroidals", MNRAS, 277, 1354.
17. Keane, M., Olszewski, E., Suntzeff, N., & Saha, A. 1993, "The Horizontal Branch and Variable Stars in Leo I: Further Evidence for an Intermediate Age Population", BAAS, 25, 844.
18. Knapp, G.R., Kerr, F.J., & Bowers, P.F. 1978, "Upper Limits to the H I Content of the Dwarf Spheroidal Galaxies", AJ, 83, 360.
19. Lee, M.G., Freedman, W., Mateo, M., Thompson, I., Roth, M., & Ruiz, M.-T. 1993, "Leo I: The Youngest Milky Way Dwarf Spheroidal Galaxy?", AJ, 106, 1420.
20. Lee, .M.G., Thompson, I., Freedman, W.L., Mateo, M., Roth, M., & Ruiz, M.-T. 1992, "Stellar Populations in the Dwarf Spheroidal Galaxy Leo I and the Mass of our Galaxy", in the Globular Cluster-Galaxy Connection, ASP Conf. Series, edited by J. Brodie and G. Smith.
21. Mateo, M., Olszewski, E.W., Lee, M.-G., Saha, A., Hodge, P., Keane, M., Suntzeff, N., Freedman, W., & Thompson, I. 1994, "Deep HST Photometry of the Leo I Dwarf Spheroidal Galaxy", BAAS, 26, 1395.

22. Reid, N., & Mould, J. 1991, "The Leo I Color-Magnitude Diagram", AJ, 101, 1299.
23. Suntzeff, N.B., Aaronson, M., Olszewski, E.W., & Cook, K.H. 1986, "The Abundances and Velocities of the Leo Dwarf Spheroidals", AJ, 91, 1091.
24. van den Bergh, S. 1994, "The Evolutionary History of Low-Luminosity Local Group Dwarf Galaxies", ApJ, 428, 617.
25. Zaritsky, D., Olszewski, E.W., Schommer, R.A., Peterson, R.C. & Aaronson, M. 1989, "Velocities of Stars in Remote Galactic Satellites and the Mass of the Galaxy", ApJ, 345, 759.

Leo II

Alternate name(s): DDO 93
Right Ascension (2000): 11h 13m 29s
Declination (2000): +22d 09.2m
Type: dE0
Apparent magnitude (V): 12.0
Color (B-V): 0.65
(U-B): 0.29
(V-R):
Color Excess, E(B-V): 0.02
Absolute magnitude (M_V): -9.6
Distance (kpc): 205
Radial velocity (solar, km/sec): 76

Objects Identified On the Atlas:

Globular clusters: none
Source:
Open clusters: none
Source:
OB associations: none
Source:
Variable stars: not shown
Source:
Carbon stars: 7; designation: C
Source: [2], [3]
HII regions: none
Source:
Planetary nebulae: none
Source:
Dust clouds: none
Source:

Leo II

Chart 139

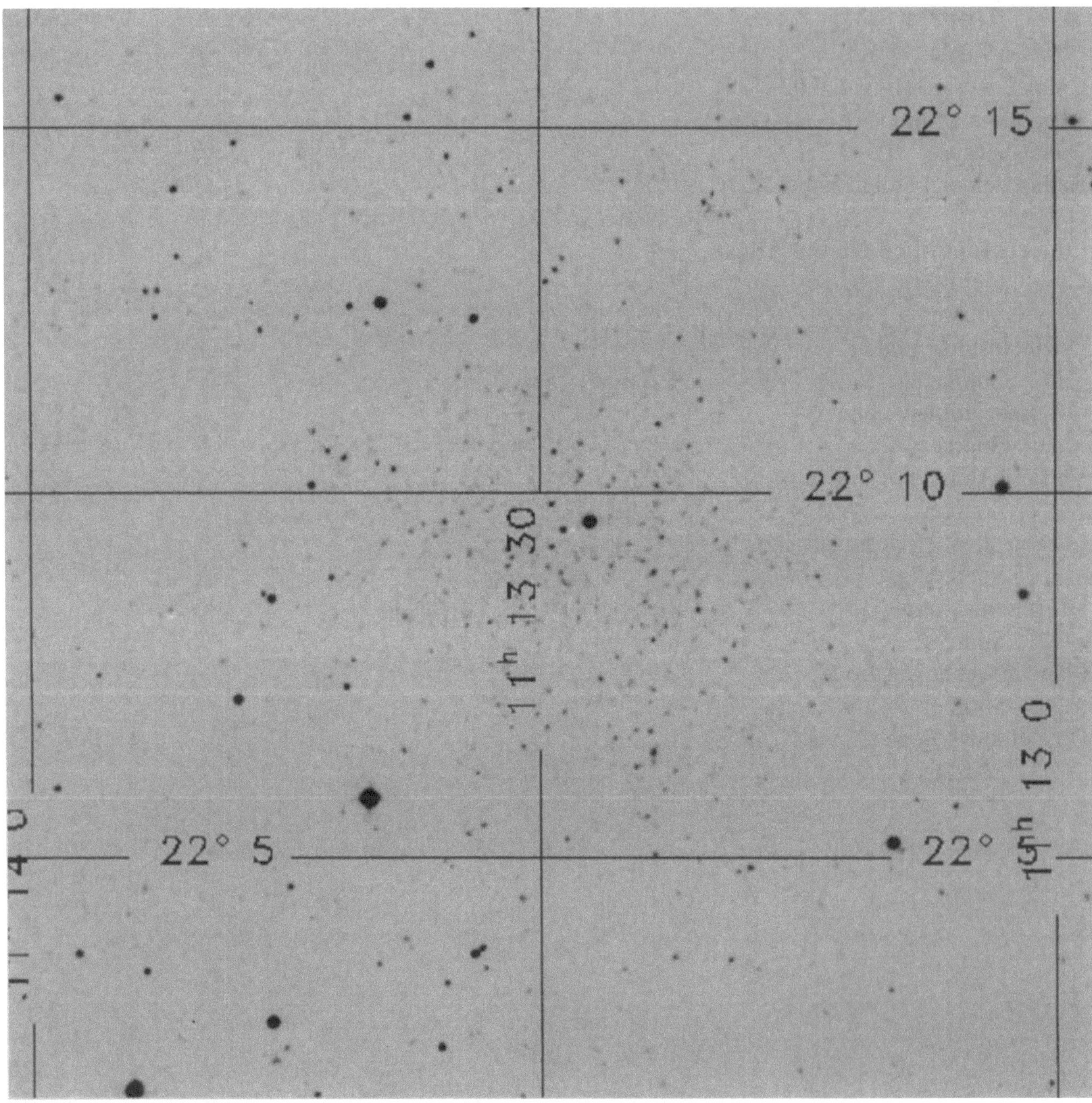

Leo II

Chart 140

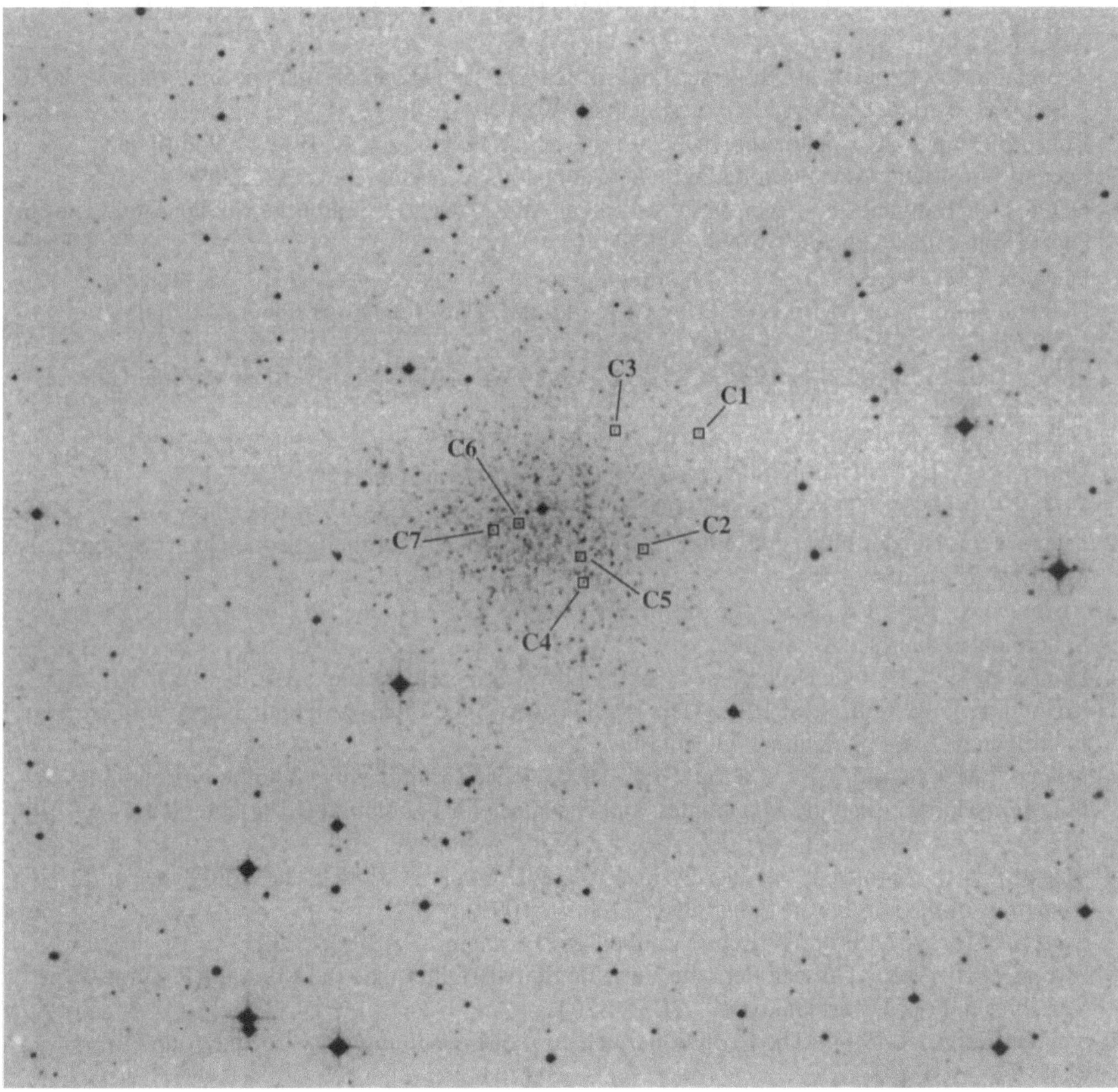

Leo II

References

1. Aaronson, M., & Mould, J. 1985, "Infrared Photometry and the Comparative Stellar Content of Dwarf Spheroidals in the Galactic Halo", ApJ, 290, 191.
2. Aaronson, M., Olszewski, E.W., & Hodge, P.W. 1983, "Carbon Stars and the Seven Dwarfs", ApJ, 267, 274.
3. Azzopardi, M., Lequeux, J., & Westerlund, B.E. 1985, "New Carbon Stars in Spheroidal Galaxies. I. Sculptor, Carina, Leo I and Leo II Systems", A&A, 144, 388.
4. Bono, G., Caputo, F., Santolamazza, P., Cassisi, S., & Piersimoni, A. 1997, "Evolutionary Scenario for Metal-Poor Pusating Stars. II. Anomalous Cepheids", AJ, 113, 2209.
5. Demers, S., Battinelli, P., Irwin, M.J., & Kunkel, W.E. 1995, "A Search for Stellar Aggregates in Dwarf Spheroidal Galaxies", MNRAS, 274, 491.
6. Demers, S., & Harris, W.E. 1983, "A Color-Magnitude Diagram for Leo II", AJ, 88, 329.
7. Demers, S., & Irwin, M.J. 1993, "Deep CCD Photometry of the Dwarf Spheroidal Galaxy Leo II", MNRAS, 261, 657.
8. Faber, S.M., & Lin, D.N.C. 1983, "Is There Nonluminous Matter in Dwarf Spheroidal Galaxies?", ApJ, 266, L17.
9. Harrington, R.G., & Wilson, A.G. 1950, "Two New Stellar Systems in Leo", PASP, 62, 118.
10. Hodge, P.W. 1962, "Distribution of Stars in the Leo II Dwarf Galaxy", AJ, 67, 125.
11. Hodge, P.W. 1982, "The Color and Surface Brightness of the Leo II Dwarf Galaxy", AJ, 87, 1668.
12. Irwin, M., & Hatzidimitriou, D. 1995, "Structural Parameters for the Galactic Dwarf Spheroidals", MNRAS, 277, 1354.
13. Knapp, G.R., Kerr, F.J., & Bowers, P.F. 1978, "Upper Limits to the H I Content of the Dwarf Spheroidal Galaxies", AJ, 83, 360.
14. Lee, M.G. 1995, "VI CCD Photometry of the Dwarf Spheroidal Galaxy Leo II", AJ, 110, 1155.
15. Mighell, K.J., & Rich, R.M. 1996, "The Leo II Dwarf Spheroidal – An Old Galaxy with a Large Intermediate-Age Population", AJ, 111, 777.
16. Nemec, J.M., Nemec, A.F.L., & Lutz, T.E. 1994, "Period-Luminosity-Metallicity Relations, Pulsation Modes, Absolute Magnitudes, and Distances for Population II Variable Stars", AJ, 108, 222.
17. Suntzeff, N.B., Aaronson, M., Olszewski, E.W., & Cook, K.H. 1986, "The Abundance and Velocities of the Leo Dwarf Spheroidals", AJ, 91, 1091.
18. Swope, H.H. 1967, "Progress Report on the Leo II System", PASP, 79, 439.
19. Swope, H.H. 1968, "Thirteen Periodic Variable Stars Brighter than the Normal RR Lyrae-Type Variables in Four Dwarf Galaxies", AJ, 73 S204.
20. van den Bergh, S. 1994, "The Evolutionary History of Low-Luminosity Local Group Dwarf Galaxies", ApJ, 428, 617.
21. Vogt, S.S., Mateo, M., Olszewski, E.W., & Keane, M.J. 1995, "Internal Kinematics of the Leo II Dwarf Spheroidal Galaxy", AJ, 109, 151.
22. Zaritsky, D., Olszewski, E.W., Schommer, R.A., Peterson, R.C., & Aaronson, M. 1989, "Velocities of Stars in Remote Galactic Satellites and the Mass of the Galaxy", ApJ, 345, 759.

LGS 3

Alternate name(s): Pisces
Right Ascension (2000): 01h 03m 43s
Declination (2000): +21d 53.1m
Type: dE/dIrr
Apparent magnitude (V): 14.26
Color (B-V): 0.73
(U-B):
(V-R): 0.42
Color Excess, E(B-V): 0.08
Absolute magnitude (M_V): -10.5
Distance (kpc): 810
Radial velocity (solar, km/sec): -272

Objects Identified On the Atlas:
Globular clusters: none
Source:
Open clusters: none
Source:
OB associations: none
Source:
Variable stars: none
Source:
Carbon stars: none
Source:
HII regions: none
Source:
Planetary nebulae: none
Source:
Dust clouds: none
Source:

LGS 3

Chart 141

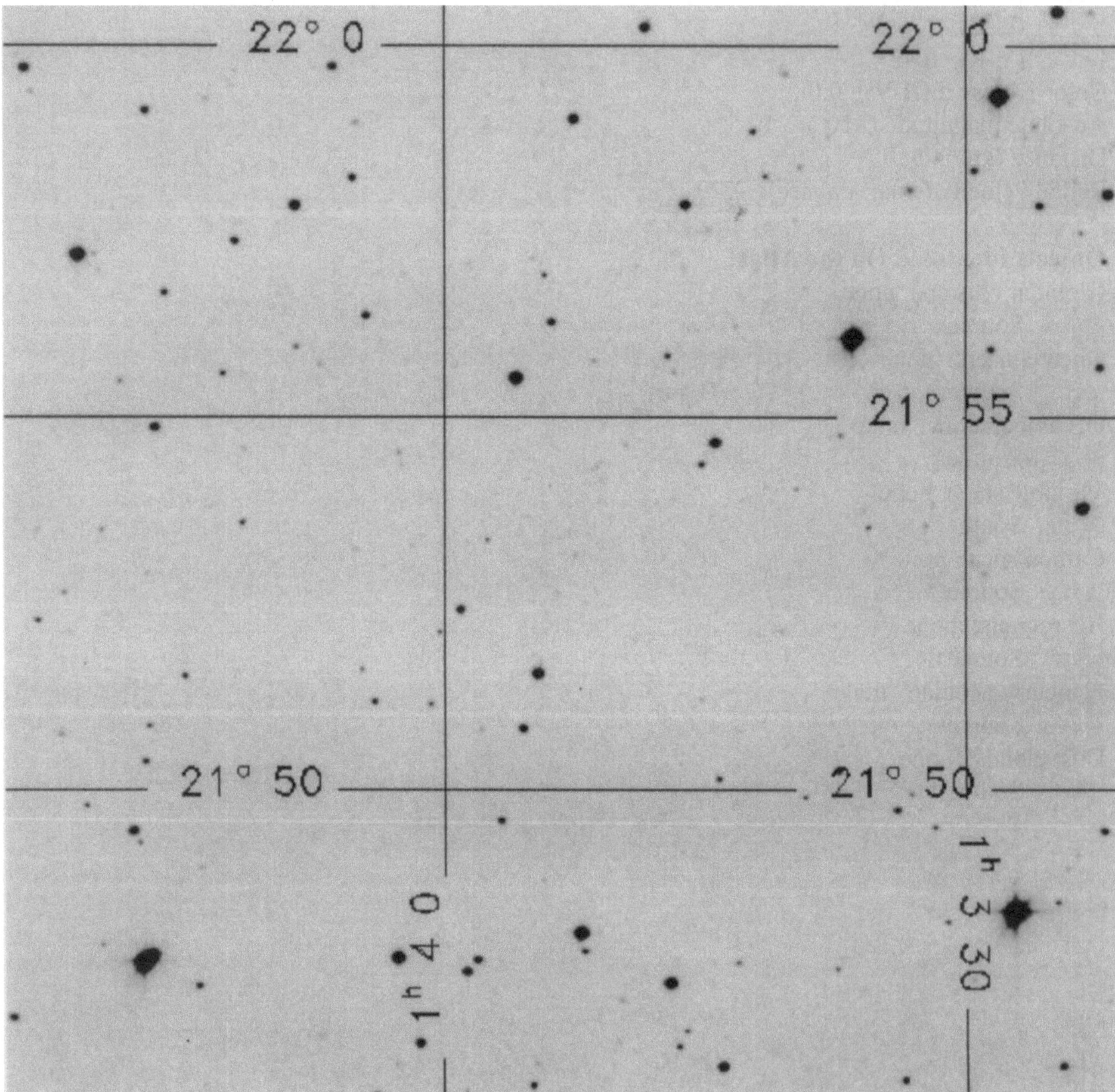

LGS 3

Chart 142

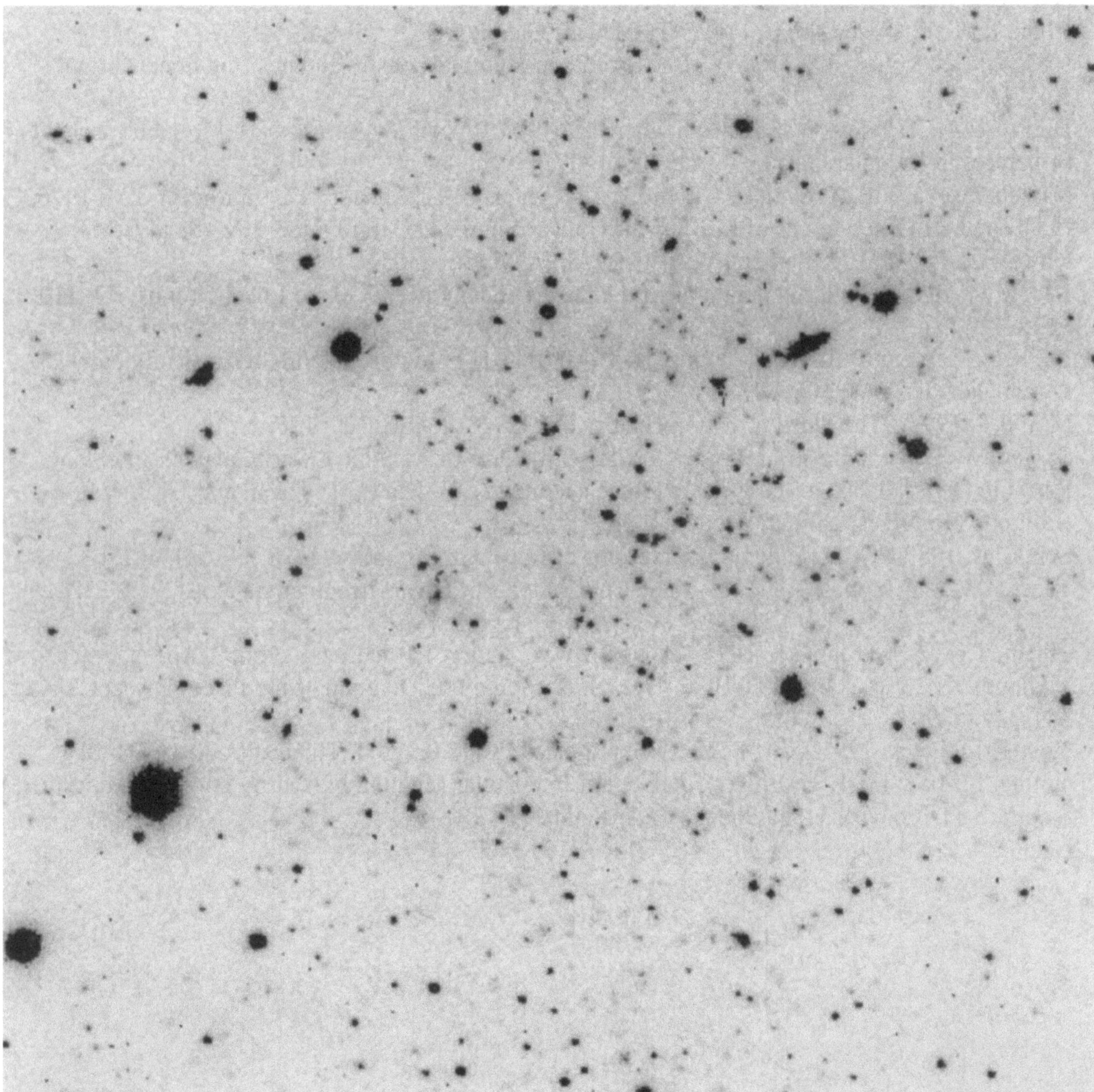

LGS 3

References

1. Aparicio, A., Gallart, C., & Bertelli, G. 1997, "The Stellar Content and the Star Formation History of the Local Group Dwarf Galaxy LGS 3", AJ, 114, 680.
2. Christian, C.A., & Tully, R.B. 1983, "The Local Group Irregular Galaxies LGS 3 and Pegasus", AJ, 88, 934.
3. Cook, K.H., & Olszewski, E. 1989, "The Dwarf Galaxy LGS 3", BAAS, 21, 775.
4. Hodge, P., & Miller, B.W. 1995, "H II Regions in Four Galaxies in and near the Local Group", ApJ, 451, 176.
5. Hunter, D.A., Hawley, W.N., & Gallagher, J.S. 1993, "A Survey for Extra-H II Region Ionized Gas Structures in Irregular Galaxies", AJ, 106, 1797.
6. Karachentseva, V.E., Prugniel, P., Vennik, J., Richter, G.M., Thuan, T.X., & Martin, J.M. 1996, "CCD and HI Observations of LSB Dwarf Galaxies in the General Field", A&AS, 117, 343.
7. Kowal, C., Lo, K.Y., & Sargent, W.L.W. 1978, IAUC No. 3305.
8. Lee, M.G. 1995, "Stellar Populations of the Dwarf Galaxy LGS 3 in the Local Group", AJ, 110, 1129.
9. Lo, K.Y., Sargent, W.L.W., & Young, K. 1993, "The H I Stucture of Nine Intrinsically Faint Dwarf Galaxies", AJ, 106, 507.
10. Mould, J. 1997, "The Star-Formation History of LGS 3", PASP, 109, 125.
11. Sargent, W.L.W., & Lo, K.Y. 1986, "The H I Structures of Extreme Dwarf Galaxies", in Star Forming Dwarf Galaxies and Related Objects, edited by D. Kunth, T. Thuan and J.T.T. Van (Editions Frontieres, Gif-sur-Yvette), p. 253.
12. Schild, R. 1980, "CCD Camera Observations of Local Group Galaxy LGS-3", ApJ, 241, 63.
13. Tacconi, L.J., & Young, J.S. 1987, "The CO Contents of Dwarf Irregular Galaxies", ApJ, 322, 681.
14. Thuan, T.X., & Martin, G.E. 1979, "A New Dwarf Galaxy in the Local Group", ApJ, 232, L11.
15. Tikhonov, N., & Makarova, L. 1996, "A Colour-Magnitude Diagram of the Pisces Dwarf Galaxy", AN, 317, 179.
16. Young, J.S., et al. 1995, "The FCRAO Extragalactic CO Survey. I. The Data", ApJS, 98, 219.
17. Young, L.M., & Lo, K.Y. 1997, "The Neutral Interstellar Medium in Nearby Dwarf Galaxies. III. Sagittarius DIG, LGS 3, and Phoenix", ApJ, 490, 710.

NGC 147

Alternate name(s): DDO 3
Right Ascension (2000): 00h 33m 12s
Declination (2000): +48d 30.5m
Type: dE4
Apparent magnitude (V): 9.41
Color (B-V): 0.92
(U-B): 0.32
(V-R): 0.55
Color Excess, E(B-V): 0.04
Absolute magnitude (M_V): -14.9
Distance (kpc): 715
Radial velocity (solar, km/sec): -193

Objects Identified On the Atlas:
Globular clusters: 4; designation: GC
Source: [10]
Open clusters: none
Source:
OB associations: none
Source:
Variable stars: not shown
Source: [16]
Carbon stars: none
Source:
HII regions: none
Source:
Planetary nebulae: 5; designation: PN
Source: [7]
Dust clouds: none
Source:

NGC 147

Chart 143

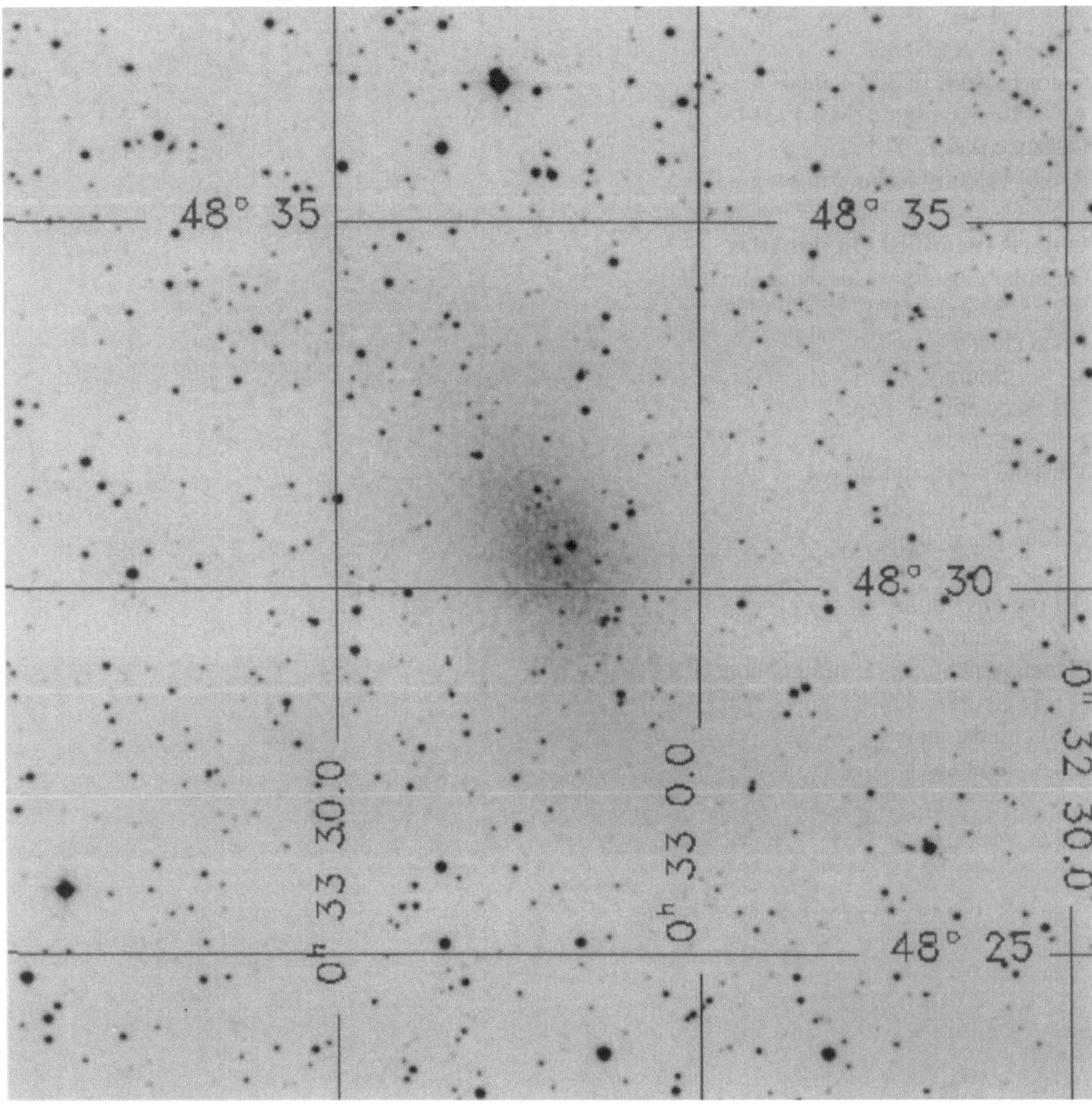

NGC 147

Chart 144

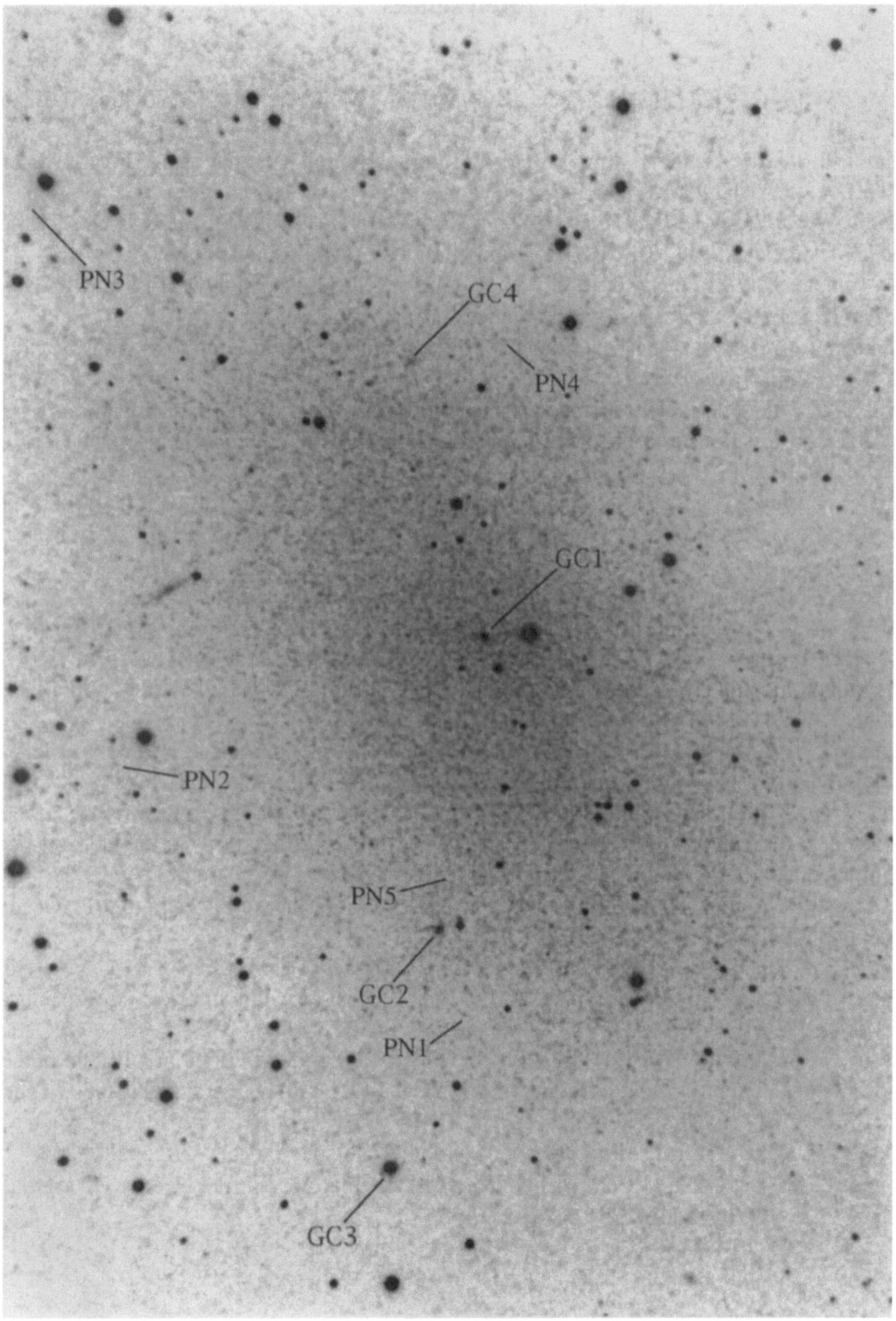

NGC 147

References

1. Baade, W. 1944, "NGC 147 and NGC 185, Two New Members of the Local Group of Galaxies", ApJ, 100, 147.
2. Bender, R. Paquet, A., & Nieto, J.-L. 1991, "Internal Stellar Kinematics of Three Dwarf Ellipticals in the Local Group", A&A, 246, 349.
3. Brandt, W.N., Ward, M. J., Fabian, A.C., & Hodge, P.W. 1997, "ROSAT HRI Observations of the Local Group Galaxies IC 10, NGC 147 and NGC 185", MNRAS, 291, 709.
4. Da Costa, G.S., & Mould, J.R. 1988, "The Globular Clusters of the Dwarf Elliptical Galaxies NGC 147, NGC 185, and NGC 205. I. Abundances", ApJ, 334, 159.
5. Davidge, T.J., 1994, "V and I Photometry of Bright Giants in the Central Regions of NGC 147", AJ, 108, 2123.
6. Ford, H. C. 1978, "Planetary Nebulae in the Andromeda Galaxy and its Companions", in Planetary Nebulae, Observations and Theory, IAU Symposium No. 76, edited by Y. Terzian (Reidel, Dordrecht), p. 19.
7. Ford, H. C., Jacoby, G., & Jenner, D. C. 1977, "Planetary Nebulae in Local Group Galaxies. IV. Identifications, Positions, and Radial Velocities of Nebulae in NGC 147 and NGC 185", ApJ, 213, 18.
8. Han, M., et al. 1997, "Stellar Populations in the Dwarf Elliptical Galaxy NGC 147", AJ, 113, 1001.
9. Ho, L. C., Filippenko, A. V., & Sargent, W. L. W. 1995, "A Search for 'Dwarf' Seyfert Nuclei. II. An Optical Spectral Atlas of the Nuclei of Nearby Galaxies", ApJS, 98, 477.
10. Hodge, P. W. 1976, "The Structure of NGC 147", AJ, 81, 25.
11. Huchra, J. P., Brodie, J. P., Caldwell, N., Christian, C., & Schommer, R. 1996, "Extragalactic Globular Clusters. IV. The Data", ApJS, 102, 29.
12. Kent, S. M. 1987, "Surface Photometry of Six Local Group Galaxies", AJ, 94, 306.
13. Mould, J. R., Kristian, J., & Da Costa, G. S. 1983, "Stellar Populations in Local Group Dwarf Elliptical Galaxies. I.. NGC 147", ApJ, 270, 471.
14. Price, J. S., Mason, S. F., & Gullixson, C. A. 1988, "CCD Surface Photometry of NGC 147, IC 1613, and NGC 6822", BAAS, 20, 701.
15. Saha, A., & Hoessel, J. G. 1987, "A Search for RR Lyrae Variables in NGC 147", AJ, 94, 1556.
16. Saha, A., Hoessel, J. G., & Mossman, A. E. 1990, "RR Lyrae Stars in Local Group Galaxies. II. NGC 147", AJ, 100, 108.
17. Sharov, A. S., & Lyuty, V. M. 1983, "Photoelectric Catalogue of Globular Clusters in the Andromeda Nebula M31 and its Companions NGC 147, NGC 185, and NGC 205", Ap&SS, 90, 371.
18. Sitko, M. L. 1984, "Infrared Photometry of Globular Clusters in M31", ApJ, 286, 209.
19. Young, L. M., & Lo, K. Y. 1997, "The Neutral Interstellar Medium in Nearby Dwarf Galaxies. II. NGC 185, NGC 205, and NGC 147", ApJ, 476, 127.

NGC 185

Alternate name(s): UGC 396
Right Ascension (2000): 00h 38m 58s
Declination (2000):+48d 20.2m
Type: dE3
Apparent magnitude (V): 9.09
Color (B-V): 1.02
(U-B): 0.34
(V-R): 0.60
Color Excess, E(B-V): 0.19
Absolute magnitude (M_V): -15.4
Distance (kpc): 620
Radial velocity (solar, km/sec): -204

Objects Identified On the Atlas:
Globular clusters: 7; designation: GC
Source: [17], [19]
Open clusters: none
Source:
OB associations: none
Source:
Variable stars: not shown
Source: [30]
Carbon stars: none
Source:
HII regions: none
Source:
Planetary nebulae: 4; designation: PN
Source: [8], [9]
Dust clouds: 2, designation: Roman numerals
Source: [17]

NGC 185

Chart 145

NGC 185

Chart 146

NGC185

References

1. Baade, W. 1944, "NGC 147 and NGC 185, Two New Members of the Local Group of Galaxies", ApJ, 100, 147.
2. Bender, R. Paquet, A., & Nieto, J.-L. 1991, "Internal Stellar Kinematics of Three Dwarf Ellipticals in the Local Group", A&A, 246, 349.
3. Brandt, W.M., Ward, M.J., Fabian, A.C., & Hodge, P.W. 1997, "ROSAT HRI Observations of the Local Group Galaxies IC 10, NGC 147 and NGC 185", MNRAS, 291, 709.
4. Ciardullo, R., Jacoby, G.H., Ford, H.C., & Neill, J.D. 1989, "Planetary Nebulae as Standard Candles. II. The Calibration in M31 and its Companions", ApJ, 339, 53.
5. Da Costa, G.S., & Mould, J.R. 1988, "The Globular Clusters of the Dwarf Elliptical Galaxies NGC 147, NGC 185, and NGC 205. I. Abundances", ApJ, 334, 159.
6. Dickel, J.R., D'Odorico, S., & Silverman, A. 1985, "VLA Observations of Three Extragalactic SNR at 20 and 6 cm", AJ, 90, 414.
7. Ford, H.C. 1978, "Planetary Nebulae in the Andromeda Galaxy and its Companions", in Planetary Nebulae, Observations and Theory, IAU Symposium No. 76, edited by Y. Terzian (Reidel, Dordrecht), p. 19.
8. Ford, H.C., Jacoby, G., & Jenner, D.C. 1977, "Planetary Nebulae in Local Group Galaxies. IV. Identifications, Positions, and Radial Velocities of Nebulae in NGC 147 and NGC 185", ApJ, 213, 18.
9. Ford, H.C., Jenner, D.C., & Epps, W.H. 1973, "Planetary Nebulae in Local Group Galaxies. I. Identifications in NGC 185, NGC 205, and NGC 221", ApJ, 183, L73.
10. Gallagher, J.S., & Hunter, D.A. 1981, "Dust Clouds in the Stellar Spheroids of Local Group Galaxies", AJ, 86, 1312.
11. Gallagher, III, J.S., Hunter, D.A., & Mould, J. 1984, "A Probable Supernova Remnant in the Dwarf Elliptical Galaxy NGC 185", ApJ, 281, L63.
12. Gottesman, S.T., & Johnson, D.W. 1983, "Neutral Hydrogen Observations of the Dwarf Elliptical Galaxies NGC 185 and NGC 205", in Internal Kinematics and Dynamics of Galaxies, IAU Symposium No. 100, edited by E. Athanassoula (Reidel, Dordrecht), p. 307.
13. Gregg, M.D. 1989, "Differential Population Synthesis of S0 Galaxies. I. The Data", ApJS, 69, 217.
14. Gregg, M.D. 1989, "Differential Population Synthesis of S0 Galaxies. II. Methods and Results", ApJ, 337, 45.
15. Held, E.V., de Zeeuw, T., Mould, J., & Picard, A. 1992, "Internal Dynamics of the Dwarf Elliptical NGC 185", AJ, 103, 851.
16. Ho, L.C., Filippenko, A.V., & Sargent, W.L.W. 1995, "A Search for 'Dwarf' Seyfert Nuclei. II. An Optical Spectral Atlas of the Nuclei of Nearby Galaxies", ApJS, 98, 477.
17. Hodge, P.W. 1963, "Distribution of Luminosity and Color in the Galaxy NGC 185", AJ, 68, 691.
18. Hodge, P.W. 1972, "The Population I Content of the Elliptical Companions of M31", in External Galaxies and Quasi-Stellar Objects, IAU Symposium No. 44, edited by D.S. Evans (Reidel, Dordrecht), p.46.
19. Hodge, P.W. 1974, "Photometry of the Globular Clusters in NGC 185", PASP, 86, 289.

20. Huchra, J.P., Brodie, J.P., Caldwell, N., Christian, C., & Schommer, R. 1996, "Extragalactic Globular Clusters. IV. The Data", ApJS, 102, 29.
21. Jenner, D.C., & Ford, H.C. 1978, "Chemical Abundances of Planetary Nebulae in NGC 185, 205, and 221", in Planetary Nebulae: Observations and Theory, IAU Symposium No. 76, edited by Y. Terzian (Reidel, Dordrecht), p. 246.
22. Johnson, D.W., & Gottesman, S.T. 1979, "Carbon Monoxide Observations of Early Type Galaxies", in Photometry, Kinematics, and Dynamics of Galaxies, edited by D.S. Evans (University of Texas, Austin), p. 57.
23. Johnson, D.W., & Gottesman, S.T. 1983, "VLA Observations of the Neutral Hydrogen Content of NGC 185 and 205", ApJ, 275, 549.
24. Kent, S.M. 1987, "Surface Photometry of Six Local Group Galaxies", AJ, 94, 306.
25. Kron, G.E., & Mayall, N.U. 1960, "Photoelectric Photometry of Galactic and Extragalactic Star Clusters", AJ, 65, 581.
26. Lee, M.G., Freedman, W.L., & Madore, B.F. 1993, "Stellar Populations in the Dwarf Elliptical Galaxy NGC 185", AJ, 106, 964.
27. Maoz, D., Filippenko, A.V., Ho, L.C., Macchetto, F.D., Rix, H.-W., & Schneider, D.P. 1996, "An Atlas of *Hubble Space Telescope* Ultraviolet Images of Nearby Galaxies", ApJS, 107, 215.
28. Price, J.S. 1985, "Optical and 2 Micron Surface Photometry of NGC 185", ApJ, 297, 652.
29. Richer, M.G., & McCall, M.L. 1995, "Oxygen Abundances in Diffuse Ellipticals and the Metallicity-Luminosity Relations for Dwarf Galaxies", ApJ, 445, 642.
30. Saha, A., & Hoessel, J.G. 1990, "RR Lyrae Stars in Local Group Galaxies. I. NGC 185", AJ, 99, 97.
31. Sharov, A.S., & Lyuty, V.M. 1983, "Photoelectric Catalogue of Globular Clusters in the Andromeda Nebula M31 and its Companions NGC 147, NGC 185, and NGC 205", Ap&SS, 90, 371.
32. Sitko, M.L. 1984, "Infrared Photometry of Globular Clusters in M31", ApJ, 286, 209.
33. Sofue, Y., & Wakamatsu, K. 1993, "CO in Elliptical Galaxies and the Universal CO-to-Dust Ratio", PASJ, 45, 529.
34. Welch, G.A., Mitchell, G.F., & Yi, S. 1996, "Observations of Molecular Gas in the Dwarf Elliptical Galaxy NGC 185", ApJ, 470, 781.
35. Wiklind, T., & Henkel, C. 1995, "Cold Dust in Elliptical Galaxies", A&A, 297, L71.
36. Wiklind, T., & Rydbeck, G. 1986, "Detection of CO ($J = 1 - 0$) in the Dwarf Elliptical Galaxy NGC 185, A&A, 164, L22.
37. Wiklind, T., & Rydbeck, G. 1986, "Detection of CO ($J = 1$-0) in the Dwarf Elliptical Galaxy NGC 185", in Star Formation in Galaxies, NAS Conference Publication 2466, edited by C.J. Lonsdale Persson (NASA Office of Space Science & Applications, Washington), p. 331.
38. Young, L.M., & Lo, K.Y. 1997, "The Neutral Interstellar Medium in Nearby Dwarf Galaxies. II. NGC 185, NGC 205, and NGC 147", ApJ, 476, 127.

NGC 6822

Alternate name(s): DDO 209
Right Ascension (2000): 19h 44m 56s
Declination (2000): -14d 48.1m
Type: dIrr
Apparent magnitude (V): 9.31
Color (B-V): 0.71
(U-B): 0.41
(V-R):
Color Excess, E(B-V): 0.26
Absolute magnitude (M_V): -14.9
Distance (kpc): 490
Radial velocity (solar, km/sec): -53

Objects Identified On the Atlas:

Globular clusters: 4; designation: H + Roman numeral
Source: [43]
Open clusters: 26; designation: C
Source: [36]
OB associations: 16; designation: A
Source: [36]
Variable stars: 34; designation: V
Source: [43], [51]
Wolf-Rayet stars: 12; designation: WR
Source: [5]
HII regions: 142; designation: H + Roman numeral, Ho, K + Greek letter, KD, HK
Source: [39]
Planetary nebulae and emission-line stars: 19; designation: KDe
Source: [53]
Dust clouds: 11; designation: Arabic numeral (triangles on Chart 171 are probable background galaxies)
Source: [36]

Table 6. NGC 6822 Emission Regions[a]

ID	Other names	Chart
H I	Ho 2, KD 1	158
H II	Ho 3, KD 2	161
H III	Ho 4, KD 4	158
H V	Ho 9, Ho 11, KD 19	160
H X	Ho 14, KD 26	159
Ho 1	KD 3	161
Ho 5	KD 7	169
Ho 6	KD 10	160
Ho 7	KD 11	163
Ho 8	KD 16, KD 17	168
Ho 10	KD 18	168
Ho 12	KD 23	165
Ho 15	KD 30	164
Ho 16	KD 31	162
K□	KD 2e	166
K□	KD 5e	166
K□	KD 11e	168
KD 5	Part of H III	158
KD 6	Part of H III	158
KD 8	--	169
KD 9	--	160
KD 12	--	160
KD 13	--	168
KD 14	--	168
KD 15	--	168
KD 17	--	168
KD 20	--	165
KD 21	--	160
KD 22	--	167
KD 24	--	165
KD 27	--	170
KD 28	--	170
KD 29	--	164

KD 9e	Part of Ho V	169
KD 12e	Part of KD 12	160
KD 13e	--	168
KD 15e	Part of KD 13	168
KD 17e	Part of KD 14	168
KD 20e	Part of Ho 8	168
KD 22e	Part of H V	160
KD 34e	Part of Ho 15	164
KD 36e	--	162
HK 1	--	166
HK 2	--	161
HK 3	--	158
HK 4	--	161
HK 5	--	158
HK 6	--	169
HK 7	--	169
HK 8	--	169
HK 9	--	169
HK 11	--	163
HK 12	--	163
HK 13	--	160
HK 14	--	163
HK 15	--	160
HK 16	--	160
HK 17	--	160
HK 18	--	163
HK 19	--	160
HK 20	--	160
HK 21	--	160
HK 22	--	163
HK 23	--	163
HK 24	--	168
HK 25	--	168
HK 26	--	168
HK 27	--	163
HK 28	--	163

HK 29	--	168
HK 30	--	168
HK 31	--	163
HK 32	--	160
HK 33	--	160
HK 34	--	163
HK 35	--	160
HK 36	--	160
HK 37	--	163
HK 38	--	163
HK 39	--	168
HK 40	--	160
HK 41	--	163
HK 42	--	160
HK 43	--	163
HK 44	--	160
HK 45	--	168
HK 46	--	168
HK 47	--	168
HK 48	--	168
HK 49	--	163
HK 50	--	160
HK 51	--	160
HK 52	--	168
HK 53	--	160
HK 54	--	168
HK 55	--	160
HK 56	--	168
HK 57	--	168
HK 58	--	165
HK 59	--	168
HK 60	--	160
HK 61	--	168
HK 62	--	160
HK 63	--	165
HK 64	--	160

HK 65	--	165
HK 66	--	165
HK 67	--	165
HK 68	--	159
HK 69	--	165
HK 70	--	165
HK 71	--	165
HK 72	--	165
HK 73	--	159
HK 74	--	159
HK 75	--	159
HK 76	--	159
HK 77	--	159
HK 78	--	159
HK 79	--	159
HK 80	--	159
HK 81	--	170
HK 82	--	159
HK 83	--	170
HK 84	--	159
HK 85	--	159
HK 86	--	164
HK 87	--	159
HK 88	--	164
HK 89	--	170
HK 90	--	159
HK 91	--	170
HK 92	--	159
HK 93	--	159
HK 94	--	170
HK 95	--	170
HK 96	--	162
HK 97	--	162
HK 98	--	162
HK 99	--	164

HK 100	--	162
HK 101	--	162
HK 102	--	162
HK 103	--	162
HK 104	--	162
HK 105	--	162
HK 106	--	162
HK 107	--	162
HK 108	--	162
HK 109	--	160
HK 110	--	160

[a]Designations: H: reference 43; Ho: reference 35; K: reference 54; KD: reference 53; KDe: unresolved emission regions from reference 53; HK: reference 39.

NGC 6822

Chart 147

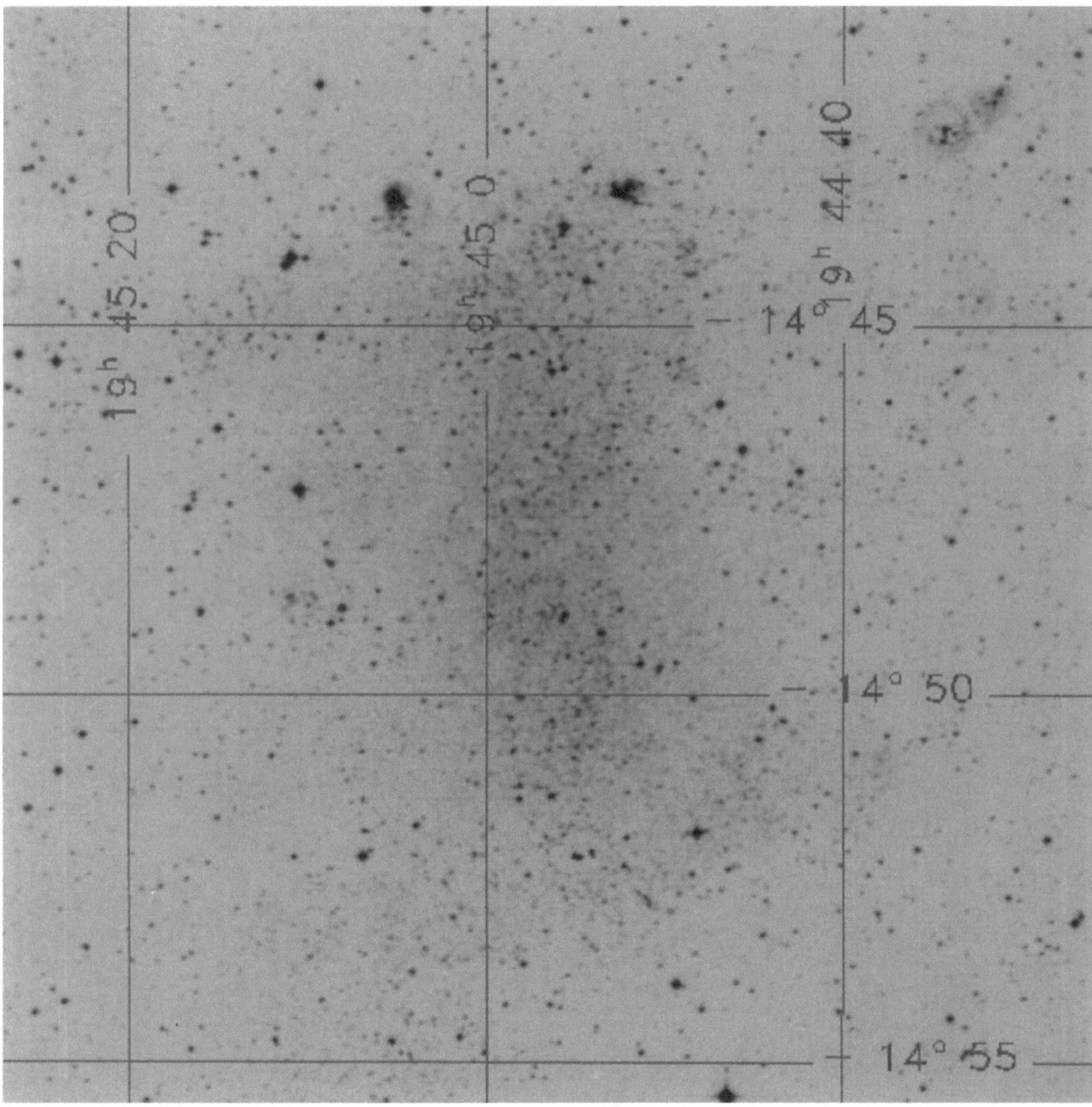

NGC 6822

Chart 148

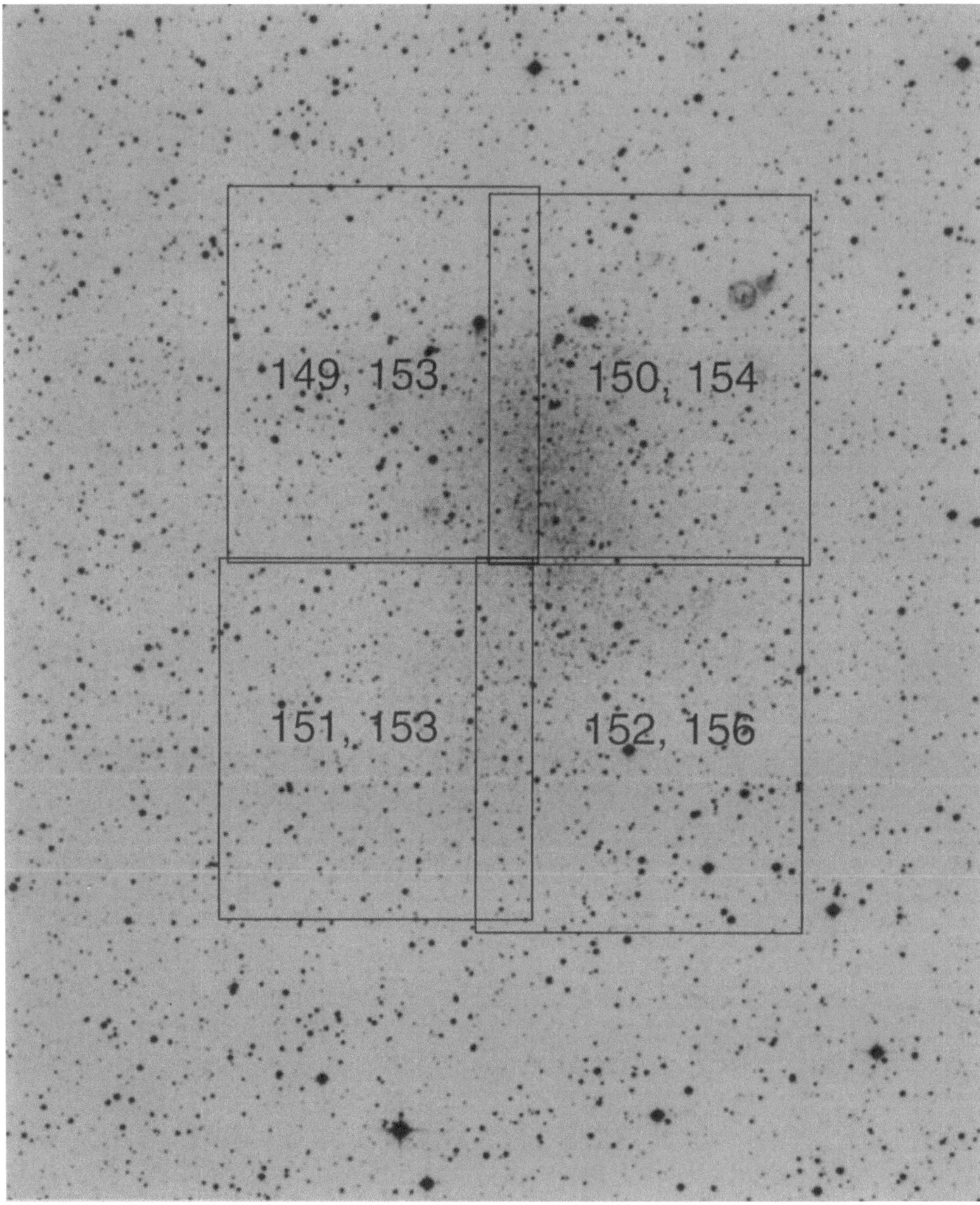

NGC 6822

Chart 149

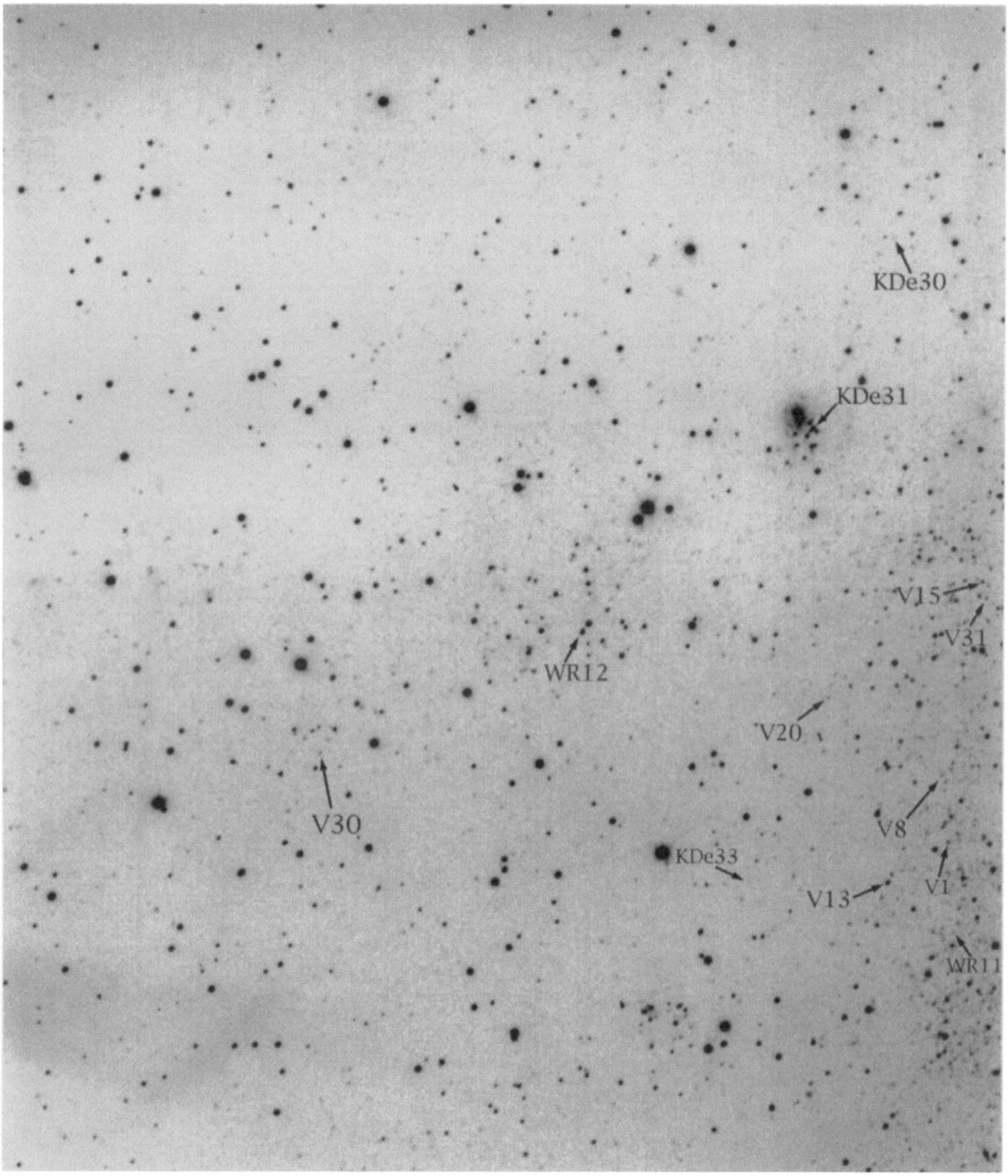

NGC 6822

Chart 150

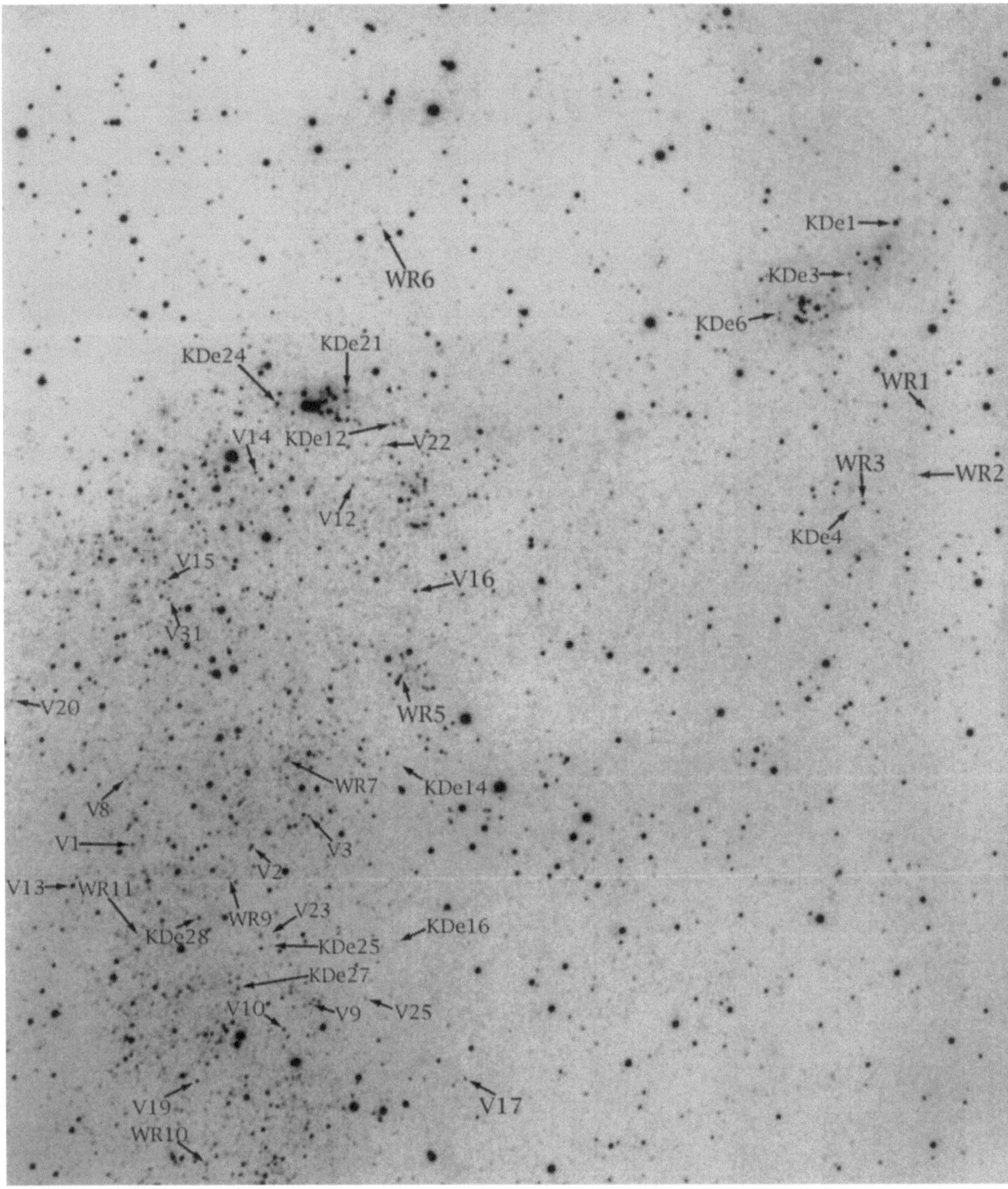

NGC 6822

Chart 151

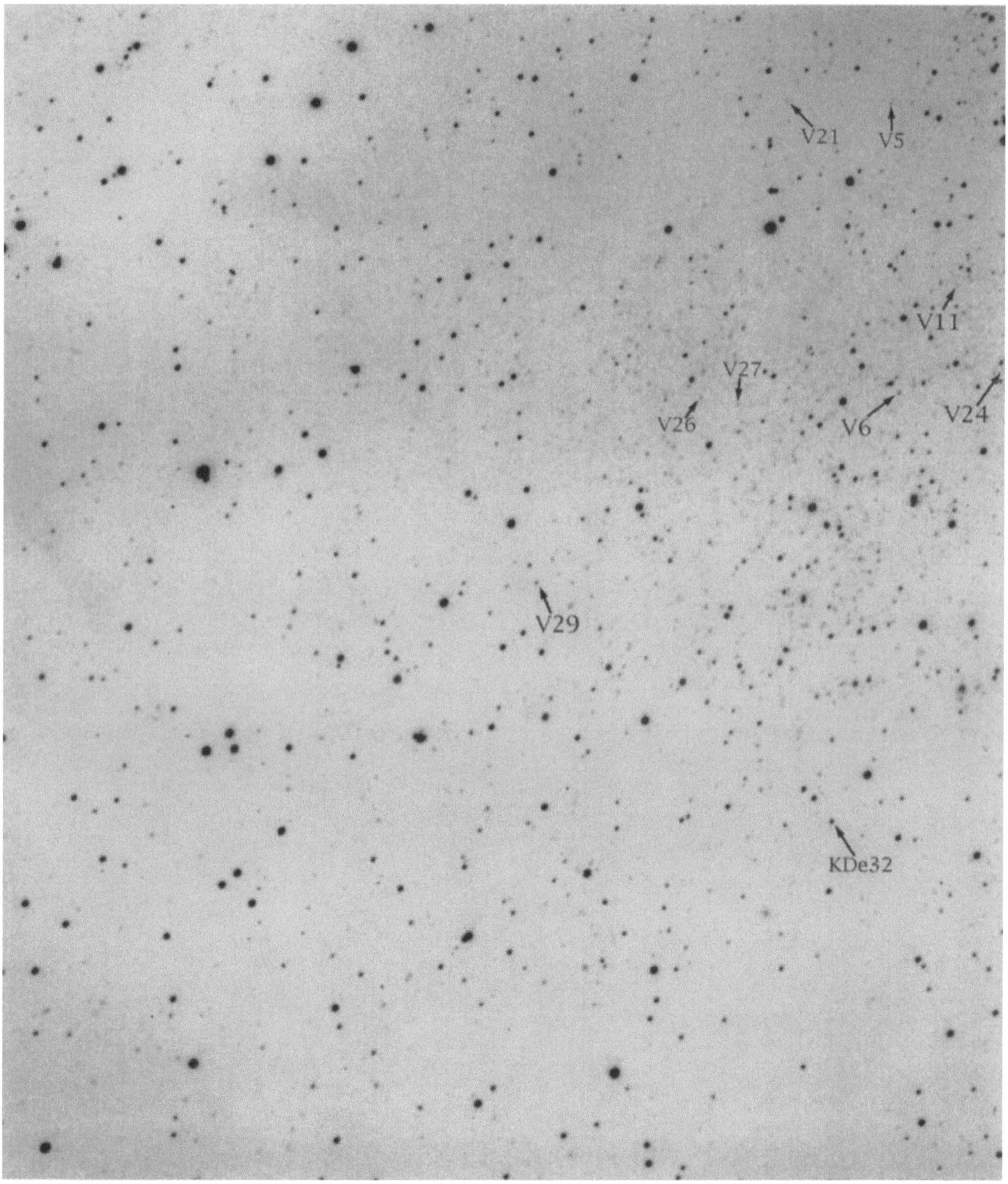

NGC 6822

Chart 152

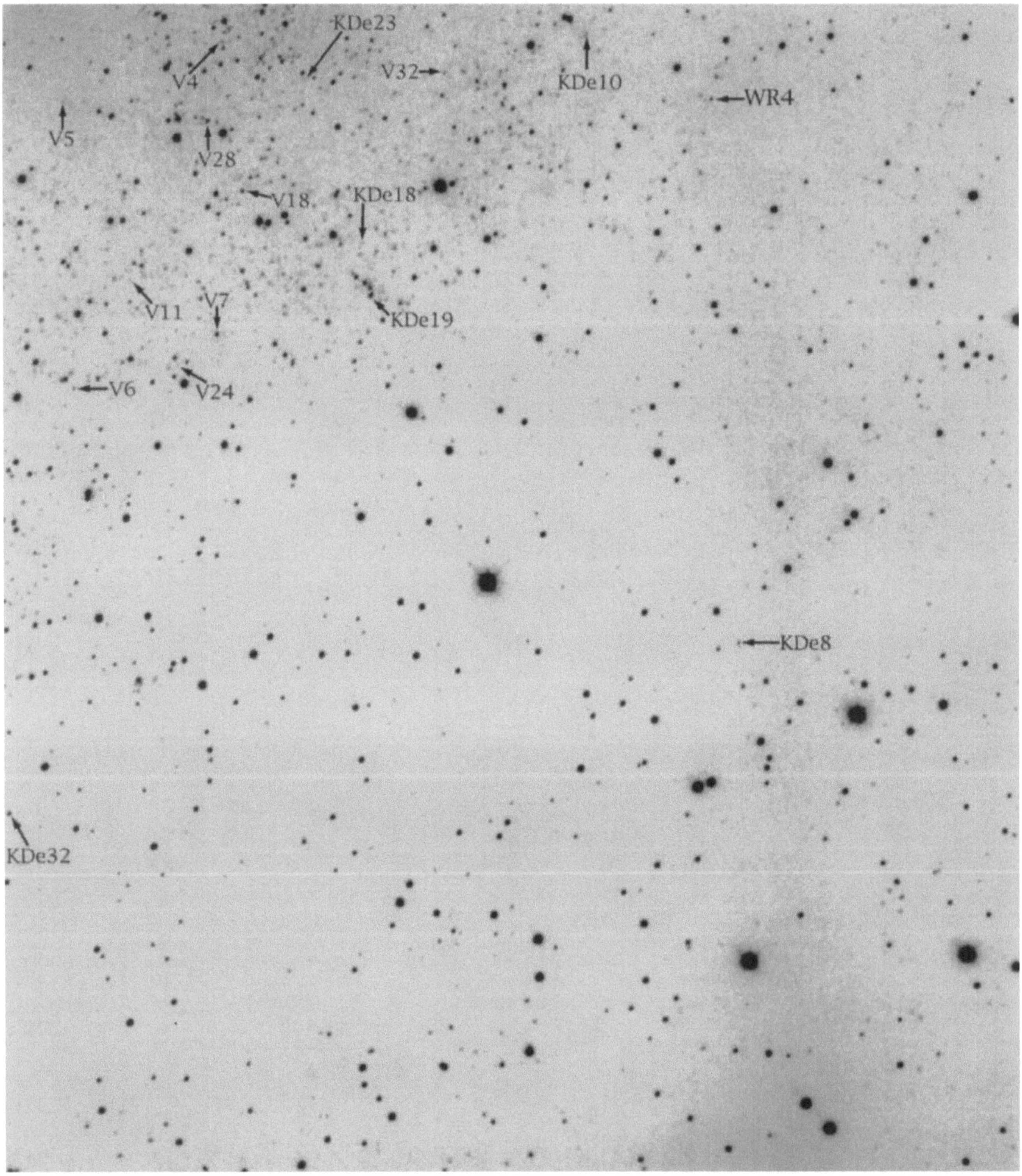

NGC 6822

Chart 153

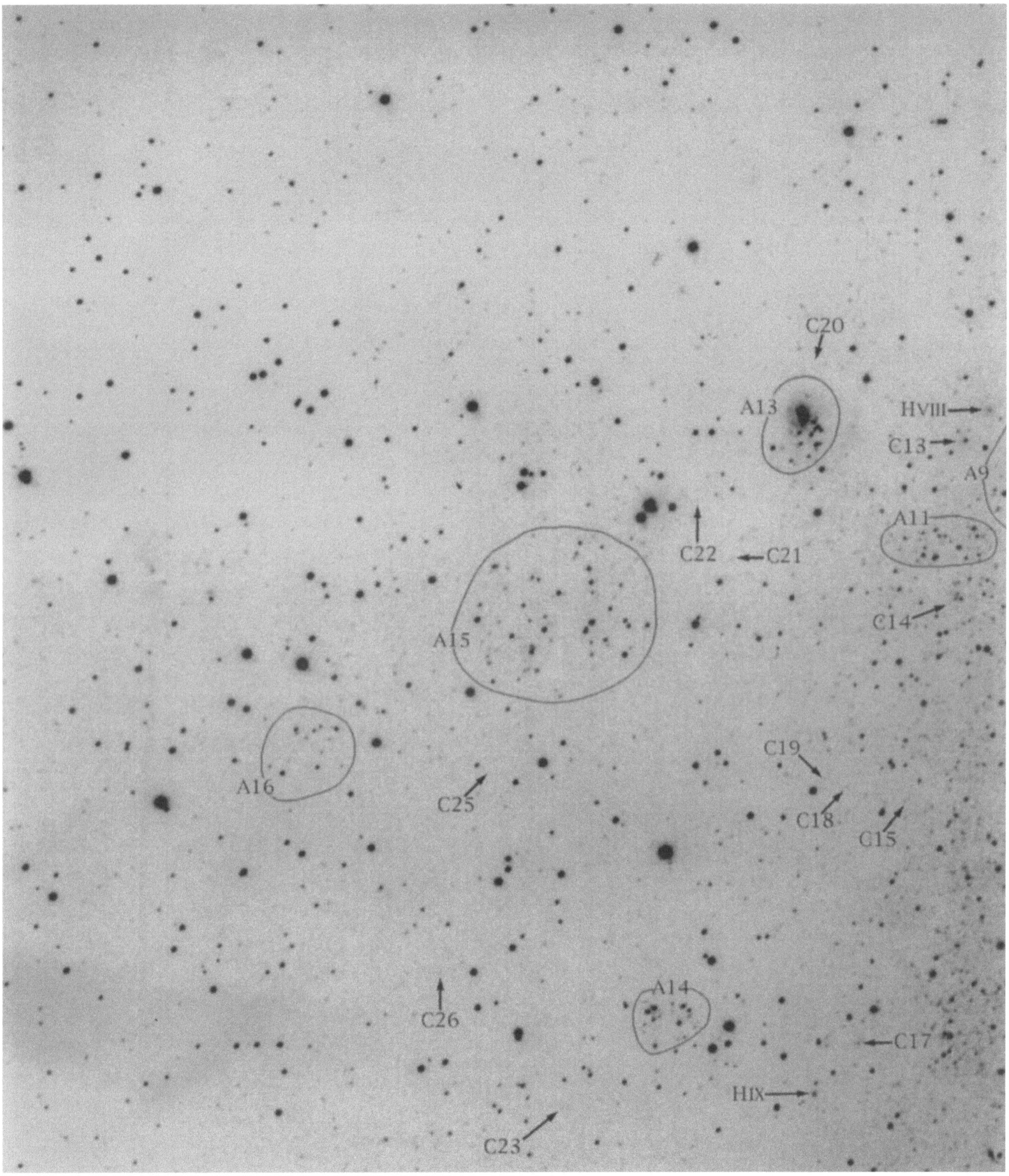

NGC 6822

Chart 154

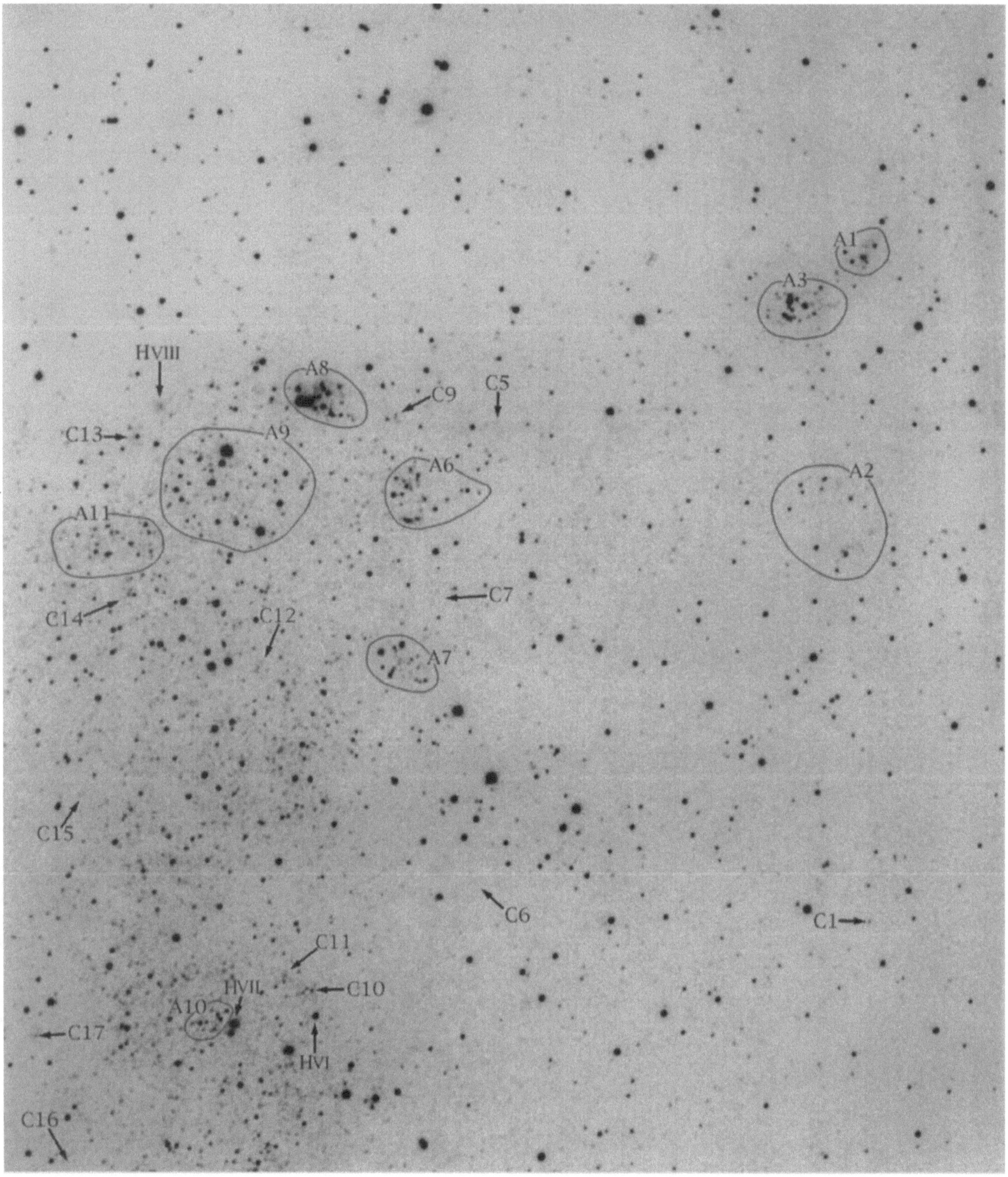

NGC 6822

Chart 155

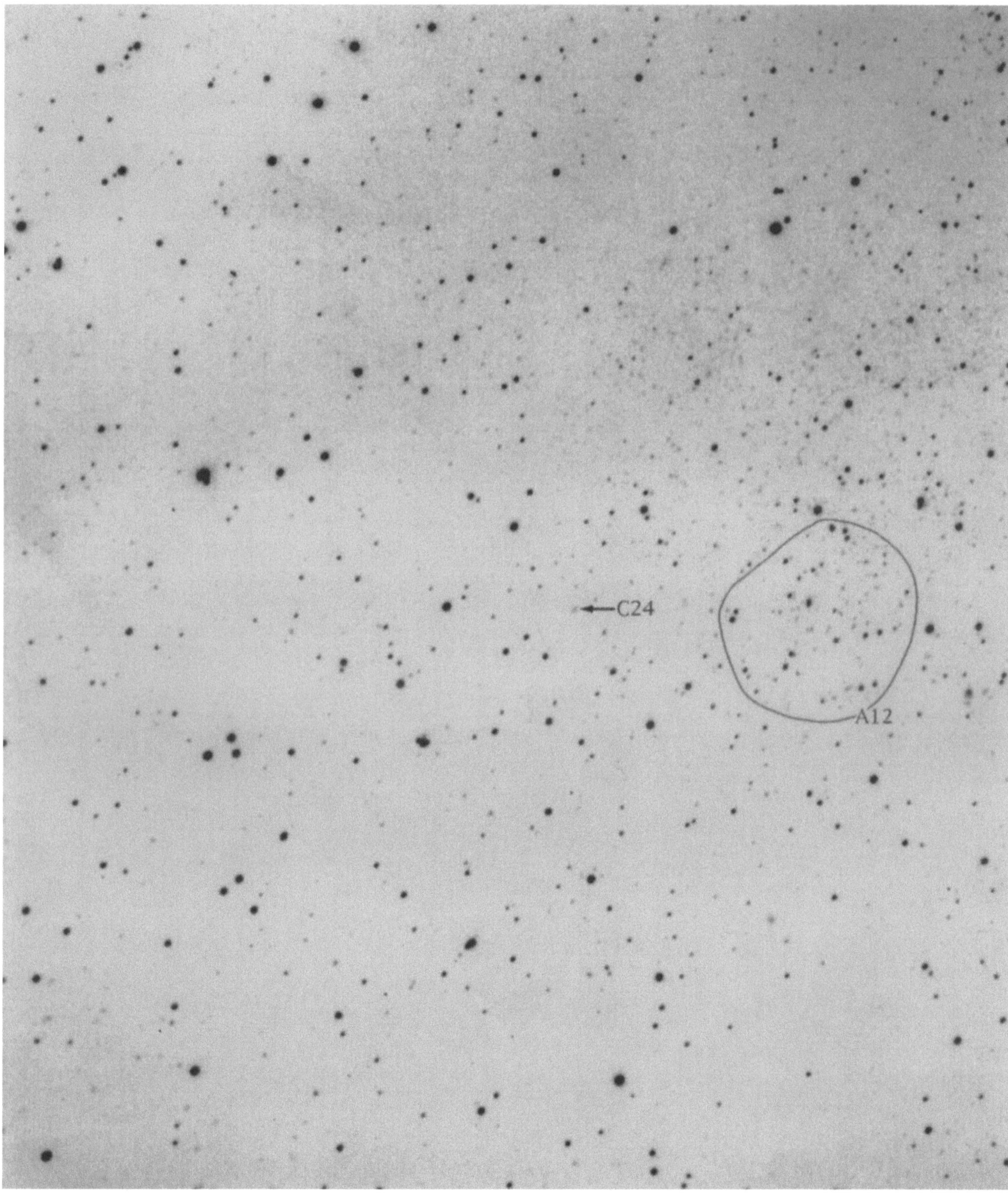

NGC 6822

Chart 156

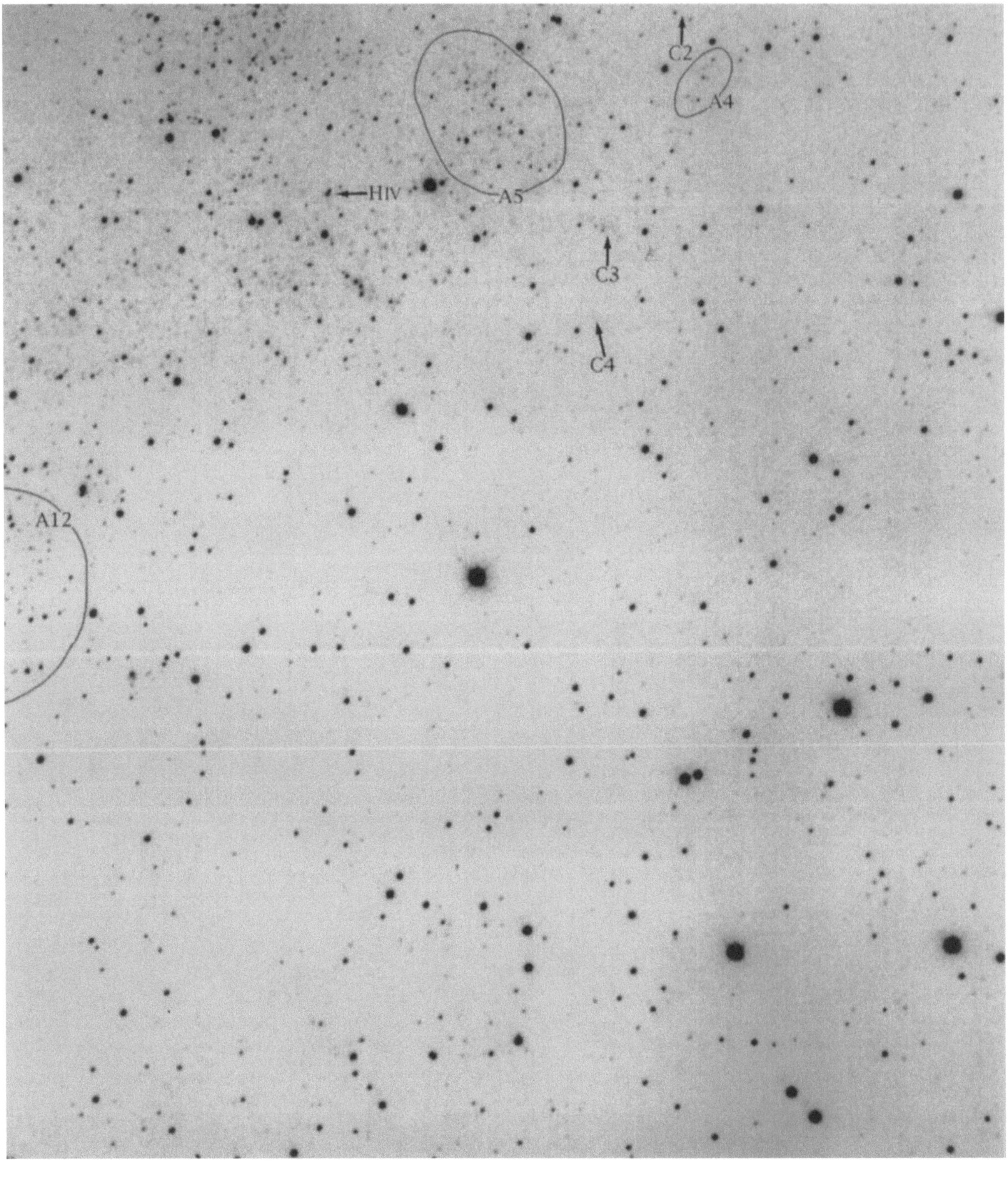

NGC 6822

Chart 157

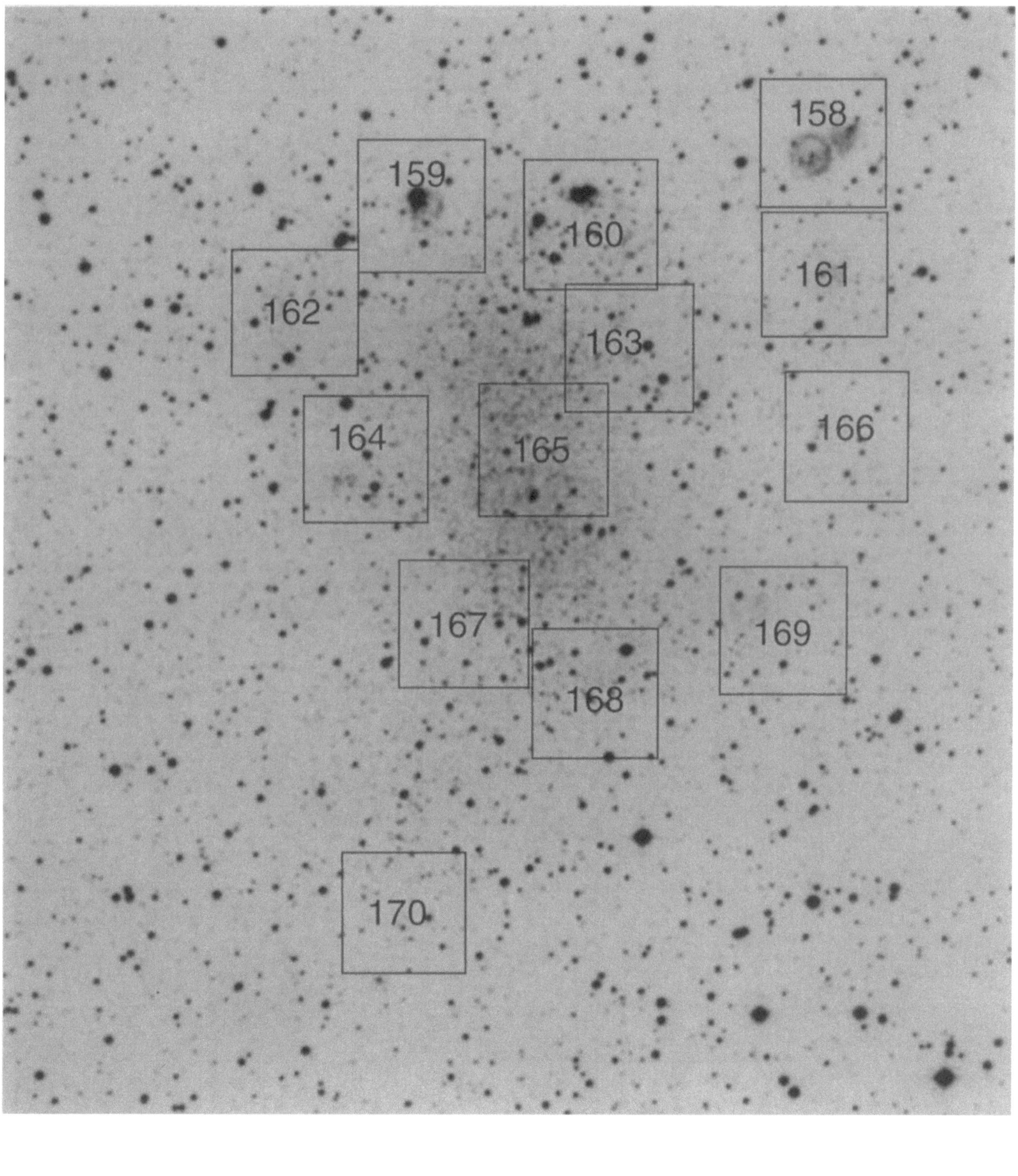

NGC 6822

Chart 158

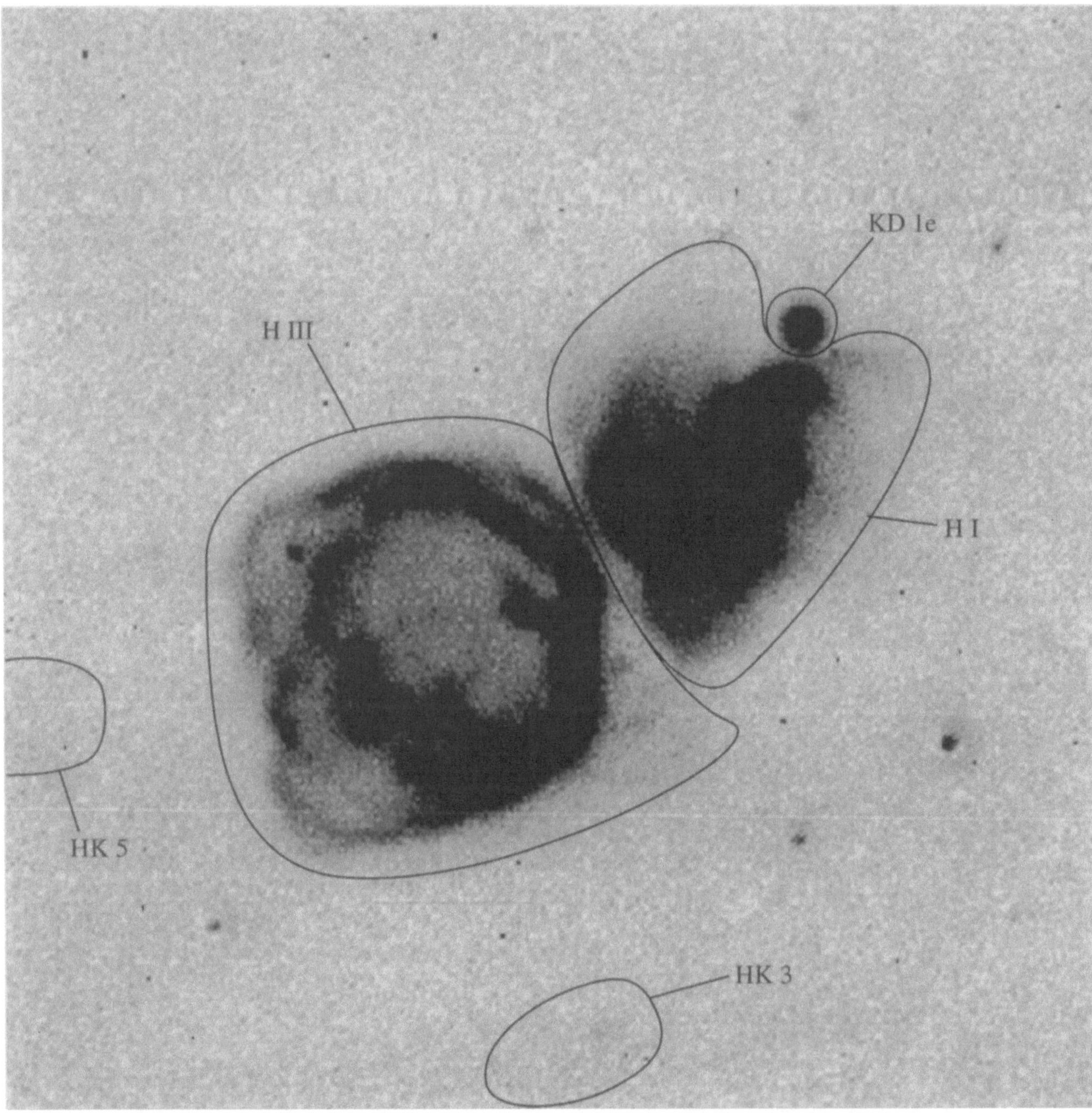

NGC 6822

Chart 159

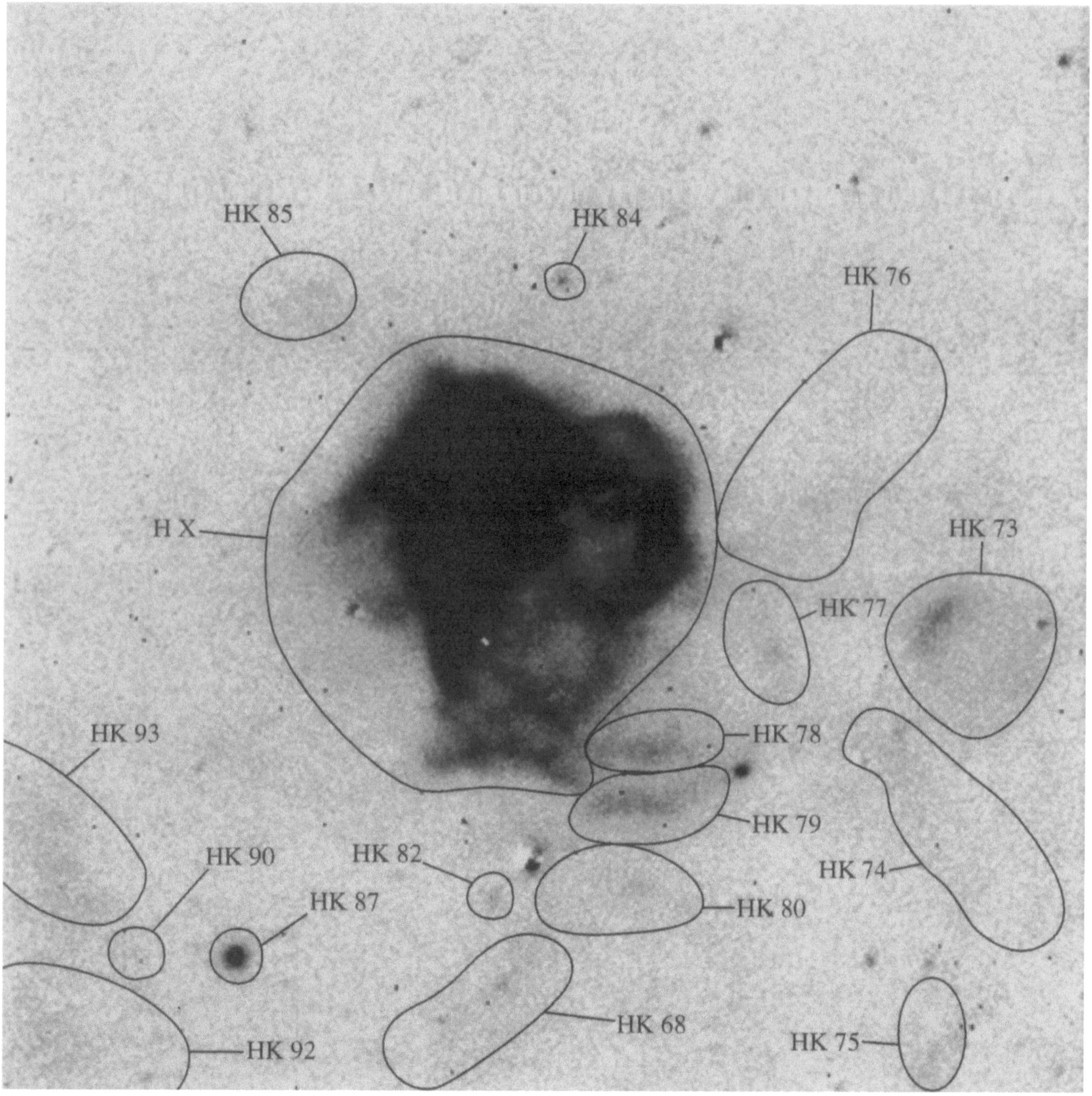

NGC 6822

Chart 160

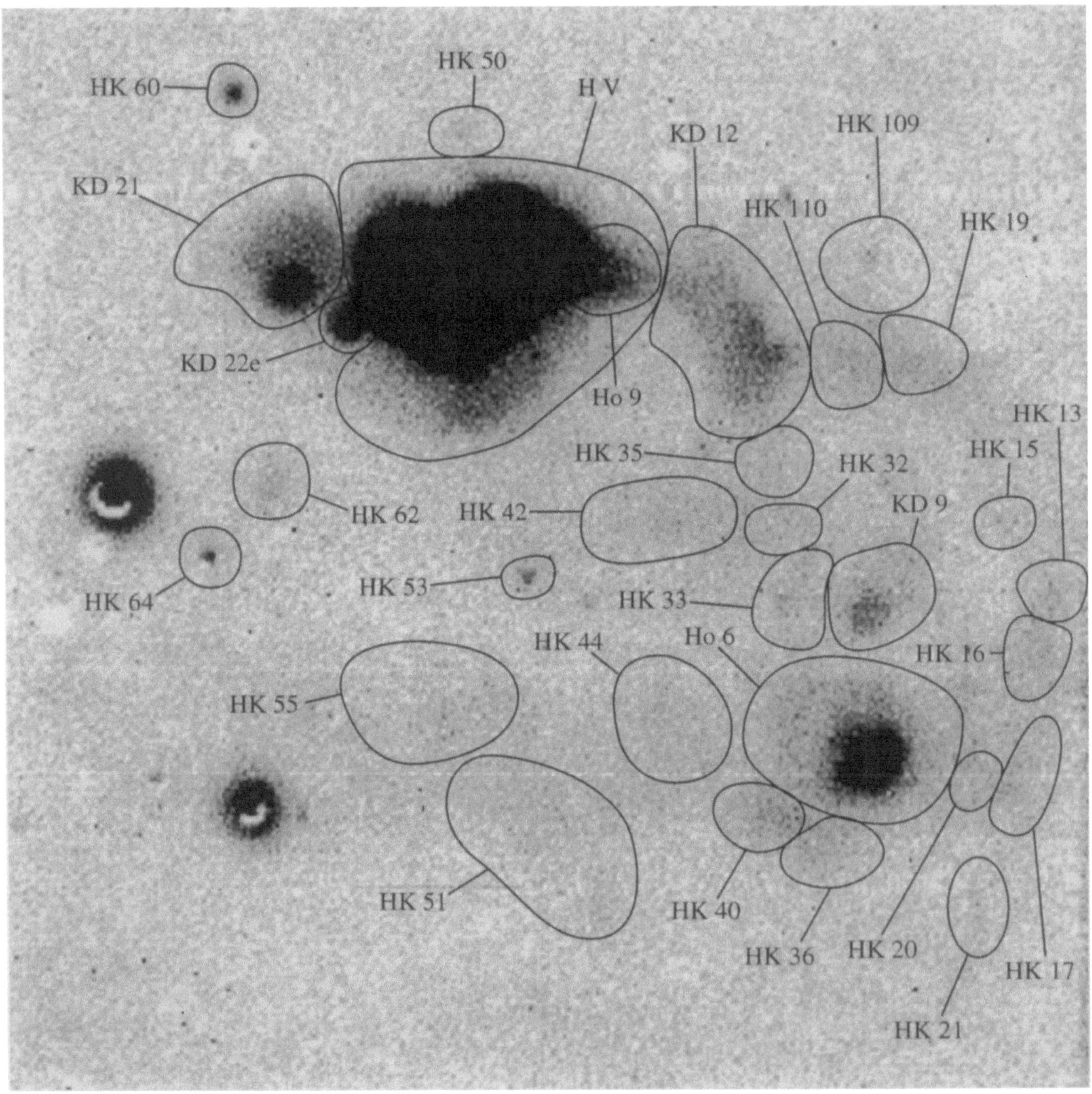

NGC 6822

Chart 161

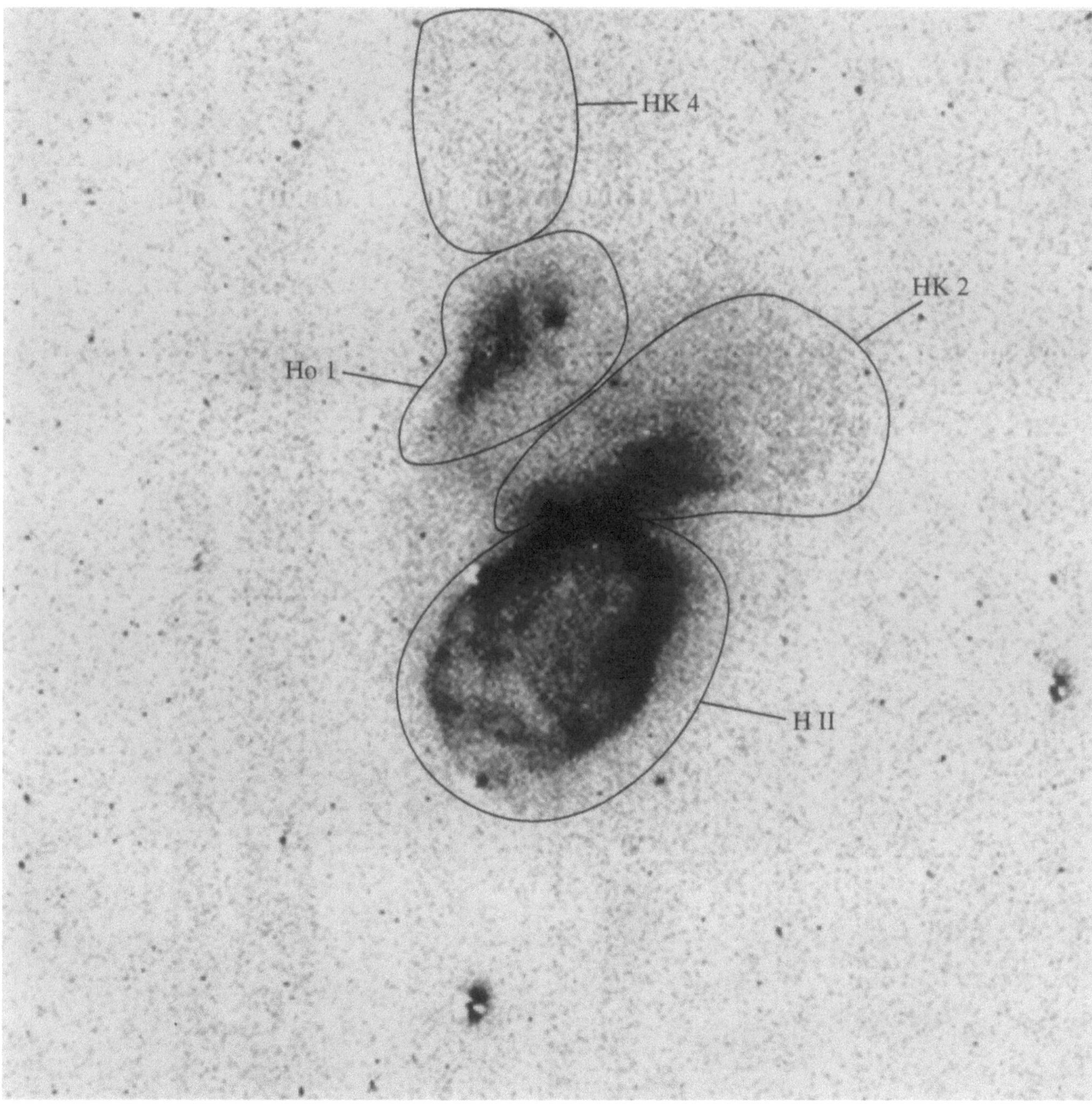

NGC 6822

Chart 162

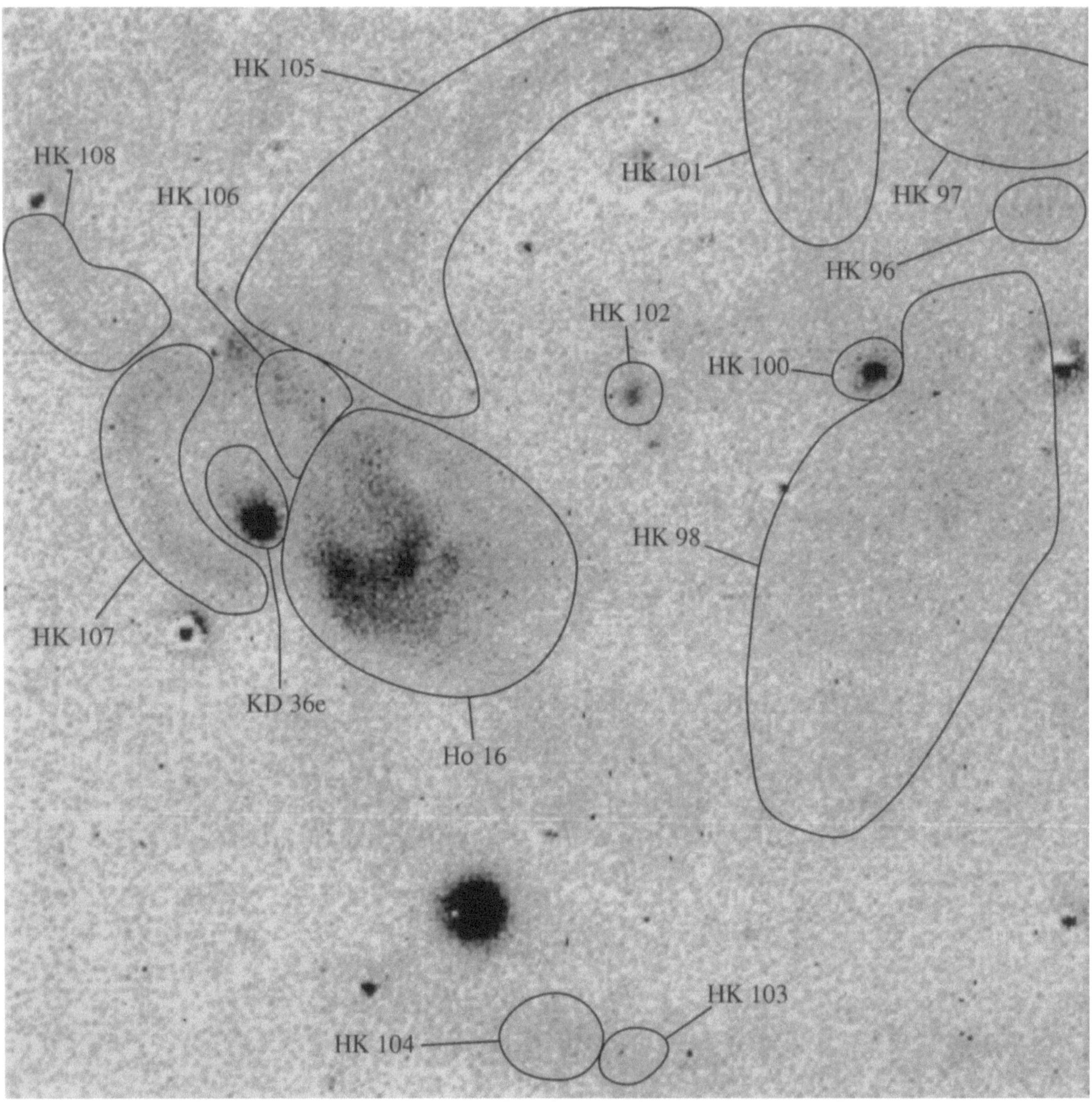

NGC 6822

Chart 163

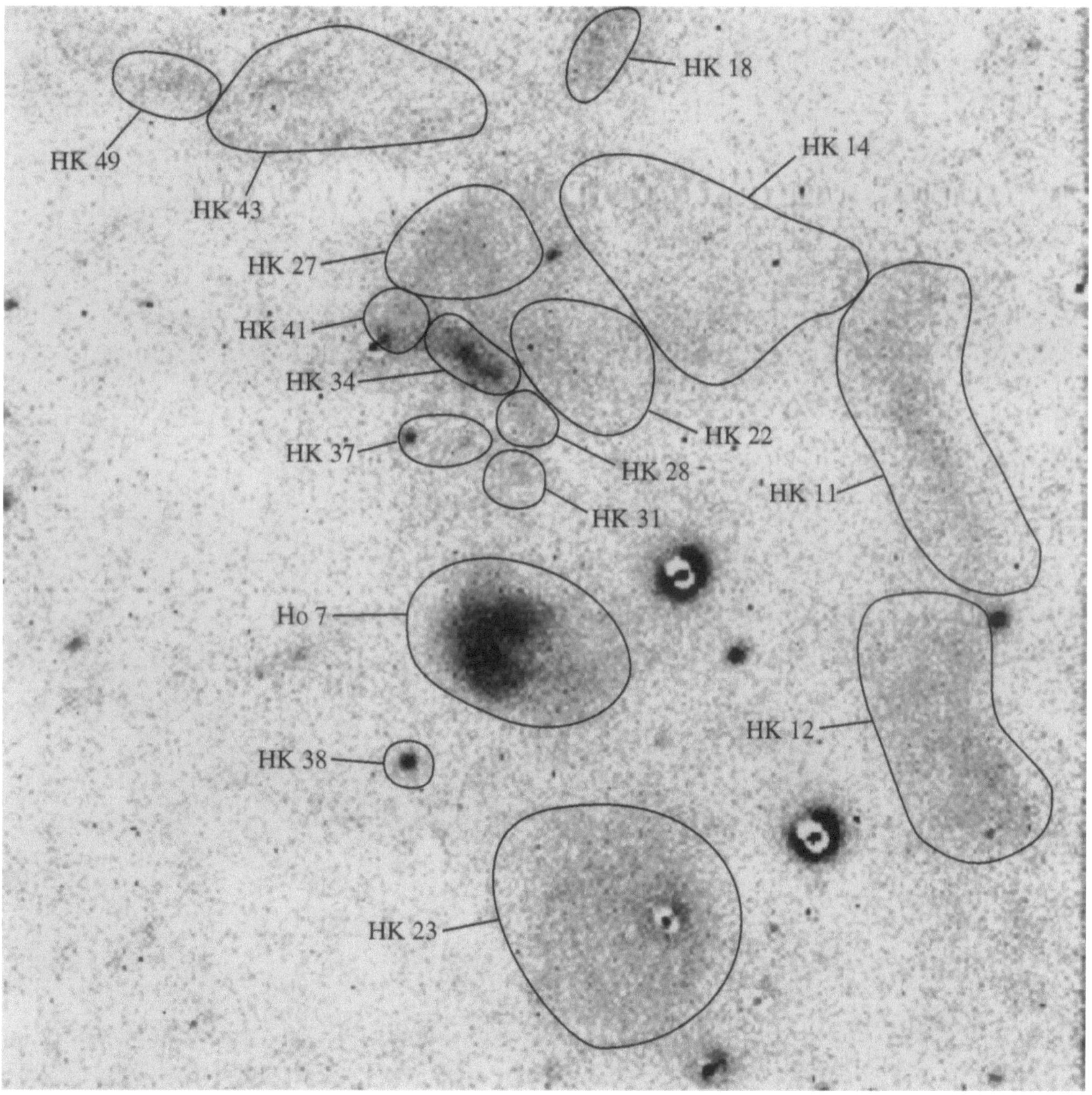

NGC 6822

Chart 164

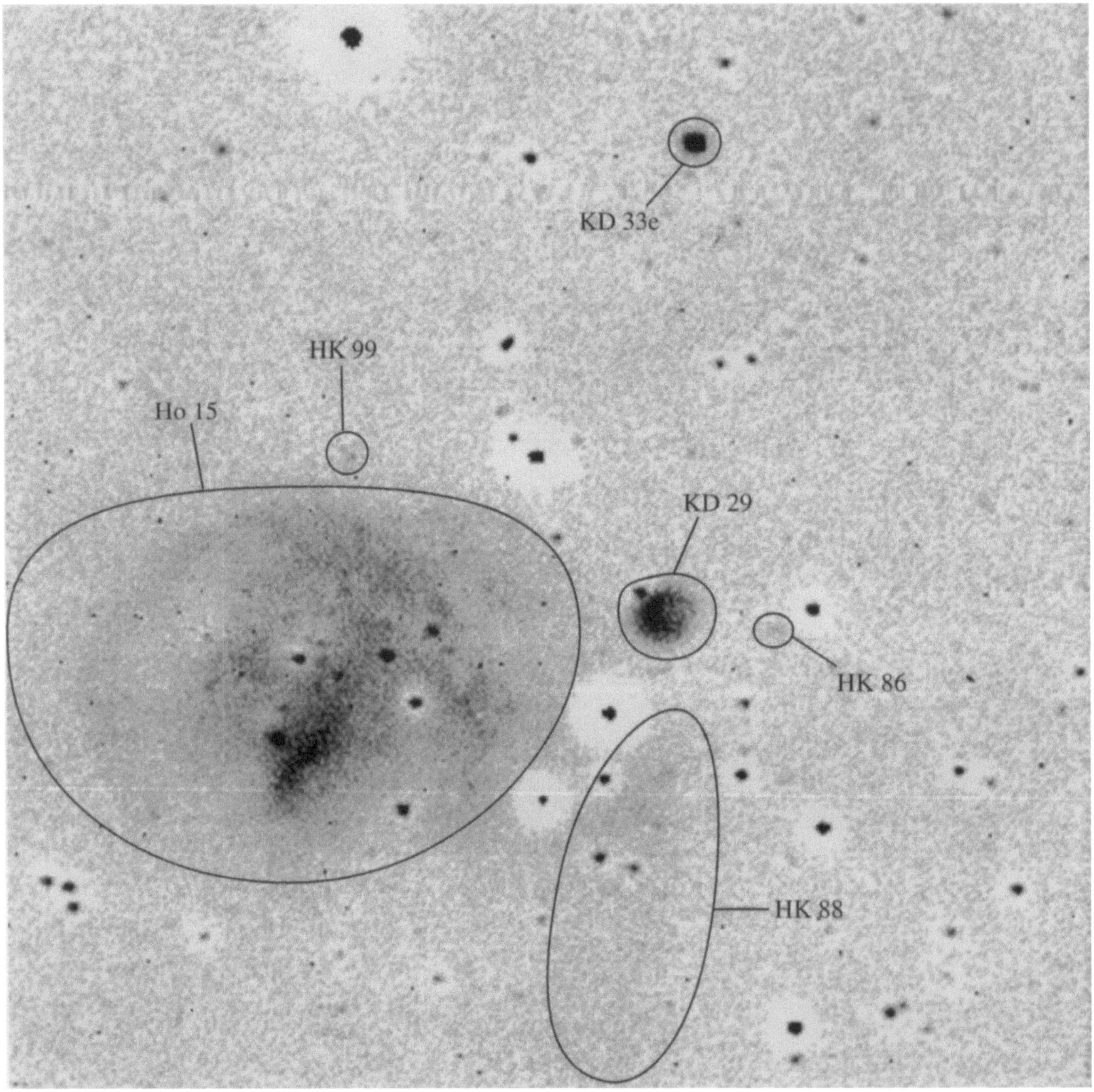

NGC 6822

Chart 165

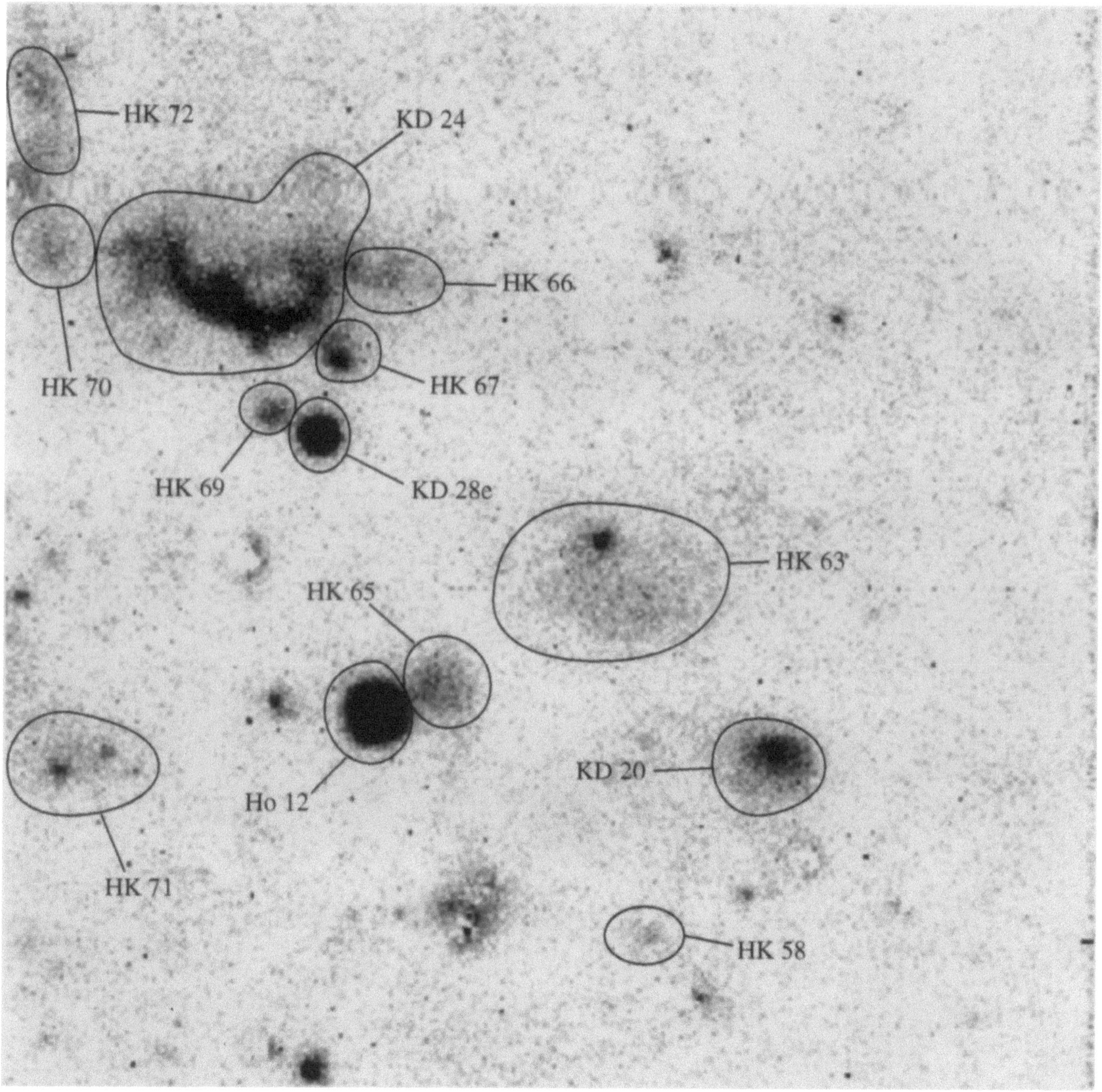

NGC 6822

Chart 166

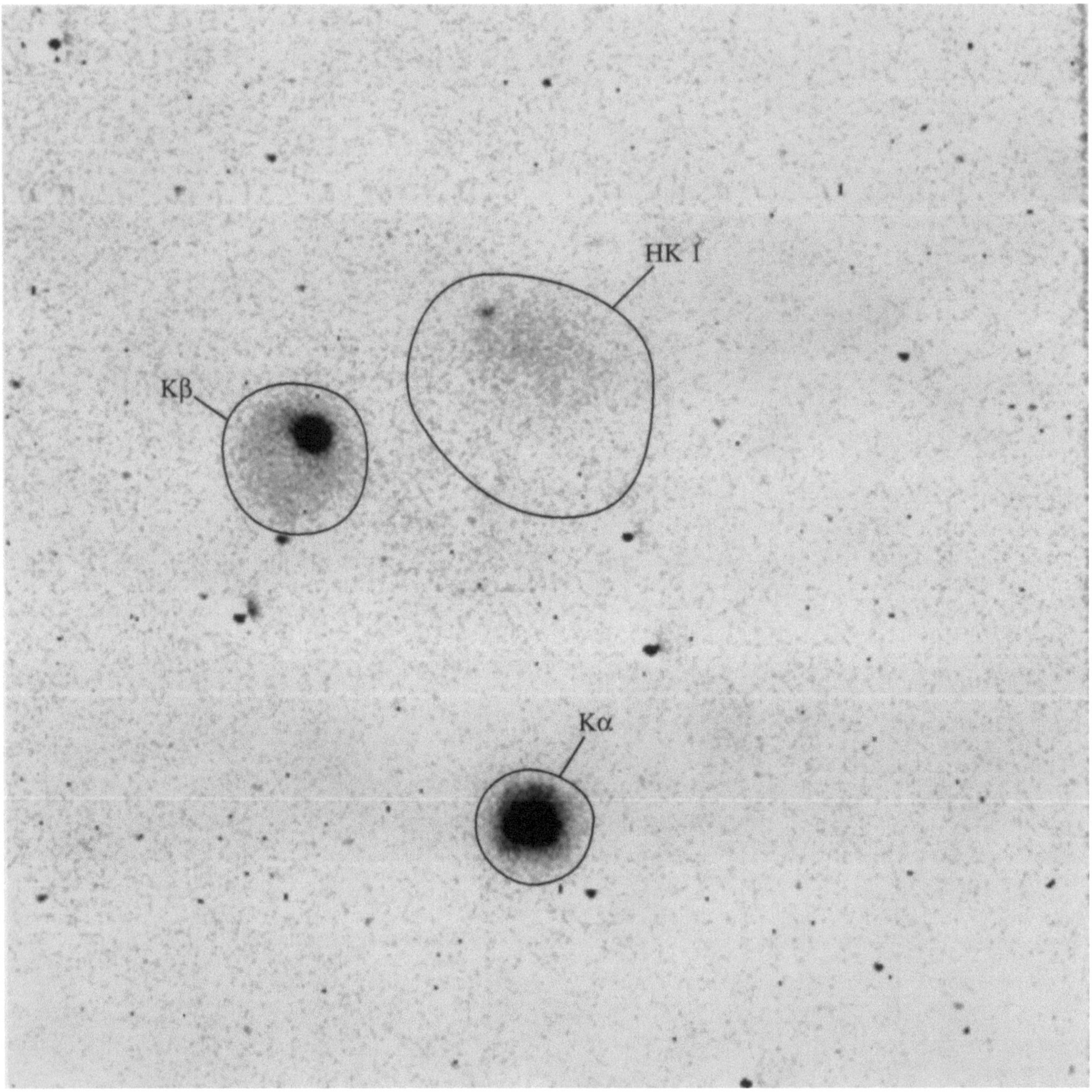

NGC 6822

Chart 167

NGC 6822

Chart 168

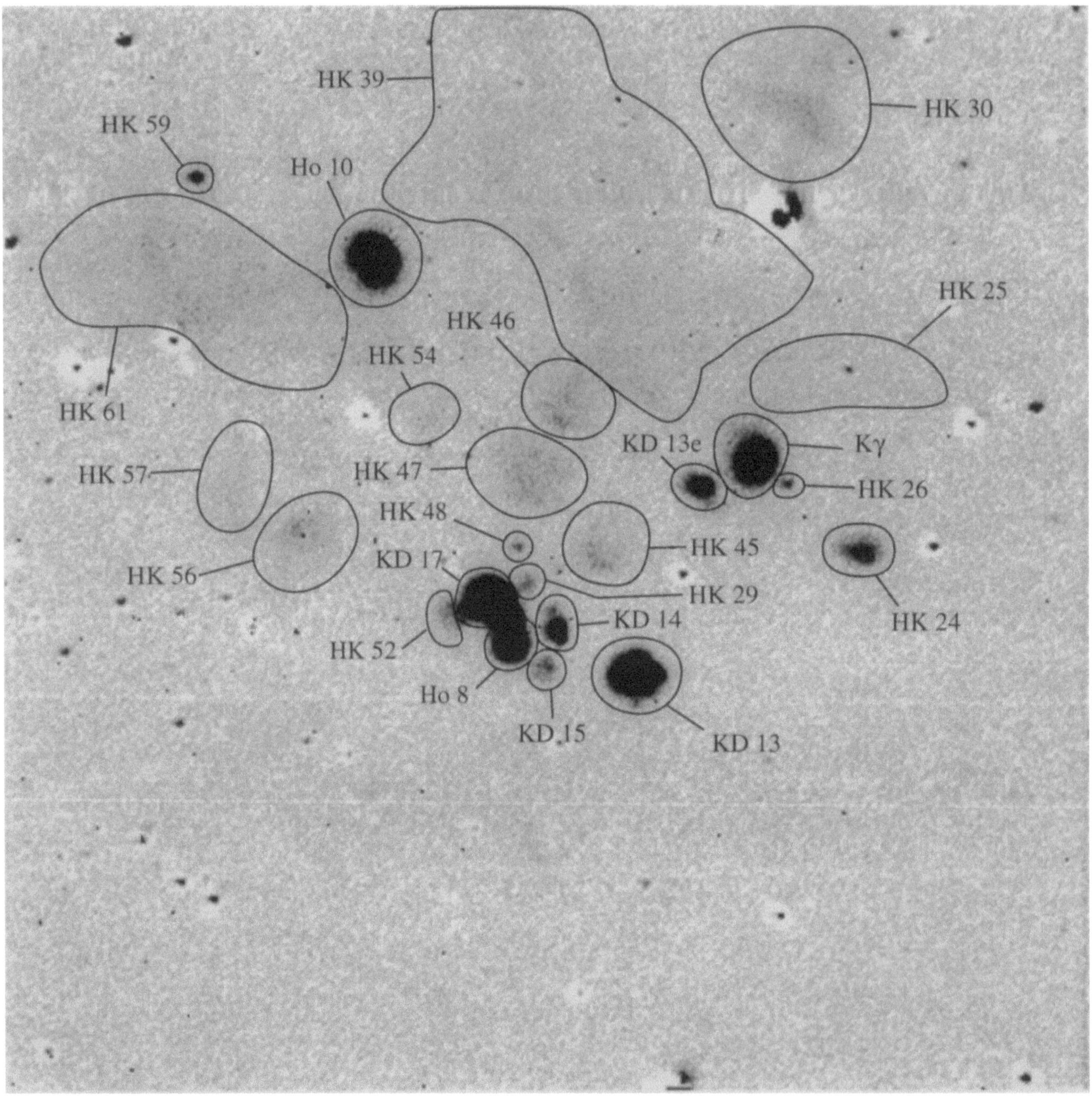

NGC 6822

Chart 169

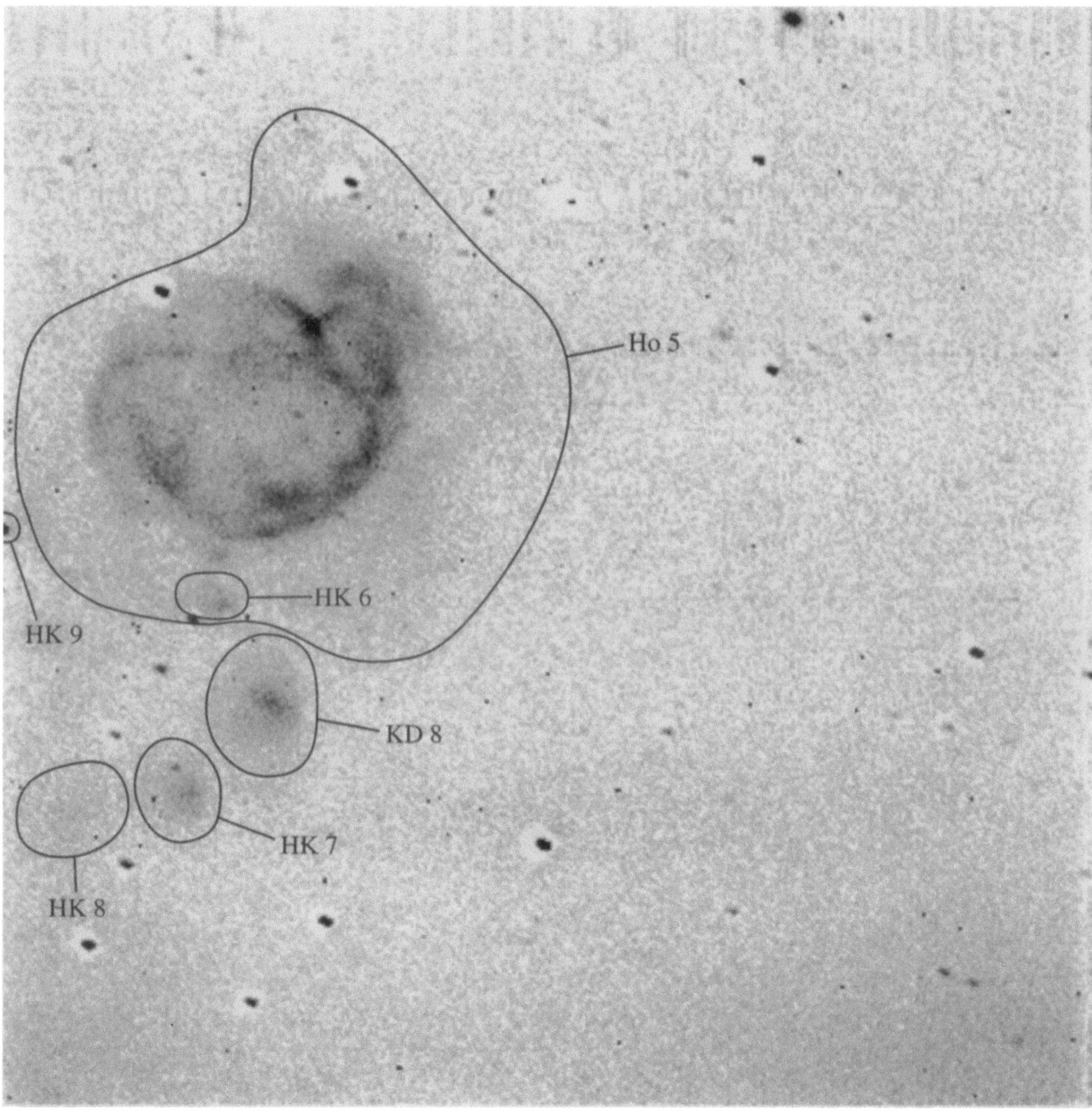

NGC 6822

Chart 170

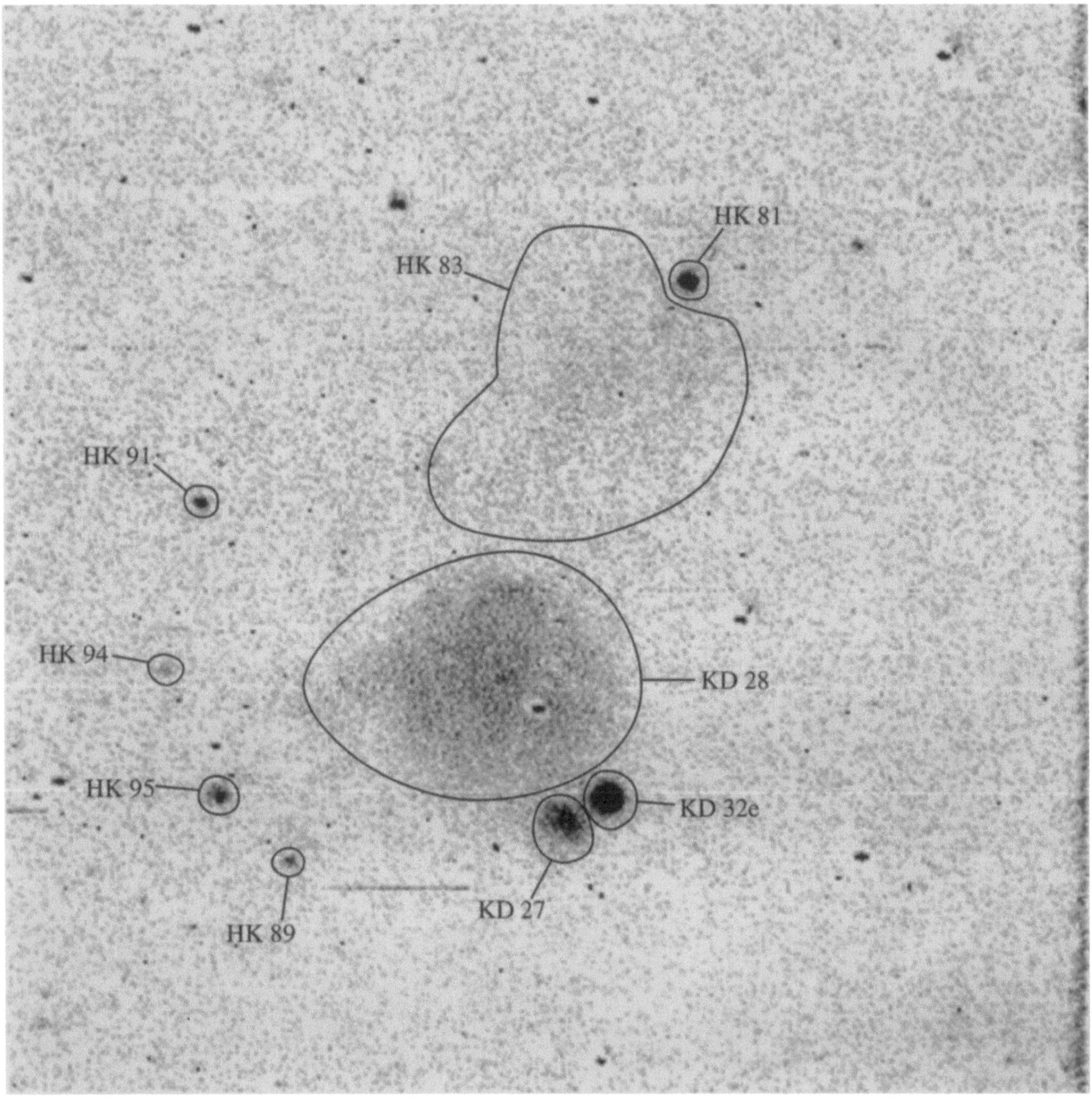

NGC 6822

Chart 171

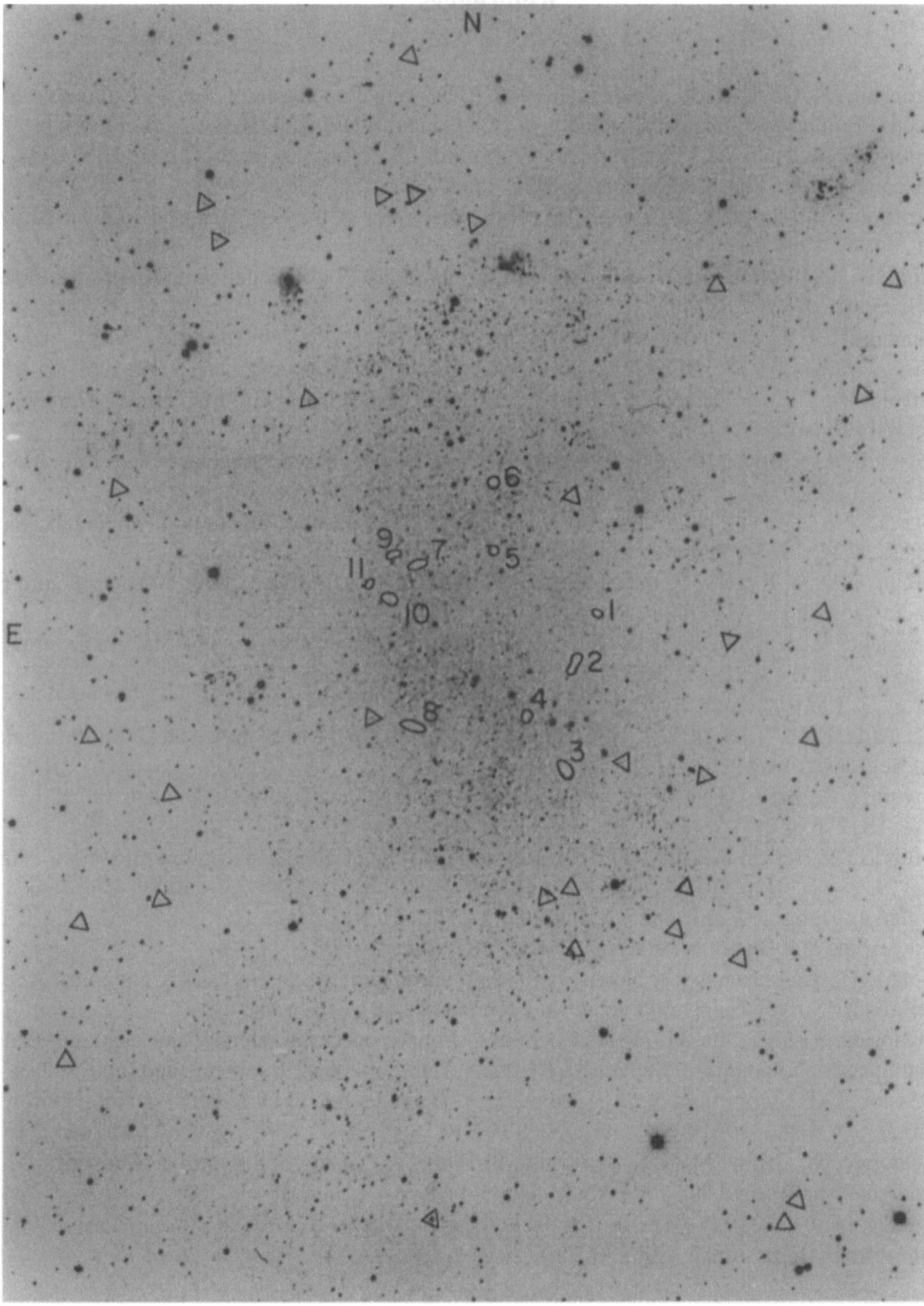

NGC 6822

References

1. Aaronson, M., Cook, K.H., & Norris, J. 1985, "The AGB Population of Nearby Galaxies", in Spectral Evolution of Galaxies, edited by C. Chiosi and A. Renzini (Reidel, Dordrecht), p. 171.
2. Aaronson, M., Da Costa, G.S., Hartigan, P., Mould, J.R., Norris, J., & Stockman, H.S. 1984, "Discovery of the First Carbon Star in NGC 6822", ApJ, 277, L9.
3. Aaronson, M., Mould, J., & Cook, K.H. 1985, "Discovery of the First S Star in NGC 6822", ApJ, 291, L41.
4. Alloin, D., Collin-Souffrin, S., Joly, M., & Vigroux, L. 1979, "Nitrogen and Oxygen Abundances in Galaxies", A&A, 78, 200.
5. Armandroff, T.E., & Massey, P. 1985, "Wolf-Rayet Stars in NGC 6822 and IC 1613", ApJ, 291, 685.
6. Armandroff, T.E., & Massey, P. 1991, "Wolf-Rayet Stars in Local Group Galaxies: Numbers and Spectral Properties", AJ, 102, 927.
7. Arsenault, R. & Roy, J.R. 1986, "Integrated Hα Profiles of Giant Extragalactic H II Regions", AJ, 92, 567.
8. Bogess, N.W. 1967, "The Structure of NGC 6822", Ph.D. thesis, University of Michigan.
9. Bogess, N.W. 1968, "Structure of NGC 6822", AJ, 73, S6.
10. Bothun, G.D., & Rogers, C. 1992, "Surface Photometry of Disk Galaxies at 60 and 100 μm: Radial Gradients in Dust Temperature in Optically Thin Disks", AJ, 103, 1484.
11. Breysacher, J., & Lequeux, J. 1983, "Search for Wolf-Rayet, Carbon and M Stars in External Galaxies with the GRISM/GRENS Technique", The Messenger, 33, 21.
12. Burley, J. 1963, "Neutral Hydrogen in NGC 6822", AJ, 68, 274.
13. Castañada, H.O., Vilchez, J.M., & Copetti, M.V.F. 1992, "Density Studies on Giant Extragalactic H II Regions", A&A, 260, 370.
14. Clayton, C.A. 1987, "The Dynamics of the Giant Ring Nebula Hubble III in NGC 6822", MNRAS, 226, 493.
15. Collier, J., & Hodge, P. 1994, "The H II Regions of NGC 6822. III. A Photometric Atlas", ApJS, 92, 119.
16. Collier, J., Hodge, P., & Kennicutt, R.C. 1995, "The H II Regions of NGC 6822. IV. A Multivariate and Bivariate Analysis", PASP, 107, 361.
17. Cook, K.H., Aaronson, M., & Norris, J. 1986, "Carbon and M Stars in Nearby Galaxies: A Preliminary Survey Using a Photometric Technique", ApJ, 305, 634.
18. D'Odorico, S., & Dopita, M. 1983, "Chemical Abundances in the Interstellar Medium of Galaxies from Spectrophotometry of Supernova Remnants", in Supernova Remnants and their X-Ray Emission, IAU Symposium No. 101, edited by J. Danziger and P. Gorenstein (Reidel, Dordrecht), p. 517.
19. D'Odorico, S., Dopita, M.A., & Benvenuti, P. 1980, "A Catalog of Supernova Remnant Candidates in Nearby Galaxies", A&AS, 40, 67.
20. de Vaucouleurs, G. 1978, "The Extragalactic Distance Scale. II. Distances of the Nearest Galaxies from Primary Indicators", ApJ, 223, 730.

21. Dickel, J.R., D'Odorico, S., & Silverman, A. 1985, "VLA Observations of Three Extragalactic SNR at 20 and 6 cm", AJ, 90, 414.
22. Dufour, R.J., & Talent, D.L. 1980, "Spectrophotometry of a Planetary Nebula in the Irregular Galaxy NGC 6822", ApJ, 235, 22.
23. Elias, J.H., & Frogel. J.A. 1985, "M Supergiants in Local Group Irregular Galaxies: Metallicities and Distances", ApJ, 289, 141.
24. Elmegreen, B.G., Elmegreen, D.M., & Morris, M. 1980, "On the Abundance of Carbon Monoxide in Galaxies: A Comparison of Spiral and Magellanic Irregular Galaxies", ApJ, 240, 455.
25. Fabbiano, G., Kim, D.-W., & Trinchieri, G. 1992, "An X-Ray Catalog and Atlas of Galaxies", ApJS, 80, 531.
26. Freedman, W.L. 1986, "Recent CCD Observations of Cepheids in Nearby Galaxies: A Progress Report", in Galaxy Distances and Deviations from Universal Expansion, edited by B.F. Madore and R. B. Tully (Reidel, Dordrecht), p. 21.
27. Gallagher, III, J.S., Hunter, D.A., Gillett, F.C., & Rice, W.L. 1991, "Far-Infrared Properties of NGC 6822", ApJ, 371, 142.
28. Gallart, C., Aparicio, A., Bertelli, G., & Chiosi, C. 1996, "The Local Group Dwarf Irregular Galaxy NGC 6822. II. The Old and Intermediate-Age Star Formation History", AJ, 112, 1905.
29. Gallart, C., Aparicio, A., Bertelli, G., & Chiosi, C. 1996, "The Local Group Dwarf Irregular Galaxy NGC 6822. III. The Recent Star Formation History", AJ, 112, 2596.
30. Gallart, C., Aparicio, A., Chiosi, C., Bertelli, G., & Vilchez, J.M. 1994, "Asymptotic Giant Branch Stars in NGC 6822: Probes of Star Formation History", ApJ, 425, L9.
31. Gallart, C., Aparicio, A., & Vilchez, J.M. 1996, "The Local Group Dwarf Irregular Galaxy NGC 6822. I. The Stellar Content", AJ, 112, 1928.
32. Gottesman, S.T., & Weliachew, L. 1977, "An Interferometer Study of the Neutral Hydrogen Associated with the Optical Core of the Irregular Galaxy NGC 6822", A&A, 61, 523.
33. Hippelein, H.H. 1986, "Turbulent Gas Motions in Giant H II Regions. II. The Luminosity-Velocity Dispersion Relation", A&A, 160, 374.
34. Hodge, P.W. 1969, "Distribution of H II Regions in Irregular Galaxies", ApJ, 156, 847.
35. Hodge, P.W. 1969, "H II Regions in Twenty Nearby Galaxies", ApJS, 18, 73.
36. Hodge, P.W. 1977, "The Structure and Content of NGC 6822", ApJS, 33, 69.
37. Hodge, P.W. 1980, "The Recent Evolutionary History of the Galaxies NGC 6822 and IC 1613", ApJ, 241, 125.
38. Hodge, P.W. 1986, "The Ages of Local Group Galaxies", MSAIt, 57, 553.
39. Hodge, P., Kennicutt, Jr., R.C., & Lee, M.G. 1988, "The H II Regions of NGC 6822. I. An Atlas of 157 H II Regions", PASP, 100, 917.
40. Hodge, P., Lee, M.G., & Kennicutt, Jr., R.C. 1989, "The H II Regions of NGC 6822. II. The Luminosity Function and Size Distribution", PASP, 101, 32.
41. Hodge, P., Smith, T., Eskridge, P., MacGillivray, H., & Beard, S. 1991, "A COSMOS Study of the Structure and Content of NGC 6822", ApJ, 379, 621.
42. Hoessel, J., & Anderson. N. 1986, "Photometry of Resolved Galaxies. V. NGC 6822", ApJS, 60, 507.
43. Hubble, E. 1925, "NGC 6822, A Remote Stellar System", ApJ, 62, 409.
44. Humphreys, R.M. 1980, "Studies of Luminous Stars in Nearby Galaxies. V. The Local Group Irregulars NGC 6822 and IC 1613", ApJ, 238, 65.

45. Hunter, D.A., & Gallagher, III, J.S. 1985, "Star-Forming Properties and Histories of Dwarf Irregular Galaxies: Down but not Out", ApJS, 58, 533.
46. Hunter, D.A., Hawley, W.N., and Gallagher, III, J.S. 1993, "A Survey for Extra- H II Region Ionized Gas Structures in Irregular Galaxies", AJ, 106, 1797.
47. Israel, F.P. 1997, "Dust and Molecules in the Local Group Galaxy NGC 6822. II. CO and Molecular Hydrogen", A&A, 317, 65.
48. Israel, F.P., Bontokoe, T.R., & Kester, D.J.M. 1996, "Dust and Molecules in the Local Group Galaxy NGC 6822. I. Dust, Star Formation, and the Origin of Infrared Emission", A&A, 317, 65.
49. Israel, F.P., Tacconi, L.G., & Baas, F. 1995, "CO Observations of 25 Dwarf Galaxies", A&A, 295, 599.
50. Jones, J.H., Christian, C.A., Crowe, R., McLaren, R., Crabtree, D., & Richer, H. 1987, "Near Infrared Photometry of Cepheid Variables in NGC 6822", BAAS, 19, 1052.
51. Kayser, S.E. 1967, "Photometry of the Nearby Irregular Galaxy, NGC 6822", AJ, 72, 134.
52. Kennicutt, Jr., R.C. 1984, "Structural Properties of Giant H II Regions in Nearby Galaxies", ApJ, 287, 116.
53. Killen, R.M., & Dufour, R.J. 1982, "Diffuse Nebulae and Emission-Line Stars in NGC 6822", PASP, 94, 444.
54. Kinman, T.D., Green, J.R., & Mahaffey, C.T.C. 1979, "New Compact H II Regions in NGC 6822", PASP, 91, 749.
55. Klein, U. & Gräve, R. 1986, "Multi-Frequency Radio Continuum Observations of Dwarf Irregular Galaxies", A&A, 161, 155.
56. Klein, U., Gräve, R., & Wielebinski, R. 1983, "A Survey of the Distribution of λ2.8 cm Radio Continuum in Nearby Galaxies. III. A Small Sample of Irregular and Blue Compact Galaxies", A&A, 117, 332.
57. Lee, M.G., Freedman, W.L., & Madore, B.F. 1992, "I-band Cepheid Distance to NGC 6822", presented at New Perspectives on Stellar Pulsation and Pulsating Variable Stars, IAU Colloquium 139.
58. Lequeux, J. 1979, "Comparison of the Rates of Formation of Massive Stars and of the Initial Mass Functions in Galaxies of the Local Group", A&A, 71, 1.
59. Lequeux, J., Peimbert, M., Rayo, J.F., Serrano, A., & Torres-Peimbert, S. 1979, "Chemical Composition and Evolution of Irregular and Blue Compact Galaxies", A&A, 80, 155.
60. Madore, B.F., & Freedman, W.L. 1991, "The Cepheid Distance Scale", PASP, 103, 933.
61. Marconi, G., Tosi, M., Greggio, L., & Focardi, P. 1995, "Star Formation in Dwarf Irregular Galaxies: NGC 6822", AJ, 109, 173.
62. Markert, T.H., & Donahue, M.E. 1985, "Observations of Four Nearby Galaxies with the *Einstein Observatory*", ApJ, 297, 564.
63. Massey, P., Armandroff, T.E., Pyke, R., Patel, K., & Wilson, C.D. 1995, "Hot Luminous Stars in Selected Regions of NGC 6822, M31, and M33", AJ, 110, 2715.
64. Massey, P., Conti, P.S., & Armandroff, T.E. 1987, "The Spectra of Extragalactic Wolf-Rayet Stars", AJ, 94, 1538.
65. McAlary, C.W., Madore, B.F., McGonegal, R., McLaren, R.A., & Welch, D.L. 1983, "The Distance to NGC 6822 from Infrared Photometry of Cepheids", ApJ, 273, 539.
66. Moffat, A.F.J., & Shara, M.M. 1983, "Wolf-Rayet Stars in the Local Group Galaxies M31 and NGC 6822", ApJ, 273, 544.
67. Mould, J., & Aaronson, M. 1986, "The Formation and Evolution of Carbon Stars", ApJ, 303, 10.

68. O'Dell, C.R., & Castañada, H.O. 1984, "[O II] Studies of Galactic Planetary Nebulae and Extragalactic H II Complexes", ApJ, 283, 158.
69. Ohta, K., Tomita, A., Saitō, M. 1993, "CO-to H_2 Conversion Factor in Dwarf Irregular Galaxies", PASJ, 45, L21.
70. Pagel, B.E.J., Edmunds, M.G., & Smith, G. 1980, "On the Composition of H II Regions in Southern Galaxies. II. NGC 6822 and NGC 1313", MNRAS, 193, 219.
71. Patel, K., & Wilson, C.D. 1995, "Are Galaxies Optically Thin to their own Lyman-continuum Radiation? II. NGC 6822", ApJ, 453, 162.
72. Peimbert, M., & Spinrad, H. 1970, "On the Chemical Abundance of NGC 6822", A&A, 7, 311.
73. Perrine, C.D. 1922, "Notes on Four Interesting Nebulae", MNRAS, 82, 486.
74. Price, J.S., Mason, S.F., & Gullixson, C.A. 1988, "CCD Surface Photometry of NGC 147, IC 1613, and NGC 6822", BAAS, 20, 701.
75. Rice, W. 1993, "An Atlas of High-Resolution *IRAS* Maps of Nearby Galaxies", AJ, 105, 67.
76. Richer, M.G., & McCall, M.L. 1995, "Oxygen Abundances in Diffuse Ellipticals and the Metallicity-Luminosity Relations for Dwarf Galaxies", ApJ, 445, 642.
77. Richer, H.B., & Westerlund, B.E. 1983, "Carbon Stars in Local Group Galaxies", ApJ, 264, 114.
78. Sandage, A., & Tamman, G.A. 1974, "Steps Toward the Hubble Constant. I. Calibration of the Linear Sizes of Extragalactic H II Regions", ApJ, 190, 525.
79. Schmidt, E.G., & Spear, G.G. 1989, "Pulsation Properties of Extragalactic Cepheids. I. Differential Light Curves for Field No. 1 in NGC 6822", MNRAS, 236, 567.
80. Skillman, E.D., Terlevich, R., & Melnick, J. 1989, "Abundances in Southern Local Group Dwarf Irregular Galaxies", MNRAS, 240, 563.
81. Smith, H.E. 1975, "Spectrophotometric Observations of Ionized Hydrogen Regions in Nearby Spiral and Irregular Galaxies", ApJ, 199, 591.
82. Tarrab, I. 1983, "Estimated Energy and Momentum Input to the Interstellar Medium for Several External Galaxies", A&A, 125, 308.
83. Tomita, A., Ohta, K., & Saitō, M. 1993, "H-Alpha Velocity Fields of Giant H II Regions in Four Nearby Dwarf Irregular Galaxies", PASJ, 45, 693.
84. van den Bergh, S., & Humphreys, R.M. 1979, "*UBV* Photometry in the Dwarf Irregular Galaxy NGC 6822", AJ, 84, 604.
85. Vangioni-Flam, E., Lequeux, J., Maucherat-Joubert, M., & Rocca-Volmerange, B. 1980, "Ultraviolet Studies of the Magellanic Clouds. II. Internal Extinction, Formation of Massive Stars, Comparison with Other Galaxies", A&A, 90, 73.
86. Visvanathan, N. 1989, "The Zero Point of the Cepheid Period-Luminosity Relation in the 1.05 Micron Wave Band", ApJ, 346, 629.
87. Volders, L., & Högbom, J.A. 1961, "Observations of Neutral Hydrogen in IC 1613, NGC 6822 and M82", Bull. Ast. Inst. Neth., 15, 307.
88. Walker, A.R. 1987, "Notes on Cepheid Variables in Galaxies More Distant Than the Magellanic Clouds", S. African Astron. Obs. Circ., 11, 125.
89. Westerlund, B.E., Azzopardi, M., Breysacher, J., & Lequeux, J. 1983, "Discovery of a Wolf-Rayet Star in NGC 6822", A&A, 123, 159.
90. Wilson, C.D. 1992, "The Stellar Populations of NGC 6822", AJ, 104, 1374.
91. Wilson, C.D. 1992, "A Comparison of the Properties of an Unbiased Sample of OB Associations in the Local Group Galaxies M33 and NGC 6822", ApJ, 384, L29.
92. Wilson, C.D. 1992, "CO Emission from the Dwarf Irregular Galaxy NGC 6822", ApJ, 391, 144.

93. Wilson, C.D. 1994, "First Observations of Individual Molecular Clouds in the Irregular Galaxy NGC 6822", ApJ, 434, L11

Pegasus

Alternate name(s): DDO 216
Right Ascension (2000): 23h 28m 34s
Declination (2000):+14d 44.8m
Type: dIrr
Apparent magnitude (V): 12.04
Color (B-V): 0.61
(U-B): 0.06
(V-R):
Color Excess, E(B-V): 0.02
Absolute magnitude (M_V): -12.9
Distance (kpc): 955
Radial velocity (solar, km/sec): -182

Objects Identified On the Atlas:
Globular clusters: none
Source:
Open clusters: 1; designation: C1
Source: [this atlas]
OB associations: 2; designation: A
Source: [this atlas]
Variable stars: 12; designation: V
Source: [6]
Carbon stars: none
Source:
HII regions: 1; designation: AG1
Source: [2]
Planetary nebulae: 1; designation: PN
Source: [11]
Dust clouds: none
Source:

Pegasus

Chart 172

Pegasus

Chart 173

Pegasus

References

1. Aparicio, A. 1994, "The Distance to the Pegasus Dwarf Irregular Galaxy: An Open Question", ApJ, 437, L27.
2. Aparicio, A., & Gallart, C. 1995, "The Stellar Content of the Pegasus Dwarf Irregular Galaxy", AJ, 110, 2105.
3. Aparicio, A., Gallart, C., & Bertelli, G. 1997, "The Star Formation History of the Pegasus Dwarf Irregular Galaxy", AJ, 114, 669.
4. Christian, C.A., & Tully, R.B. 1983, "The Local Group Irregular Galaxies, LGS 3 and Pegasus", AJ, 88, 934.
5. Fisher, J.R., & Tully, R.B. 1979, "Photographs of Dwarf Irregular Galaxies in the Vicinity of the Local Group and the M81 Group", AJ, 84, 62.
6. Hoessel, J.G., Abbot, M.J., Saha, A., Mossman, A.E., & Danielson, G.E. 1990, "Variable Stars in the Pegasus Dwarf Galaxy (DDO 216)", AJ, 100, 1151.
7. Hoessel, J.G., & Mould, J.R. 1982, "Photometry of Resolved Galaxies. I. The Pegasus Dwarf Irregular", ApJ, 254, 38.
8. Hoffman, G.L., Salpeter, E.E., Farhat, B., Roos, T., Williams H., & Helou, G. 1996, "Arecibo H I Mapping of a Large Sample of Dwarf Irregular Galaxies", ApJS, 105, 269.
9. Hunter, D.A., Hawley, W.N., & Gallagher, III, J.S. 1993, "A Survey for Extra-H II Region Ionized Gas Structures in Irregular Galaxies", AJ, 106, 1797.
10. Israel, F.P., & Burton, W.B. 1986, "Search for ^{12}CO Emission from Twelve Irregular Dwarf Galaxies", A&A, 168, 369.
11. Jacoby, G.H., & Lesser, M.P. 1981, "Distances to Nearby Galaxies Derived from Planetary Nebulae", AJ, 86, 185.
12. Lo, K.Y., Sargent, W.L.W., & Young, K. 1993, "The H I Structure of Nine Intrinsically Faint Dwarf Galaxies", AJ, 106, 507.
13. Rowan-Robinson, M., Phillips, T.G., & White, G. 1980, "A Survey of Nearby Galaxies for CO", A&A, 82, 381.
14. Sandage, A. 1986, "The Brightest Stars in Nearby Galaxies. VII. The Pegasus and Leo A Im V Dwarfs", AJ, 91, 796.
15. Sargent, W.L.W., & Lo, K.-Y. 1986, "The H I Structures of Extreme Dwarf Galaxies", in Star Forming Dwarf Galaxies and Related Objects, edited by D. Kunth, T. Thuan, and J.T.T. Van (Editions Frontieres, Gif-sur-Yvette), p. 253.
16. Skillman, E.D., Bomans, D.J., & Kobulnicky, H.A. 1997, "Interstellar Medium Abundances in the Pegasus Dwarf Irregular Galaxy", ApJ, 474, 205.

Phoenix

Alternate name(s):
Right Ascension (2000): 01h 51m 06s
Declination (2000):-44d 26.7m
Type: dE3/dIrr
Apparent magnitude (V): 13.2
Color (B-V): 0.61
(U-B): -0.21
(V-R):
Color Excess, E(B-V): 0.02
Absolute magnitude (M_V): -10.1
Distance (kpc): 445
Radial velocity (solar, km/sec): 56

Objects Identified On the Atlas:
Globular clusters: 3 (suspected); designation: GC
Source: [2]
Open clusters: none
Source:
OB associations: 1; designation: A1
Source: [this atlas]
Variable stars: none
Source:
Carbon stars: 3 candidates (not shown)
Source: [8]
HII regions: none
Source:
Planetary nebulae: none
Source:
Dust clouds: none
Source:

Phoenix

Chart 174

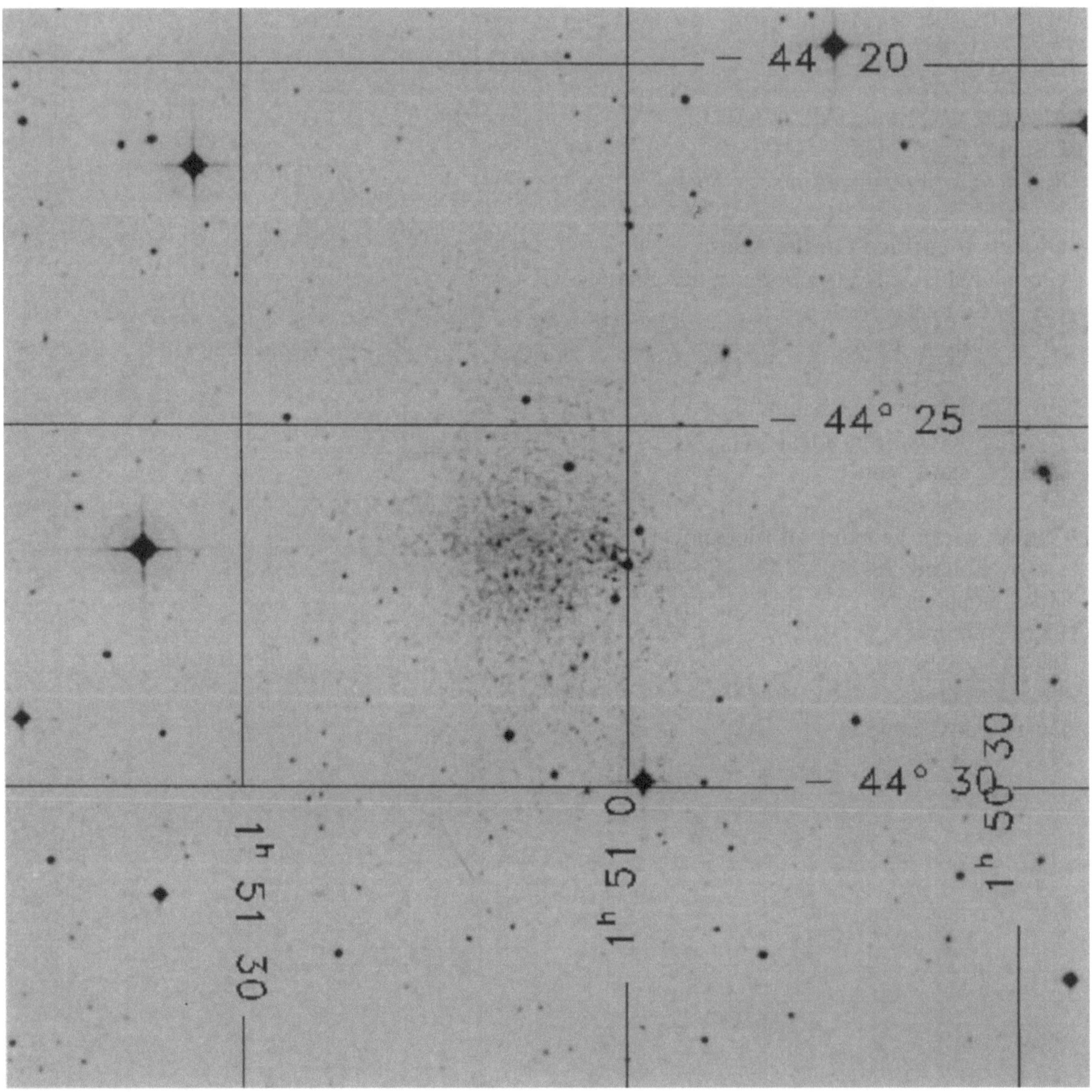

Phoenix

Chart 175

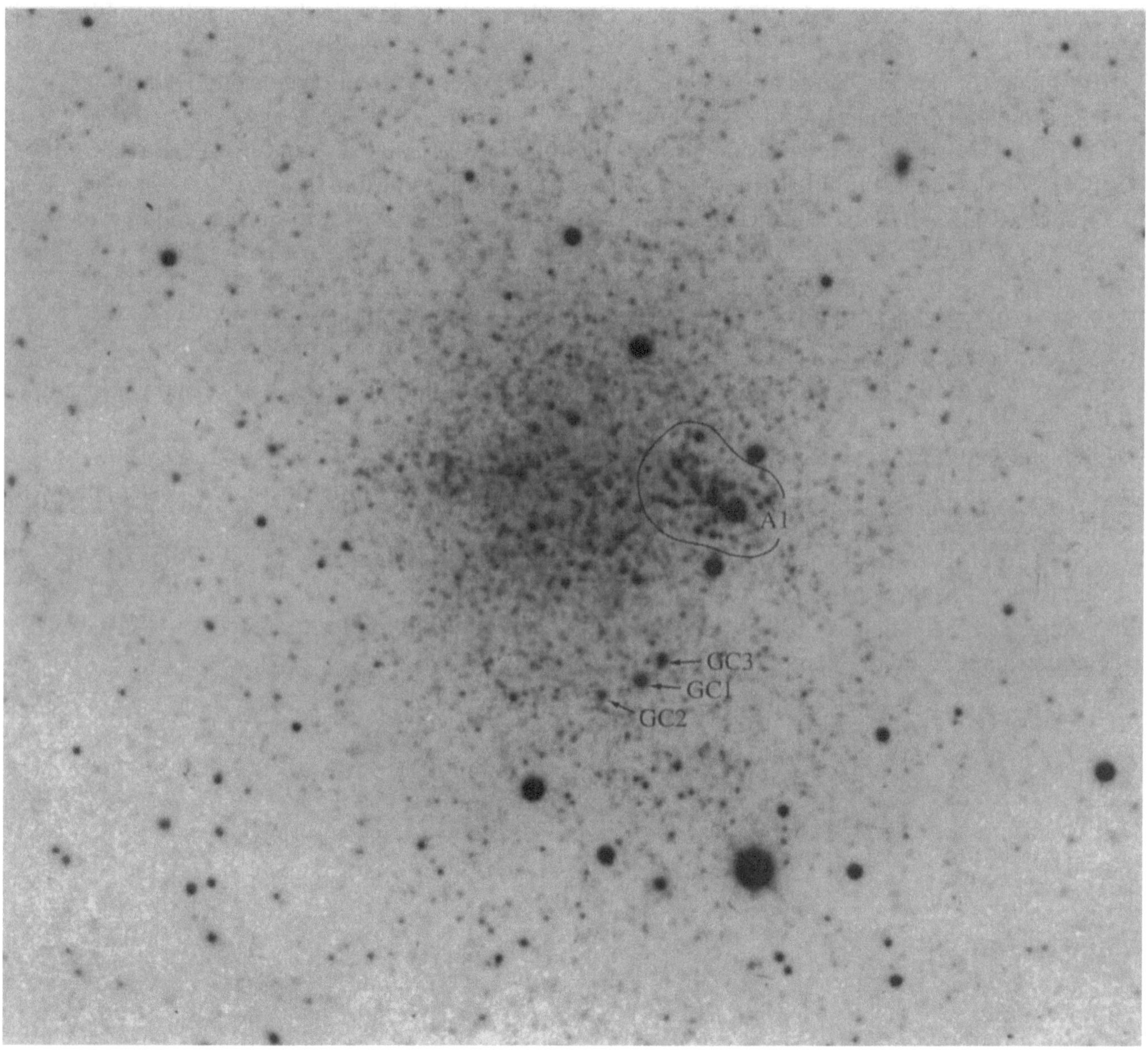

Phoenix

References

1. Caldwell, N., & Schommer, R. 1988, "Cepheid Variables in the Phoenix Galaxy and IC 5152", in the Extragalactic Ditance Scale, ASP Conf. Series v.4, edited by S. van den Bergh and C.J. Pritchet (Astronomical Society of the Pacific, San Francisco), p. 77.
2. Canterna, R. & Flower, P.J. 1977, "A New Dwarf Irregular Galaxy in the Constellation Phoenix", ApJ, 212, L57.
3. Carignan, C., Demers, S., & Côté, S. 1991,"The H I Content of the Local Group Dwarf (Spheroidal or Irregular?) Galaxy Phoenix", ApJ, 381, L13.
4. Da Costa, G.S. 1994, "New Results for Local Group Dwarf Galaxies", in Dwarf Galaxies, edited by G. Meylan an P. Prugniel (European Southern Observatory, Munich), p. 221.
5. Gratton, R.G., Ortolani, S., & Richter, O. 1986, "The Dwarf Galaxies in Fornax and in Phoenix", MSAIt, 57, 561.
6. Madore, B.F., & Freedman, W.L. 1991, "The Cepheid Distance Scale", PASP, 103, 933.
7. Morras, R., & Bajaja, E. 1986, "H I Around the Dwarf Galaxy in Phoenix", Rev. Mex. Astron. Astrof., 13, 69.
8. Ortolani, S., & Gratton, R.G. 1988, "Stellar Photometry in the Phoenix Dwarf Galaxy", PASP, 100, 1405.
9. Schuster, H.E., & West, R.M. 1976, "A Very Distant Globular Cluster?", A&A, 49, 129.
10. van de Rydt, F., Demers, S., & Kunkel, W.E. 1991, "Phoenix: An Intermediate Dwarf Galaxy in the Local Group", AJ, 102, 130.
11. van den Bergh, S. 1994, "The Evolutionary History of Low-Luminosity Local Group Dwarf Galaxies", ApJ, 428, 617.
12. Young, L.M., & Lo, K.Y. 1997, "The Neutral Interstellar Medium in Nearby Dwarf Galaxies. III. Sagittarius DIG, LGS 3, and Phoenix", ApJ, 490,710.

Sag DIG

Alternate name(s): UKS 1927-177
Right Ascension (2000): 19h 29m 59s
Declination (2000): -17d 40.7m
Type: dIrr
Apparent magnitude (V): 13.9
Color (B-V):
(U-B):
(V-R):
Color Excess, E(B-V): 0.22
Absolute magnitude (M_V): -11.8
Distance (kpc): 1060
Radial velocity (solar, km/sec): -80

Objects Identified On the Atlas:
Globular clusters: none
Source:
Open clusters: none
Source:
OB associations: 2; designation: A
Source: [this atlas]
Variable stars: none
Source:
Carbon stars: none
Source:
HII regions: 3 (Nos. 1 and 2 may be foreground emission-line stars) designation: Arabic numerals
Source: [7]
Planetary nebulae: none
Source:
Dust clouds: none
Source:

Sag DIG

Chart 176

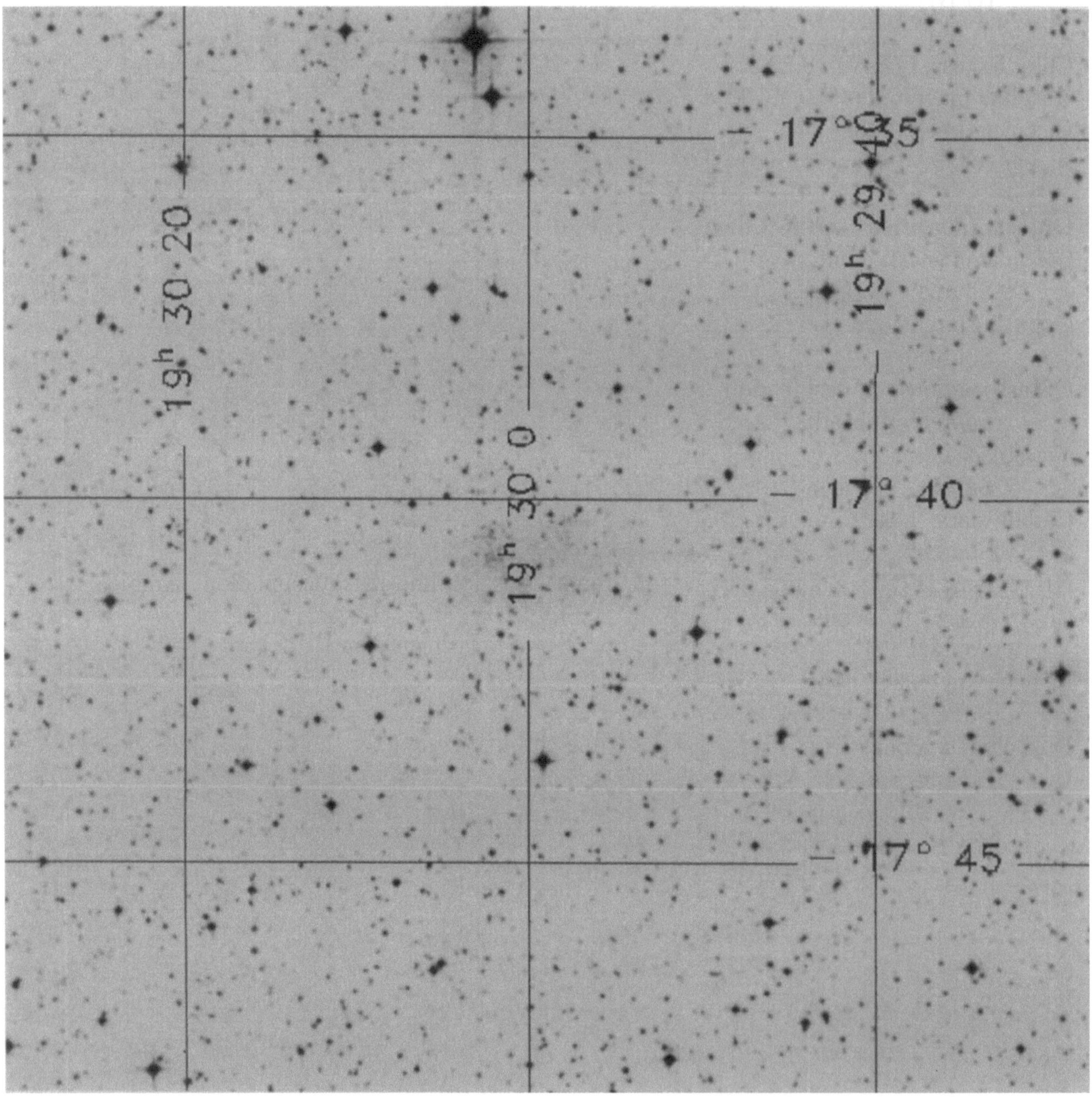

Sag DIG

Chart 177

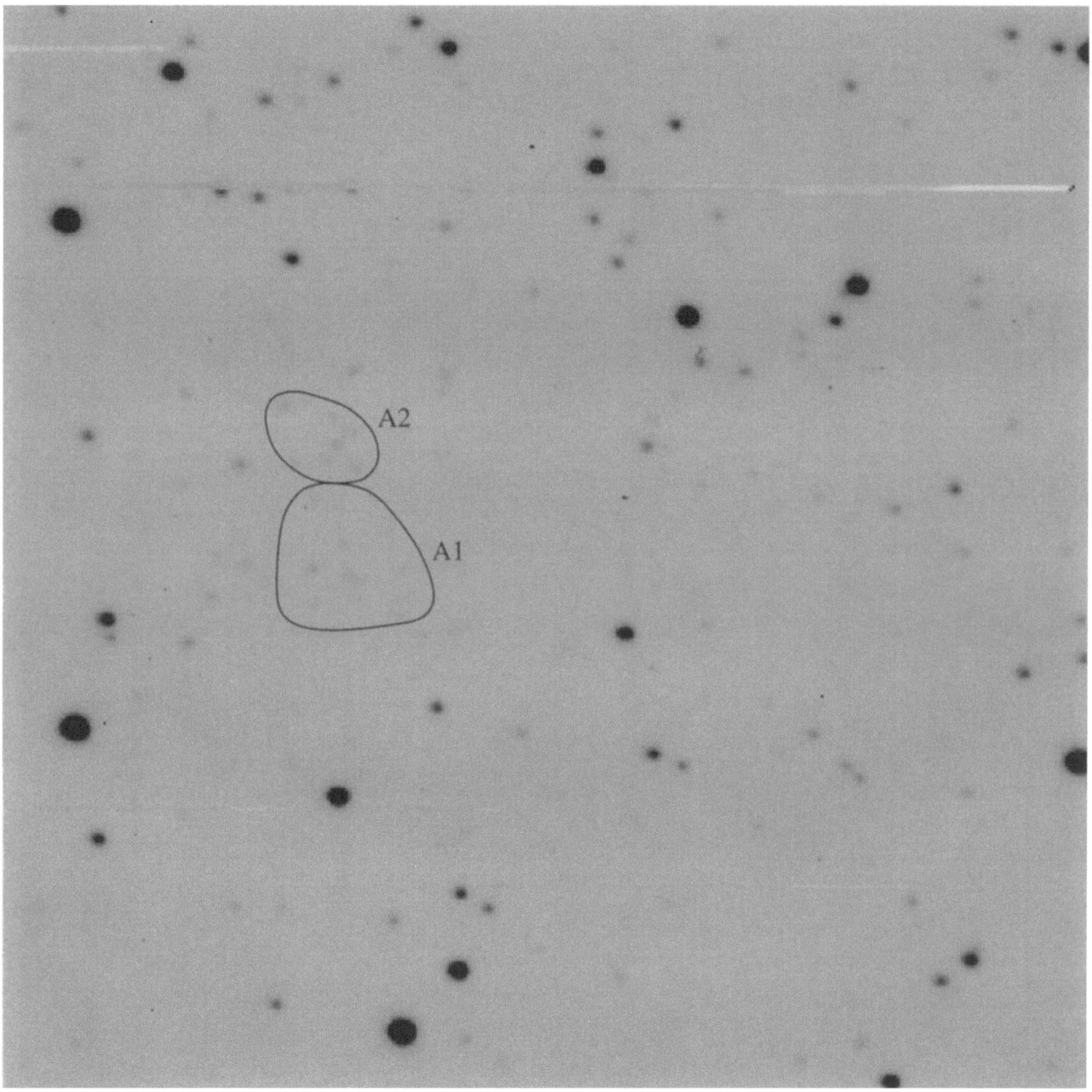

Sag DIG

Chart 178

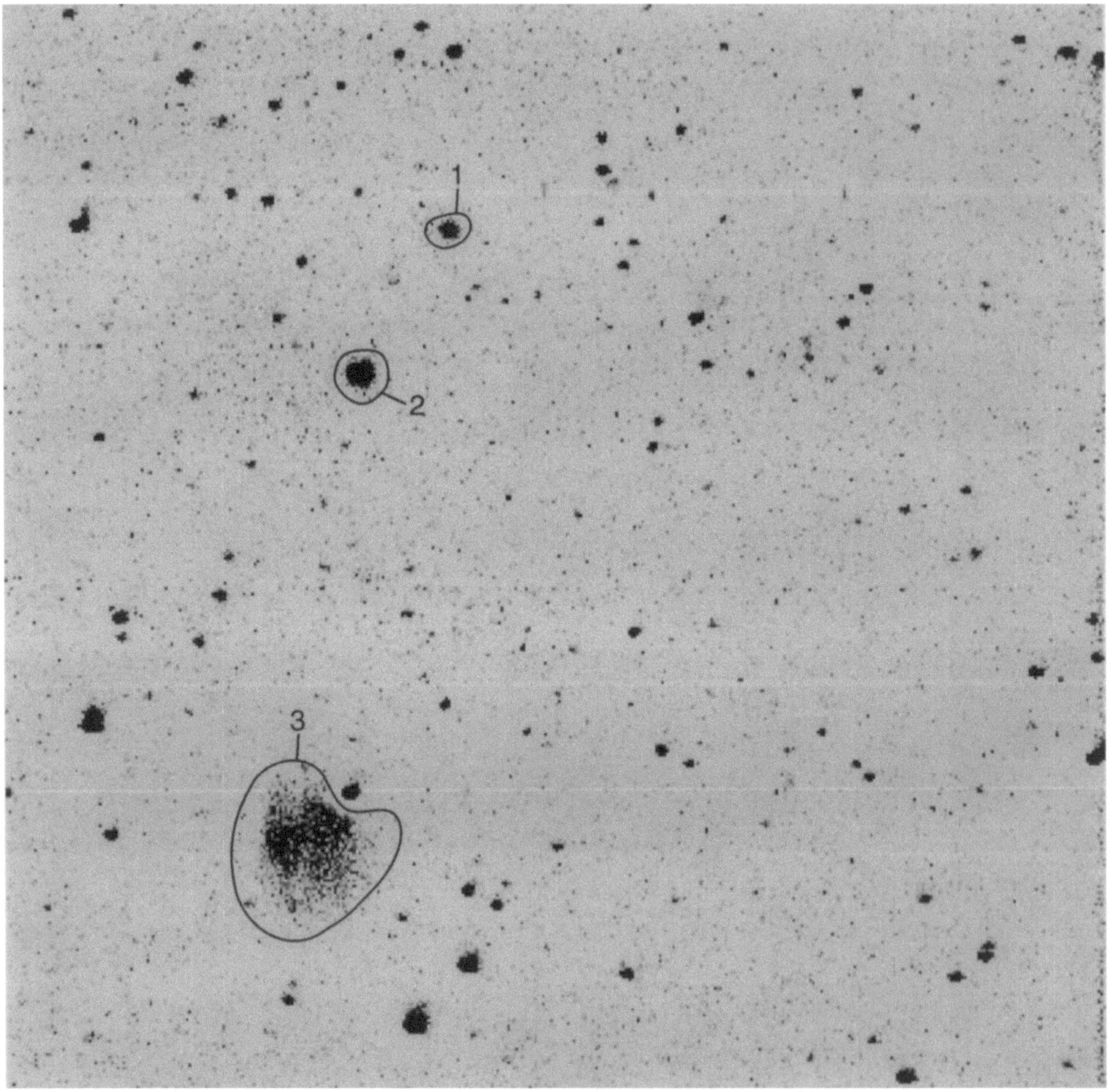

Sag DIG

References

1. Cesarsky, D.A., Laustsen, S., Lequeux, J., Schuster, H.-E., & West, R.M. 1977, "Two New Faint Stellar Systems Discovered on ESO Schmidt Plates", A&A, 61, L31.
2. Cook, K.H., "Asymptotic Giant Branch Populations in Composite Stellar Systems", BAAS, 20, 742.
3. Lo, K.Y., Sargent, W.L.W., & Young, K. 1993, "The H I Structure of Nine Intrinsically Faint Dwarf Galaxies", AJ, 106, 507.
4. Longmore, A.J., Hawarden, T.G., Webster, B.L., Goss, W.M., & Mebold, U. 1978, "Two New Dwarf Galaxies in the Local Group: UKS 1927 – 177 and UKS 2323 – 326", MNRAS, 184, 97P.
5. Sargent, W.L.W., & Lo, K.-Y. "The H I Structures of Extreme Dwarf Galaxies", in Star-Forming Dwarf Galaxies and Related Objects, edited by D. Kunth, T.X. Thuan, and J.T.T. Van (Editions Frontieres, Gif-sur-Yvette), p. 253.
6. Skillman, E.D., Terlevich, R., & Melnick, J. 1989, "Abundances in Southern Local Group Dwarf Irregular Galaxies", MNRAS, 240, 563.
7. Strobel, N.V., Hodge, P., & Kennicutt, R.C. 1991, "H II Regions in Seven Dwarf Irregular Galaxies", ApJ, 383, 148.
8. Young, L.M., & Lo, K.-Y. 1997, "The Neutral Interstellar Medium in Nearby Dwarf Galaxies. III. Sagittarius DIG, LGS 3, and Phoenix", ApJ, 490, 710.

Sagittarius

Alternate name(s): Sgr
Right Ascension (2000): 18h 55m 03s
Declination (2000): -30d 28.7m
Type: dE5
Apparent magnitude (V): 4.0
Color (B-V):
(U-B):
(V-R):
Color Excess, E(B-V): 0.15
Absolute magnitude (M_V): -13.4
Distance (kpc): 24
Radial velocity (solar, km/sec): 140

Objects Identified On the Atlas:

Globular clusters: 5; designation: traditional names
Source: [9]
Open clusters: none
Source:
OB associations: none
Source:
Variable stars: not shown
Source: [2], [15], [19]
Carbon stars: none
Source:
HII regions: none
Source:
Planetary nebulae: not shown
Source: [24]
Dust clouds: none
Source:

Sagittarius

Chart 179

Sagittarius

References

1. Alard, C. 1996, "Evidence for the Sagittarius Dwarf Galaxy at Low Galactic Latitudes", ApJ, 458, L17.
2. Alcock, C., et al. 1997, "MACHO Project Photometry of RR Lyrae Stars in the Sagittarius Dwarf Galaxy", ApJ, 474, 217.
3. Catelan, M. 1997, "Color-Magnitude Diagrams of Merged Globular Clusters: Metallicity Effects", ApJ, 478, L99.
4. Da Costa, G.S., & Armandroff, T.W. 1995, "Abundances and Kinematics of the Globular Cluster Systems of the Galaxy and of the Sagittarius Dwarf", AJ, 109, 2533.
5. Edelsohn, D.J., & Elmegreen, B.G. 1997, "Computer Models of the Sagittarius Dwarf Interaction with the Milky Way", MNRAS, 290, 7.
6. Fahlman, G.G., Mandsuhev, G., Richer, H.B., Thompson, I.B., & Sivaramakrishnan, A. 1996, "The Main-Sequence Stars of the Sagittarius Dwarf Galaxy", ApJ, 459, L65.
7. Harding, P. 1994, "The Sagittarius Dwarf", BAAS, 26, 1395.
8. Ibata, R.A., Gilmore, G., & Irwin, M.J. 1994, "A Dwarf Satellite Galaxy in Sagittarius", Nature, 370, 194.
9. Ibata, R.A., Gilmore, G., & Irwin, M.J. 1995, "Sagittarius: The Nearest Dwarf Galaxy", MNRAS, 277, 781.
10. Ibata, R.A., Wyse, R.F.G., Gilmore, G., Irwin, M.J., & Suntzeff, N.B. 1997, "The Kinematics, Orbit, and Survival of the Sagittarius Dwarf Spheroidal Galaxy", AJ, 113, 634.
11. Koribalski, B., Johnston, S., & Otrupcek, R. 1994, "H I Observations of the New Sagittarius Dwarf", MNRAS, 270, L43.
12. Johnston, K.V., Spergel, D.N., & Hernquist, L. 1995, "The Disruption of the Sagittarius Dwarf Galaxy", ApJ, 451, 598.
13. Layden, A.C., & Sarajedini, A. 1997, "The Globular Cluster M54 and the Star Formation History of the Sagittarius Dwarf Galaxy", ApJ, 486, L107.
14. Marconi, G., Bounanno, R., Castellani, M., Iannicola, G., Molaro, P., Pasquini, L., & Pulone, L. 1998, "The Stellar Content of the Sagittarius Dwarf Galaxy", A&A, 330, 453.
15. Mateo, M., Kubiak, M., Szymańsky, M., Kalużny, J., Krzemiński, W., & Udalski, A. 1995, "The Optical Gravitational Lensing Experiment: Variable Stars in the Sagittarius Dwarf Spheroidal Galaxy", AJ, 110, 1141.
16. Mateo, M., Mirabal, N., Udalski, A., Szymańsky, M., Kalużny, J., Kubiak, M., Krzemiński, W., & Stanek, K.Z. 1996, "Discovery of a Tidal Extension of the Sagittarius Dwarf Spheroidal Galaxy", ApJ, 458, L13.
17. Mateo, M., Udalski, A., Szymańsky, M., Kalużny, J., Kubiak, M., & Krzemiński, W. 1995, "The Optical Gravitational Lensing Experiment: Deep Photometry of the Sagittarius Dwarf Spheroidal Galaxy", AJ, 109, 588.
18. Ng, Y.K. 1997, "Are the Bulge C-stars Related to the Sagittarius Dwarf Galaxy?", A&A, 328, 211.
19. Ng, Y.K., & Schultheis, M. 1997, "Field # 3 of the Palomar-Groningen Survey. I. Variable Stars in the Outer Edge of the Sagittarius Dwarf Galaxy", A&AS, 123, 115.

20. Sarajedini, A., & Layden, A.C. 1995, "A Photometric Study of the Globular Cluster M54 and the Sagittarius Dwarf Galaxy: Evidence for Three Distinct Populations", AJ, 109, 1086.
21. Velazquez, H., & White, S.D.M. 1995, "On the Dynamics of the Sagittarius Dwarf Galaxy", MNRAS, 275, L23.
22. Walsh, J.R., Dudziak, G., Minniti, D., & Zijlstra, A.A. 1997, "Chemical Abundances of Planetary Nebulae in the Sagittarius Dwarf Elliptical Galaxy", ApJ, 487, 651.
23. Whitelock, P.A., Irwin, M., & Catchpole, R.M. 1996, "Bright Giants in the Sagittarius Dwarf Galaxy", New Astronomy, 1, 57.
24. Zijlstra, A.A., & Walsh, J.R. 1996, "Two Planetary Nebulae in the Sagittarius Dwarf Galaxy", A&A, 312, L21.

Sculptor

Alternate name(s):
Right Ascension (2000): 01h 00m 09s
Declination (2000): -33d 42.5m
Type: dE3
Apparent magnitude (V): 8.5
Color (B-V): 0.7
(U-B):
(V-R):
Color Excess, E(B-V): 0.02
Absolute magnitude (M_V): -11.0
Distance (kpc): 79
Radial velocity (solar, km/sec): 108

Objects Identified On the Atlas:
Globular clusters: none
Source:
Open clusters: none
Source:
OB associations: none
Source:
Variable stars: 3 anomalous Cepheids; designation: V; >600 RR Lyraes (not shown)
Source: [32], [34]
Carbon stars: 8; designation: C
Source: [4], [5], [28]
HII regions: none
Source:
Planetary nebulae: none
Source:
Dust clouds: none
Source:

Sculptor

Chart 180

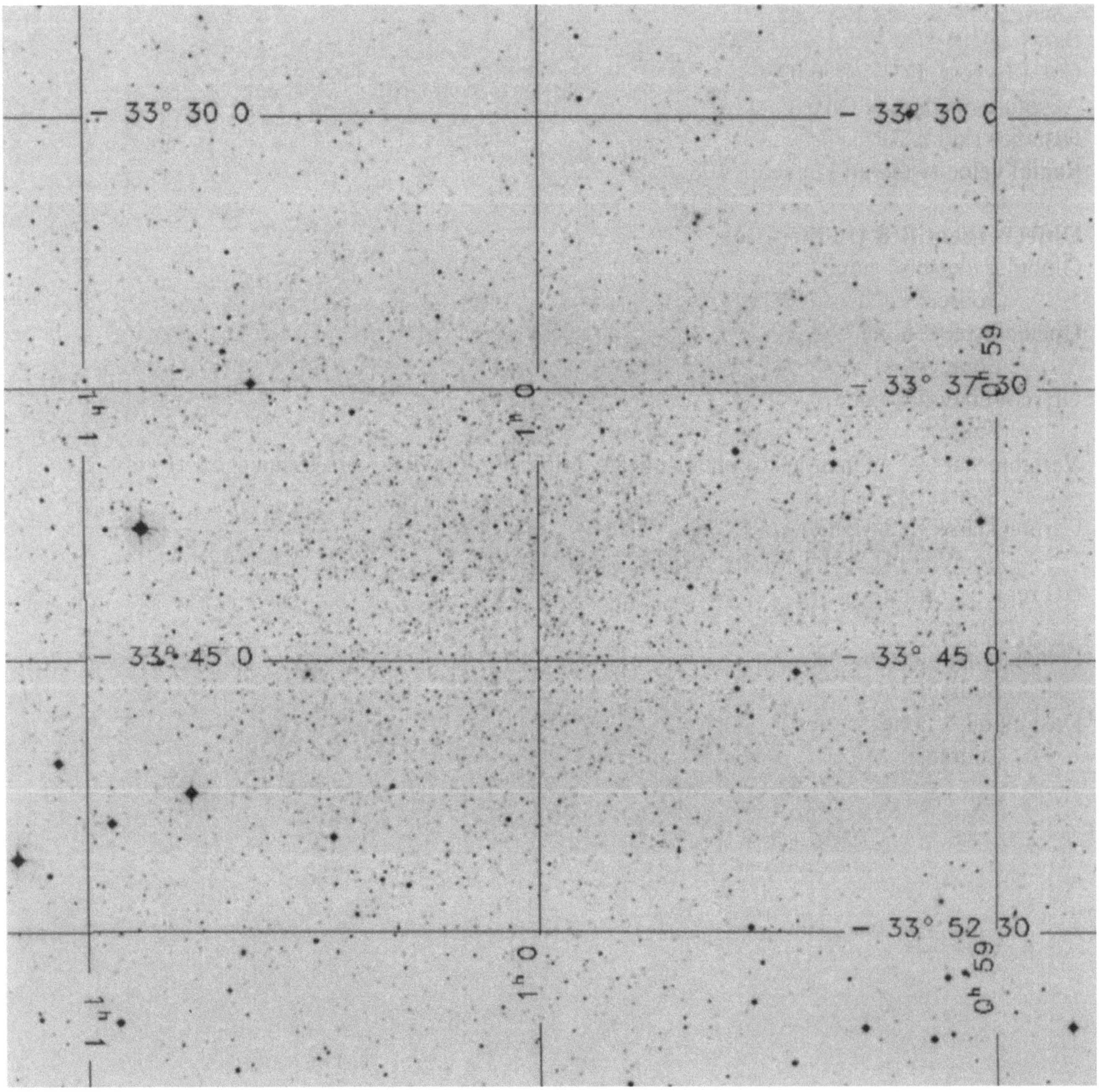

Sculptor

Chart 181

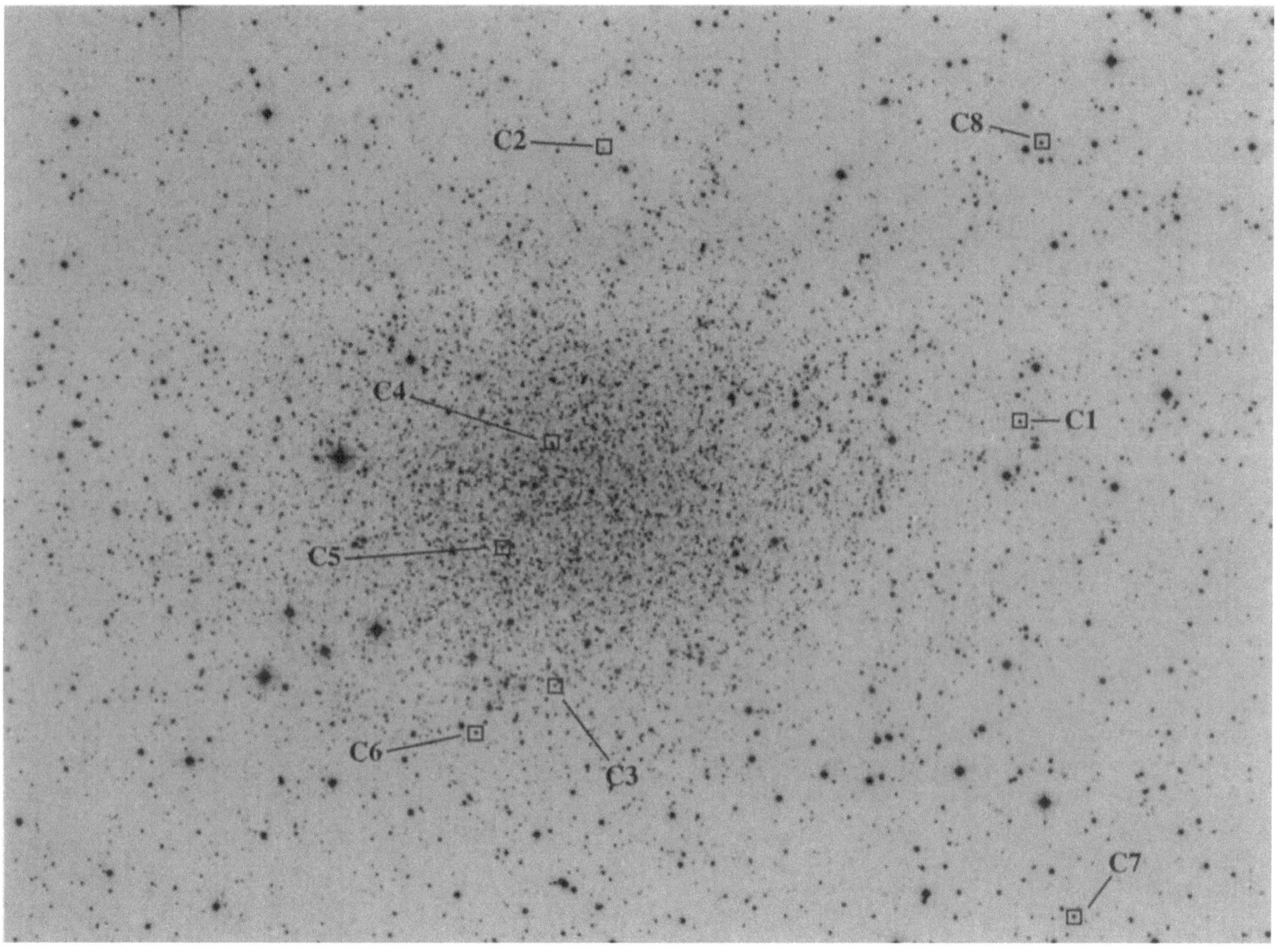

Sculptor

Chart 182

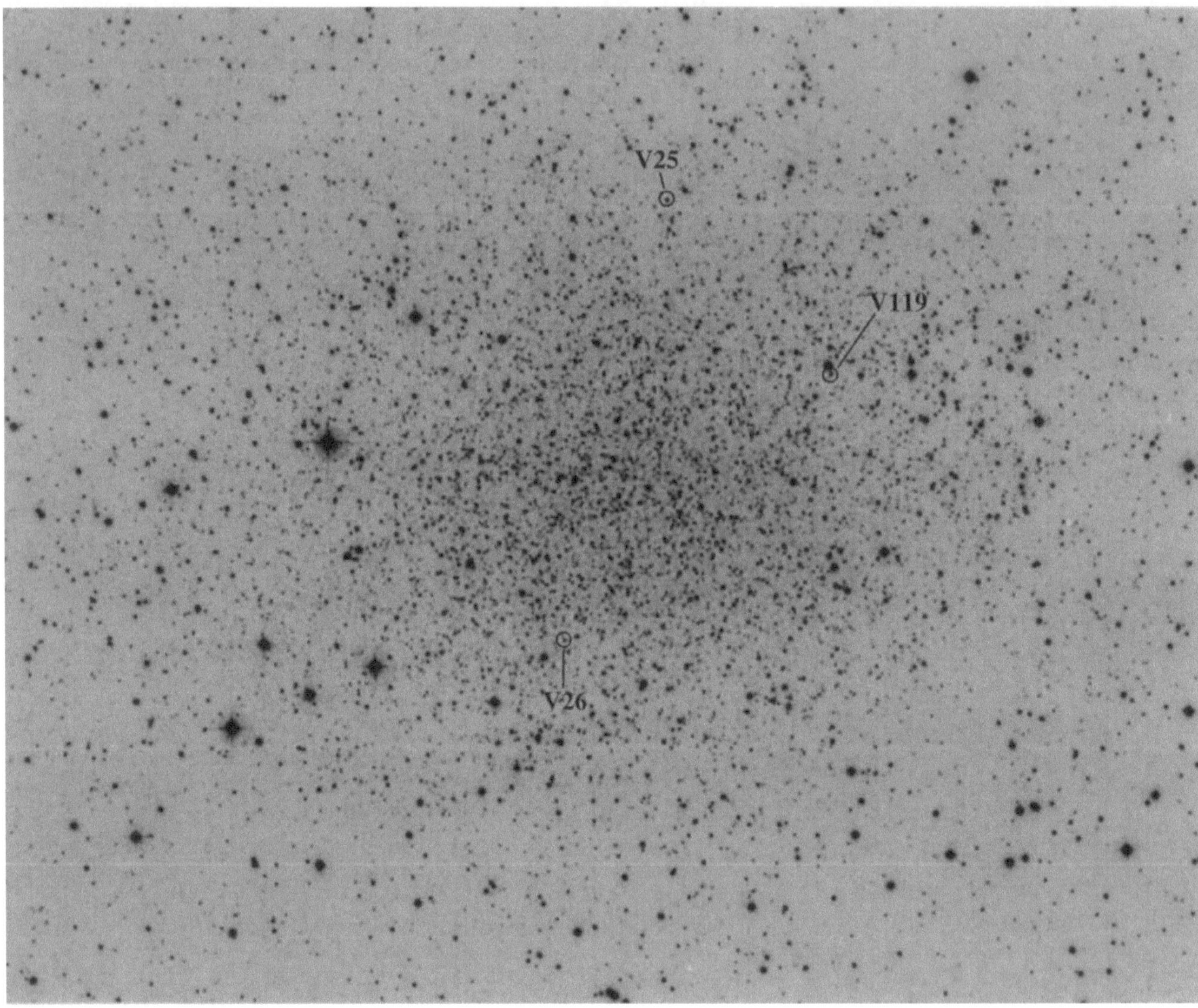

Sculptor

References

1. Aaronson, M., & Mould, J. 1980, "Carbon Stars in the Fornax Dwarf Spheroidal Galaxy", ApJ, 240, 804.
2. Aaronson, M. & Olszewski, E.W. 1987, "Accurate Radial Velocities for Carbon Stars in the Sculptor Dwarf Spheroidal", AJ, 94, 657.
3. Armandroff, T.E., & Da Costa, G.S. 1986, "The Radial Velocity, Velocity Dispersion, and Mass-to-Light Ratio of the Sculptor Dwarf Galaxy", AJ, 92, 777.
4. Azzopardi, M., Lequeux, J., & Westerlund, B.E. 1985, "New Carbon Stars in Spheroidal Galaxies. I. Sculptor, Carina, Leo I and Leo II Systems", A&A, 144, 388.
5. Azzopardi, M., Lequeux J., & Westerlund, B.E. 1986, "New Carbon Stars in Spheroidal Galaxies. II. Draco, Ursa Minor, and New Data on Sculptor, Carina, and Leo I", A&A, 161, 232.
6. Baade, W., & Hubble, E. 1939, "The New Stellar Systems in Sculptor and Fornax", PASP, 51, 6.
7. Bono, G., Caputo, F., Santolamazza, P., Cassisi, S., & Piersimoni, A. 1997, "Evolutionary Scenario for Metal-Poor Pulsating Stars. II. Anomalous Cepheids", AJ, 113, 2209.
8. Cowley, A.P., Hartwick, F.D.A., & Sargent, W.L.W. 1997, "Spectra of Giant Stars in Distant Satellites of the Galaxy", ApJ, 220, 453.
9. Da Costa, G.S. 1984, "The Age(s?) of the Sculptor Dwarf Galaxy", ApJ, 285, 483.
10. Demers, S., Kunkel, W.E., & Krautter, A. 1980, "The Structure of the Sculptor System", AJ, 85, 1587.
11. Eskridge, P.B. 1988, "The Global Luminosity Function of the Sculptor Dwarf Elliptical Galaxy", AJ, 95, 445.
12. Eskridge, P.B. 1988, "The Structure of the Sculptor Dwarf Elliptical Galaxy. I. The Radial Profile", AJ, 95, 1706.
13. Eskridge, P.B. 1988, "The Structure of the Sculptor Dwarf Elliptical Galaxy. II. Two Dimensial Structure", AJ, 96, 1336.
14. Faber, S.M., & Lin, D.N.C. 1983, "Is there Nonluminous Matter in Dwarf Spheroidal Galaxies?", ApJ, 266, L17.
15. Frogel, J.A., Blanco, V.M., & McCarthy, M.F. 1982, "The Late Type Stellar Content of the Fornax and Sculptor Dwarf Galaxies", ApJ, 252, 133.
16. Grebel, E.K., Roberts, W.J., & van de Rydt, F. 1994, "Population Studies in Dwarf Spheroidal Galaxies: Washington Photometry of Sculptor and Fornax", in The Local Group: Comparative and Global Properties, edited by A. Layden, R.C. Smith, and J. Storm (European Southern Observatory, Munich), p. 148.
17. Hodge, P.W. 1961, "The Distribution of Stars in the Sculptor Dwarf Galaxy", AJ, 66, 384.
18. Hodge, P.W. 1965, "A Color-Magnitude Diagram for the Sculptor Dwarf Galaxy", ApJ, 142, 1390.
19. Hodge, P.W. 1966, "Surface Photometry of the Sculptor Dwarf Galaxy", AJ, 71, 204.
20. Innanen, K.A., & Papp, K.A. 1979, "Extratidal Variables and the Dynamics of the Sculptor Dwarf Galaxy", AJ, 84, 601.
21. Irwin, M., & Hatzidimitriou, D. 1995, "Structural Parameters for the Galactic Dwarf Spheroidals", MNRAS, 277, 1354.

22. Kalużny, J., Kubiak, M., Szymański, M., Udalski, A., Krzemiński, W., & Mateo, M. 1995 OGLE Catalogue of Variable Stars in the Sculptor Dwarf Spheroidal", A&AS, 112, 407.
23. Kunkel, W.E., & Demers, S. 1977, "The Sculptor Dwarf Galaxy: Photoelectric Sequence and a Preliminary Color-Magnitude Diagram", ApJ, 214, 21.
24. Light, R.M., 1987, "Measurement of Abundances and their Variations in Dwarf Spheroidal Galaxies", BAAS, 19, 1033.
25. Nemec, J.M., Nemec, A.F.L., & Lutz, T.E. 1994, "Period-Luminosity-Metallicity Relations, Pulsation Modes, Absolute Magnitudes, and Distances for Population II Variable Stars", AJ, 108, 222.
26. Norris, J., & Bessell, M.S. 1978, "Abundance Variations on the Giant Branches of the Sculptor and Ursa Minor Dwarf Spheroidal Galaxies and the ω Centauri Anomaly", ApJ, 225, L49.
27. Queloz, D., Dubath, P., & Pasquini, L. 1995, "A Kinematic Study of the Sculptor Dwarf Spheroidal Galaxy", A&A, 300, 31.
28. Richer, H.B., & Westerlund, B.E. 1983, "Carbon Stars in Local Group Galaxies", ApJ, 264, 114.
29. Schweitzer, A.E., Cudworth, K.M., Majewski, S.R., & Suntzeff, N.B. 1995, "The Absolute Proper Motion and a Membership Survey of the Sculptor Dwarf Spheroidal Galaxy", AJ, 110, 2747.
30. Seitzer, P., & Frogel, J.A. 1985, "Radial Velocities of Carbon Stars in Three Dwarf Spheroidal Galaxies", AJ, 90, 1796.
31. Shapley, H. 1938, "Two Stellar Systems of a New Kind", Nature, 142, 715.
32. Shapley, H. 1939, "Galactic and Extragalactic Studies. II. Notes on the Peculiar Stellar Systems in Sculptor and Fornax", Proceedings of the Natl. Acad. of Sci., 25, 565.
33. Smith, G.H., & Dopita, M.A. 1983, "The Chemical Inhomogeneity of the Sculptor Dwarf Spheroidal Galaxy", ApJ, 271, 113.
34. Smith, H.A., & Stryker, L.L. 1986, "Anomalous Cepheids in the Sculptor Dwarf Galaxy", AJ, 92, 328.
35. Swope, H.H. 1968, "Thirteen Periodic Variable Stars Brighter than the Normal RR Lyrae-Type Variables in Four Dwarf Galaxies", AJ, 73, S204.
36. van Agt, S. 1978, "The Sculptor Dwarf Spheroidal Galaxy. I. Discovery and Identification of Variable Stars", Pub. DDO, 3, 205.
37. van den Bergh, S. 1994, "The Evolutionary History of Low-Luminosity Local Group Dwarf Galaxies", ApJ, 428, 617.
38. Zinnecker, H. 1994, "ROSAT Observations Towards the Carina and Sculptor Dwarf Galaxies", in Dwarf Galaxies, edited by G. Meylan and P. Prugniel (European Southern Observatory, Munich), p. 231.

Sextans I

Alternate name(s):
Right Ascension (2000): 10h 13m 03s
Declination (2000): -01d 36.9m
Type: dE4
Apparent magnitude (V): 10.3
Color (B-V):
(U-B):
(V-R):
Color Excess, E(B-V): 0.03
Absolute magnitude (M_V): -9.4
Distance (kpc): 86
Radial velocity (solar, km/sec): 227

Objects Identified On the Atlas:
Globular clusters: none
Source:
Open clusters: none
Source:
OB associations: none
Source:
Variable stars: 46; designation: V
Source: [9]
Carbon stars: none
Source:
HII regions: none
Source:
Planetary nebulae: none
Source:
Dust clouds: none
Source:

Sextans I

Chart 183

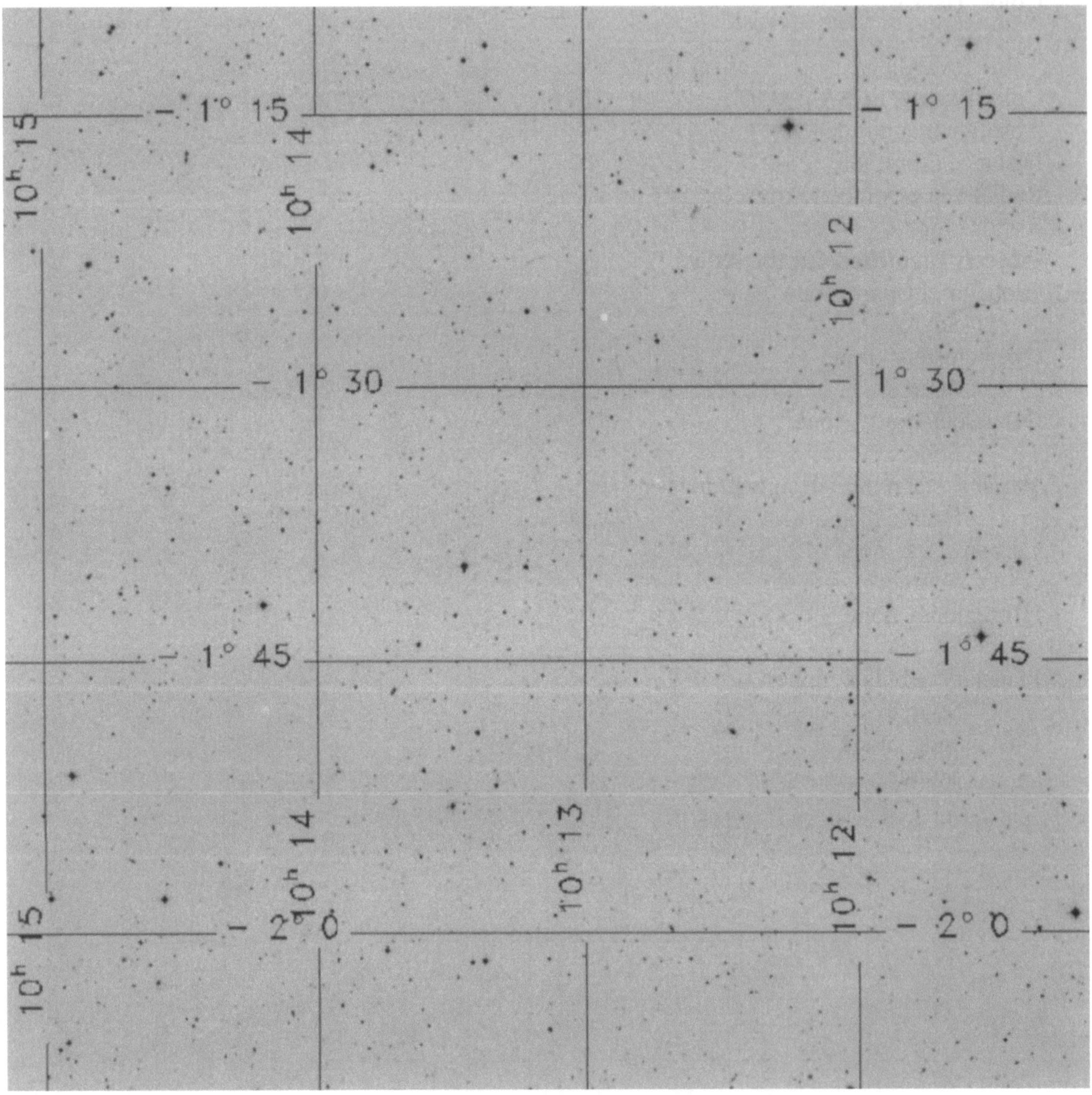

Sextans I

Chart 184

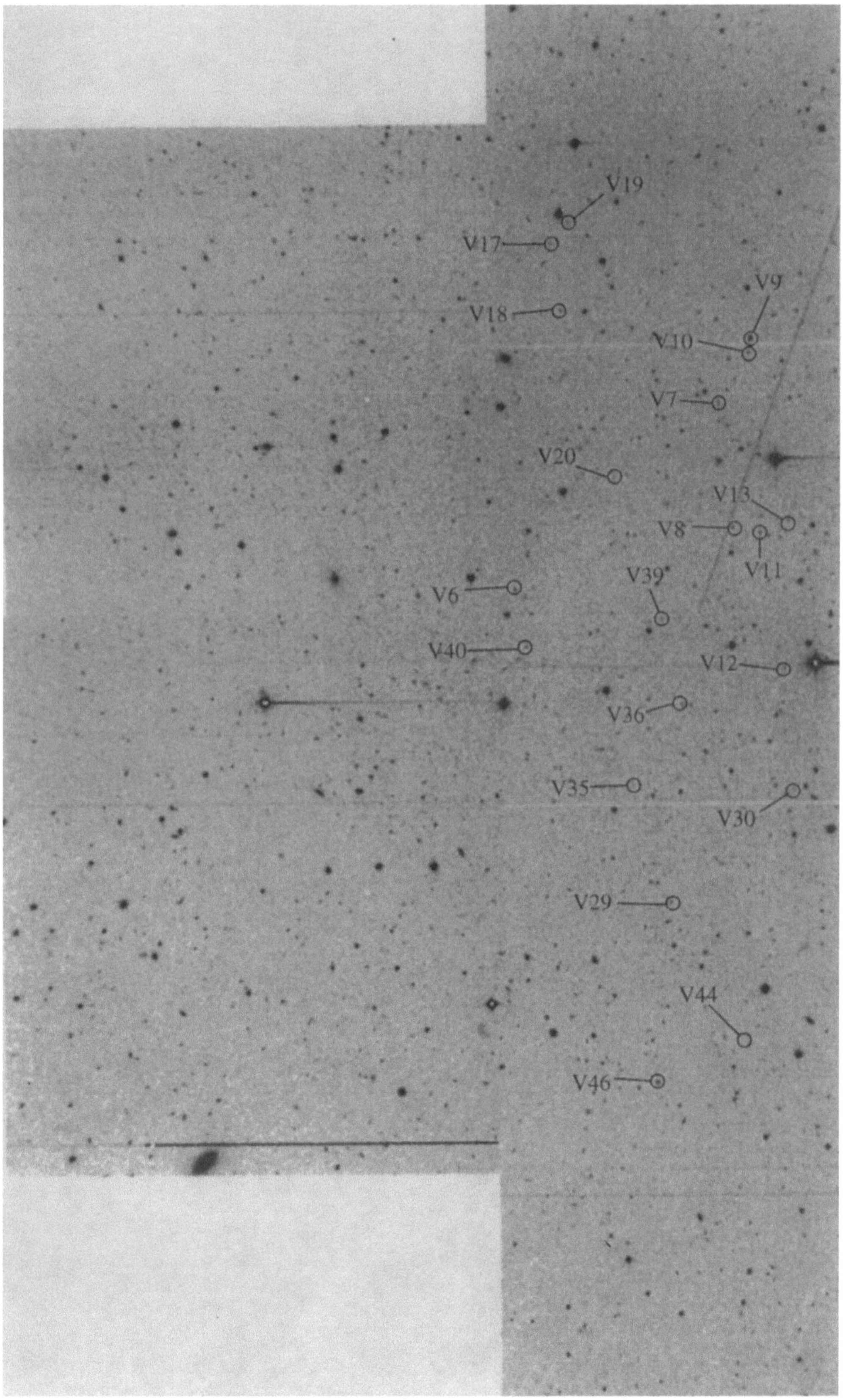

Sextans I

Chart 185

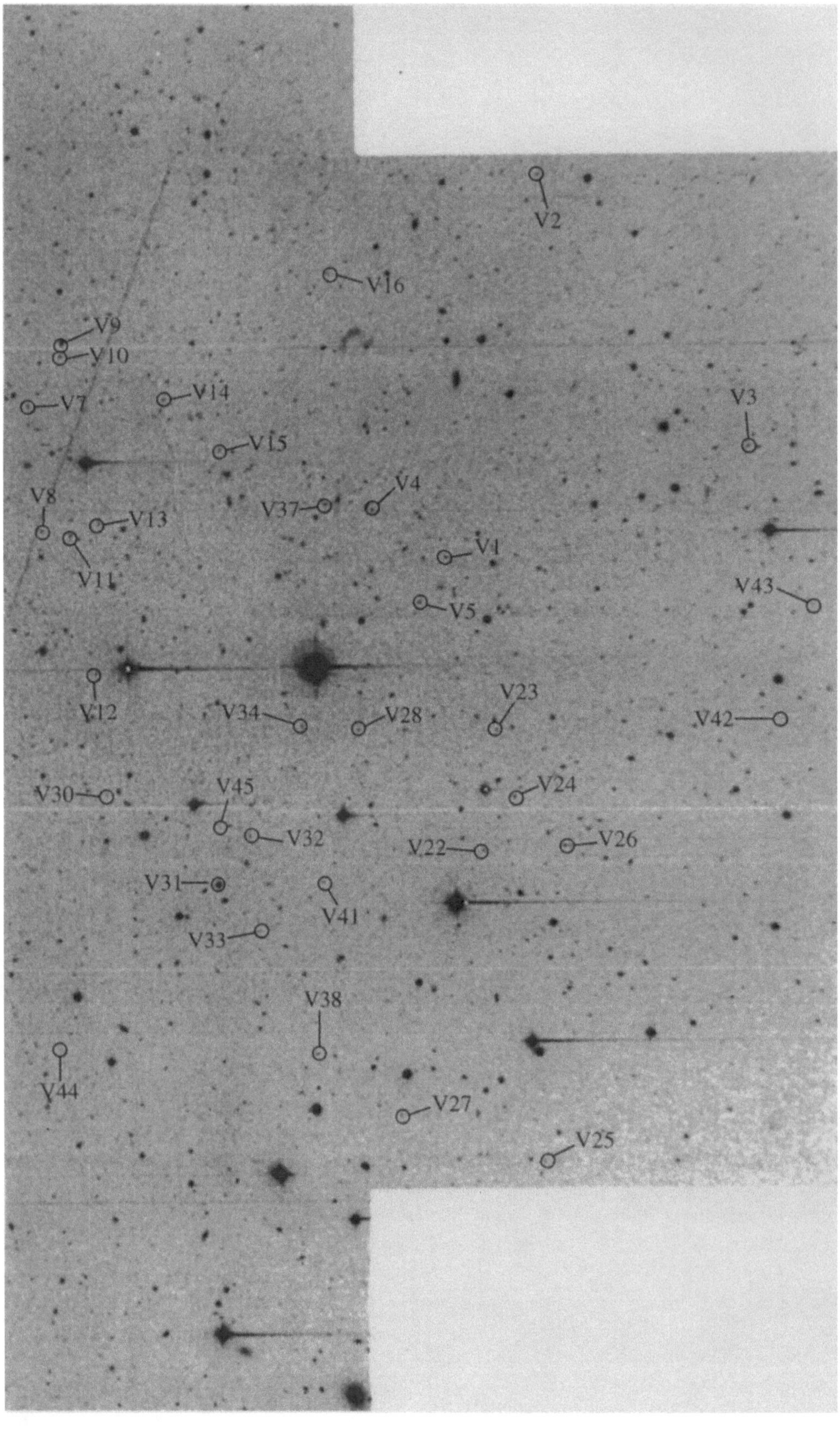

Sextans I

References

1. Bono, G., Caputo, F., Santolamazza, P., Cassisi, S., & Piersimoni, A. 1997, "Evolutionary Scenario for Metal-Poor Pulsating Stars. II. Anomalous Cepheids", AJ, 113, 2209.
2. Da Costa, G.S. 1992, "Dwarf Spheroidal Galaxies", in The Stellar Populations of Galaxies, IAU Symposium No. 149, edited by B. Barbuy and A. Renzini (Kluwer, Dordrecht).
3. Da Costa, G.S., Hatzidimitriou, D., Irwin, M.J., & McMahon, R.G. 1991, "The Radial Velocity and Metal Abundance of the Sextans Dwarf Spheroidal Galaxy", MNRAS, 249, 473.
4. Geisler, D., & Sarajedini, A. 1996, "Standards Giant Branches in the Washington System: Calibration and Application to the Draco, Ursa Minor, and Sextans Dwarf Spheroidal Galaxies", in Formation of the Galactic Halo…Inside and Out, edited by H. Morrison, and A. Sarajedini (Astronomical Society of the Pacific, San Francisco), p. 524.
5. Gould, A., Guhathakurta, P., Richstone, D., & Flynn, C. 1992, "Evidence for Dwarf Stars at $D \sim$ 100 Kiloparsecs near the Sextans Dwarf Spheroidal Galaxy", ApJ, 388, 345.
6. Hargreaves, J.C., Gilmore, G., Irwin, M.J., & Carter, D. 1994, "A Dynamical Study of the Sextans Dwarf Spheroidal Galaxy", MNRAS, 269, 957.
7. Irwin, M.J., Bunclark, P.S., Bridgeland, M.T., & McMahon, R.G. 1990, "A New Satellite Galaxy of the Milky Way in the Constellation of Sextans", MNRAS, 244, 16P.
8. Irwin, M., & Hatzidimitriou, D. 1995, "Structural Parameters for the Galactic Dwarf Spheroidals", MNRAS, 277, 1354.
9. Mateo, M., Fischer, P., & Krzemiński, W. 1995, "Variable Stars in the Sextans Dwarf Spheroidal Galaxy", AJ, 110, 2166.
10. Mateo, M., Nemec, J., Irwin, M., & McMahon, R. 1991, "Deep CCD Photometry of the Sextans Dwarf Galaxy", AJ, 101, 892.
11. Suntzeff, N.B., Mateo, M., Terndrup, D.M., Olszewski, E.W., Geisler, D., & Weller, W. 1993, "Spectroscopy of Giants in the Sextans Dwarf Spheroidal Galaxy", ApJ, 418, 208.
12. van den Bergh, S. 1994, "The Evolutionary History of Low-Luminosity Local Group Dwarf Galaxies", ApJ, 428, 617.
13. Weller, W., Mateo, M., & Krzemiński, W. 1991, "The RR Lyr Stars of the Sextans Dwarf Spheroidal Galaxy", BAAS, 23, 877.

Sextans A

Alternate name(s): DDO 75
Right Ascension (2000): 10h 11m 06S
Declination (2000): -04d 42.5m
Type: dIrr
Apparent magnitude (V): 11.30
Color (B-V): 0.38
(U-B): -0.32
(V-R):
Color Excess, E(B-V): 0.03
Absolute magnitude (M_V): -14.6
Distance (kpc): 1440
Radial velocity (solar, km/sec): 325

Objects Identified On the Atlas:

Globular clusters: none
Source:
Open clusters: none
Source:
OB associations: 6; designation: A
Source: [this atlas]
Variable stars: 8; designation: V
Source: [24]
Carbon stars: none
Source:
HII regions: 25; designation: Arabic numerals
Source: [7]
Planetary nebulae: 1; designation PN1
Source: [17]
Dust clouds: none
Source:

Sextans A

Chart 186

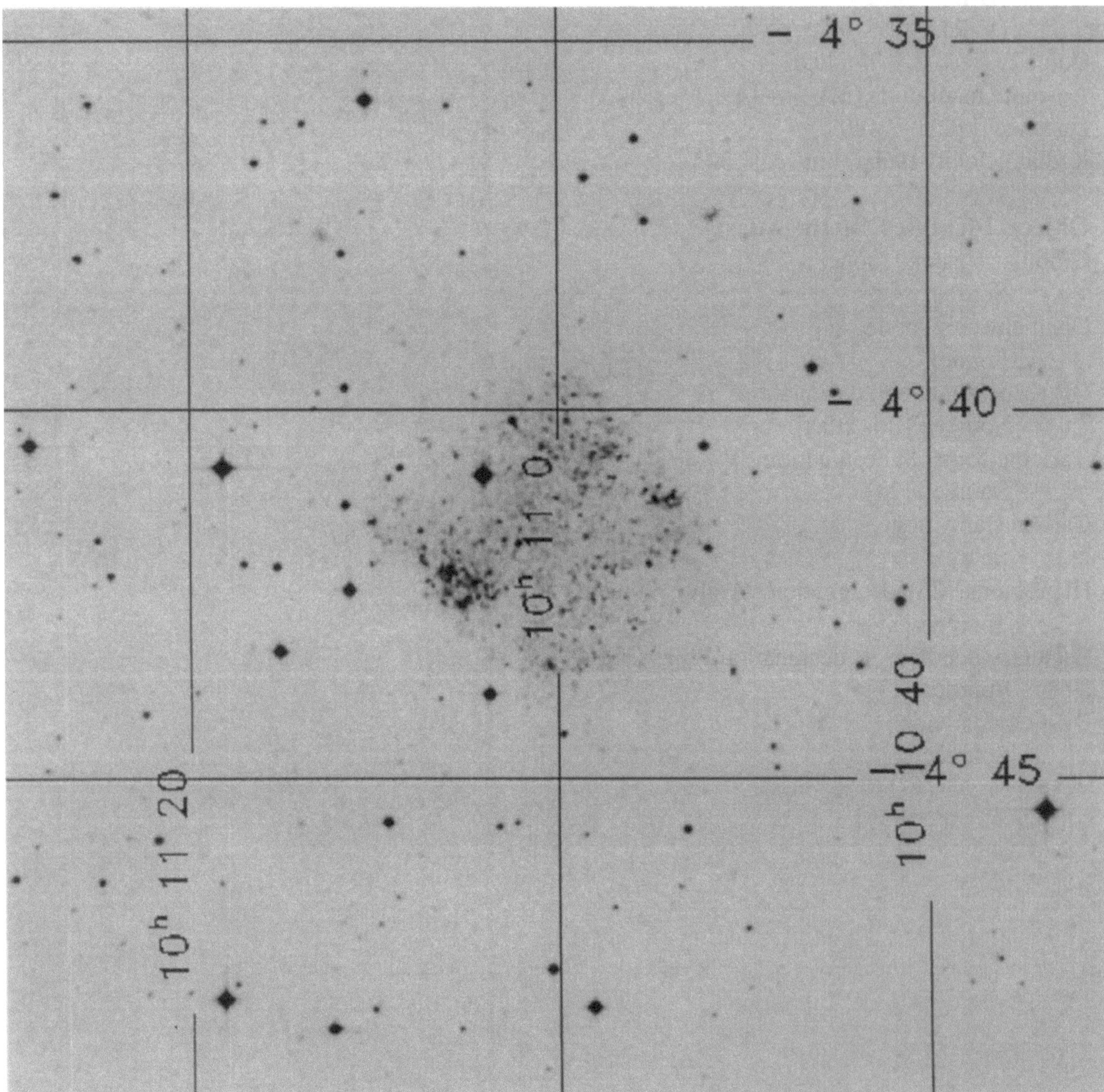

Sextans A

Chart 187

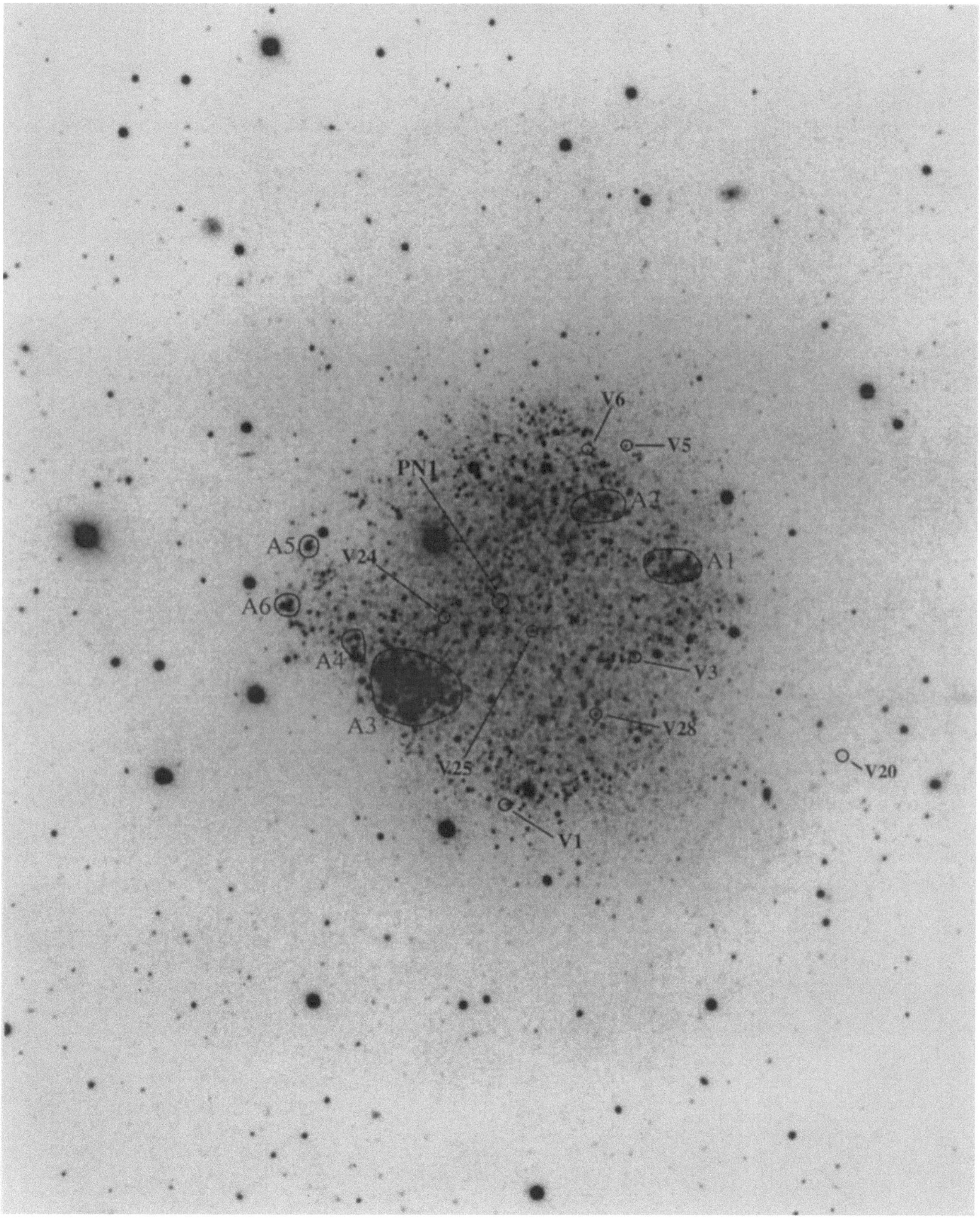

Sextans A

Chart 188

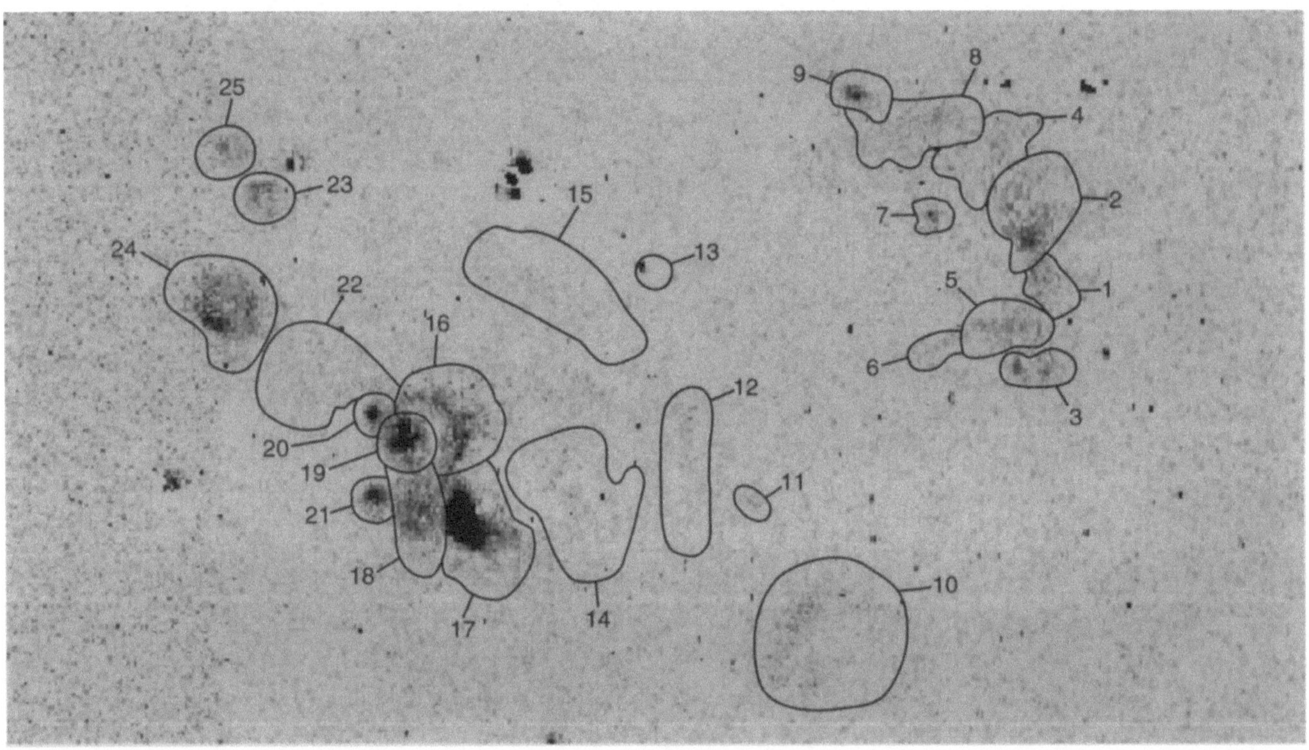

Sextans A

References

1. Ables, H.D. 1971, "Optical Study of Nearby Galaxies", Pub. of the U.S. Naval Observatory, Vol. 20, Part 4, 7.
2. Aparicio, A., García-Pelayo, J.M., Moles, M., & Melnick, J. 1987, "CCD Photometry of Resolved Dwarf Irregular Galaxies. I. Sextans A", A&AS, 71, 297.
3. Aparicio, A., Rodriguez-Ulloa, J.A. 1992, "H II Regions in Sextans A", A&A, 260, 77.
4. Elias, J.H., & Frogel, J.A. 1985, "M Supergiants in Local Group Irregular Galaxies: Metallicities and Distances", ApJ, 289, 141.
5. Fisher, J.R., & Tully, R.B. 1979, "Photographs of Dwarf Irregular Galaxies in the Vicinity of the Local Group and the M81 Group", AJ, 84, 62.
6. Freedman, W.L. 1986, "Recent CCD Observations of Cepheids in Nearby Galaxies: A Progress Report", in Galaxy Distances and Deviations from Universal Expansion, edited by B.F. Madore and R.B. Tully (Reidel, Dordrecht), p. 21.
7. Hodge, P., Kennicutt, R.C., & Strobel, N. 1994, "The H II Regions of Sextans A", PASP, 106, 765.
8. Hoessel, J.G., Schommer, R.A., & Danielson, G.E. 1983, "Photometry of Resolved Galaxies. II. Sextans A", ApJ, 274, 577.
9. Huchtmeier, W.K., Seiradakis, J.H., & Materne, J. 1981, "The Neutral Hydrogen Distribution of Irregular Galaxies", A&A, 102, 134.
10. Hunter, D.A., & Gallagher, III, J.S. 1985, "Star-Forming Properties and Histories of Dwarf Irregular Galaxies: Down but not Out", ApJS, 58, 533.
11. Hunter, D.A., & Gallagher, III, J.S. 1990, "Properties of Ionized Interstellar Froth in Irregular Galaxies", ApJ, 362, 480.
12. Hunter, D.A., & Gallagher, II, J.S. 1992, "Supergiant Filaments of Ionized Gas in Irregular Galaxies", ApJ, 391, L9.
13. Hunter, D.A., & Gallagher, III, J.S. 1997, "An Emission-Line Study of Supergiant Ionized Filaments in Irregular Galaxies", ApJ, 475, 65.
14. Hunter, D.A., Gallagher, III, J.S., Rice, W.L., & Gillett, F.C. 1989, "*IRAS* Observations of a Large Sample of Normal Irregular Galaxies", ApJ, 336, 152.
15. Hunter, D.A. Hawley, W.N., & Gallagher, III, J.S. 1993, "A Survey for Extra-H II Region Ionized Gas Structures in Irregular Galaxies", AJ, 106, 1797.
16. Hunter, D.A., & Plummer, J.D. 1996, "Sextans A: A Case of Star Formation and Gas Densities in Irregular Galaxies", ApJ, 462, 732.
17. Jacoby, G.H., & Lesser, M.P. 1981, "Distances to Nearby Galaxies Derived from Planetary Nebulae", AJ, 86, 185.
18. Madore, B.F., & Freedman, W.L. 1991, "The Cepheid Distance Scale", PASP, 103, 933.
19. Ohta, K., Tomita, A., & Saitō, M. 1993, "CO-to-H_2 Conversion factor in Dwarf Irregular Galaxies", PASJ, 45, L21.
20. Piotto, G., Capaccioli, M., & Pellegrini, C. 1994, "On the Cepheid Variables of Nearby Galaxies. II. Sextans A and Sextans B", A&A, 287, 371.
21. Rowan-Robinson, M., Phillips, T.G., & White, G. 1980, "A Survey of Nearby Galaxies for CO", A&A, 82, 381.

22. Sakai, S., Madore, B.F., & Freedman, W.L. 1996, "Tip of the Red Giant Branch Distances to Galaxies. III. The Dwarf Galaxy Sextans A", ApJ, 461, 713.
23. Sandage, A., & Carlson, G. 1982, "Distance and Absolute Magnitudes of the Brightest Stars in the Dwarf Galaxy Sextans A", ApJ, 258, 439.
24. Sandage, A., & Carlson, G. 1985, "The Brightest Stars in Nearby Galaxies. V. Cepheids and the Brightest Stars in the Dwarf Galaxy Sextans B Compared to those in Sextans A", AJ, 90, 1019.
25. Skillman, E.D., Kennicutt, R.C., & Hodge, P.W. 1989, "Oxygen Abundances in Nearby Dwarf Irregular Galaxies", ApJ, 347, 875.
26. Skillman, E.D., Terlevich, R., Teuben, P.J., & van Woerden, H. 1988, "H I Synthesis Observations of the Dwarf Irregular Galaxy Sextans A", A&A, 198, 33.
27. Skillman, E.D., Terlevich, R., & van Woerden, H. 1986, "H I Observations of Sextans A", in Star Forming Dwarf Galaxies and Related Objects, edited by D. Kunth, T. Thuan, and J. Van (Editions Frontieres, Gif-sur-Yvette), p. 263.
28. Tomita, A., Ohta, K., & Saitō, M. 1993, "H-Alpha Velocity Fields of Giant H II Regions in Four Nearby Dwarf Irregular Galaxies", PASJ, 45, 693.
29. Visvanathan, N. 1989, "The Zero Point of the Cepheid Period-Luminosity Relation in the 1.05 Micron Wave Band", ApJ, 346, 629.
30. Walker, A.R. 1987, "Studies of the Dwarf Irregular Galaxies. I. *B* and *V* CCD Photometry of the Brighter Stars in Sextans A", MNRAS, 224, 935.
31. Walker, A.R. 1987, "Notes on Cepheid Variables in Galaxies More Distant Than the Magellanic Clouds", S. African Astron. Obs. Circ., 11, 125.

Sextans B

Alternate name(s): DDO 70
Right Ascension (2000): 10h 00m 00s
Declination (2000): +5d 19.7m
Type: dIrr
Apparent magnitude (V): 11.43
Color (B-V): 0.48
(U-B): -0.16
(V-R):
Color Excess, E(B-V): 0.01
Absolute magnitude (M_V): -14.2
Distance (kpc): 1345
Radial velocity (solar, km/sec): 303

Objects Identified On the Atlas:
Globular clusters: none
Source:
Open clusters: none
Source:
OB associations: 3; designation: A
Source: [this atlas]
Variable stars: 7; designation: V
Source: [7], [9]
Carbon stars: none
Source:
HII regions: 12; designation: Arabic numerals
Source: [12]
Planetary nebulae: none
Source:
Dust clouds: none
Source:

Sextans B

Chart 189

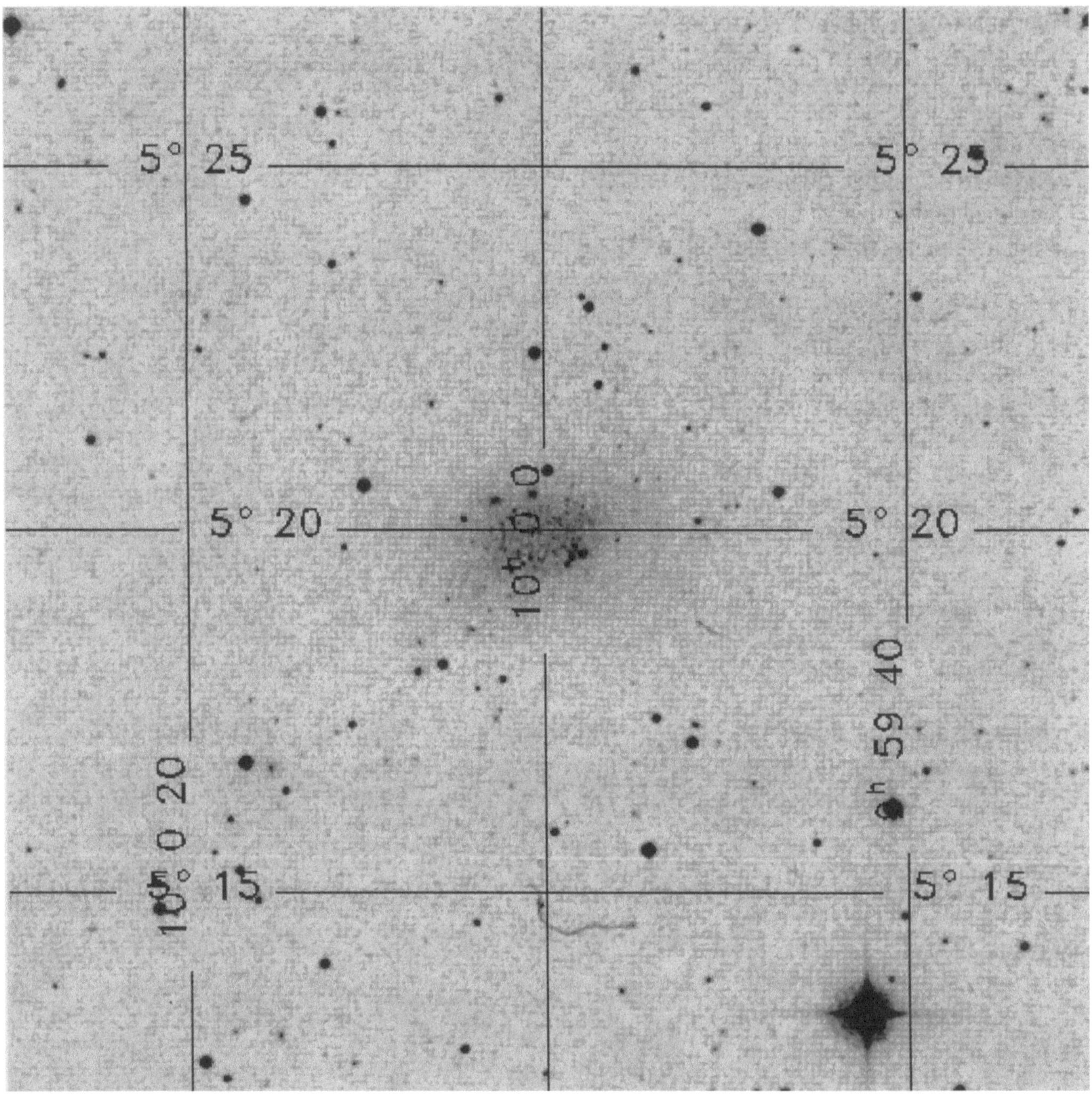

Sextans B

Chart 190

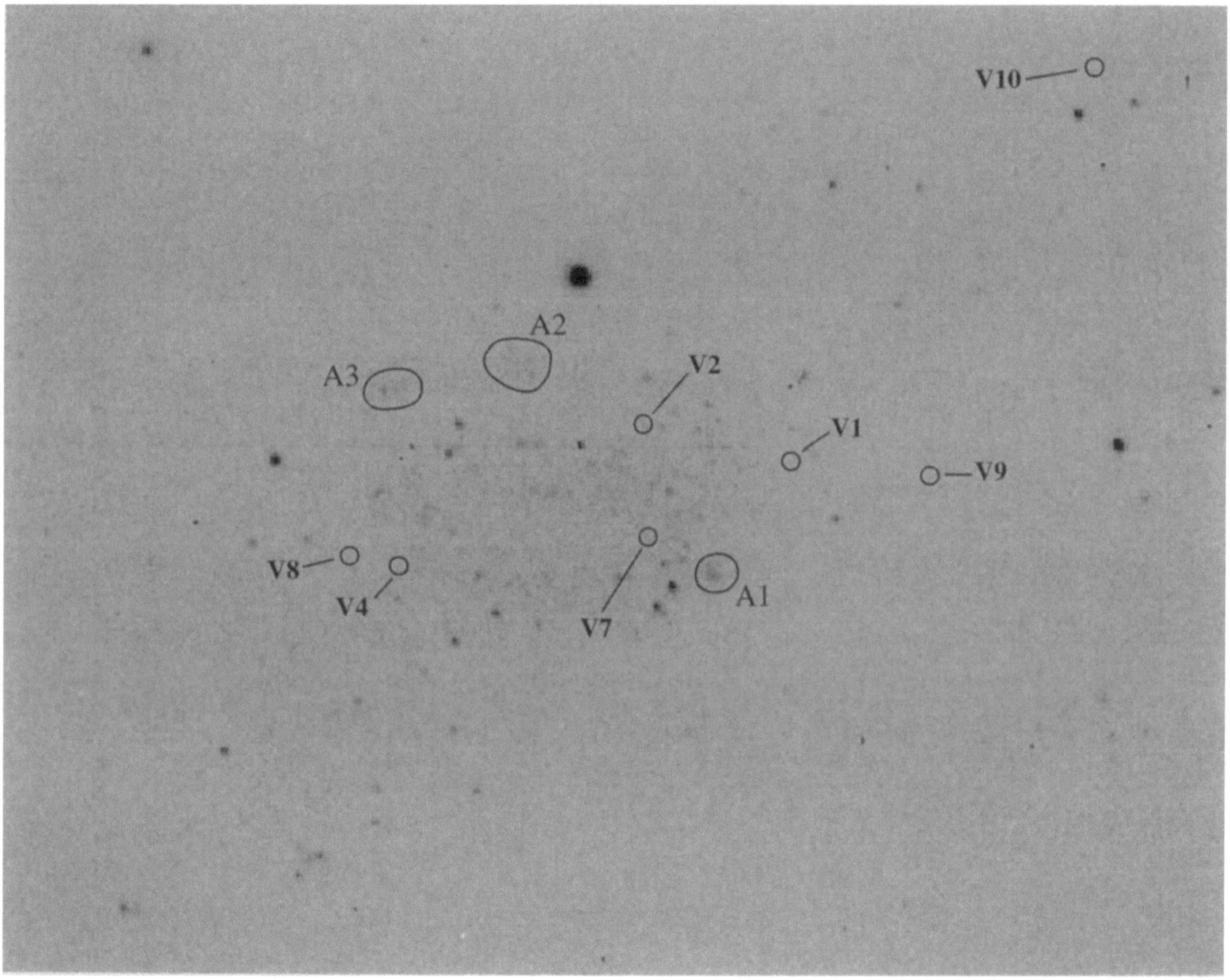

Sextans B

Chart 191

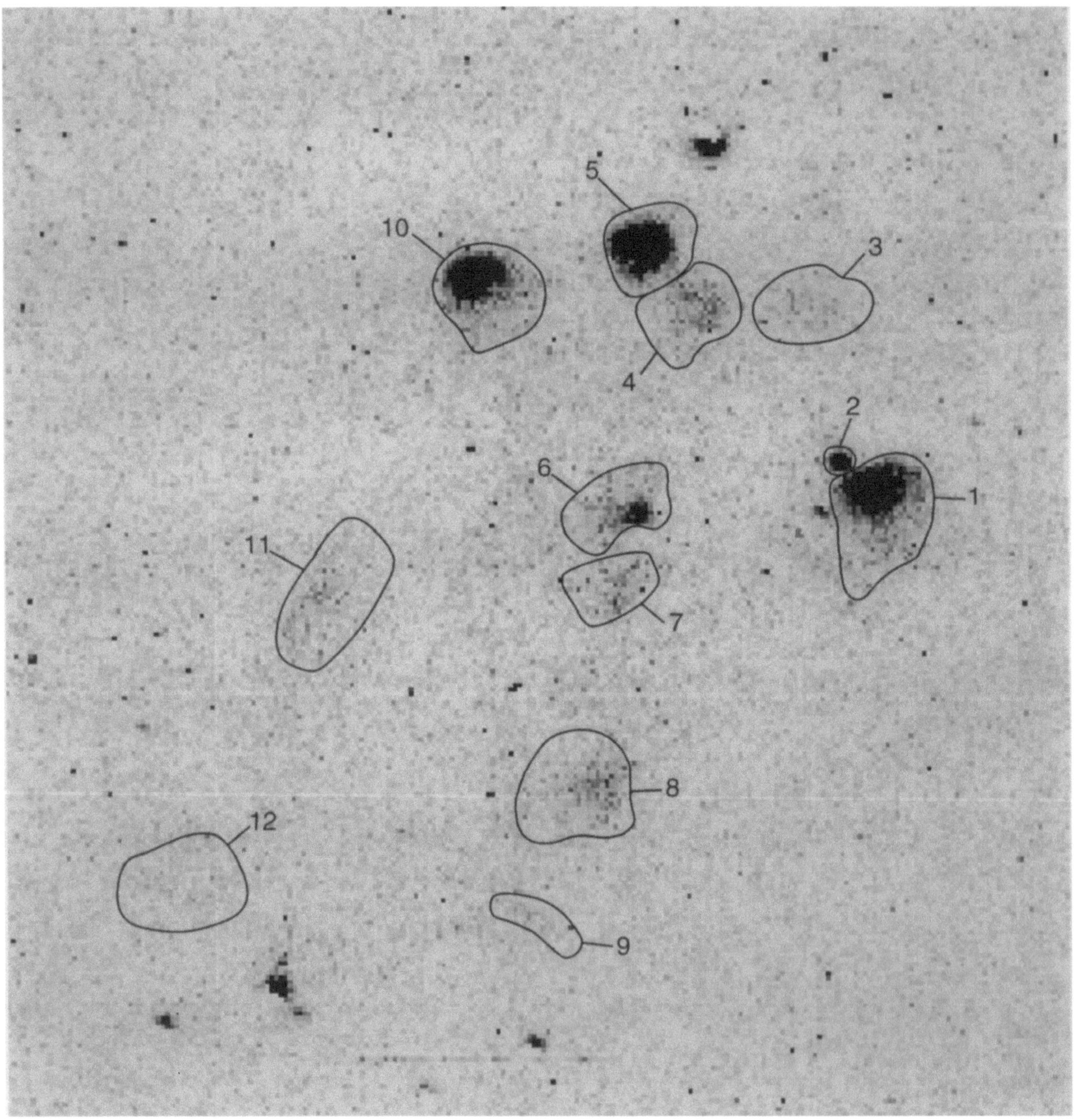

Sextans B

References

1. Comte, G., Lequeux, J., & Viallefond, F. 1985, "VLA Observations of Dwarf Galaxies", in Star-Forming Dwarf Galaxies and Related Objects, edited by D. Kunth, T.X. Thuan, and J.T.T. Van (Editions Frontiers, Gif-sur-Yvette), p. 273.
2. Fisher, J.R., & Tully, R.B. 1979, "Photographs of Dwarf Irregular Galaxies in the Vicinity of the Local Group and the M81 Group", AJ, 84, 62.
3. Hoffman, G.L., Salpeter, E.E., Farhat, B., Roos, T., Williams, H., & Helou, G. 1996, "Arecibo H I Mapping of a Large Sample of Dwarf Irregular Galaxies", ApJS, 105, 269.
4. Hunter, D.A., Hawley, W.N., and Gallagher, III, J.S. 1993, "A Survey for Extra-H II Region Ionized Gas Structures in Irregular Galaxies", AJ, 106, 1797.
5. Madore, B.F., & Freedman, W.L. 1991, "The Cepheid Distance Scale", PASP, 103, 933.
6. Moles, M., Aparicio, A., and Masegosa, J. 1990, "The Metal Content of GR8 and Sextans B", A&A, 228, 310.
7. Piotto, G., Capaccioli, M., & Pellegrini, C. 1994, "On the Cepheid Variables of Nearby Galaxies. II. Sextans A and Sextans B", A&A, 287, 371.
8. Sakai, S., Madore, B.F., & Freedman, W.L. 1996, "Tip of the Red Giant Branch Distances to Galaxies. IV. Sextans B", ApJ, 480, 589.
9. Sandage, A., & Carlson, G. 1985, "The Brightest Stars in Nearby Galaxies. V. Cepheids and the Brightest Stars in the Dwarf Galaxy Sextans B Compared with those in Sextans A", AJ, 90, 1019.
10. Skillman, E.D., Kennicutt, R.C., & Hodge, P.W. 1989, "Oxygen Abundances in Nearby Dwarf Irregular Galaxies", ApJ, 347, 875.
11. Stasińska, G., Comte, G., & Vigroux, L. 1986, "Spectrophotometry of H II Regions in Irregular Galaxies", A&A, 154, 352.
12. Strobel, N.V., Hodge P.W., & Kennicutt, R.C. 1991, "H II Regions in Seven Dwarf Irregular Galaxies", ApJ, 383, 148.
13. Tosi, M., Greggio, L., & Focardi, P. 1989, "Star Formation in Dwarf Irregular Galaxies: Preliminary Results", Ap&SS, 156, 295.
14. Tosi, M., Greggio, L., Marconi, G., & Focardi. P. 1991, "Star Formation in Dwarf Irregular Galaxies: Sextans B", AJ, 102, 951.
15. Walker, A.R. 1987, "Notes on Cepheid Variables in Galaxies More Distant Than the Magellanic Clouds", S. African Astron. Obs. Circ., 11, 125.

Tucana

Alternate name(s): SGC 2238.5-6441
Right Ascension (2000): 22h 41m 50s
Declination (2000): -64d 25.2m
Type: dE5
Apparent magnitude (V): 15.15
Color (B-V): 0.7
(U-B):
(V-R): 0.55
Color Excess, E(B-V): 0.00
Absolute magnitude (M_V): -9.5
Distance (kpc): 880
Radial velocity (solar, km/sec):

Objects Identified On the Atlas:
Globular clusters: none
Source:
Open clusters: none
Source:
OB associations: none
Source:
Variable stars: none
Source:
Carbon stars: none
Source:
HII regions: none
Source:
Planetary nebulae: none
Source:
Dust clouds: none
Source:

Tucana

Chart 192

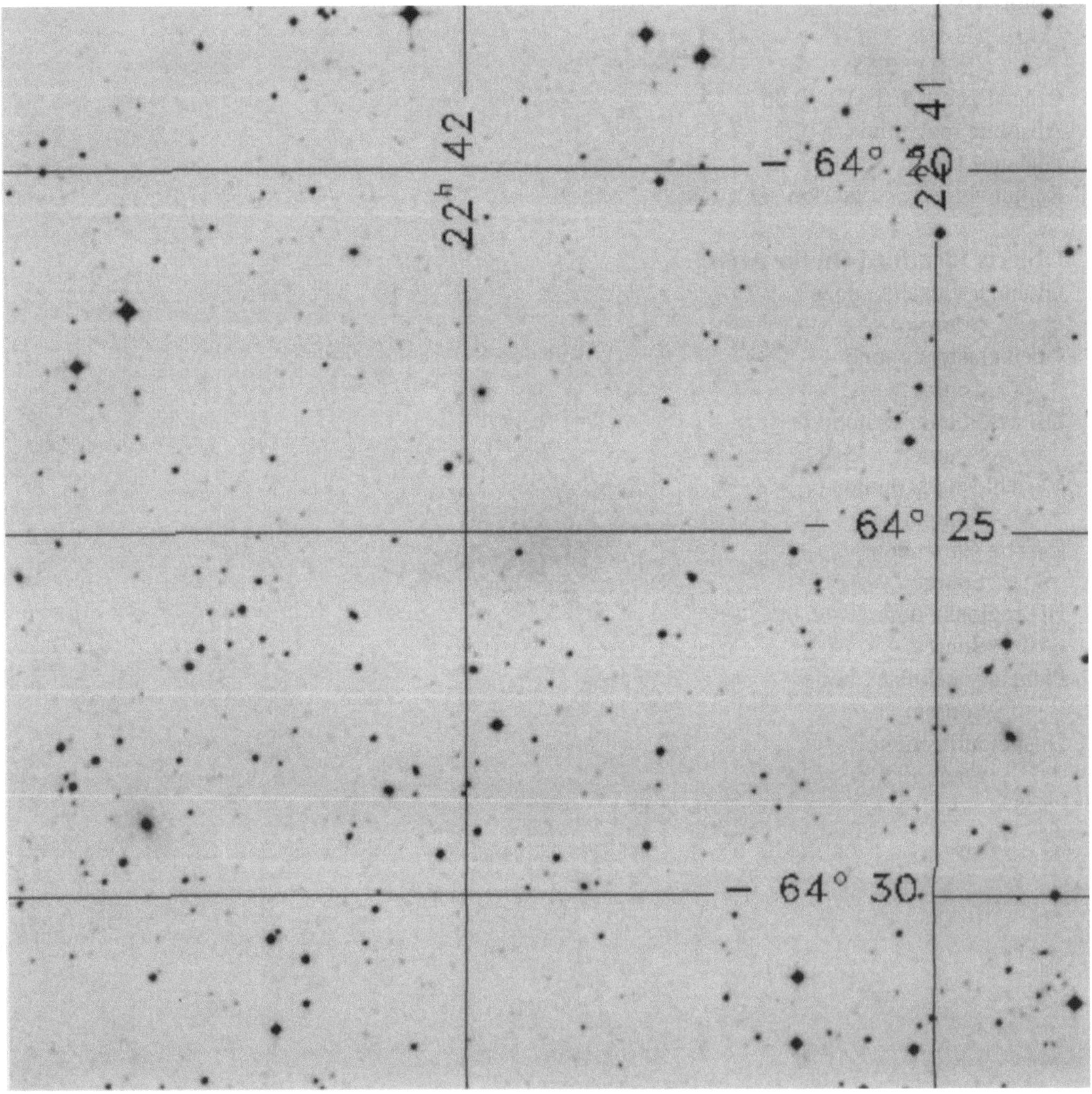

Tucana

Chart 193

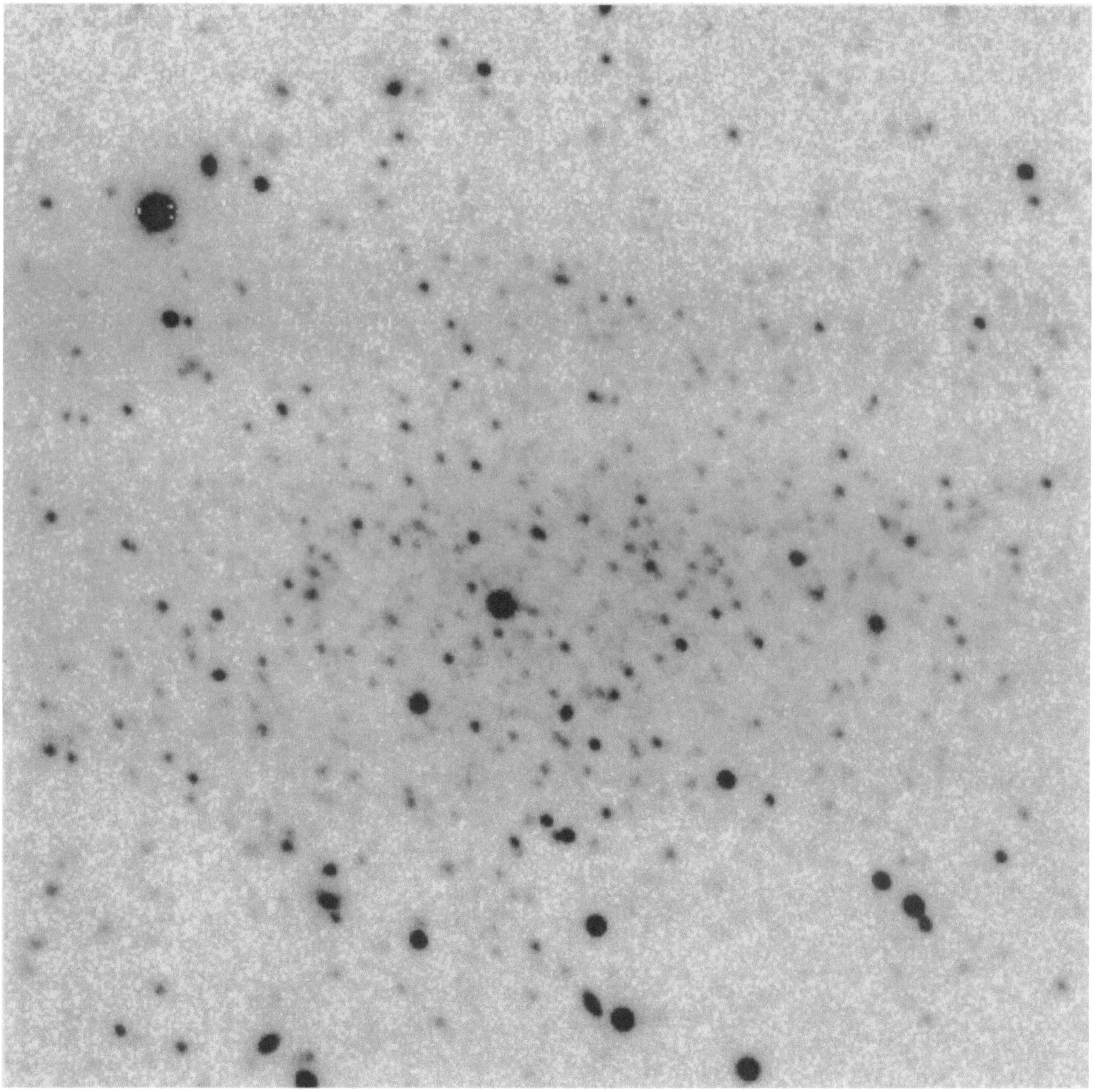

Tucana

References

1. Castellani, M., Marconi, G., & Buonanno, R. 1996, "The Tucana Dwarf Galaxy", A&A, 310, 715.
2. Da Costa, G.S. 1994, "New Results for Local Group Dwarf Galaxies", in Dwarf Galaxies, edited by G. Meylan, and P. Prugniel (European Southern Observatory, Munich), p. 221.
3. Lavery, R.J., & Mighell, K.J. 1992, "A New Member of the Local Group: The Tucana Dwarf Elliptical Galaxy", AJ, 103, 81.
4. Osterloo, T., Da Costa, G.S., & Staveley-Smith, L. 1996, "H I Observations of the Tucana Dwarf Elliptical Galaxy", AJ, 112, 1969.
5. Saviane, I., Held, E.V., & Piotto, G. 1996, "CCD Photometry of the Tucana Dwarf Galaxy", A&A, 315, 40.

UGC A92

Alternate name(s): EGB 0427+63
Right Ascension (2000): 04h 32m 01s
Declination (2000): +63d 36.4m
Type: dIrr
Apparent magnitude (V):
Color (B-V):
(U-B):
(V-R):
Color Excess, E(B-V): 0.30
Absolute magnitude (M_V):
Distance (kpc): 796
Radial velocity (solar, km/sec): -99

Objects Identified On the Atlas:
Globular clusters: none
Source:
Open clusters: none
Source:
OB associations: none
Source:
Variable stars: none
Source:
Carbon stars: none
Source:
HII regions: 25; designation: Arabic numerals
Source: [2]
Planetary nebulae: none
Source:
Dust clouds: none
Source:

UGC A92

Chart 194

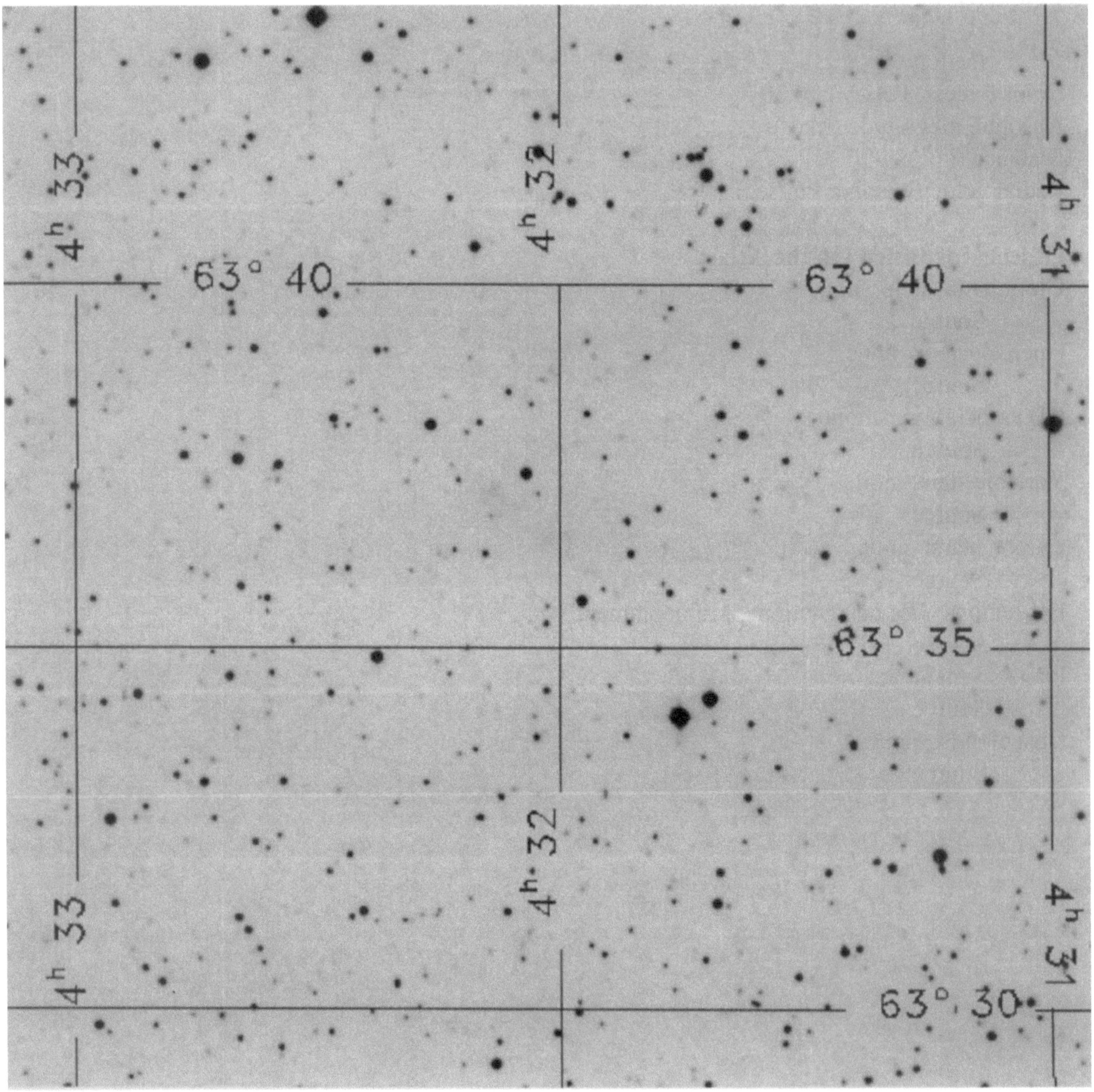

UGC A92

Chart 195

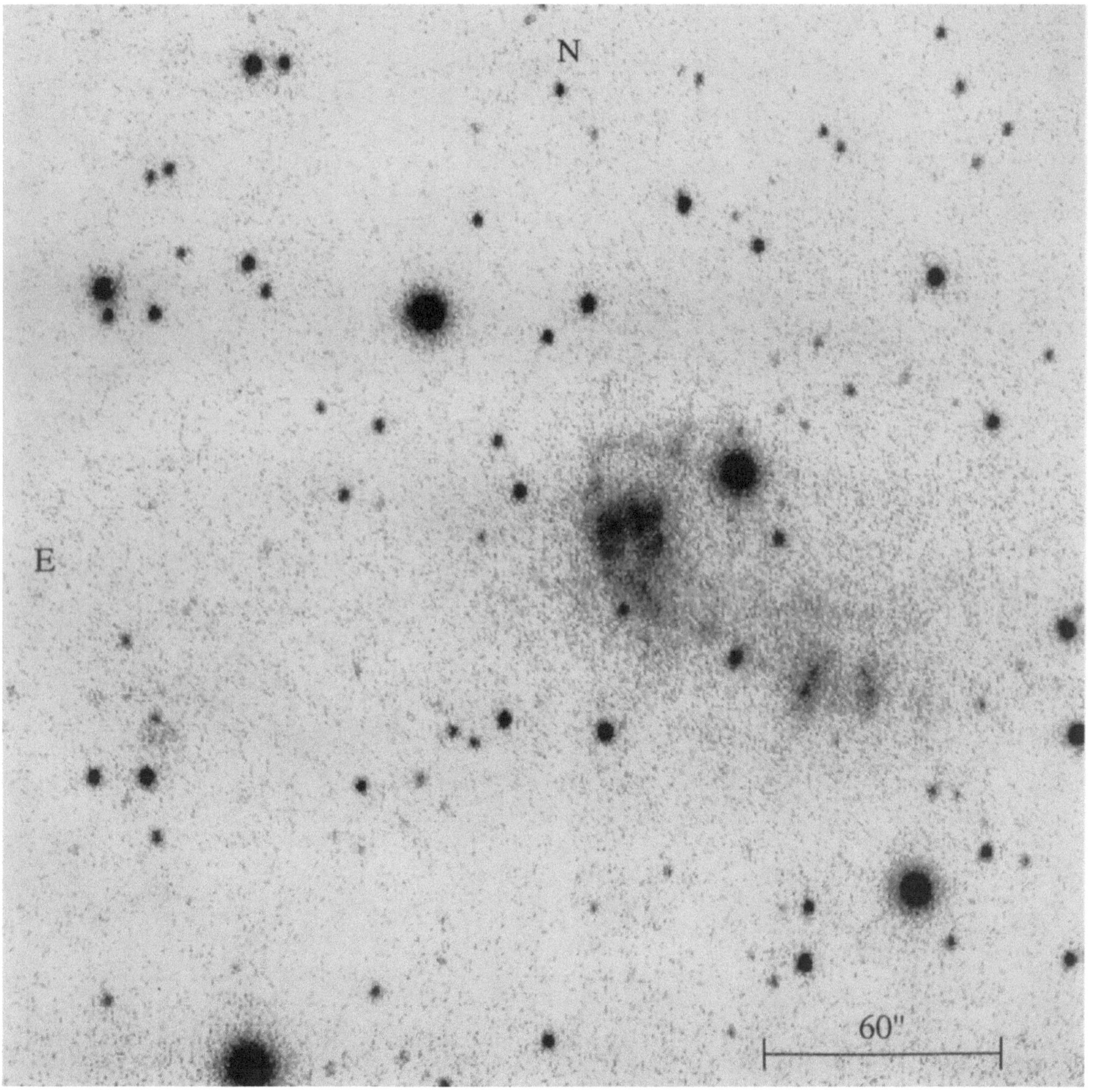

UGC A92

Chart 196

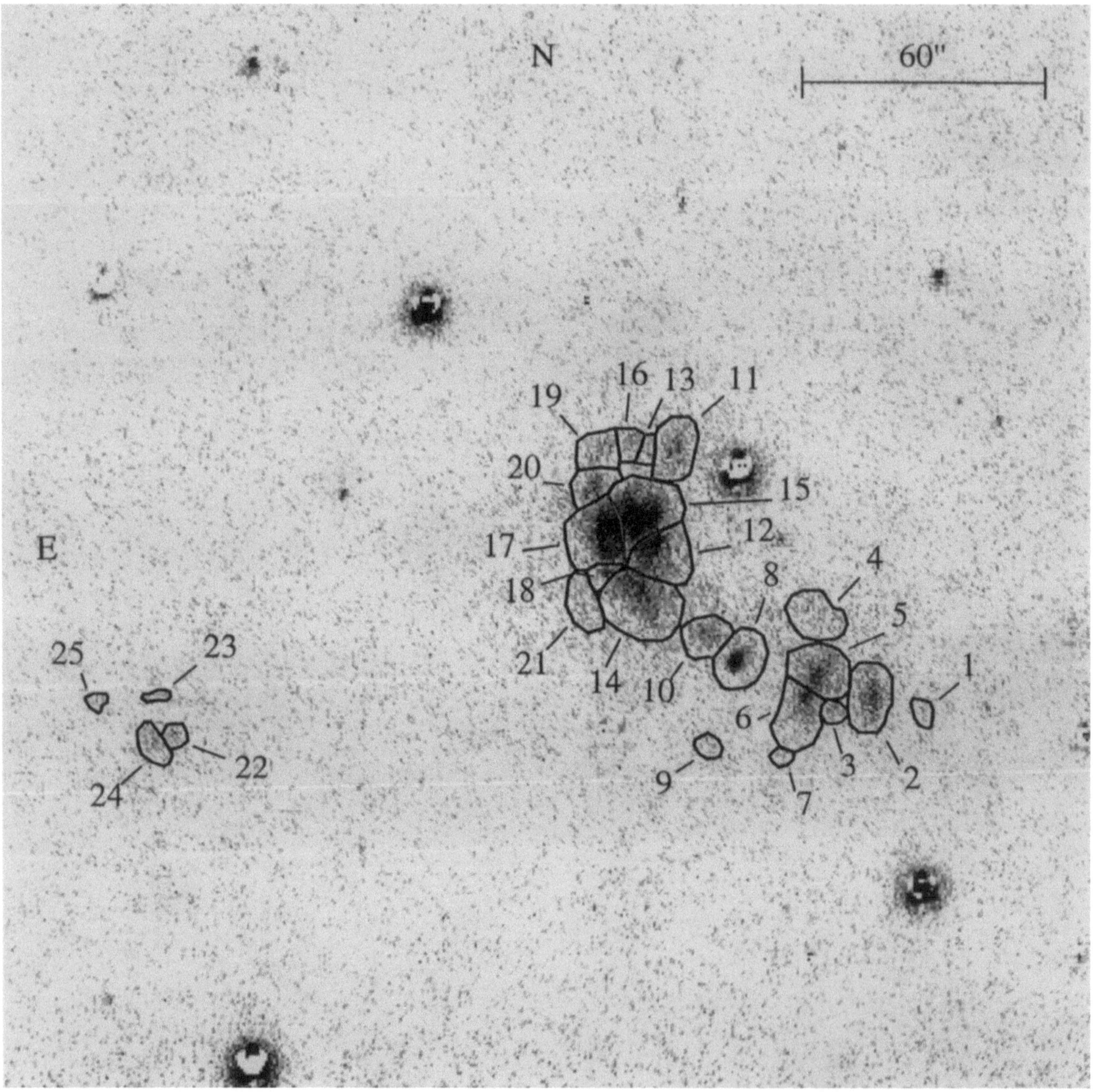

UGC A92

References

1. Ellis, G.L., Grayson, E.T., & Bond, H.E. 1984, "A Search for Faint Planetary Nebulae on Palomar Sky Survey Prints", PASP, 96, 283.
2. Hodge, P., & Miller, B.W. 1995, "H II Regions in Four Galaxies in and near the Local Group", ApJ, 451, 176.
3. Hoessel, J.G., Saha, A., & Danielson, G.E. 1988, "Deep CCD Observations of Nearby Dwarf Galaxy Candidates", PASP, 100, 680.
4. Karachentsev, I.D., Tikhonov, N.A., & Sazonova, L.N. 1994, "NGC 1569 and UGCA 92 - A Nearby Pair of Galaxies in the Milky Way", Astron. Letters, 20, 84.
5. Karachentseva, V.E., Prugniel, P., Vennik. L., Richter, G.M., Thuan, T.X., & Martin, J.M. 1996, "CCD and H I Observations of LSB Dwarf Galaxies in the General Field", A&AS, 117, 343.

Ursa Minor

Alternate name(s): DDO 199
Right Ascension (2000): 15h 09m 11s
Declination (2000): +67d 12.9m
Type: dE5
Apparent magnitude (V): 10.3
Color (B-V): 1.25
(U-B): -0.33
(V-R):
Color Excess, E(B-V): 0.03
Absolute magnitude (M_V): -9.0
Distance (kpc): 66
Radial velocity (solar, km/sec): -248

Objects Identified On the Atlas:
Globular clusters: none
Source:
Open clusters: none
Source:
OB associations: none
Source:
Variable stars: 93; designations: Arabic numerals
Source: [26]
Carbon stars: 2; designations: star K, V335
Source: [4], [6]
HII regions: none
Source:
Planetary nebulae: none
Source:
Dust clouds: none
Source:

Ursa Minor

Chart 197

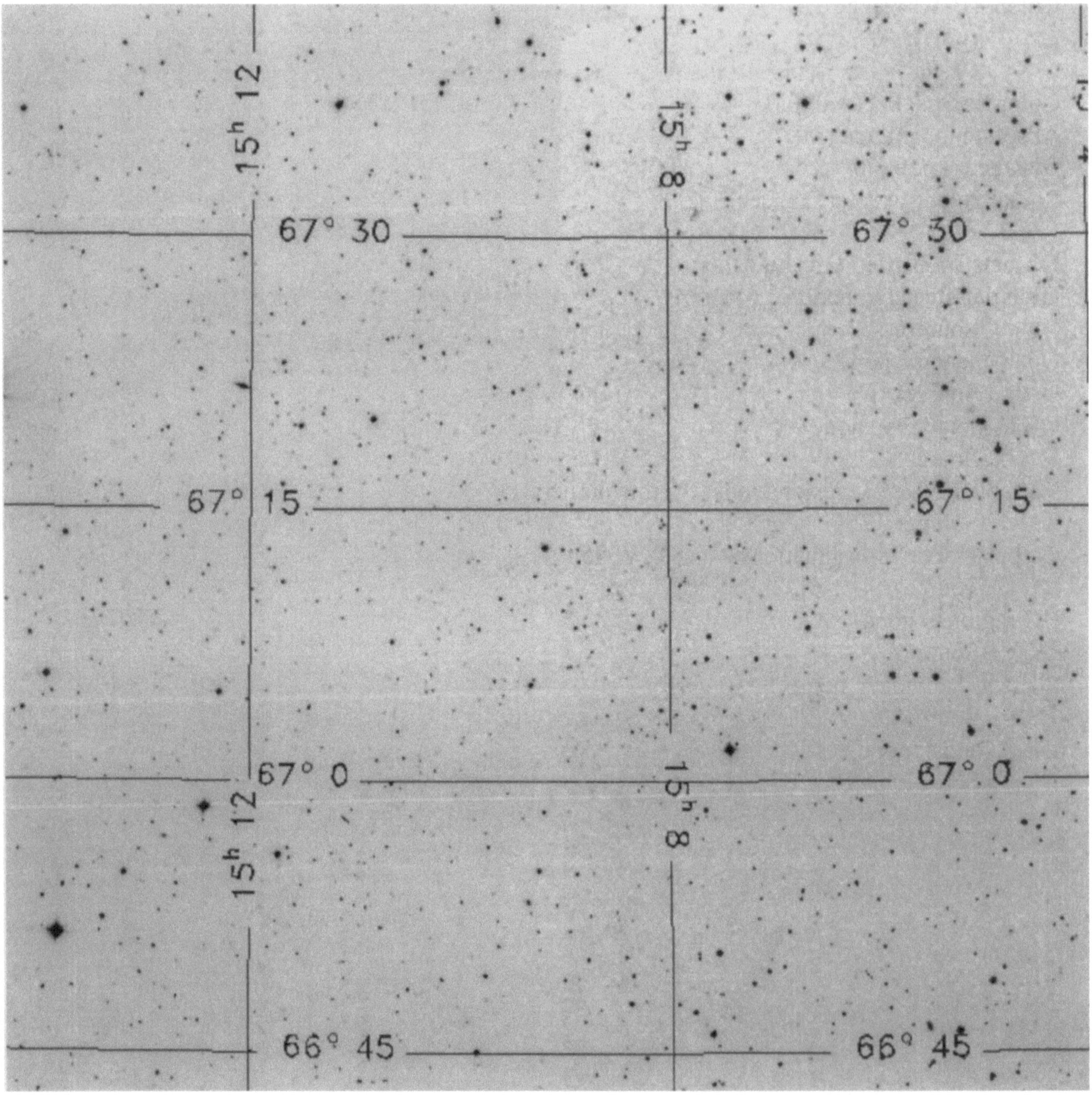

Ursa Minor

Chart 198

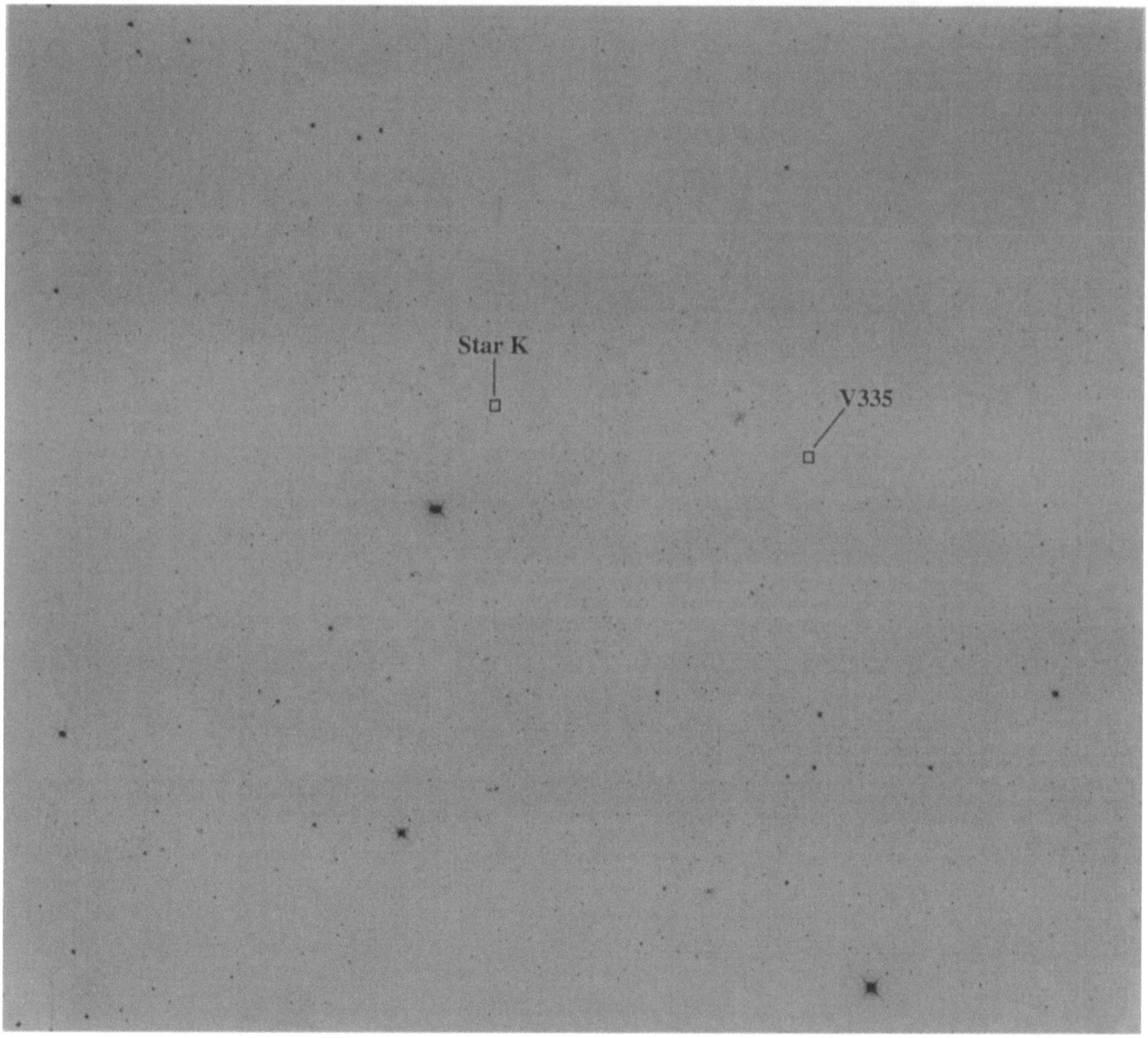

Ursa Minor

Chart 199

Ursa Minor

Chart 200

Ursa Minor

References

1. Aaronson, M. 1983, "Accurate Radial Velocities for Carbon Stars in Draco and Ursa Minor: The First Hint of a Dwarf Spheroidal Mass-to-Light Ratio", ApJ, 266, L11.
2. Aaronson, M., & Mould, J. 1985, "Infrared Photometry and the Comparative Stellar Content of Dwarf Spheroidals in the Galactic Halo", ApJ, 290, 191.
3. Aaronson, M., & Olszewski, E.W. 1987, "The Search for Dark Matter in Draco and Ursa Minor: A Three Year Progress Report", in Dark Matter in the Universe, IAU Symposium No. 117, edited by J. Kormendy and G.R. Knapp (Reidel, Dordrecht), p. 153
4. Aaronson, M., Olszewksi, E.W., & Hodge, P.W. 1983, "Carbon Stars and the Seven Dwarfs", ApJ, 267, 274.
5. Armandroff, T.E., Olszewski, E.W., & Pryor, C. 1995, "The Mass-to-Light Ratios of the Draco and Ursa Minor Dwarf Spheroidal Galaxies. I. Radial Velocities from Multifiber Spectroscopy", AJ, 110, 2131.
6. Azzopardi, M., Lequeux, J., & Westerlund, B.E. 1986, "New Carbon Stars in Spheroidal Galaxies. II. Draco, Ursa Minor, and New Data on Sculptor, Carina, and Leo I", A&A, 161, 232.
7. Bell, R.A. 1985, "Abundances of Giant Stars in the Draco and Ursa Minor Dwarf Galaxies", PASP, 97, 219.
8. Bono, G., Caputo, F., Santolamazza, P., Cassisi, S., & Piersimoni, A. 1997, "Evolutionary Scenario for Metal-Poor Pulsating Stars. II. Anomalous Cepheids", AJ, 113, 2209.
9. Canterna, R., & Schommer, R.A. 1978, "The Chemical Abundances of Extreme Population II Systems", ApJ, 219, L119.
10. Cowley, A.P., Hartwick, F.D.A., & Sargent, W.L.W. 1978, "Spectra of Giant Stars in Distant Satellites of the Galaxy", ApJ, 220, 453.
11. Cudworth, K.M., Olszewski, E.W., & Schommer, R.A. 1986, "Proper Motions and Bright-Star Photometry in the Ursa Minor Dwarf Galaxy", AJ, 92, 766.
12. Demers, S., Battinelli, P., Irwin, M.J., & Kunkel, W.E. 1995, "A Search for Stellar Aggregates in Dwarf Spheroidal Galaxies", MNRAS, 274, 491.
13. Faber, S.M., & Lin, D.N.C. 1983, "Is there Nonluminous Matter in Dwarf Spheroidal Galaxies?", 266, L17.
14. Geisler, D., & Sarajedini, A. 1996, "Standards Giant Branches in the Washington System: Calibration and Application to the Draco, Ursa Minor, and Sextans Dwarf Spheroidal Galaxies", in Formation of the Galactic Halo…Inside and Out, edited by H. Morrison and A. Sarajedini (Astronomical Society of the Pacific, San Francisco), p. 524.
15. Gerhard, O.E., & Spergel, D.N. 1992, "Dwarf Spheroidal Galaxies and the Mass of the Neutrino", ApJ, 389, L9.
16. Hargreaves, J.C., Gilmore, G., Irwin, M.J., & Carter, D. 1994, "A Dynamical Study of the Ursa Minor Spheroidal Galaxy", MNRAS, 271, 693.
17. Hill, J.M., Olszewski, E.W., Lesser, M.P., Aaronson, M., & Eisenhamer, J.D. 1986, "New Member Giants in Dwarf Spheroidal Galaxies Using the MX Spectrometer", BAAS, 18, 996.
18. Hodge, P.W. 1966, : Distribution of Stars in the Ursa Minor Dwarf Galaxy", AJ, 69, 438.

19. Irwin, M., & Hatzidimitriou, D. 1995, "Structural Parameters for the Galactic Dwarf Spheroidals", MNRAS, 277, 1354.
20. Knapp, G.R., Kerr, F.J., & Bowers, P.F. 1978, "Upper Limits to the H I Content of the Dwarf Spheroidal Galaxies", AJ, 83, 360.
21. Lake, G. 1990, "The Distribution of Dark Matter in Draco and Ursa Minor", MNRAS, 244, 701.
22. Light, R.M. 1987, "Measurement of Abundances and their Variations in Dwarf Spheroidal Galaxies", BAAS, 19, 1033.
23. Lynden-Bell, D. 1982, "The Ursa Minor Dwarf Galaxy is a Member of the Magellanic Stream", Observatory, 102, 7.
24. Markert, T.H., & Donahue, M.E. 1985, "Observations of Four Nearby Galaxies with the *Einstein Observatory*", ApJ, 297, 564.
25. Nemec, J.M., Nemec, A.F.L., & Lutz, T.E. 1994, "Period-Luminosity-Metallicity Relations, Pulsation Modes, Absolute Magnitudes, and Distances for Population II Variable Stars", AJ, 108, 222.
26. Nemec, J.M., Wehlau, A., & Mendes de Oliveira, C. 1988, "Variable Stars in the Ursa Minor Dwarf Galaxy", AJ, 96, 528.
27. Norris, J., & Bessell, M.S. 1978, "Abundance Variations on the Giant Branches of the Sculptor and Ursa Minor Dwarf Spheroidal Galaxies and the ω Centauri Anomaly", ApJ, 225, L49.
28. Olszewski, E.W., & Aaronson, M. 1985, "The Ursa Minor Dwarf Galaxy: Still an Old Stellar System", AJ, 90, 2221.
29. Olszewski, E.W., Aaronson, M., & Hill, J.M. 1995, "Nine Seasons of Velocity Measurements in the Draco and Ursa Minor Dwarf Spheroidal Galaxies with the MMT Echelle", AJ, 110, 2120.
30. Olszewski, E.W., Pryor, C., & Armandroff, T.E. 1996, "The Mass-to-Light Ratios of the Draco and Ursa Minor Dwarf Spheroidal Galaxies. II. The Binary Population and its Effect on the Measured Velocity Dispersions of Dwarf Spheroidals", AJ, 111, 750.
31. Pryor, C., & Kormendy, J. 1990, "The Dark Matter Halos of Draco and Ursa Minor, AJ, 100, 127.
32. Schommer, R.A., Olszewski, E.W., & Cudworth, K.M. 1981, "The Ursa Minor Dwarf Galaxy", in Astrophysical Parameters for Globular Clusters, IAU Colloquium No. 68, edited by A.G.D. Philip and D.S. Hayes (L. Davis Press Inc., Schenectady), p. 453.
33. Schommer, R.A., Olszewski, E.W., & Kunkel, W.E. 1978, "A Preliminary Color-Magnitude Diagram for the Ursa Minor Dwarf Galaxy", in The HR Diagram, IAU Symposium, No. 80, edited by A.G.D. Philip and D.S. Hayes (Reidel, Dordrecht), p. 269.
34. Schweitzer, A.E., & Cudworth, K.M. 1996, "The Absolute Proper Motion of the Ursa Minor Dwarf Spheroidal Galaxy", BAAS, 28, 835.
35. Stetson, P.B. 1984, "Spectroscopy of Giant Stars in the Draco and Ursa Minor Dwarf Galaxies", PASP, 96, 128.
36. Strobel, N.V., & Lake, G. 1994, "Black Hole Models for the Dark Matter in Draco and Ursa Minor", ApJ, 424, L83.
37. Suntzeff, N.B., Olszewski, E., Kraft, R.P., Friel, E., Aaronson, M., & Cook, K. 1983, "The Ursa Minor Dwarf Galaxy", BAAS, 15, 907.
38. Swope, H.H. 1968, "Thirteen Periodic Variable Stars Brighter than the Normal RR Lyrae-Type Variables in Four Dwarf Galaxies", AJ, 73, S204.
39. van Agt, S. 1967, "A Discussion of the Ursa Minor Dwarf Galaxy Based on Plates Obtained by Walter Baade", Bull. Astr. Inst. Neth., 19, 275.

40. van den Bergh, S. 1994, "The Evolutionary History of Low-Luminosity Local Group Dwarf Galaxies", ApJ, 428, 617.
41. Wilson, A.G. 1955, "Sculptor-Type Systems in the Local Group of Galaxies", PASP, 67, 27.
42. Zinn, R. 1981, "The Metal Abundance Range in the Ursa Minor Dwarf Galaxy", ApJ, 251, 52.

WLM

Alternate name(s): DDO 221
Right Ascension (2000): 00h 01m 58s
Declination (2000): -15d 27.8m
Type: dIrr
Apparent magnitude (V): 10.85
Color (B-V): 0.42
(U-B): -0.21
(V-R):
Color Excess, E(B-V): 0.02
Absolute magnitude (M_V): -14.0
Distance (kpc): 925
Radial velocity (solar, km/sec): -123

Objects Identified On the Atlas:

Globular clusters: 1; designation: GC1
Source: [2]
Open clusters: none
Source:
OB associations: 4; designation: A
Source: [this atlas]
Variable stars: 33; designation: V
Source: [15]
Carbon stars: none
Source:
HII regions: 21; designation: Arabic numerals
Source: [6]
Planetary nebulae: none
Source:
Dust clouds: none
Source:

WLM

Chart 201

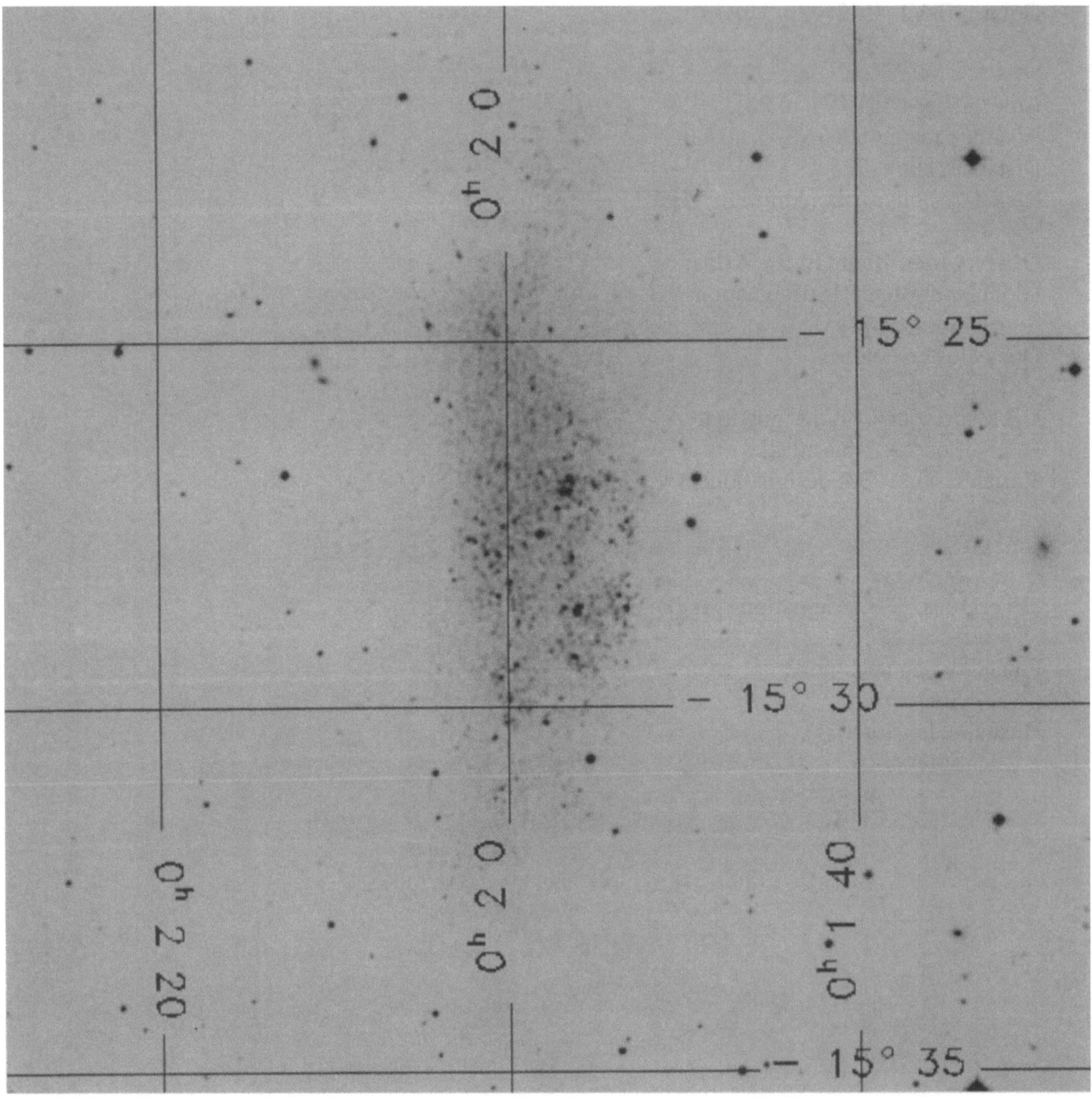

WLM

Chart 202

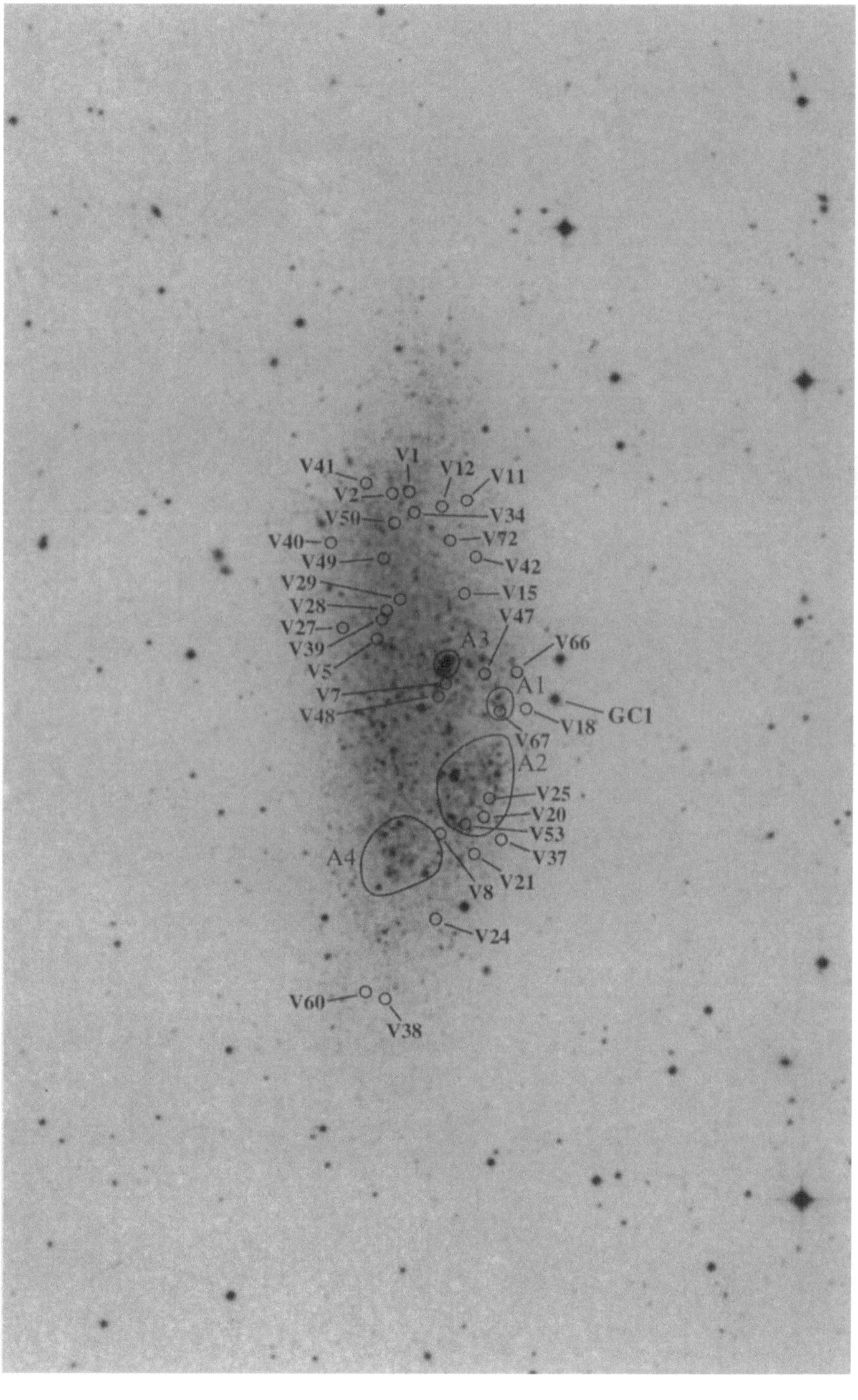

WLM

Chart 203

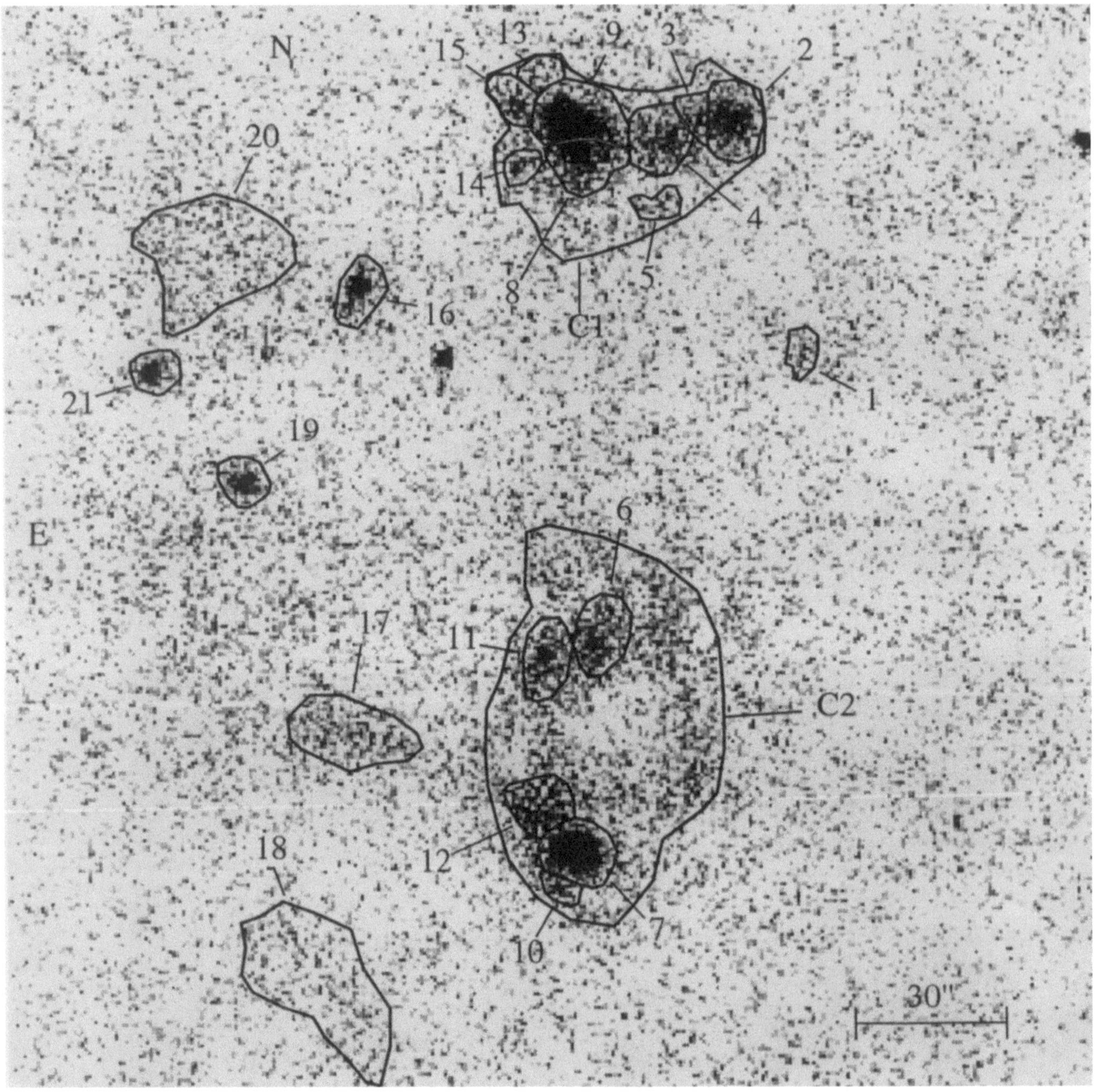

WLM

References

1. Aaronson, M., Cook, K.H., & Norris, J. 1985, "The AGB Population of Nearby Galaxies", in Spectral Evolution of Galaxies, edited by C. Chiosi and A. Renzini (Reidel, Dordrecht), p. 171.
2. Ables, H.D., & Ables, P.G. 1977, "*B, V* Photometry of the WLM Galaxy", ApJS, 34, 245.
3. Cook, K.H., Aaronson, M., & Norris, J. 1986, "Carbon and M Stars in Nearby Galaxies: A Preliminary Survey Using a Photometric Technique", ApJ, 305, 634.
4. Ferraro, F.R., Fusi Pecci, F., Tosi, M., & Buonanno, R. 1989, "A Method to Study the Star Formation History of Dwarf Irregular Galaxies. I. CCD Photometry of WLM", MNRAS, 241, 433.
5. Freedman, W.L. 1986, "Recent CCD Observations of Cepheids in Nearby Galaxies: A Progress Report", in Galaxy Distances and Deviations from Universal Expansion, edited by B.F. Madore and R.B. Tully (Reidel, Dordrecht), p. 21.
6. Hodge, P., & Miller, B.W. 1995, "H II Regions in Four Galaxies in and near the Local Group", ApJ, 451, 176.
7. Huchtmeier, W.K., Seiradakis, J.H., & Materne, J. 1981, "The Neutral Hydrogen Distribution of Irregular Galaxies", A&A, 102, 134.
8. Hunter, D.A., Hawley, W.N., & Gallagher, III, J.S. 1993, "A Survey for Extra- H II Region Ionized Gas Structures in Irregular Galaxies", AJ, 106, 1797.
9. Jacoby, G.H., & Lesser, M.P. 1981, "Distances to Nearby Galaxies Derived from Planetary Nebulae", AJ, 86, 185.
10. Lee, M.G., Freedman, W.L., & Madore, B.F. 1992, "I-band Cepheid Distance to WLM", presented at New Perspectives on Stellar Pulsation and Pulsating Variable Stars, IAU Colloquium 139.
11. Madore, B.F., & Freedman, W.L. 1991, "The Cepheid Distance Scale", PASP, 103, 933.
12. Minniti, D., & Zijlstra, A.A. 1997, "Stellar Populations of the Dwarf Irregular Galaxy WLM", AJ, 114, 147.
13. Minniti, D., & Zijlstra, A.A. 1996, "Dwarf Galaxies Also Have Stellar Halos", ApJ, 467, L13.
14. Mould, J., & Aaronson, M. 1986, "The Formation and Evolution of Carbon Stars", ApJ, 303, 10.
15. Sandage, A., & Carlson, G. 1985, "The Brightest Stars in Nearby Galaxies. VI. Cepheids and the Brightest Stars in WLM", AJ, 90, 1464.
16. Skillman, E.D., Terlevich, R., & Melnick, J. 1989, "Abundances in Southern Local Group Dwarf Irregular Galaxies", MNRAS, 240, 563.
17. Tosi, M., Greggio, L., & Focardi, P. 1989, "Star Formation in Dwarf Irregular Galaxies: Preliminary Results", Ap&SS, 156, 295.
18. Walker. A.R. 1987, "Notes on Cepheids Variables in Galaxies More Distant Than the Magellanic Clouds", S. African Astron. Obs. Circ., 11, 125.

Part III

Charts of four additional Local Group Members

And V

Alternate name(s):
Right Ascension (2000): 01h 10m 17s
Declination (2000): 47d 37m 41s
Type: dSph
Apparent magnitude (V):
Color (B-V):
(U-B):
(V-R):
Color Excess, E(B-V):
Absolute magnitude (M_V): -9.1
Distance (kpc): 810
Radial velocity (solar):

Objects Identified On the Atlas:
Globular clusters:
Source:
Open clusters:
Source:
OB associations:
Source:
Variable stars:
Source:
Carbon stars:
Source:
HII regions:
Source:
Planetary nebulae:
Source:
Dust clouds:
Source:

And V

Chart 204

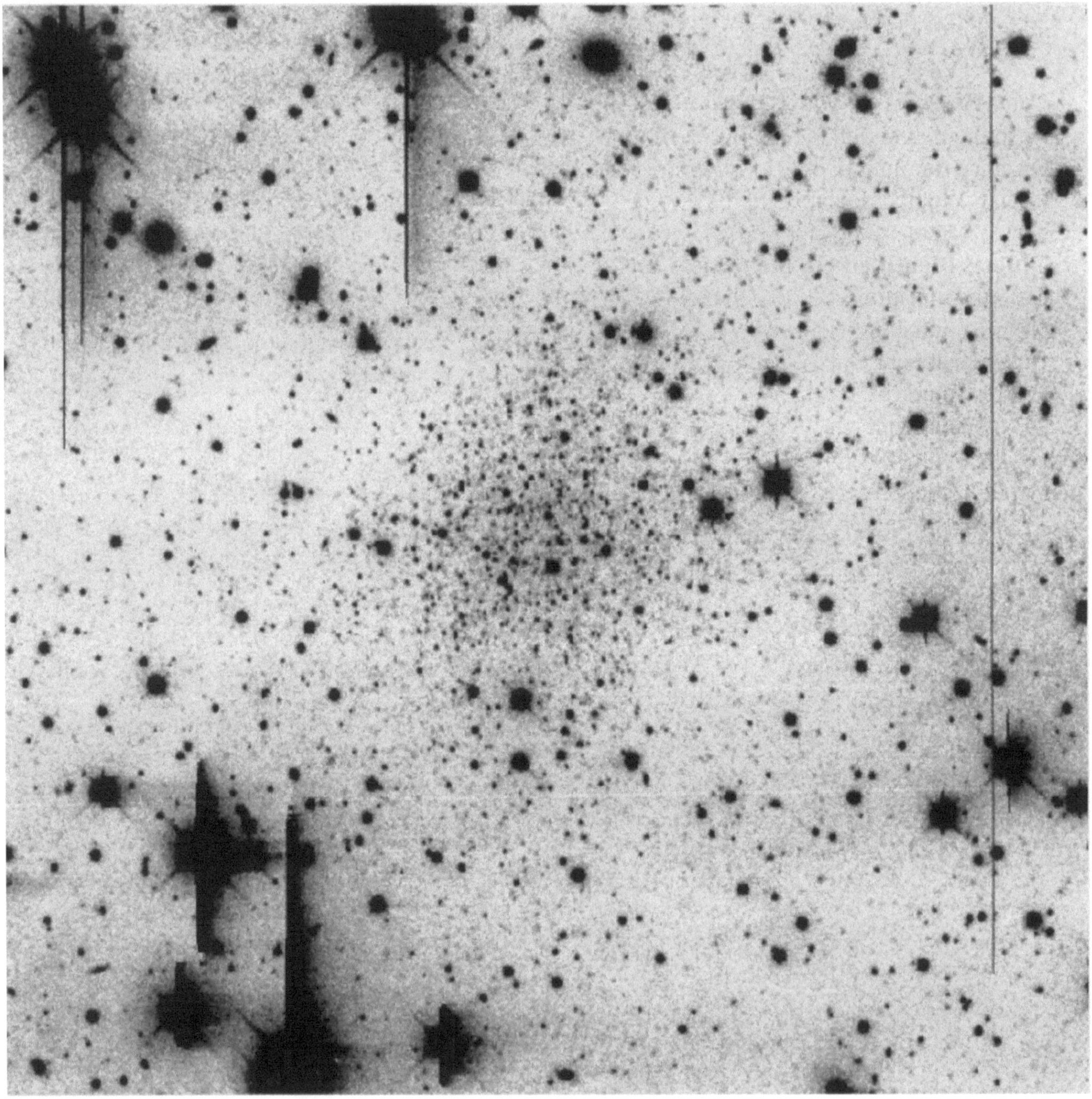

And VI

Alternate name(s): Pegasus
Right Ascension (2000): 23h 51m 39s
Declination (2000): 24d 35m 42s
Type: dSph
Apparent magnitude (V): 14.1
Color (B-V):
(U-B):
(V-R):
Color Excess, E(B-V):
Absolute magnitude (M_V): -11.3
Distance (kpc): 780
Radial velocity (solar):

Objects Identified On the Atlas:
Globular clusters:
Source:
Open clusters:
Source:
OB associations:
Source:
Variable stars:
Source:
Carbon stars:
Source:
HII regions:
Source:
Planetary nebulae:
Source:
Dust clouds:
Source:

And VI

Chart 205

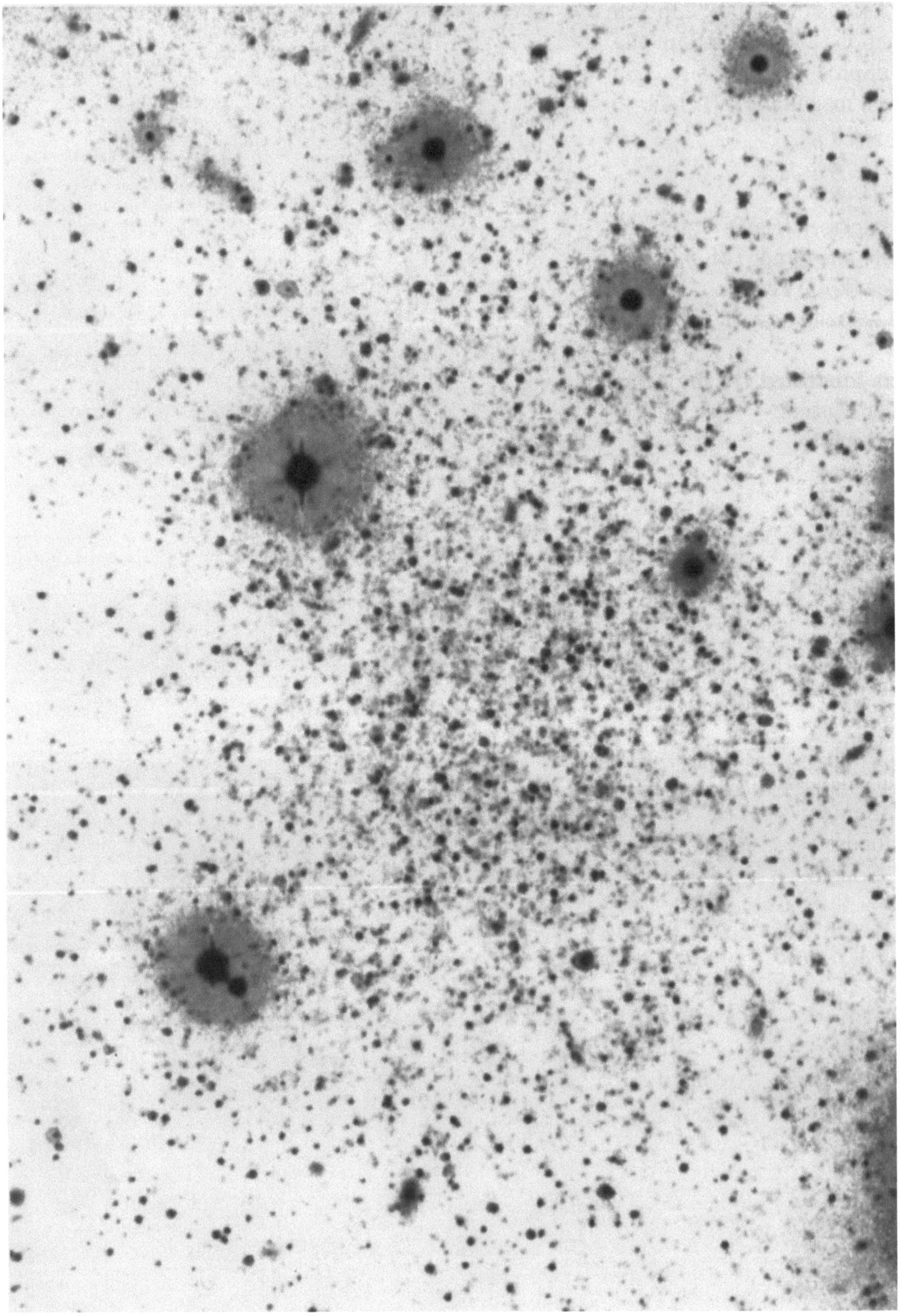

And VII

Alternate name(s): Cassiopeia
Right Ascension (2000): 23h 26m 31s
Declination (2000): 50d 41m 31s
Type: dSph
Apparent magnitude (V):
Color (B-V):
(U-B):
(V-R):
Color Excess, E(B-V):
Absolute magnitude (M_V): -12.0
Distance (kpc): 690
Radial velocity (solar):

Objects Identified On the Atlas:
Globular clusters:
Source:
Open clusters:
Source:
OB associations:
Source:
Variable stars:
Source:
Carbon stars:
Source:
HII regions:
Source:
Planetary nebulae:
Source:
Dust clouds:
Source:

And VII

Chart 206

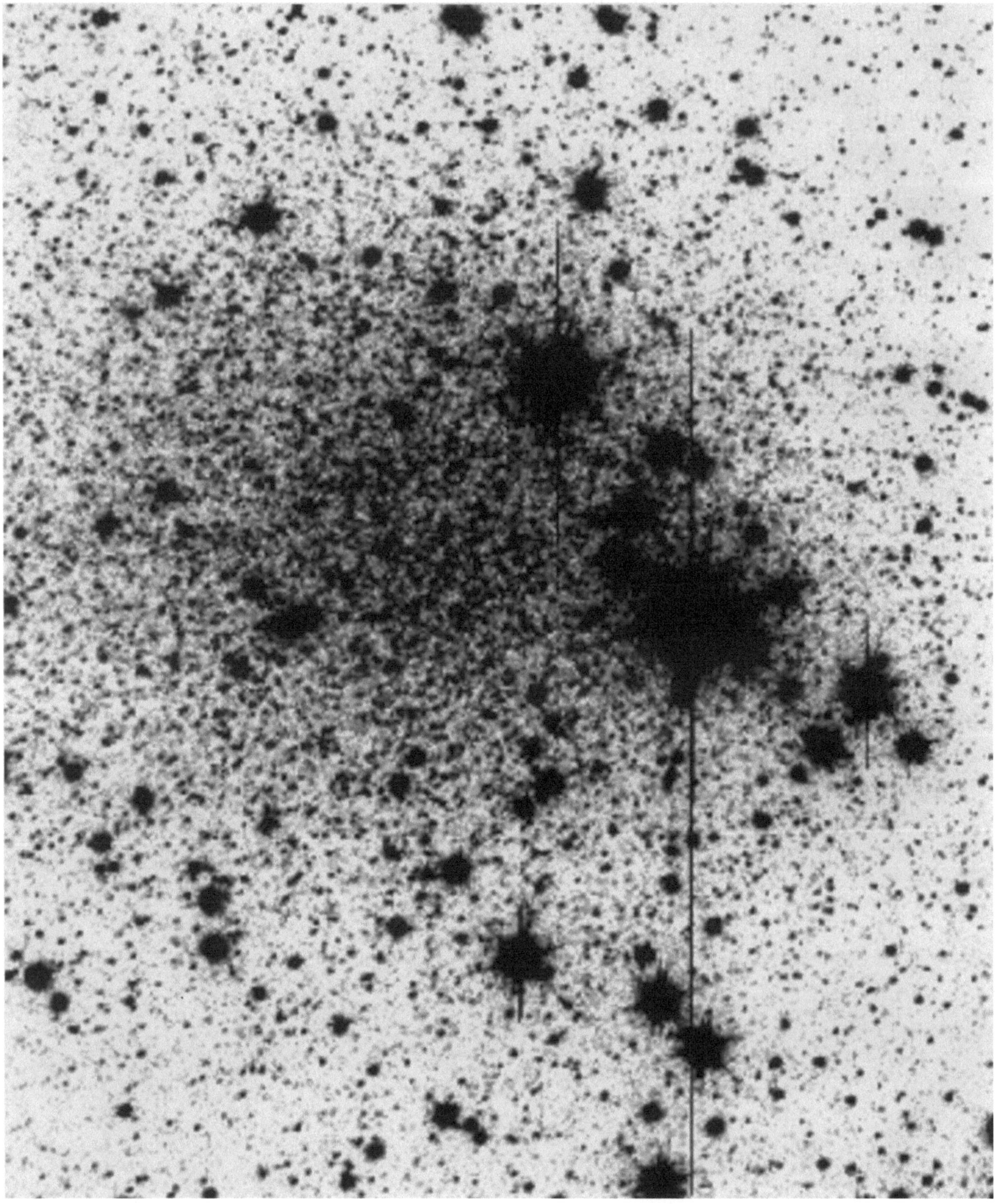

Cetus

Alternate name(s):
Right Ascension (2000): 00h 26m 11s
Declination (2000): -11d 02m 40s
Type: dSph
Apparent magnitude (V): 14.4
Color (B-V):
(U-B):
(V-R):
Color Excess, E(B-V):
Absolute magnitude (M_V): -10.1
Distance (kpc): 775
Radial velocity (solar):

Objects Identified On the Atlas:
Globular clusters:
Source:
Open clusters:
Source:
OB associations:
Source:
Variable stars:
Source:
Carbon stars:
Source:
HII regions:
Source:
Planetary nebulae:
Source:
Dust clouds:
Source:

Cetus

Chart 207

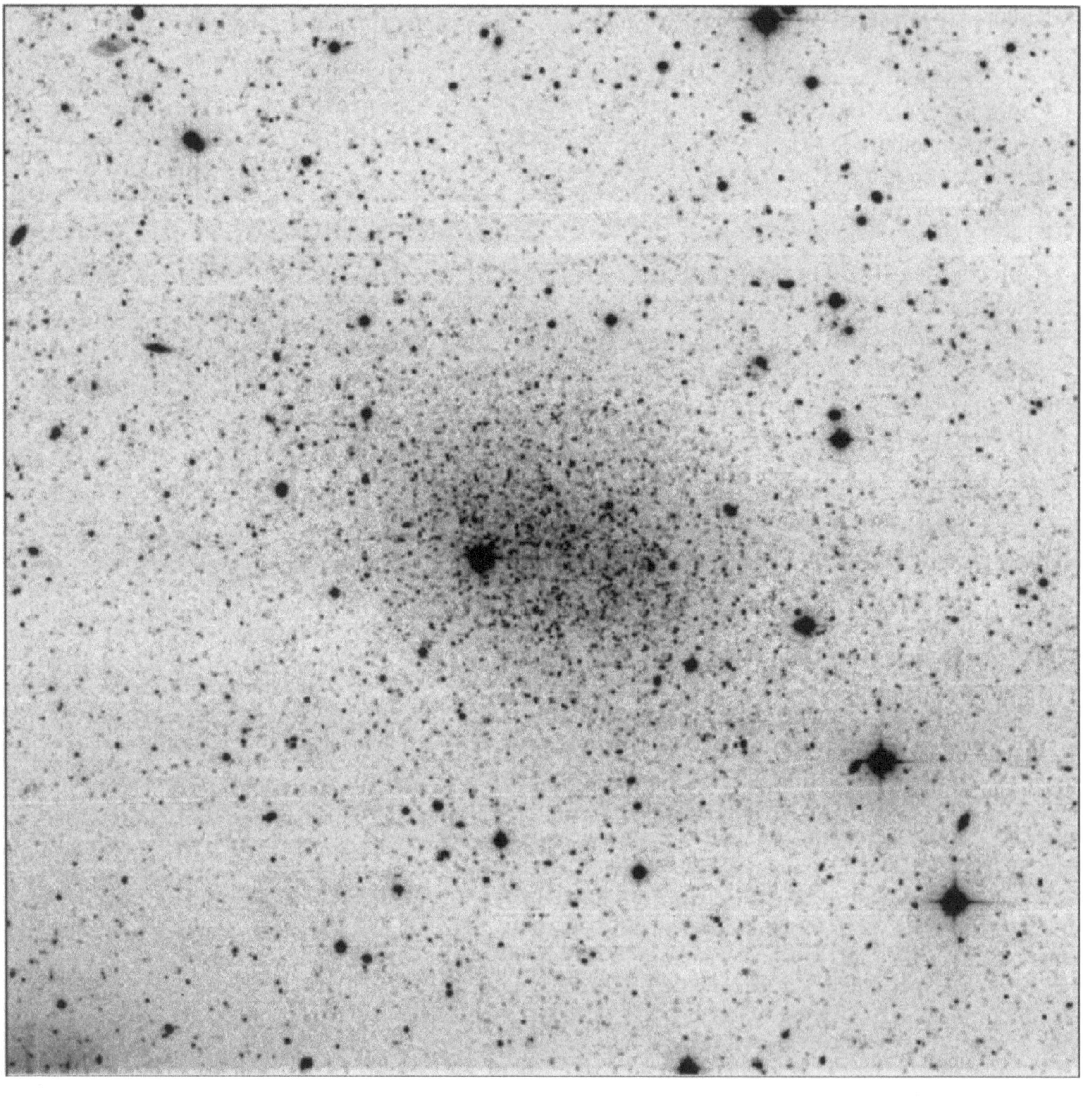

MIX
Papier aus verantwortungsvollen Quellen
Paper from responsible sources
FSC® C105338

If you have any concerns about our products,
you can contact us on
ProductSafety@springernature.com

In case Publisher is established outside the EU,
the EU authorized representative is:
Springer Nature Customer Service Center GmbH
Europaplatz 3, 69115 Heidelberg, Germany

Printed by Libri Plureos GmbH
in Hamburg, Germany